# XII International Conference on Beauty, Charm, and Hyperons in Hadronic Interactions (BEACH 2016)



Journal of Physics: Conference Series Volume 770

Fairfax, Virginia, USA
12 – 18 June 2016

## Editors:

**Marco Bozzo**
**David S. Armstrong**
**Patrizia Cenci**
**Kevin Giovanetti**
**Christina Lazzeroni**
**Janis McKenna**
**Philip Rubin**
**Elton S. Smith**
**Nick Solomey**
**Igor Strakowsky**

ISBN: 978-1-5108-3528-3
ISSN: 1742-6588

**Printed from e-media with permission by:**

Curran Associates, Inc.
57 Morehouse Lane
Red Hook, NY 12571

**Some format issues inherent in the e-media version may also appear in this print version.**



Printed by Curran Associates, Inc. (2017)

For permission requests, please contact the Institute of Physics
at the address below.

Institute of Physics
Dirac House, Temple Back
Bristol BS1 6BE UK

Phone: 44 1 17 929 7481
Fax: 44 1 17 920 0979

techtracking@iop.org

**Additional copies of this publication are available from:**

Curran Associates, Inc.
57 Morehouse Lane
Red Hook, NY 12571 USA
Phone: 845-758-0400
Fax: 845-758-2633
Email: curran@proceedings.com
Web: www.proceedings.com

# Table of contents

## Volume 770

**XII International Conference on Beauty, Charm, and Hyperons in Hadronic Interactions (BEACH 2016)**

**12–18 June 2016, George Mason University, Fairfax, Virginia, USA**

**Accepted papers received: 8 November 2016**
**Published online: 9 December 2016**

### Preface



### Papers

#### Overview

**Session 1**



**Session 2**



**Session 3**

**Session 4**



**Session 5**

**Session 6**



**Session 7**

**Session 8**



**Session 9**



**Session 10**

**Session 11**



**Session 12**

**Session 15**



**Session 16**

**Session 17**



**Session 18**

**Poster Session**



**Closing session**



**Poster Session**

**Proceedings of the**

# XII Conference on Beauty, Charm, Hyperons in Hadronic Interactions BEACH

George Mason University - $12^{th}$ to $18^{th}$ of June 2016

**Main Editor:**
Marco Bozzo - University of Genoa and INFN Genoa, Italy

**Editors:**
David S. Armstrong - College of William & Mary, Virginia, USA
Patrizia Cenci - Istituto Nazionale di Fisica Nucleare, Perugia, Italy
Kevin Giovanetti - James Madison University, Harrisonburg, Virginia, USA
Cristina Lazzeroni - School of Physics and Astronomy, University of Birmingham, UK
Janis McKenna - University of British Columbia, Vancouver, Canada
Philip Rubin - George Mason University, Fairfax, Virginia, USA
Elton S. Smith - Jefferson Lab, Virginia, USA
Nick Solomey - Wichita University, Kansas, USA
Igor Strakowsky - The George Washington University, Washington, DC 20052, USA

Journal of Physics: Conference Series **770** (2016) 011001 doi:10.1088/1742-6596/770/1/011001

# Preface

This volume contains the contributed papers presented at the $12^{th}$ *International Conference on Hyperons, Charm and Beauty Hadrons*, currently known as the BEACH Conferences. The BEACH conferences cover a broad range of physics topics in the field of Hyperon and heavy-flavor physics. This conference continues the BEACH series, which began with a meeting in Strasbourg in 1995 and since then offers a biennial opportunity for both theorists and experimentalists from the high-energy physics community to discuss all aspects of flavour physics.

The $12^{th}$ Conference took place in the Research Hall Room 163 of the Geoege Mason University in Fairfax, Virginia from June $12^{th}$ to June $18^{th}$ and was attended by 65 participants. All of the sessions were plenary sessions accommodating review talks and shorter contributions discussing both theory and recent experiments. At the end of the conference Alan Schwarz summarized and put in context all the presentations of the conference giving a very interesting Summary talk.

I want to thank here the Local Organizing Committee that organized the Conference. Many from the University Staff have contributed to the smooth running of the conference. I would like also to thank the Local Scientific Secretariat for their invaluable help in making the conference a truly enjoyable event; a special thanks goes to Maria Hobbs, our local secretary, who worked tirelessly in the organization of every detail. Finally we would like thank the European Organization for Nuclear Research and the George Mason University for their generous support.

The next BEACH Conference will be held at Peniche, north of Lisbon, Portugal at the beginning of summer 2018 and I hope that we will all meet again there.

Marco Bozzo,
Chairman
Università di Genova and INFN Genova (Italy)

Journal of Physics: Conference Series **770** (2016) 011001 doi:10.1088/1742-6596/770/1/011001

## Advisory Committee

| | |
|---|---|
| Federico Antinori | INFN Padova |
| Marco Bozzo | Università di Genova and INFN Genova (chair) |
| Stan Brodsky | SLAC and Stanford University |
| Antonio Ferrer | IFIC, Valencia |
| Valerie Gibson | University of Cambridge |
| Francis Halzen | University of Wisconsin–Madison |
| David Hitlin | California Institute of Technology, Pasadena |
| Calvin Kalman | Concordia University, Montreal |
| Vera Luth | SLAC and Stanford University |
| Tatsuya Nakada | EPFL, Lausanne |
| Gigi Rolandi | CERN, Geneva |
| Patrick Roudeau | Université Paris-Sud, Orsay |
| Chris Sachrajda | University of Southampton |
| Marjorie Shapiro | University of California, Berkeley |
| Teodor Siemiarczuk | National Centre for Nuclear Studies, Warsaw |
| Maria Smizanska | Lancaster University |

## Organizing Committee

| | |
|---|---|
| David Armstrong | College of William and Mary |
| Marco Bozzo | Università di Genova and INFN Genova |
| Patrizia Cenci | INFN Perugia |
| Kevin Giovanetti | James Madison University |
| Cristina Lazzeroni | University of Birmingham |
| Janis McKenna | University of British Columbia |
| Philip Rubin | George Mason University (Chair) |
| Elton Smith | Jefferson Laboratory |
| Nick Solomey | Wichita State University |
| Igor Strakovsky | George Washington University |

### Conference Assistants

| | | |
|---|---|---|
| Adam Conovaloff | Maria D'souza | Kathleen Enos |
| Concorde Habineza | Stephanie Kuhta | Andy Won |

George Mason University

Figure 1: BEACH 2016 - The participants.

Journal of Physics: Conference Series **770** (2016) 011001 doi:10.1088/1742-6596/770/1/011001

# BEACH 2014 – List of Participants

1 Walter Ikegami Andersson
Uppsala University , Sweden
walter.ikegami_andersson
@physics.uu.se

2 Artur Ankowski
Virginia Tech , USA
artank@vt.edu

3 Cioli Barazandeh
Northern Virginia Community College , USA
clb29910@email.vccs.edu

4 Adam Barton
Lancaster University , United Kingdom
abarton@cern.ch

5 Violaine Bellee
EPFL , Lausanne , Switzerland
violaine.bellee@epfl.ch

6 Jake Bennett
Carnegie Mellon University , USA
jvbennett@cmu.edu

7 Cesare Bini
University of Rome La Sapienza and INFN Roma , Italy
cesare.bini@roma1.infn.it

8 Austin Bradley
George Mason University ,
abradl12@gmu.edu

9 Marco Bozzo
INFN Genoa , Italy
marco.bozzo@cern.ch

10 Raul Briceno
Jefferson Lab , USA
bricneo.raul@gmail.com

11 Angelo Carbone
University and INFN Bologna , Italy
carbone@bo.infn.it

12 Andrea Celentano
INFN Genova , Italy
andrea.celentano@ge.infn.it

13 Patrizia Cenci
INFN Perugia , Italy
patrizia.cenci@pg.infn.it

14 Racha Cheaib
University of Mississippi , USA
rcheaib@olemiss.edu

15 Hok-Chuen (Tom) Cheng
University of Michigan, Ann Arbor , USA
hccheng@umich.edu

16 Adam Conovaloff
George Mason University ,
aconovaloff@gmail.com

17 Rafael S. Coutinho
University of Zurich , Switzerland
rsilvaco@cern.ch

18 Sudipan De
Universidade de São Paulo , Brasil
Sudipan.De@cern.ch

19 Umberto De Sanctis
University of Sussex , United Kingdom
umberto.de.sanctis@cern.ch

20 Antonio De Santis
INFN - Laboratori Nazionali di Frascati , Italy
antonio.desantis@lnf.infn.it

21 Biplab Dey
INFN Milano , Italy
biplab.dey@cern.ch

22 Fabio Ferrari
INFN - Bologna , Italy
fabio.ferrari@cern.ch

23 Salvatore Fiore
INFN Roma , Italy
salvatore.fiore@roma1.infn.it

24 Anjan Giri
IIT Hyderabad , Iran
giria@iith.ac.in

25 Angelo Gutarra-Leon
Northern Virginia Community College , USA
ajg269@email.vccs.edu

26 Brian Hamilton
University of Maryland, College Park , USA
bkham@umd.edu

27 Tristan O.S. Haseler
Georgia State University , USA
thaseler1@student.gsu.edu

28 Jenny Holzbauer
University of Mississippi , USA
jholzbau@fnal.gov

29 En-Chuan Huang
University of Illinois at Urbana-Champaign , USA
gratain@gmail.com

30 David Hutchcroft
University of Liverpool , United Kingdom
David.Hutchcroft@cern.ch

31 Walter Jaronski
Radford University , USA
wjaronsk@radford.edu

32 Bai-Cian Ke
Carnegie Mellon University , USA
baiciank@andrew.cmu.edu

33 Zachary King
University of Cincinnati , USA
kingze@mail.uc.edu

34 Jochen Krempel
ETH Zurich , Switzerland
jochen.krempel@phys.ethz.ch

35 Cristina Lazzeroni
University of Birmingham , United Kingdom
cristina.lazzeroni@cern.ch

36 Alexey Luchinsky
Institute for High Energy Physics, Protvino , Russia
alexey.luchinsky@gmail.com

37 Niels Madsen
Swansea University , United Kingdom
niels.madsen@cern.ch

38 Janis McKenna
University of British Columbia , Canada
janis@physics.ubc.ca

39 Iliescu Mihail
INFN-Laboratori Nazionali di Frascati , Italy
mihai@lnf.infn.it

40 Rukmani Mohanta
University of Hyderabad , Iran
tt rukmani98@gmail.com

41 Irina Nasteva
Universidade Federal do Rio de Janeiro , Brasil
tt irina@if.ufrj.br

42 Scarlet Norberg
University of Puerto Rico , Puerto Rico
scarletnorberg2014@gmail.com

43 Toshio Numao
TRIUMF , Canada
toshio@triumf.ca

44 Agnieszka Oblakowska-Mucha
AGH - UST , Poland
amucha@agh.edu.pl

45 Andrea Pattori
University of Zurich , Switzerland
pattori@physik.uzh.ch

46 Marc Pelizaeus
Ruhr-University Bochum , Germany
pattori@physik.uzh.ch

47 Mathieu Perrin-Terrin
CERN , Switzerland
mathieu.perrin-terrin@cern.ch

48 Alessandro Pilloni
Jefferson Lab , USA
pillaus@jlab.org

49 Christoph Redmer
Johannes Gutenberg-University Mainz , Germany
redmer@uni-mainz.de

50 Patrizia Rossi
Jefferson Lab , USA
rossi@jlab.org

51 Philip Rubin
George Mason University ,
prubin@gmu.edu

52 Saurabh Sandilya
University of Cincinnati , USA
saurabhsandilya@gmail.com

53 Alan Schwartz
University of Cincinnati , USA
alan.j.schwartz@uc.edu

54 Maria Sevilla
University of Liverpool , United Kingdom
Maria.Cerda-Sevilla@liv.ac.uk

55 Nick Solomey
Wichita State University , USA
nick.solomey@wichita.edu

56 Tomasz Szumlak
AGH - UST , Poland
szumlak@agh.edu.pl

57 Fuquan Wang
University of Wisconsin , USA
fuquan.wang@cern.ch

58 Peter Weidenkaff
University Mainz , Germany
weidenka@uni-mainz.de

59 Christian Weiss
Jefferson Lab , USA
weiss@jlab.org

60 Mark Whitehead
CERN , Switzerland
mark.peter.whitehead@cern.ch

61 Jack Wimberley
University of Maryland, College Park , USA
jtwimb@umd.edu

62 Joanna Zalipska
National Centre for Nuclear Research , Poland
joanna.zalipska@ncbj.gov.pl

63 Zhengqiao Zhang
Brookhaven National Lab , USA
qiao@rcf.rhic.bnl.gov

64 Jianhui Zhu
Central China Normal University , China
zhu@subatech.in2p3.fr

# Peer review statement

All papers published in this volume of *Journal of Physics: Conference Series* have been peer reviewed through processes administered by the proceedings Editors. Reviews were conducted by expert referees to the professional and scientific standards expected of a proceedings journal published by IOP Publishing.

Journal of Physics: Conference Series **770** (2016) 012001 doi:10.1088/1742-6596/770/1/012001

# BEACH 2016: an Overview (of sorts)

**P D Rubin**

Department of Physics & Astronomy, MSN 3F3
George Mason University
Fairfax, VA 22030 USA

E-mail: prubin@gmu.edu

**Abstract.** A brief history of the conference series and its contemporary context are presented.

## 1. Introduction

The first *Conference on Hyperons, Charm and Beauty Hadrons* met in Strasbourg, France, in September 1995, and was considered by attendees to have been a success. Many supported a continuing series of such conferences that would focus on hyperons and charm and beauty hadrons, distinct from existing heavy flavor meetings. A second conference was then held in Montreal, Canada, in August 1996. Subsequent editions of the conference, now known as BEACH, have been held biennially, with venues alternating between Europe and North America. George Mason University is happy to host this twelfth edition.

## 2. An "evolution" of topics

At the third edition in Genoa, Italy, in June - July 1998, the organizers already referred to BEACH's plenary talks-only policy as a "tradition." Topics covered included [1]

- CP violation in B decays
- Top physics
- Charmonium production and decay
- Heavy-quark effective theory
- Non-relativistic QCD
- Electroweak measurements; tests of the Standard Model
- Spectroscopy
- New detectors and related technologies

At our twelfth edition, we cover

- Hadron production and decays
- Heavy quarks, quarkonium, and hyperons
- Electroweak measurements
- Neutrinos
- Symmetry violations

- Lattice and non-relativistic QCD
- Physics beyond the Standard Model
- New facilities and projects

## 3. The evolution of results

Because BEACH conference programs remain close to their origins, the conferences have documented the growth and maturation of its purview.

As an example, consider the Cabibbo-Kobayashi-Maskawa matrix unitarity triangle apex position [2] in terms of the Wolfenstein parameters, $\bar{\eta}$ and $\bar{\rho}$ [see Figure 1]. Thus, Nicola Cabibbo's presentation [3] of the latest fit result [see Figure 2] in his introductory overview of the sixth edition in Chicago, in July 2004, was subsequently updated [see Figure 3] by Jernej Kamenik's [4] of the eleventh edition in Birmingham, England, in July 2014.

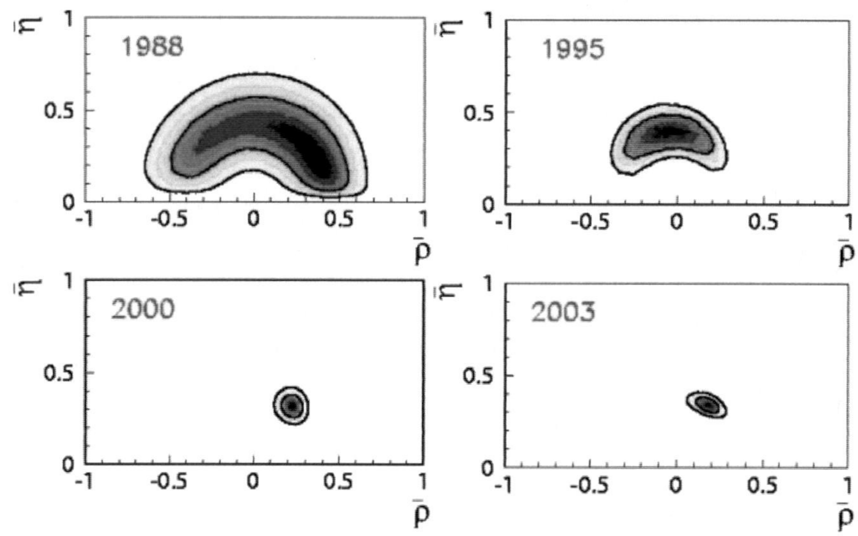


**Figure 1.** History of unitarity triangle apex, 1988 - 2003

## 4. BEACH's place

In November - December 2011, the U.S. Department of Energy, Office of Science, sponsored a workshop [5] intended to make the case, at the dawn of the LHC era, for the continuing importance of a broad approach to what at that point had been called flavor physics. The workshop was called "Fundamental Physics at the Intensity Frontier." Six working groups–heavy quarks, charged leptons, neutrinos, photons, proton decay, and nucleons, nuclei, and atoms–were charged with elucidating "the full spectrum of Intensity Frontier physics" and identifying the facilities required to investigate it.

The heavy quark working group identified CP asymmetries, rare decays, and precision measurement of key distributions in the strange, charm, and bottom sector as its primary foci. The charged lepton group identified lepton flavor violation with muons and taus and $g-2$ as its foci. The neutrinos group identified lepton flavor violation with neutrino oscillations and double beta decay. The photons group, now dubbed the hidden sectors group, set its sights on new particles searches. The proton decay group vowed to keep looking for protons to decay, and

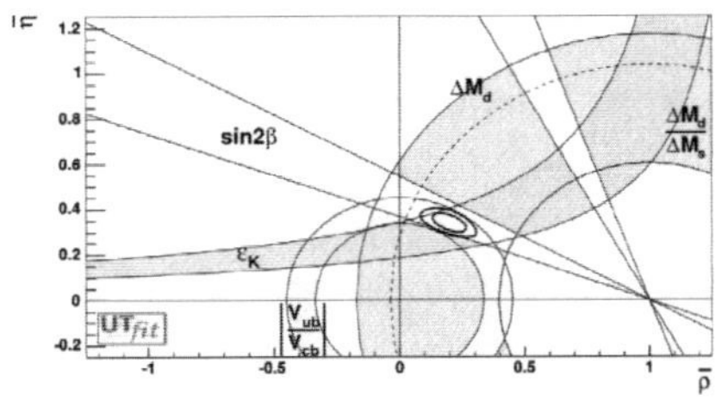


**Figure 2.** Unitarity triangle apex, 2004

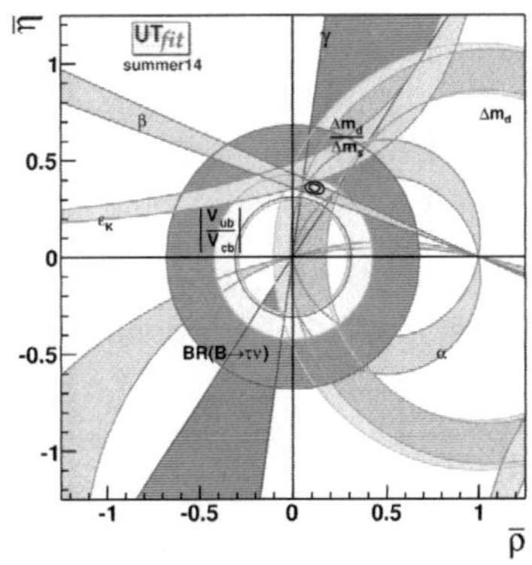


**Figure 3.** Unitarity triangle apex, 2014

the nucleons and atoms group would search for electric dipole moments and make additional precision measurements of parity violation.

The program of this twelfth edition of the BEACH conferences, which addresses nearly every item on the Intensity Frontier's "broad and diverse, yet connected, set of science opportunities," shows that BEACH remains not just relevant, but central, to these endeavors.

Here, we see where we are and where we are going with respect to this science.

## References

[1] Hyperons Conference (n.d) https://www.ge.infn.it/hyperons98/
[2] *15 YEARS of* $(\rho - \eta)$ *predictions* (n.d.) http://www.utfit.org/foswiki/pub/UTfit/Old/ckm-history/ckm-history.html#15yearhisto
[3] *BEACH 2004 - An Introduction* (June 2004) http://capp.iit.edu/beach04/talks/cabibbo.pdf
[4] *The Flavorful Road to New Physics* (July 2014) http://indico.cern.ch/event/283978/contributions/648565/attachments/524082/722823/Kamenik_BEACH_2014_1.pdf
[5] H Weertz, *et al Fundamental Physics at the Intensity Frontier* May 2012 http://www.ipd.anl.gov/anlpubs/2012/06/73380.pdf

# Measurement of W- and Z-boson production in p–Pb collisions at $\sqrt{s_{\mathrm{NN}}} = 5.02$ TeV with ALICE at the LHC

**J Zhu**[1,2]**, on behalf of the ALICE collaboration**
[1]Key Laboratory of Quark and Lepton Physics (MOE) and Institute of Particle Physics, Central China Normal University, Wuhan, China
[2]SUBATECH, Ecole des Mines de Nantes, Université de Nantes, CNRS-IN2P3, Nantes, France

E-mail: Jianhui.Zhu@cern.ch

**Abstract.** In hadronic collisions, electroweak bosons (W and Z) are produced in initial hard scattering processes and they are not affected by the strong interaction. In proton–proton collisions, they have been suggested as standard candles for luminosity monitoring and their measurement can improve the evaluation of detector performances. In nucleus-nucleus and proton–nucleus collisions, electroweak bosons provide a check at first order of the validity of binary collision scaling, while small deviations allow studying the nuclear modifications of parton distribution functions. In ALICE, the production of W and Z bosons is measured via the contribution of W-boson decays to the inclusive $p_{\mathrm{T}}$-differential muon yield and the invariant mass of unlike-sign muon pairs from Z-boson decays, respectively, at large rapidities. This paper reports on the production cross sections of muons from W muonic decay and of Z boson in p–Pb collisions at $\sqrt{s_{\mathrm{NN}}} = 5.02$ TeV, which are compared to the theoretical calculations, and on the W-boson yields normalized to the average number of binary nucleon-nucleon collisions as a function of the event activity.

## 1. Introduction

The high collision energies available at the Large Hadron Collider (LHC) at CERN allow for an abundant production of hard probes, such as heavy quarks, quarkonia, high $p_{\mathrm{T}}$ jets and electroweak bosons ($W^{\pm}$, $Z^{0}$). Due to large masses of electroweak bosons, they are produced in initial hard parton scattering processes, with a formation time of the order of $1/M \sim 0.002 - 0.003$ fm/c and, having a lifetime of 0.08-0.10 fm/c, thus decay before the formation of the quark-gluon plasma (QGP), which is a deconfined phase of QCD matter that is formed in high-energy heavy-ion collisions. Furthermore, their leptonic decay products do not interact strongly with the hot and dense QCD medium. In proton-proton collisions, precise theoretical predictions of the W- and Z-boson production make them good standard candles for the luminosity measurement. Moreover, they can be used to constrain the parton distribution functions (PDFs) at high momentum transfer ($Q^2$). In proton-nucleus and nucleus-nucleus collisions, precise measurements of W- and Z-boson production can constrain the modification of the parton distribution functions in the nucleus (nPDFs) [1] and at first order they can be used to test the scaling of hard particle production with the number of binary nucleon-nucleon collisions [2].

## 2. Experimental apparatus and data sample

The ALICE experiment is equipped with two magnets and various detectors for triggering, tracking and particle identification [3, 4]. Muons are reconstructed in the muon spectrometer, which covers the pseudo-rapidity range $-4 < \eta_{\rm lab} < -2.5$. The spectrometer consists of a thick front absorber, a dipole magnet, five tracking stations, a muon filter and two trigger stations. The inner tracking system consists of six layers of silicon detectors covering the range $|\eta_{\rm lab}| < 0.9$. In this analysis, the interaction vertex is measured with the two innermost layers, which are equipped with Silicon Pixel Detectors (SPD). The number of clusters in the outer layer of the SPD (CL1) can be used to characterize the event activity. Two arrays of scintillators placed at both sides of the interaction point, the VZERO detector (V0A ($2.8 < \eta_{\rm lab} < 5.1$) and V0C ($-3.7 < \eta_{\rm lab} < -1.7$)), provide the trigger and an additional event activity estimation. A third estimation of the event activity is obtained with the Zero Degree Calorimeters (ZNA and ZNC) located at both sides of the detector at 112.5 m from the main interaction point along the beam line. Due to the energy asymmetry of the proton and lead colliding beams at the LHC, the centre-of-mass system of nucleon-nucleon collisions is shifted by $\Delta y = 0.465$ in the direction of the proton beam, with respect to the laboratory frame. Experimental data have been collected with two beam configurations (p-Pb and Pb-p) by inverting the orbits of the two particle species. The rapidity regions covered by the muon spectrometer in the two cases are $2.03 < y^{\mu}_{\rm cms} < 3.53$ (forward, p-going direction) and $-4.46 < y^{\mu}_{\rm cms} < -2.96$ (backward, Pb-going direction).

The data have been collected at $\sqrt{s_{\rm NN}} = 5.02$ TeV with minimum bias (MB), high-$p_{\rm T}$ single-muon (MSH) and low-$p_{\rm T}$ di-muon (MUL) triggers. The MB trigger is defined requiring hits in both sides of the VZERO detector in coincidence with the beam counters. The MSH trigger is defined by asking, in addition to the MB condition, for a high transverse momentum muon with $p_{\rm T} \gtrsim 4.2$ GeV/c in the trigger system of the spectrometer. The MUL is a coincidence of MB trigger and an unlike-sign muon pair selected with $p_{\rm T} \gtrsim 0.5$ GeV/c for each muon. The MSH and MUL data samples are used for W- and Z-boson analysis, respectively. The integrated luminosity for the forward and backward rapidity measurements are $5.03\pm0.18$ nb$^{-1}$ and $5.81\pm0.20$ nb$^{-1}$ [5], respectively. The track in the muon tracking system is required to be in the geometrical acceptance ($-4 < \eta_{\rm lab} < -2.5$), to have a polar angle measured at the end of the front absorber of $170° < \theta_{\rm abs} < 178°$ and match a tracklet in the trigger stations in order to reject the remaining contamination from hadrons. Furthermore, the correlation between momentum and distance of closest approach (DCA) to the interaction vertex was used to remove beam-gas collisions and secondary particles produced in the absorber.

## 3. W-boson measurement

The present analysis is based on the extraction of the W-boson contribution from the transverse momentum distribution of inclusive muons. At high $p_{\rm T}$, the main contributions to the yield of inclusive muons come from the muonic decays of W bosons, the di-muon decays of $Z^0/\gamma^*$ bosons and the semi-muonic decays of beauty hadrons. The yield of muons from W-boson decays can be obtained through a fit based on suitable parameterizations of the different components. The distribution of muons from intermediate vector boson decays is described by Monte Carlo (MC) templates obtained from POWHEG simulations [6]. The isospin dependence of the W-boson differential cross-section [7] is accounted for by simulating separately p-p and p-n collisions and then summing the results together according to:

$$\frac{1}{N_{\text{p-Pb}}} \cdot \frac{dN_{\text{p-Pb}}}{dp_{\rm T}} = \frac{Z_{\rm Pb}}{A_{\rm Pb}} \cdot \frac{1}{N_{\text{p-p}}} \cdot \frac{dN_{\text{p-p}}}{dp_{\rm T}} + \frac{A_{\rm Pb} - Z_{\rm Pb}}{A_{\rm Pb}} \cdot \frac{1}{N_{\text{p-n}}} \cdot \frac{dN_{\text{p-n}}}{dp_{\rm T}} \tag{1}$$

where $A_{\rm Pb} = 208$ and $Z_{\rm Pb} = 82$. The generation of muons from $Z^0/\gamma^*$ decays is performed in an equivalent way. For the description of the background at low $p_{\rm T}$ ($10 < p_{\rm T} < 40$ GeV/c), which originates from the decays of beauty hadrons, three approaches are adopted: MC templates based on pQCD calculations with FONLL [8], functional form, which was already successfully adopted for analogous measurements by the ATLAS experiment at the LHC [9] and another function derived from the second term of the parameterization used by ATLAS, which can be written as $f(p_{\rm T}) = a \cdot \frac{e^{b \cdot \sqrt{p_{\rm T}}}}{p_{\rm T}^c}$. The different contributions are summed together in the fit function:

$$f(p_{\rm T}) = N_{\mu\leftarrow \rm HF} \cdot f_{\mu\leftarrow \rm HF}(p_{\rm T}) + N_{\mu\leftarrow \rm W} \cdot f_{\mu\leftarrow \rm W}(p_{\rm T}) + N_{\mu\leftarrow \rm Z^0/\gamma^*} \cdot f_{\mu\leftarrow \rm Z^0/\gamma^*}(p_{\rm T}) \quad (2)$$

where $f_{\mu\leftarrow \rm HF}$ can be either a functional form or the MC template for muons from heavy-flavour decays, and $f_{\mu\leftarrow \rm W}$ and $f_{\mu\leftarrow \rm Z^0/\gamma^*}$ are the MC templates for muons from W and $\rm Z^0/\gamma^*$ boson decays, respectively. The number of muons from W decays ($N_{\mu\leftarrow \rm W}$) is a free parameter, while the ratio $N_{\mu\leftarrow \rm Z^0/\gamma^*}/N_{\mu\leftarrow \rm W}$ is fixed to the value obtained with POWHEG. Figure 1 presents three examples of the combined fit for $\mu^+$ by varying heavy-flavour background description.

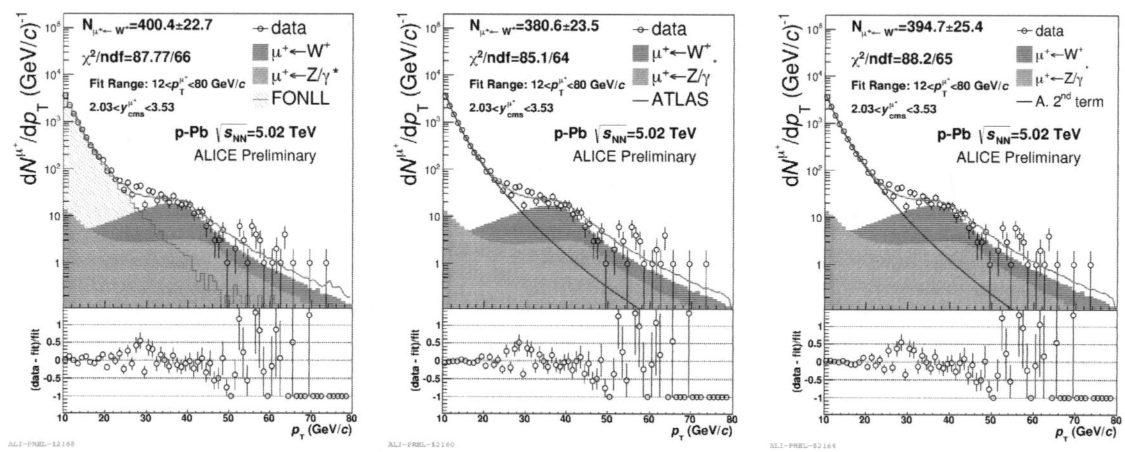


**Figure 1.** Examples of the combined fit to the inclusive $p_{\rm T}$-spectrum of $\mu^+$ at forward rapidity. Heavy-flavour templates based on FONLL [8] (left), the ATLAS function [9] (middle) and the $2^{nd}$ term (second addend) of the ATLAS function (right). The errors on the yields represent the statistical uncertainties.

The uncertainty on the background description is determined by using either FONLL-based MC templates or functional forms and by varying the $p_{\rm T}$ range in the fit. The relative contribution of $\rm Z^0/\gamma^*$ is varied based on POWHEG and PYTHIA [10] predictions. The effect of the detector alignment is taken into account by producing the templates using different residual alignment assumptions. The signal was extracted several times by varying the above conditions and the final result is chosen as the weighted average of all of the trials. The systematic uncertainties are given by the RMS of the trials. It is worth noting that POWHEG does not account for the nuclear modification of the PDFs. Hence, the signal extraction using PYTHIA with EPS09 parameterization of the nuclear PDFs [11] was included in the calculation of the systematic uncertainties.

After signal extraction, the yields of muons from W-boson decays are corrected for acceptance times efficiency (88% for forward, 76% for backward). The production cross-sections of muons from W-boson decays at forward and backward rapidity are shown in Figure 2 and compared with theoretical predictions based on NLO pQCD calculations with CT10 PDFs [12] and modified CT10 PDFs with the EPS09 parametrization of nuclear effects. Both the theoretical predictions with and without EPS09 agree with the measurement within uncertainties.

The yield of muons from W-boson decays can also be extracted per event activity interval. It is corrected for acceptance times efficiency and normalized to the number of equivalent MB event. Thus, the normalized yield can be divided by the average number of binary nucleon-nucleon collisions $\langle N_{\rm coll}\rangle$ in each event activity interval, in order to obtain the yield per binary collision. The average number of binary collisions $\langle N_{\rm coll}\rangle$ is determined from a Glauber-model based analysis in the case of V0A, V0C and CL1 estimators, while for the ZN estimator it is computed with a hybrid approach, assuming that the particle density at mid-rapidity is proportional to the number of nucleons participating in the collision, $N_{\rm part}$ [2]. The measured yields of muons from W-boson decays over $\langle N_{\rm coll}\rangle$ in p–Pb collisions at forward and backward rapidities are shown in Figure 3 as a function of event activity. The results present a flat trend as function of event activity, and are similar for the three estimators within uncertainties.

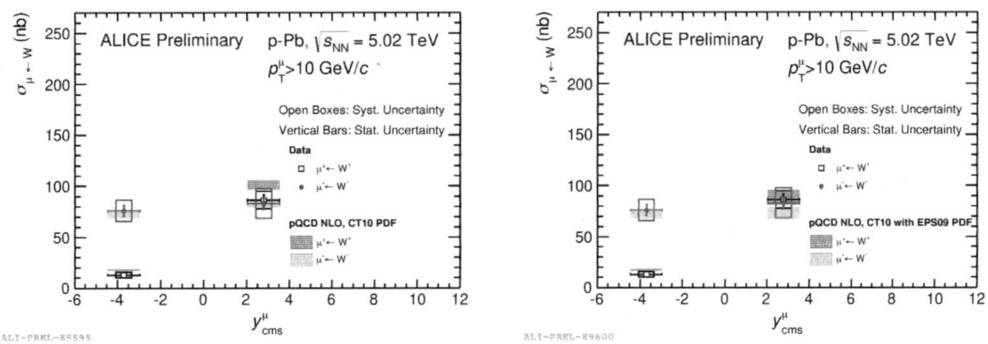


**Figure 2.** The cross section of muons from W-boson decays compared with pQCD [1] calculations with CT10 PDFs [12] (left) and CT10 PDFs including nuclear effects EPS09 [11] (right) at forward and backward rapidity. The cut of $p_{\rm T} > 10$ GeV/c is added.

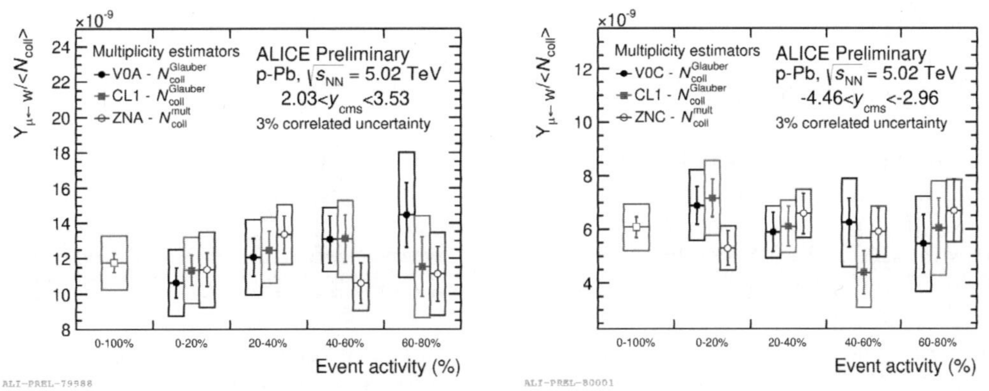


**Figure 3.** Muon yields from W-boson decays normalized to the average number of binary collisions as a function of event activity at forward (left) and backward (right) rapidity. The green points represent the values in minimum-bias collisions.

## 4. Z-boson measurement

The signal of Z boson is seen as a peak around 90 GeV/$c^2$ in the invariant-mass distribution of unlike-sign muon pairs, selected according to the criterion described in section 2 and with

$p_{\rm T} > 20$ GeV/c in order to reduce background from lower mass resonances decay and from the semi-leptonic decay of charm and beauty hadrons. It was verified that loosening the requirement on the minimum $p_{\rm T}$ of the muons to 10 GeV/c does not introduce any additional unlike-sign muon pair with $m_{\mu\mu} > 40$ GeV/c$^2$. The invariant mass distribution is shown in Figure 4. 22 (2) candidates with $m_{\mu\mu} > 60$ GeV/c$^2$ were reconstructed at forward (backward) rapidity in p–Pb collisions. At forward rapidity, the distribution is compared with expectations from the POWHEG simulations with CT10 PDFs and EPS09 nPDFs. The parameters were extracted from a fit using the Crystal-Ball function [13], which consists of a Gaussian core with power-law tails on both sides. A good agreement within the experimental uncertainties between data and POWHEG is obtained. The contribution to the invariant-mass distribution from the combinatorial background can be estimated using the like-sign dimuon distribution: no candidates were found in the region $60 < m_{\mu\mu} < 120$ GeV/c$^2$. A 0.1% upper limit for this contribution is obtained by extrapolating the like-sign dimuon distribution at low mass to the region of interest. Contributions from other physics processes, like the semi-leptonic decay of $c\bar{c}$, $b\bar{b}$ and $t\bar{t}$ pairs and the muonic decay of $\tau$ pairs in the process ($Z \to \tau\tau \to \mu\mu$), is estimated using MC simulations to be less than 0.7% (0.4%) at forward and backward rapidity, respectively.

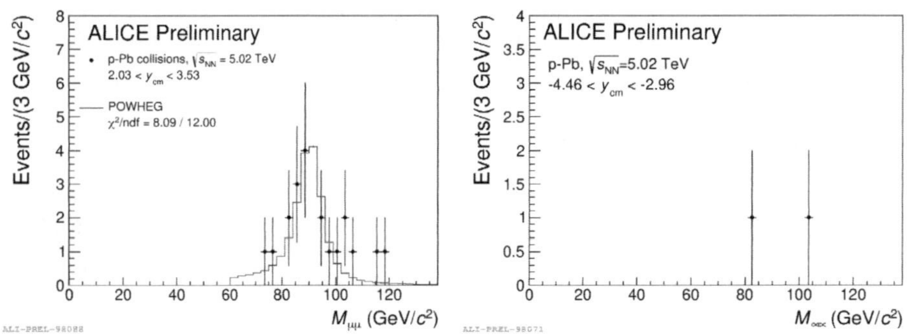


**Figure 4.** Invariant-mass distribution of unlike-sign muon pairs with $p_{\rm T} > 20$ GeV/c in the p-going (left panel) and Pb-going (right panel) direction. The blue line represents the distribution obtained using POWHEG simulation with CT10 PDFs and EPS09 nPDFs and normalized to the number of Z candidates in the data.

The measured number of candidates is corrected by the acceptance times efficiency (78% for forward, 61% for backward) with a relative systematic uncertainty (1% for forward, 2% for backward). The measured production cross section of muons from Z-boson decays at forward and backward rapidity in p-Pb collisions is shown in Figure 5 (left) and compared with next-to-next-to leading order (NNLO) Fully Exclusive W and Z (FEWZ) [14] predictions with CT10nlo [15], CTEQ6m [16], JR09NNLO [17] and MSTW2008NNLO [18] with and without nPDFs. Since there are few candidates at backward rapidity, limiting the measurement statistical precision, an upper limit of the cross section is also evaluated. The ratio of the measured cross sections (obtained by ALICE and LHCb [19]) to FEWZ are shown in Figure 5 (right). The measurements of ALICE are in agreement with theoretical calculations at both rapidity regions.

## 5. Conclusion

The production cross section of muons from W- and Z-boson decays at forward and backward rapidity has been measured in p-Pb collisions at $\sqrt{s_{\rm NN}} = 5.02$ TeV with the ALICE muon spectrometer. Theoretical predictions based on NLO pQCD and FEWZ calculations with CT10 PDFs agree with the measurement within uncertainties for W and Z boson, respectively. Taking into account the EPS09 parametrization of nuclear effects on the PDFs further improves the

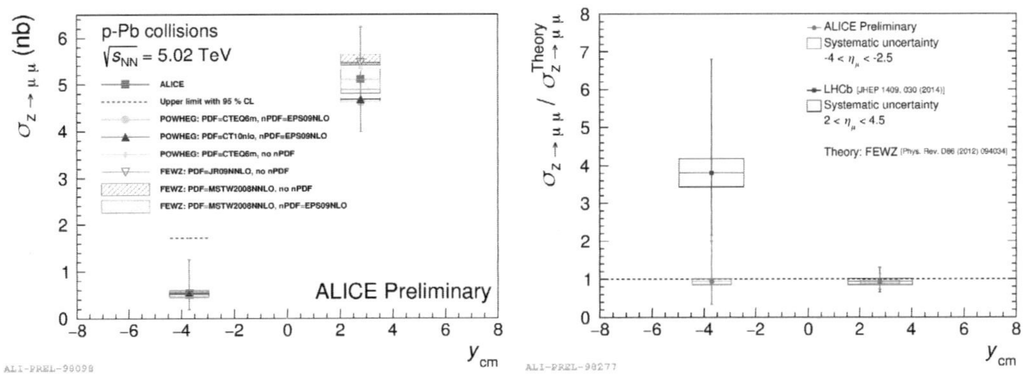


**Figure 5.** (Left) The cross section of muons from Z-boson decays compared to FEWZ [14] calculations with different PDFs with and without nuclear effects EPS09 [11]. (Right) The ratio of cross section to FEWZ calculation measured by the ALICE collaboration is compared with those obtained by the LHCb collaboration [19].

agreement between theoretical predictions and the measurements at forward rapidity where shadowing is expected to be important. The ALICE measurement on Z-boson agrees with the results from LHCb [19] at forward and backward rapidities, whereas CMS [20, 21] and ATLAS [22] results indicate that electroweak bosons favour the modification of the PDFs at mid-rapidity. The yield of muons from W-boson decays in different event activity scales with $\langle N_{\rm coll} \rangle$ within the experimental uncertainties.

## Acknowledgments

This work is supported partly by the Chinese Ministry of Science and Technology 973 grant 2013CB837803, NSFC Grant 11375071, 11475068, IRG 11521064 and CSC grant No. 201306770033.

## References

[1] Paukkunen H and Salgado C A 2011 *JHEP* **03** 071
[2] Adam J *et al.* (ALICE Collaboration) 2015 *Phys. Rev.* C **91** 064905
[3] Aamodt K *et al.* (ALICE Collaboration) 2008 *JINST* **3** S08002
[4] Abelev B *et al.* (ALICE Collaboration) 2014 *Int.J.Mod.Phys.* A **29** 1430044
[5] Adam J *et al.* (ALICE Collaboration) 2015 *JHEP* **11** 127
[6] Alioli S, Nason P, Oleari C and Re E 2008 *JHEP* **07** 060
[7] Conesa del Valle Z, CERN-THESIS-2007-102
[8] Cacciari M, Greco M and Nason P 1998 *JHEP* **05** 007
[9] Aad G *et al.* (ATLAS Collaboration) 2011 ATLAS-CONF-2011-078
[10] Sjostrand T, Mrenna S and Skands P 2006 *JHEP* **05** 026
[11] Eskola K J, Paukkunen H and Salgado C A 2009 *JHEP* **04** 065
[12] Lai H L *et al.* 2010 *Phys.Rev.* D **82** 074024
[13] Gaiser J 1982 PhD thesis (http://www.slac.stanford.edu/cgi-wrap/getdoc/slac-r-255.pdf)
[14] Gavin R, Li Y, Petriello F and Quackenbush S 2011 *Comput. Phys. Commun.* **182** 2388-2403
[15] Lai H L, Guzzi M, Huston J, Li Z, Nadolsky P M, Pumplin J and Yuan C P 2010 *Phys. Rev.* D **82** 074024
[16] Nadolsky P M *et al.* 2008 *Phys. Rev.* D **78**(1) 013004
[17] Jimenez-Delgado P and Reya E 2009 *Phys. Rev.* D **79** 074023
[18] Martin A D, Stirling W J, Thorne R S and Watt G 2009 *Eur. Phys. J.* C **63** 189-285
[19] Aaij R *et al.* (LHCb Collaboration) 2014 *JHEP* **09** 030
[20] Khachatryan V *et al.* (CMS Collaboration) 2015 *Phys. Lett.* B **750** 565-586
[21] Khachatryan V *et al.* (CMS Collaboration) 2016 *Phys. Lett.* B **759** 36-57
[22] Aad G *et al.* (ATLAS Collaboration) 2015 *Phys. Rev.* C **92** 044915

# Production of exotic and conventional quarkonia and open beauty/open charm at ATLAS

**C Bini (on behalf of the ATLAS collaboration.)**
Sapienza Universitá and INFN Roma, Italy
E-mail: cesare.bini@roma1.infn.it

**Abstract.** The ATLAS experiment at LHC is carrying on a wide programme to study the production properties of conventional and exotic quarkonium, beauty, and charm bound states. The latest results on $J/\psi$, $\psi(2s)$ and X(3872) production at 7, 8, and 13 TeV, together with D meson production with Run-1 are presented. Studies of associated production of charmonium with vector bosons, searches for exotic states in the bottomonium sector and a new measurement of the ratio of b-quark fragmentation functions are also briefly presented.

## 1. Introduction

The production of heavy flavors (HF) in high energy hadron collisions is characterized by a large cross-section. Since the first measurements at the Tevatron [1] it has been clear that models based on perturbative QCD (pQCD) were not able to account for such a large production cross-section. The theoretical estimate of these cross-section is a challenge for QCD. In fact many ingredients are needed to obtain consistent estimates to be compared with experimental values, including proton Parton Density Functions (PDFs), hard-scattering cross-sections and hadronization models. Several approaches have been developed in recent years to provide theoretical environments for these calculations. Among them the Color Singlet Model (CSM) and the Non-Relativistic QCD (NRQCD including Color Singlet and Color Octet production models) are considered to account for quarkonia production; GM-VFNS (General Mass Variable Flavour Number Scheme), FONLL (Fixed Order + Next to Leading Logs) and MC@NLO and POWHEG complemented by Pythia/Herwig for hadronization are used to describe open HF production. A recent comprehensive review of these models can be found in Ref. [2].

In the following we present a summary of the most recent measurements performed by the ATLAS experiment at LHC on quarkonia and open HF production.

## 2. The ATLAS experiment

The results presented here are based on the data taken by the ATLAS experiment [3] at LHC during the Run-1 in 2011 and 2012 (7 and 8 TeV center of mass (c.o.m.) energy corresponding to integrated luminosities of $\sim$ 5 and $\sim$ 21 $fb^{-1}$ respectively) and on the first data of Run-2 taken in 2015 (13 TeV, $\sim$ 4 $fb^{-1}$).

The possibility to perform HF physics in ATLAS is based on two main ingredients: low $p_T$ muon triggers and track reconstruction in the inner detector. The muon spectrometer provides di-muon triggers with muons of $p_T$ down to 4÷6 GeV and up to $|\eta|$ <2.4, in an invariant mass region between the $J/\psi$ and the $\Upsilon$ using the RPC (in the barrel) and TGC (in the endcap) trigger detectors. The inner detector provides tracking in a 2T magnetic field using silicon pixels, silicon strips and transition radiation detectors. In Run-2 an additional pixel layer (the so-called Insertable B-layer (IBL)) has been added closer to the beam interaction point. Tracks are detected up to $|\eta|$ <2.5. The typical resolution in di-muon invariant mass ranges between 50 MeV (at the $J/\psi$ peak) and 150 MeV (at the $\Upsilon$ peak).

In the case of charmonium, two distinct production mechanisms are possible at LHC: prompt and non-prompt production. Prompt particles are directly produced in the primary pp interaction or through feed-down from decays of heavier (directly produced) states; non-prompt particles are produced in the decays of b-hadrons. The two categories can be separated experimentally exploiting the long b-hadron lifetime. The discrimination is based on the so called "pseudo-proper time" $\tau$:

$$\tau = \frac{L_{xy} m(\mu\mu)}{|\vec{p}_T(\mu\mu)|} \tag{1}$$

where $L_{xy}$ is the travel distance of the quarkonium in the transverse plane, $m(\mu\mu)$ and $\vec{p}_T(\mu\mu)$ are the mass and the transverse momentum of the muon pair respectively.

## 3. Quarkonia production

For each considered charmonium final state prompt and non-prompt production cross-sections are obtained separately by counting the events through a combined fit of invariant mass and pseudo-proper time. An example of fit for the $\mu\mu$ final state involving $J/\psi$ and $\psi(2S)$ [4] is shown in figure 1. All the different contributions are shown. This kind of analysis has been performed for $J/\psi$ [4], $\psi(2S)$ [5], [6], $\chi_{c1}$ and $\chi_{c2}$ [7]. In general the prompt charmonium production turns out to be well described by NLO NRQCD. Predictions based on NNLO* colour-singlet model calculations clearly underestimate the data, especially at high transverse momenta. The non-prompt charmonium production is reasonably well described by FONLL. The comparison between data and theory for the $J/\psi$ data at 7 TeV c.o.m. energy is shown as a function of $p_T(\mu\mu)$ for different $|\eta|$ ranges in figure 2.

The same analysis has been applied on the first data from Run-2. Figure 3 compares the non-prompt fraction as a function of $p_T(\mu\mu)$ of ATLAS data at different center of

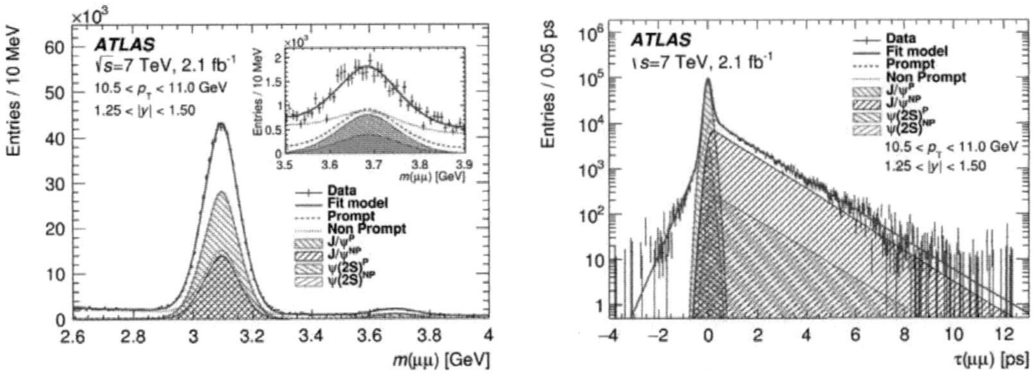


**Figure 1.** Invariant mass (left plot) and pseudo-proper decay time (right plot) spectra of $\mu\mu$ final states for a given region in $p_T(\mu\mu)$ and $y(\mu\mu)$. The signals of the $J/\psi$ and $\psi(2S)$ (also in the insert of the left plot) can be easily seen. The result of the combined fit is also shown with details of individual components.(From Ref. [4])

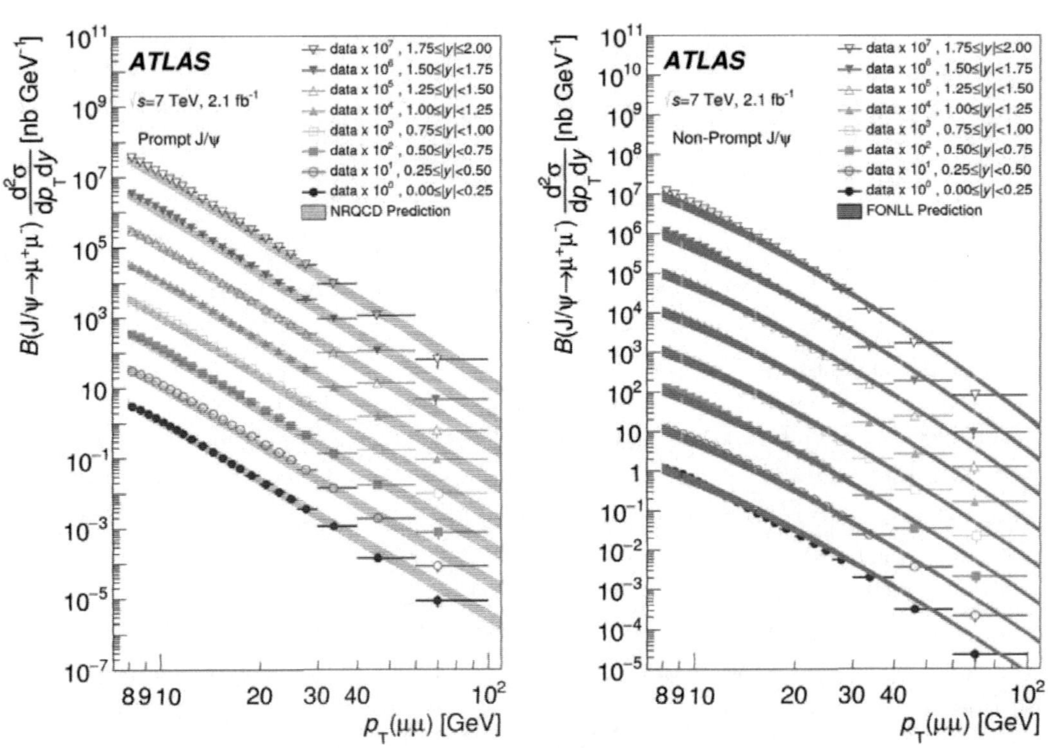


**Figure 2.** Prompt (left plot) and non-prompt (right plot) $J/\psi$ production cross-section as a function of $p_T(\mu\mu)$ up to 100 GeV in slices of $y(\mu\mu)$. The data are compared with predictions from NRQCD (left) and FONLL (right). Reasonable agreement is found in both cases in a large kinematical range.(From Ref. [4])

mass energies [8]. The plot includes data from the Tevatron [9] at 1.96 TeV c.o.m. energy. The comparison shows that the non-prompt fraction behavior with $p_T(\mu\mu)$ doesn't depend on the c.o.m. energy. Only some discrepancy is observed in the absolute scale with respect to lower energy data.

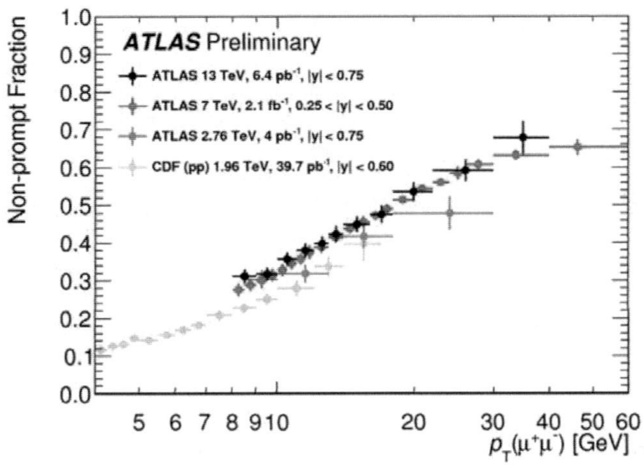


**Figure 3.** Non-prompt $J/\psi$ production fraction compared to previous measurements from ATLAS in pp collisions at 2.76 and 7 TeV, and from CDF in $p\bar{p}$ collisions at 1.96 TeV. (From Ref. [8])

In the analysis of the $J/\psi\pi^+\pi^-$ final state [6] we find (see figure 4), in addition to the $\psi$(2S) peak, a clear signal from the X(3872), the well-known exotic candidate. A good description of the prompt X(3872) production cross-section is obtained assuming that this meson is a mixture of a molecular state and a $\chi_c$ state [10]. On the other hand the non-prompt production cross-section is overestimated by all models.

A search for possible exotic $X_b$ states decaying to $\Upsilon(1S)\pi^+\pi^-$, similar to the X(3872) in the beauty sector with masses in the range $10.5 \div 10.7$ GeV, has also been performed [11]. No signal is found. Upper limits ranging between 0.02 and 0.03 at the different masses are obtained for the ratio between $X_b$ and $\Upsilon$(2S) production.

Finally it is worth mentioning the measurement of associated production of the $J/\psi$ with vector bosons W and Z [12], [13]. In both cases ATLAS finds a cross-section significantly larger than the one expected by NLO NRQCD calculations. Results with increased statistical significance are expected soon from Run-2 data.

## 4. Open Charm/Beauty production

A systematic study of the production of $D^{*\pm}$, $D^{\pm}$ and $D_s^{\pm}$ has been performed [14] using the final states $K\pi\pi$ and $KK\pi$. D mesons are reconstructed for $p_T$ up to 100 GeV. Figure 5 shows the mass peaks corresponding to the three mesons.

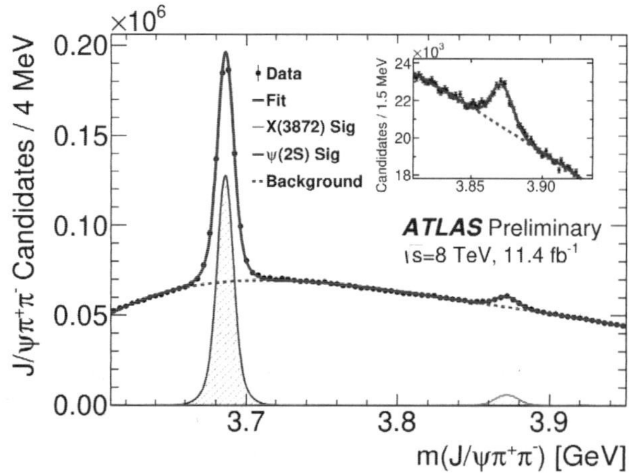


**Figure 4.** Invariant mass spectrum of $J/\psi\pi^+\pi^-$ with $J/\psi \to \mu\mu$. In addition to the $\psi(2S)$ peak, a clear signal (see also insert) is observed corresponding to the X(3872) meson.(From Ref. [6])

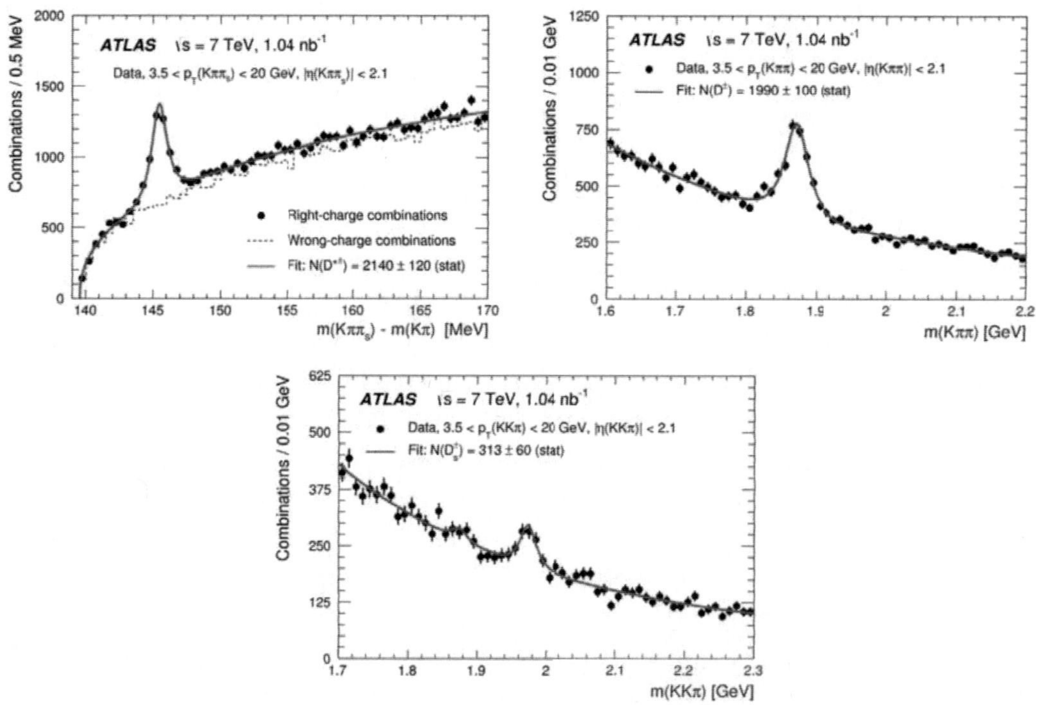


**Figure 5.** From left to right, examples of $D^{*\pm}$, $D^{\pm}$ and $D_s^{\pm}$ peaks in the $p_T$ and $\eta$ regions specified in each plot. The number of resulting candidates is also given (from Ref. [14]).

The measured $D^*$ and D meson production cross-sections integrated in $\eta$ are shown in figure 6 as a function of $p_T(D)$ and are compared to different models. The best agreement is obtained in both cases for the GM-VFNS model, that is able to describe both shape and normalization of the distributions.

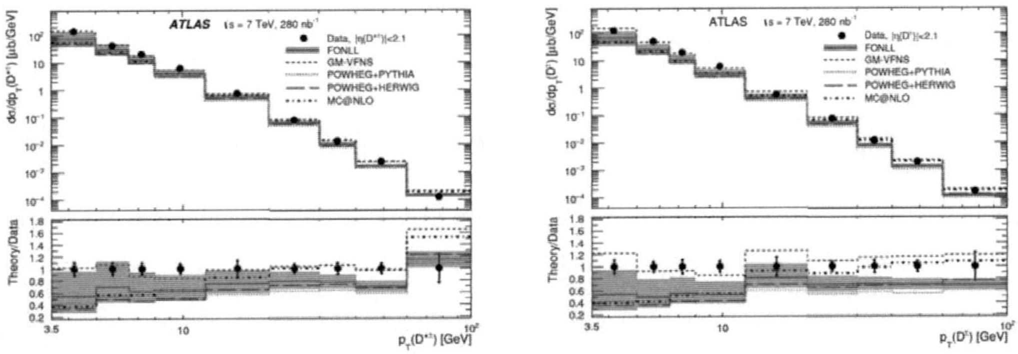


**Figure 6.** $D^{*\pm}$ (left) and $D^{\pm}$ production cross-section as a function of $p_T(D)$ and integrated in $\eta$. The data are compared with several different predictions. The lower plots show the ratio between theory and experiment.(From Ref. [14])

An important ingredient in the interpretation of rare B decays is the ratio of b-quark fragmentation functions $f_s/f_d$. This quantity can be extracted in pp collisions from the ratio of the decays $B_d^0 \to J/\psi K^{*0}$ with $K^{*0} \to K^+\pi^-$ to $B_s^0 \to J/\psi\phi$ with $\phi \to K^+K^-$. Both decays are observed with high statistics in ATLAS [15]. The resulting value of $f_s/f_d$ is:

$$\frac{f_s}{f_d} = 0.240 \pm 0.004(stat) \pm 0.010(syst) \pm 0.017(th) \tag{2}$$

in agreement with results from other experiments. This result extends the knowledge of this quantity to higher $p_T(B)$ values.

## References

[1] CDF collaboration 1997, *Phys. Rev. Lett.* **79** 572-577
[2] Andronic A et al. 2015, arXiv:1506.03981v2 [nucl-ex]
[3] ATLAS collaboration 2008, *J. Instrum.* **3** S08003
[4] ATLAS collaboration 2016, *Eur. Phys. J.* C **76** 5,283
[5] ATLAS collaboration 2014, *JHEP* **09** 079
[6] ATLAS collaboration 2016, *ATLAS-CONF* **028**
[7] ATLAS collaboration 2014, *JHEP* **07** 154
[8] ATLAS collaboration 2015, *ATLAS-CONF* **030**
[9] CDF collaboration 2005, *Phys. Rev.* D **71** 032001
[10] Meng C, Han H and Chao K 2013, arXiv:1304.6710[hep-ph]
[11] ATLAS collaboration 2015, *Phys. Lett.* B **740** 199-217
[12] ATLAS collaboration 2014, *JHEP* **04** 172
[13] ATLAS collaboration 2015, *Eur. Phys. J.* C **75** 229
[14] ATLAS collaboration 2016, *Nucl. Phys.* B **907** 717
[15] ATLAS collaboration 2015, *Phys. Rev. Lett.* **115** 262001

Journal of Physics: Conference Series **770** (2016) 012004 doi:10.1088/1742-6596/770/1/012004

# Double Charmonia Production at LHCb

**A V Luchinsky**

Institute for High Energy Physics, Protvino, Russia

E-mail: alexey.luchinsky@ihep.ru

**Abstract.** In this note next-to-leading order corrections to double charmonium production at LHCb are considered. It is shown that real gluon emission leads to significant change of the cross sections of considered processes. For example, forbidden at leading order production of $C$-odd state $\chi_c J/\psi$ is possible at NLO. Distributions over various kinematical variables are analyzed and the question of separating contributions of different mechanisms (direct $J/\psi J/\psi$ production, radiative feed-down from $\chi_c J/\psi$, double parton scattering) is considered.

## 1. Introduction

Presented article is devoted to theoretical study of real gluon emission effects to double charmonia production at LHC.

Charmonium production in high energy hadronic interaction was studied in details both theoretically and in experiment. In the case of single charmonium production the main theoretical approaches include:

- Leading order (LO) processes $gg \to \mathcal{Q}$. In the collinear gluon approximation this approach cannot describe final meson's transverse momentum distribution. In addition, due to charge parity conservation production of $C$-odd states (e.g. $J/\psi$) is forbidden.
- $k_T$ factorization [1]. In this model transverse momentum dependence of initial gluons' structure functions is taken into account, so $p_T$ distributions can be described. Charge parity selection rules, however, still hold and $J/\psi$ meson cannot be produced.
- Next-to-Leading-Order (NLO) processes $gg \to \mathcal{Q}g$. Due to the presence of additional final gluon all mentioned above problems can be solved. This approach was used, for example, in papers [2, 3] and good agreement with experiment was observed.

Recently the era of double charmonium production began [4–7]. It is clear, that in this case the same theoretical methods can be used and they have the same problems. For example, in LO collinear gluon approximation [8–10] one cannot describe distribution over the transverse momentum of the pair. Due to charge parity conservation $C$-odd final states (e.g. $J/\psi\chi_{cJ}$, $J/\psi\eta_c$, etc.) cannot be produced. As in the case of single charmonium production all these problems can be solved if one consider NLO processes $gg \to \mathcal{Q}_1\mathcal{Q}_2 g$.

Next to leading order corrections to double charmonium production in high energy hadronic interactions were studied already theoretically [11–13]. In these articles, however, only production of $S$-wave charmonia states was considered and theoretical predictions were made only for LHC experiment at $\sqrt{s} = 7\,\text{TeV}$. In the presented article we are going to consider $\sqrt{s} = 13\,\text{TeV}$ experiment and study also production of $P$-wave charmonium states $\chi_{cJ}$.

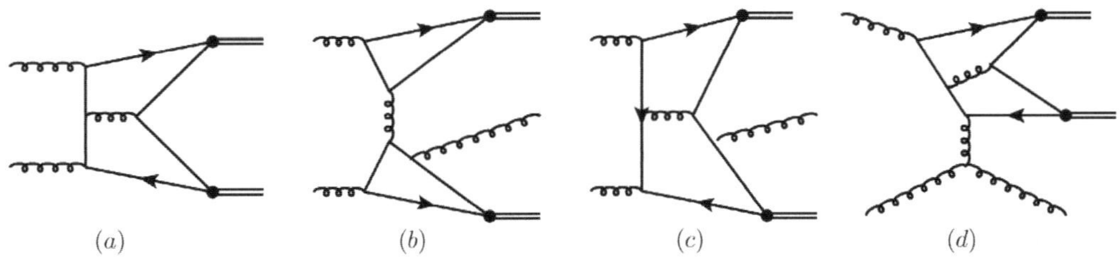


**Figure 1.** Typical Feynman diagrams of double charmonium production in gluon interaction at LO (a) and NLO* (b,c,d)

The rest of the paper is organized as follows. In the next section we study partonic cross sections of $gg \to \psi\psi g$ and $gg \to \psi\chi_{cJ}g$ processes. In section III hadronic reactions are considered. The last section is reserved for brief discussion.

## 2. Partonic Processes

In our paper we will use Non-relativistic quantum (NRQCD) [14, 15] and restrict ourselves to leading order in relative quarks' velocity colour-singlet (CS) approximation. Under these assumptions the matrix element of the considered process can be factorized into short distance and long distance parts. The first is easily calculated in perturbation theory, while for the latter one, that describes quark-antiquark pair hadronization into physical meson, some nonperturbative methods should be used. We prefer to determine the long distance matrix elements from experimental widths of the corresponding $J/\psi$ and $\chi_c$ meson decays.

Let us first discuss briefly leading order contribution to $gg \to J/\psi J/\psi$ reaction. In total there are 72 Feynman diagrams that can contribute to this process with the typical one shown in Fig.1(a). Few comments should be made. First of all, since each quark carries one half of meson's momentum, significant suppression is present and the corresponding cross section falls rapidly with the increase of the partonic energy $\sqrt{\hat{s}}$. In addition, due to back-to-back kinematics, in this process neither $p_T^{\mathrm{pair}}$ nor azimuthal distribution can be described. Finally, no $C$-odd final states (e.g. $J/\psi\chi_c$) can be produced due to charge parity conservation. Since experimentally both these distributions and final states are observed it is clear that LO description is unsatisfactory and NLO contributions should be taken into account.

There are two significant ingredients that should be taken into account at NLO. The first one is 1-loop diagrams. It is clear, however that for these diagrams all mentioned above drawbacks still hold: due to back-to-back kinematics some important distributions cannot be described and charge parity selection rules are still present. For this reason we will not consider 1-loop corrections in the following. In addition at NLO there are diagrams with real gluon emission, that can both generate transverse momentum and azimuthal distributions and allow the production of $C$-odd final state. Typical diagrams of this type are shown in Fig.1(b-d). In total there are 438 such diagrams, but charge parity conservation leads to cancellation of some sets of these diagrams. It is important to notice that for different final states ($J/\psi J/\psi$ or $J/\psi\chi_c$) contributions of different diagrams cancel.

Let us consider first direct production of $J/\psi J/\psi$ pair at NLO. First of all, it is clear that colour-singlet vector meson cannot couple to two (possible virtual) gluons, so 106 diagrams similar to those shown in Fig.1(b) do not contribute. Next, at the leading order in relative quarks' momentum all final state radiation diagrams (see Fig.1c) give finite result. Thus, the only divergent diagrams are initial state radiation (ISR), like the one shown in Fig.1(d). This divergence can be compensated by 1-loop corrections to hadron distribution functions, but we chose to put a cut-off on the transverse momentum of final charmonium pair $p_T^{\mathrm{pair}} > \Delta$. It is

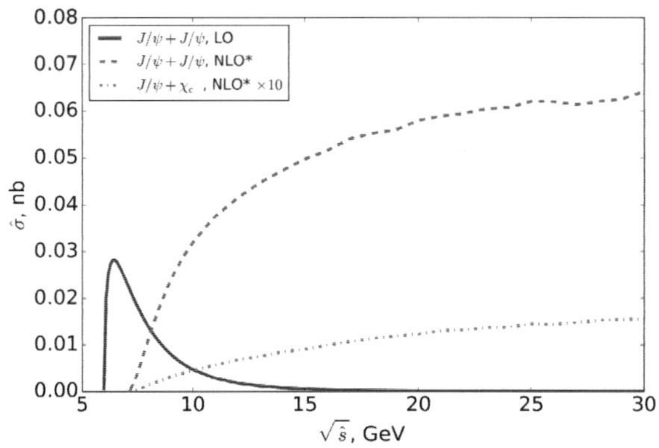


**Figure 2.** Partonic cross sections of LO $J/\psi J/\psi$ (solid line), NLO $J/\psi J/\psi$ (dashed line) and $J/\psi\chi_c$ (dotted line) reactions

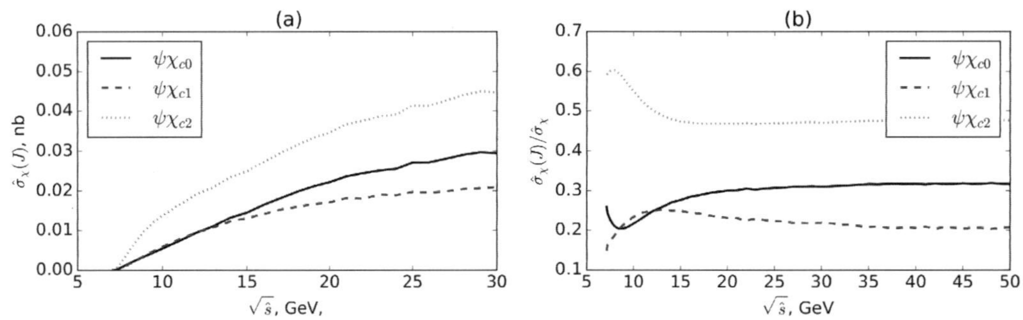


**Figure 3.** Partonic cross sections $\hat{\sigma}_\chi(J) = \hat{\sigma}(gg \to J/\psi\chi_{cJ}g)$. The value $\hat{\sigma}_\chi$ in the right figure stands for total partonic cross section $\sum_J \hat{\sigma}_\chi(J)$

also interesting to note that contributions of symmetric color structure $d_{abc}$ cancel and only antisymmetric one $f_{abc}$ is left. These cancellation can be explained by assigning negative and positive charge parities to $f_{abc}$ and $d_{abc}$ respectively.

The situation is completely different in the case of $J/\psi\chi_c$ final state. First of all, $\chi_c$ meson can couple to gluon pair, so only half of the processes shown in Fig.1(b) diagrams cancel. Since we have to keep some quarks' relative momentum to produce $P$-wave meson state some of Final State Radiation (FSR) diagrams are collinearly divergent. This divergence can be absorbed if one takes into account contributions of colour-octet components, but we prefer to regularize them by imposing a restriction $k_g p_\chi > \Delta^2$. As for shown in Fig.1(d) ISR diagrams, there contribution completely cancel due to charge parity conservation in $gg \to J/\psi\chi_c$ block. Finally, our calculations show that in the case of $J/\psi\chi_c$ final state only symmetric colour structure $d_{abc}$ is left.

To summarize, we can see, that by changing the final state one can switch on and off different sets of the underlying Feynman diagrams and change the overall colour structure of the corresponding amplitude.

In Fig.2 we show partonic cross sections of $J/\psi J/\psi$ at LO, NLO, and feed-down from radiative $J/\psi\chi_c$ decays. It is clearly seen, that the leading order cross section falls rapidly with the increase of partonic energy, while NLO ones grow. It should be also noted that NLO corrections to double $J/\psi$ production cross section are comparable with LO. In Fig.3 partonic energy dependence of

Journal of Physics: Conference Series **770** (2016) 012004 doi:10.1088/1742-6596/770/1/012004

| Cut | $J/\psi + J/\psi$, LO | $J/\psi + J/\psi$, NLO* | $J/\psi + \chi_{cJ}$, f.-d. | DPS |
|---|---|---|---|---|
| $\Delta = 1$ GeV | $1.29 \pm 0.02\,\mathrm{nb}$ | $4.47 \pm 0.79\,\mathrm{nb}$ | $10.8 \pm 1.54\,\mathrm{pb}$ | $6.2 \pm 2.1\,\mathrm{nb}$ |
| $\Delta = 3$ GeV | | $1.68 \pm 0.33\,\mathrm{nb}$ | $2.41 \pm 0.4\,\mathrm{pb}$ | $4.0 \pm 0.6\,\mathrm{nb}$ |

**Table 1.** Hadronic cross sections in the LHCb acceptance at $\sqrt{s} = 13\,\mathrm{TeV}$.

$J/\psi\chi_{cJ}$ partonic cross sections is shown. From the right panel of this figure it is clear that in low energy region cross sections of scalar and tensor meson production increase, while in the case of axial meson the situation is infrared safe. Possible explanation of this behaviour is that due to Landau-Yang theorem axial meson cannot couple to two massless gluons. The same situation was observed in the case of single charmonium production [2, 3].

## 3. Hadronic Results

Let us now discuss theoretical predictions for double charmonium production in hadronic experiments. In the following results for LHCb detector (with the cut $2 < y < 4.5$ set on rapidity of final charmonia) are presented.

It turns out, that there are two possible mechanisms that can contribute to the considered process. The first one is usual single parton scattering (SPS), when cross section of the hadronic process is written as convolution of the discussed in previous section partonic cross sections and gluons' distribution functions (in the following CT10, CT14 parametrizations [16, 17] are used):

$$d\sigma = \int_0^1 dx_1 dx_2 f_g(x_1, \mu) f_g(x_2, \mu) d\hat{\sigma}(\mu). \tag{1}$$

It is important to note that both partonic distribution functions and strong coupling constant in partonic cross section depend on scale parameter. In order to study dependence of final results on scale choice we vary the value of this parameter from $\mu/2$ to $\mu$, where $\mu^2 = 16\,m_c^2 + (p_T^\psi)^2 + (p_T^\chi)^2$ is the transverse mass of the pair. The other possible mechanism is so called double parton scattering (DPS), where it is assumed that final charmonia are produced in two almost independent partonic reactions. In this case the formula for hadronic cross section is surprisingly simple:

$$\sigma = \frac{1}{2}\frac{\sigma_\psi \sigma_\psi}{\sigma_{\mathrm{eff}}}, \tag{2}$$

where $\sigma_\psi = 15.3 \pm 1\,\mu b$ and $\sigma_{\mathrm{eff}} = 18 \pm 1.8\,\mathrm{mb}$ are the cross section of single $J/\psi$ production at LHCb [18] and the effective cross section [19]respectively. Our theoretical predictions for hadronic cross sections are shown in table 1. It is worth noting that in paper [12] the value $\sigma_{\mathrm{eff}} = 8.2 \pm 2.0 \pm 2.9\,\mathrm{mb}$ was proposed. It is clear, that with this value of the effective cross section the contribution of DPS mechanism is approximately doubled.

From presented table it is evident that contributions of different mechanisms are comparable to each other. To separate them one can study distributions over different kinematical variables. In Fig. 4, for example, we show distributions over invariant mass and transverse momentum of $J/\psi J/\psi$ pair with the cut-off parameter $\Delta = 1\,\mathrm{GeV}$ (it turns out that the form of these distributions only slightly depends on $\Delta$). It can be seen that in both cases distributions of different mechanisms are nearly the same. The situation is completely different for distribution over azimuthal asymmetry $\Delta\phi = |\phi_1 - \phi_2|$. From Fig. 5 it is clear that in this case the distribution form for different mechanisms are different and strongly depends on the choice of the cut-off parameter.

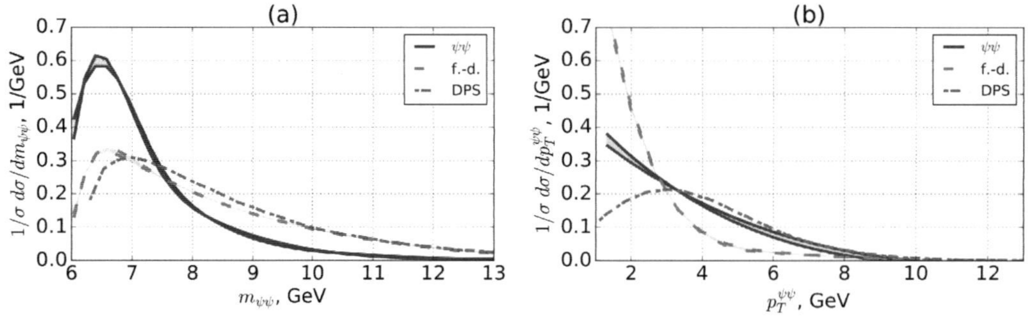


**Figure 4.** Distributions over invariant mass (left panel) and transverse momentum (right panel) of the charmonium pair

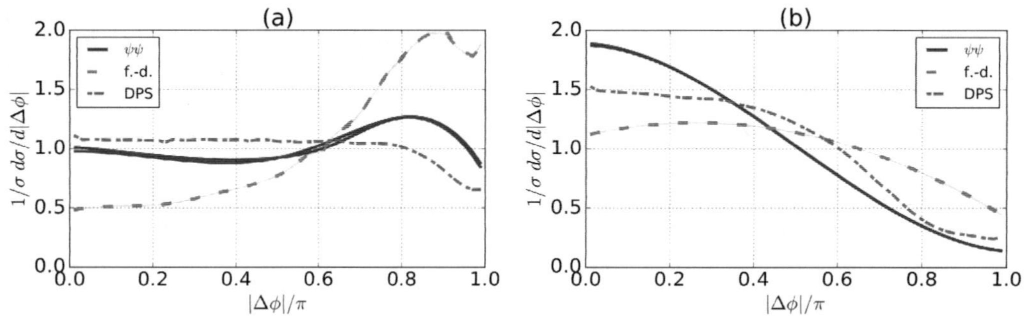


**Figure 5.** Distribution over $\Delta\phi = |\phi_1 - \phi_2|$ for $\Delta = 1\,\mathrm{GeV}$ (left panel) and $\Delta = 3\,\mathrm{GeV}$ (right panel)

We have analyzed also distributions over other kinematical variables. To be clear but concise it is desirable to quantitize in some way difference between them. In order to do this one can use, for example, the parameter connected with the Pearson correlator:

$$A_a^{(i,j)} \;=\; 1 - \left\langle \frac{d\sigma_i}{da}, \frac{d\sigma_j}{da} \right\rangle, \tag{3}$$

where $a$ is some kinematical variable ($m_{\psi\psi}$, $p_T^{\psi\psi}$ are invariant mass and trasverse momentum of final charmonium pair, $\Delta\phi = \phi_1 - \phi_2$, $\Delta y = y_1 - y_2$, and $A^T = (p_T^{\psi 1} - p_T^{\psi 2})/(p_T^{\psi 1} + p_T^{\psi 2})$ are azimuthal, rapidity, and transsvers momentum assymetries, $y_{J/\psi}$ is the rapidity of single charmonium), and $i$, $j$ stand for different subprocesses. From table 2, where the values of this parameter are shown, it is clear, for example, that $y_{J/\psi}$ distribution is post suitable for separation of direct $J/\psi J/\psi$ and feed-down from $J/\psi\chi_c$ signals.

## 4. Conclusion

Let me summarize the brief results of the talk.

It was shown, that in contrast to leading order next-to-leading-order approach is satisfactory for description of double charmonium production at LHC. Our calculations show that in the case of $J/\psi J/\psi$ final state NLO contributions are comparable or even larger then LO ones. It also turns out that production mechanisms on the partonic level depend strongly on the final

| | Combination | $m_{\psi\psi}$ | $p_T^{\psi\psi}$ | $\lvert\Delta\phi\rvert$ | $\Delta y$ | $A_T$ | $y_{J/\psi}$ |
|---|---|---|---|---|---|---|---|
| $\Delta = 1\,\mathrm{GeV}$ | $J/\psi + J/\psi$ vs $J/\psi + \chi_c$ | 0.10 | 0.02 | 0.26 | 0.10 | 0.01 | **0.54** |
| | $J/\psi + J/\psi$ vs DPS | 0.21 | 0.21 | **0.79** | 0.09 | 0.00 | 0.55 |
| | $J/\psi + \chi_c$ vs DPS | 0.03 | **0.34** | 0.26 | 0.01 | 0.01 | 0.01 |
| $\Delta = 3\,\mathrm{GeV}$ | $J/\psi + J/\psi$ vs $J/\psi + \chi_c$ | 0.07 | 0.01 | 0.14 | 0.03 | 0.05 | 0.16 |
| | $J/\psi + J/\psi$ vs DPS | **0.20** | 0.02 | 0.04 | 0.11 | 0.00 | 0.15 |
| | $J/\psi + \chi_c$ vs DPS | 0.04 | 0.06 | 0.04 | 0.03 | 0.04 | 0.01 |

**Table 2.** Correlation parameters $\mathcal{P}^a_{ij}$. Numbers shown in bold signals that corresponding kinematical distribution can be potentially used to discriminate two corresponding channels.

state. For example, for $J/\psi\chi_c$ leading order processes are forbidden completely. By switching the final state one can also turn on or off different sets of the underlying Feynman diagrams.

At the hadronic level we have calculated cross section and various differential distributions for $J/\psi J/\psi$ and $\chi_c J/\psi$ pair production at LHCb, $\sqrt{s} = 13\,\mathrm{TeV}$. The effect of double parton scattering mechanism was also estimated. In the talk it was shown that azimuthal and $|\Delta y$ asymmetries are most suitable to separate different kinds of contributions.

In our future work we are going to consider different polarization asymmetries in NLO $J/\psi J/\psi$ and $\chi_c J/\psi$ production.

More detailed description of the discussed processes can be found in [20]. The author would like to thank coauthors of this paper A.K. Likhoded and S.V. Poslavsky. I would also like to thank I. Belyaev for fruitful discussions. The work was financially supported by he Russian Foundation of Basic Research grant #14-02-00096.

## References

[1] S. P. Baranov and A. H. Rezaeian, Phys. Rev. **D93**, 114011 (2016), arXiv:1511.04089[hep-ph].
[2] A. K. Likhoded, A. V. Luchinsky, and S. V. Poslavsky, Phys. Rev. **D86**, 074027 (2012), arXiv:1203.4893[hep-ph].
[3] A. K. Likhoded, A. V. Luchinsky, and S. V. Poslavsky, Phys. Rev. **D90**, 074021 (2014), erXiv:1409.0693[hep-ph].
[4] R. Aaij et al. (LHCb), Phys. Lett. **B707**, 52 (2012), arXiv:1109.0963[hep-ex].
[5] V. M. Abazov et al. (D0), Phys. Rev. **D90**, 111101 (2014), arXiv:1406.2380[hep-ex].
[6] P. Ronchese, in *12th Conference on the Intersections of Particle and Nuclear Physics (CIPANP 2015) Vail, Colorado, USA, May 19-24, 2015* (2015), arXiv:1509.09276[hep-ex], URL http://inspirehep.net/record/1395464/files/arXiv:1509.09276.pdf.
[7] C. Collaboration (CMS) (2016).
[8] V. G. Kartvelishvili and S. M. Esakiya, Yad. Fiz. **38**, 722 (1983).
[9] B. Humpert and P. Mery, Z. Phys. **C20**, 83 (1983).
[10] A. V. Berezhnoy, A. K. Likhoded, A. V. Luchinsky, and A. A. Novoselov, Phys. Rev. **D84**, 094023 (2011), arXiv:1101.5881[hep-ph].
[11] J.-P. Lansberg and H.-S. Shao, Phys. Rev. Lett. **111**, 122001 (2013), arXiv:1308.0474[hep-ph].
[12] J.-P. Lansberg and H.-S. Shao, Phys. Lett. **B751**, 479 (2015), arXiv:1410.8822[hep-ph].
[13] L.-P. Sun, H. Han, and K.-T. Chao (2014), arXiv:1404.4042[hep-ph].
[14] G. T. Bodwin, E. Braaten, and G. P. Lepage, Phys. Rev. **D51**, 1125 (1995), [Erratum: Phys. Rev.D55,5853(1997)], hep-ph/9407339.
[15] E. Braaten and J. Lee, Phys. Rev. **D67**, 054007 (2003), [Erratum: Phys. Rev.D72,099901(2005)], hep-ph/0211085.
[16] S. Dulat, T.-J. Hou, J. Gao, M. Guzzi, J. Huston, P. Nadolsky, J. Pumplin, C. Schmidt, D. Stump, and C. P. Yuan, Phys. Rev. **D93**, 033006 (2016), arXiv:1506.07443[hep-ph].
[17] H.-L. Lai, M. Guzzi, J. Huston, Z. Li, P. M. Nadolsky, J. Pumplin, and C. P. Yuan, Phys. Rev. **D82**, 074024 (2010), arXiv:1007.2241[hep-ph].
[18] R. Aaij et al. (LHCb), JHEP **10**, 172 (2015a), arXiv:1509.00771[hep-ex].
[19] R. Aaij et al. (LHCb) (2015b), arXiv:1510.05949[hep-ex].

[20] A. K. Likhoded, A. V. Luchinsky and S. V. Poslavsky, Phys. Rev. D **94**, no. 5, 054017 (2016) [arXiv:1606.06767 [hep-ph]].

Journal of Physics: Conference Series **770** (2016) 012005 doi:10.1088/1742-6596/770/1/012005

# Production asymmetries of $b$ and $c$ hadrons at LHCb

**F Ferrari on behalf of the LHCb Collaboration**
[1]Università di Bologna, Dipartimento di Fisica ed Astronomia, via Irnerio 46, (40126) Bologna, Italy
[2]INFN - Sezione di Bologna, viale B. Pichat 6/2, (40126) Bologna, Italy
E-mail: fabio.ferrari@cern.ch

**Abstract.** Using a data sample corresponding to an integrated luminosity of 1.0 fb$^{-1}$, collected by LHCb in $pp$ collisions at a centre-of-mass energy of 7 TeV, the $D_s^+$, $D^+$, $B^0$ and $B_s^0$ production asymmetries are determined. These quantities are measured by means of $D_s^+ \to \phi(K^+K^-)\pi^+$, $D^+ \to K_S^0(\pi^+\pi^-)\pi^+$, $B^0 \to J/\psi(\mu^+\mu^-)K^{*0}(K^+\pi^-)$, $B^0 \to D^-(K^+\pi^-\pi^-)\pi^+$ and $B_s^0 \to D_s^-(K^+K^-\pi^-)\pi^+$ decays. Moreover, using the full LHCb Run 1 integrated luminosity, corresponding to 1.0 fb$^{-1}$ at 7 TeV and 2.0 fb$^{-1}$ at 8 TeV, the production and decay asymmetry of the $\Lambda_b^0$ baryon, $A_{\mathrm{P+D}}(\Lambda_b^0)$, is determined by means of $\Lambda_b^0 \to J/\psi(\mu^+\mu^-)pK^-$ decays. All asymmetries are measured as a function of $p_{\mathrm{T}}$ and $y$ (or $\eta$ in the $D^+$, $B^0$ and $B_s^0$ cases) of the involved hadrons, in order to check if there is a dependence on kinematics.

## 1. Introduction

The production rates of $b(c)$- and $\bar{b}(\bar{c})$-hadrons in $pp$ collisions at the LHC are not expected to be identical. This phenomenon, commonly referred to as production asymmetry, is related to the fact that the $\bar{b}(\bar{c})$ quark produced in the hard scattering of the two colliding protons could coalesce with a $u$ or $d$ spectator quark in the proton remnants, whereas the opposite is not possible for a $b(c)$ quark. Some theoretical models also predict an enhancement of this effect in the forward region, where the produced $b\bar{b}(c\bar{c})$ pair is closer to the remnants of the colliding protons; finally, there are also predictions of other effects coming into play at high transverse momentum [1],[2],[3].

The determination of production asymmetries is very important, since in *CP* violation measurements one needs to disentagle the physical asymmetry from other spurious effects, that could resemble the effect of *CP* violation. The production asymmetries of $D_s^+$, $D^+$, $B^0$ and $B_s^0$ mesons are defined as

$$A_{\mathrm{P}}(B^0_{(s)}) = \frac{\sigma(\overline{B}^0_{(s)}) - \sigma(B^0_{(s)})}{\sigma(\overline{B}^0_{(s)}) + \sigma(B^0_{(s)})}, \quad A_{\mathrm{P}}(D^+_{(s)}) = \frac{\sigma(D^+_{(s)}) - \sigma(D^-_{(s)})}{\sigma(D^+_{(s)}) + \sigma(D^-_{(s)})}, \tag{1}$$

where $\sigma$ stands for the production cross section.

The production and decay asymmetry for the $\Lambda_b^0 \to J/\psi pK^-$ mode can be written as

$$A_{\mathrm{P+D}}(\Lambda_b^0) = \frac{\sigma(\Lambda_b^0) - \sigma(\overline{\Lambda}_b^0)}{\sigma(\Lambda_b^0) + \sigma(\overline{\Lambda}_b^0)} + A_{CP}(\Lambda_b^0 \to J/\psi pK^-) \tag{2}$$

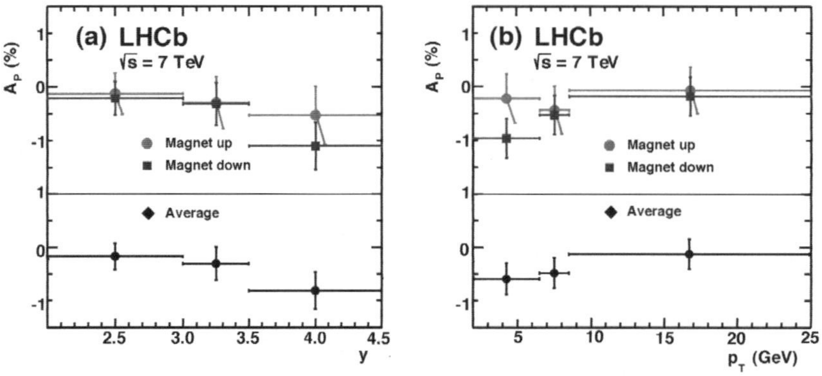


**Figure 1.** Production asymmetry of $D_s^+$ mesons as a function of (a) rapidity and (b) transverse momentum for (red circles) magnet up and (blue squares) magnet down data. The results obtained for both magnet polarities are compatible. No significant trend is observed within the current experimental precision.

where the first term on the right hand side is the analogue of the terms in Equation (1) and the second term is the *CP* asymmetry in the decay. If one assumes that *CP* violation in the $\Lambda_b^0 \to J/\psi p K^-$ decay is zero, then $A_{\mathrm{P+D}}(\Lambda_b^0) = A_{\mathrm{P}}(\Lambda_b^0)$.

## 2. $D_s^+$ production asymmetry

The $D_s^+$ production asymmetry [4] can be written as the sum of its various contributions

$$A_{\mathrm{P}}(D_s^+) = A_{\mathrm{RAW}}(D_s^+ \to \phi\pi^+) - A_{CP}(D_s^+ \to \phi\pi^+) - A_{\mathrm{D}}(\pi^+) + A_{\mathrm{D}}(K^+K^-), \quad (3)$$

where $A_{\mathrm{RAW}}(D_s^+ \to \phi\pi^+)$ is defined as the difference between $D_s^+$ and $D_s^-$ signal yields divided by their sum. The *CP* asymmetry in the decay is neglected, as this mode is Cabibbo favoured and no significant *CP* violation is expected [5],[6]. As the final state is symmetric with respect to charges of the kaons, the detection asymmetry of the kaon pair, $A_{\mathrm{D}}(K^+K^-)$, is zero. Hence the only correction that needs to be applied is due to the $\pi^\pm$ detection asymmetry. This latter quantity can be derived by taking the ratio of the pion detection efficiencies $\varepsilon(\pi^+)/\varepsilon(\pi^-)$.

In order to measure the pion efficiencies, the decay sequence $D^{*+} \to \pi_s^+ D^0(K^-\pi^+\pi^+\pi^-)$ is employed. Assuming that the $D^{*+}$ comes from the primary vertex, there are enough constraints to allow for the detection of this decay even if one pion from the $D^0$ decay is missed. One can also fully reconstruct this decay, if no charged track is missed. The ratio of fully to partially reconstructed $D^0$ decays gives a measurement of the pion detection efficiency.

The signal yield is obtained performing binned maximum likelihood fits to the invariant mass spectrum. The signal model employed is composed of triple Gaussians where all parameters are free to vary, except that two of the three Gaussians that are required to have a common mean. The background is parameterised by means of a second order polynomial.

The overall production asymmetry in the ranges $2.0 < y < 4.5$ and $p_{\mathrm{T}} > 2$ GeV is

$$A_{\mathrm{P}}(D_s^+) = (-0.33 \pm 0.22 \pm 0.10)\%$$

where the first error is statistical and the second systematic. The production asymmetry as a function of $y$ and $p_{\mathrm{T}}$ of the $D_s^+$ meson is shown in Figure 1. No significant dependence on these variables is observed within the current experimental uncertainties.

Journal of Physics: Conference Series **770** (2016) 012005 doi:10.1088/1742-6596/770/1/012005

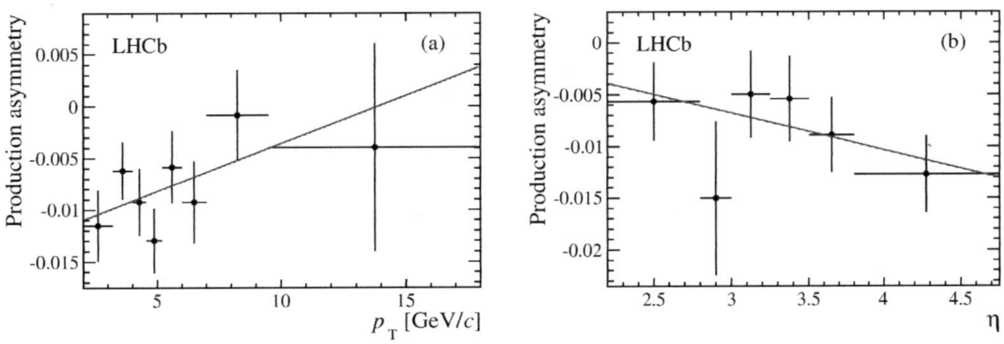


**Figure 2.** Production asymmetry of $D^+$ mesons as a function of (a) transverse momentum and (b) pseudorapidity. Although non-zero slopes are found as central values, the uncertainties are still too large to draw any conclusion on the presence of a kinematic dependence.

## 3. $D^+$ production asymmetry

The $D^+$ production asymmetry [7] can be obtained by

$$A_{\mathrm{P}}(D^+) = A_{\mathrm{RAW}}(D^+ \to K^0_S\pi^+) - A_{CP}(D^+ \to K^0_S\pi^+) + A_{\mathrm{D}}(\pi^+) + A_\epsilon, \tag{4}$$

where the raw asymmetry is defined taking the difference of $D^+$ and $D^-$ signal yields divided by their sum. The number of signal candidates is obtained by means of binned maximum likelihood fits to the $K^0_S\pi^+$ invariant mass distribution. The signal component is parameterised using Crujiff functions [8], while the background is described by a linear function plus a Gaussian function accounting for $D^+_s \to K^0_S\pi^+\pi^0$ decays, where the neutral pion is not reconstructed.

*CP* violation in the decay, $A_{CP}(D^+ \to K^0_S\pi^+)$, is estimated to be at most $1 \times 10^{-4}$ [9] and thus neglected. The pion detection asymmetry, $A_{\mathrm{D}}(\pi^+)$, is corrected for using the results obtained in the $D^+_s$ production asymmetry measurement.

The asymmetry due to *CP* violation in the neutral kaon system, $A_\epsilon$, depends on the decay time acceptance for $K^0_S$ mesons. This quantity is obtained by fitting the $K^0_S$ decay time distribution with an empirical function. The value obtained for the asymmetry is $A_\epsilon = (2.831^{+0.003}_{-0.004}) \times 10^{-4}$, where the quoted uncertainty is statistical only.

The $D^+$ production asymmetry is obtained in the kinematic range $2.0 < p_{\mathrm{T}} < 18.0$ GeV and $2.20 < \eta < 4.75$, excluding the region with $2.0 < p_{\mathrm{T}} < 3.2$ GeV and $2.20 < \eta < 2.80$, and is found to be

$$A_{\mathrm{P}}(D^+) = (-0.96 \pm 0.26 \pm 0.18)\%.$$

The production asymmetry as a function of $D^+$ meson $p_{\mathrm{T}}$ and $\eta$ is shown in Figure 2. No dependence is observed within the current experimental uncertainties.

## 4. $B^0$ and $B^0_s$ production asymmetries

The $B^0$ and $B^0_s$ production asymmetries [10] are obtained by means of two dimensional maximum likelihood invariant mass and decay time fits to the relevant distributions.

The signal mass component is parameterised by two Gaussians with the same mean and different widths. The partially reconstructed contributions are described by shapes obtained from a kernel estimation technique [11] based on invariant mass distributions obtained from fully simulated events. The combinatorial component is fitted using an exponential function.

As for the decay time, the signal is parameterized by a theoretical function describing the time dependent decay rate of a $B^0_{(s)}(\overline{B}^0_{(s)})$ to a flavour specific final state $f(\overline{f})$ convolved with a

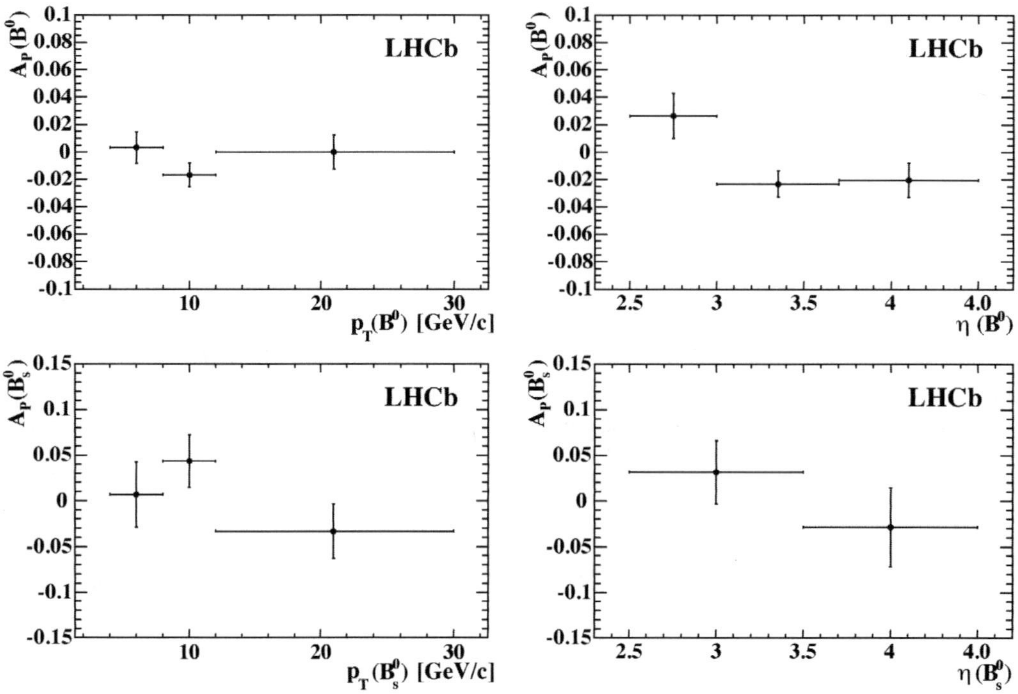


**Figure 3.** Production asymmetry of (top ) $B^0$ and (bottom) $B_s^0$ mesons as a function of (left) transverse momentum and (right) pseudorapidity. No significant trend is observed within the current experimental precision, although the first bin in $\eta$ hints at a possible feature.

resolution function and multiplied by an acceptance function. Concerning the combinatorial and partially reconstructed background, the shapes are obtained from empirical parameterizations of the decay time spectra in high and low invariant mass sidebands in data.

The production asymmetries as a function of the *b*-mesons $p_T$ and $\eta$ are shown in Figure 3. The $B^0$ and $B_s^0$ production asymmetries, integrated in the ranges $4 < p_T < 30$ GeV/$c$ and $2.5 < \eta < 4.5$, are found to be

$$A_P(B^0) = (-0.35 \pm 0.76 \pm 0.28)\% \text{ and } A_P(B_s^0) = (\quad 1.09 \pm 2.61 \pm 0.66)\%,$$

where the first uncertainties are statistical, and the second uncertainties are systematic. No evidence of production asymmetry is found within the current precision.

## 5. $\Lambda_b^0$ production asymmetry

The $\Lambda_b^0$ production and decay asymmetry [12] can be obtained as the sum of various terms

$$A_{P+D}(\Lambda_b^0) = A_{RAW}(\Lambda_b^0 \to J/\psi p K^-) + A_{PID}(\Lambda_b^0 \to J/\psi p K^-) + A_D(p) + A_D(K), \qquad (5)$$

where $A_{RAW}(\Lambda_b^0 \to J/\psi p K^-)$ is the raw asymmetry defined as the difference between $\Lambda_b^0$ and $\overline{\Lambda}_b^0$ signal yields divided by their sum, $A_{PID}(\Lambda_b^0 \to J/\psi p K^-)$ is the asymmetry between the particle identification asymmetries of $\Lambda_b^0$ and $\overline{\Lambda}_b^0$, and $A_D(p)(A_D(K))$ is the proton (kaon) detection asymmetry.

The signal yields are obtained from unbinned extended maximum likelihood fits to the invariant mass distributions of $\Lambda_b^0$ and $\overline{\Lambda}_b^0$. The signal component is parameterised using an

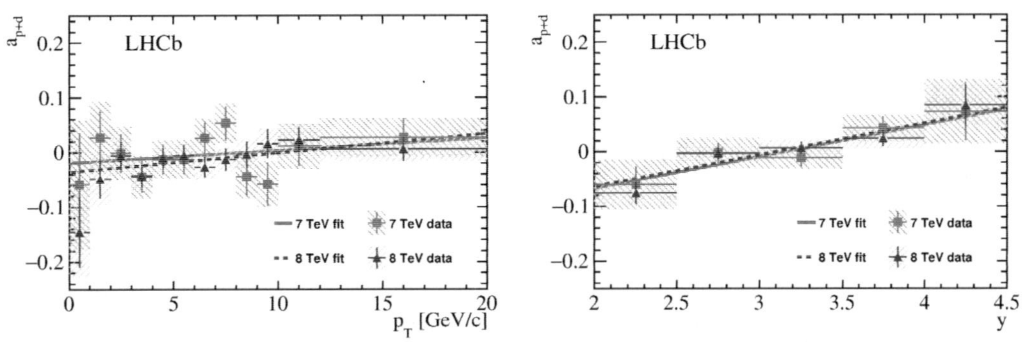


**Figure 4.** Production and decay asymmetry as a function of (left) transverse momentum and (right) rapidity of the $b$-hadrons.

double sided Crystal Ball function, that is composed by a Gaussian shape with power-law tails. The combinatorial background is modelled using an exponential function, while the contribution from $\Lambda_b^0 \to J/\psi K^+\overline{p}$ decays is taken into account using the same shape as for the signal.

The particle identification correction is calculated using samples of data with tracks from the decays $J/\psi \to \mu^+\mu^-$, $D^{*+} \to D^0(K^-\pi^+)\pi^+$ and $\Lambda_c^+ \to pK^-\pi^+$, which can be identified without requiring any particle identification cut on the daughters.

The kaon detection asymmetry is obtained from a previous LHCb publication [13], while the proton detection asymmetry is estimated from simulated events.

The quantity $A_{\rm P+D}$ as a function of $p_{\rm T}$ and $y$ is shown in Figure 4. The $p_{\rm T}$ result is consistent with a null slope, whereas the combined result for $A_{\rm P+D}(y)$ gives

$$A_{\rm P+D}(y) = (-0.001 \pm 0.007) + (0.058 \pm 0.014)(y - \langle y \rangle),$$

where $\langle y \rangle = 3.1$ is the average rapidity of the $\Lambda_b^0$ in the sample. The quoted error is the combination of the statistical and systematic contributions to the uncertainty.

## 6. Conclusion

Thanks to the unprecented amount of $b$- and $c$-hadrons produced in the LHC $pp$ collisions, LHCb is in an excellent position to perform high-precision *CP* violation measurements. One key ingredient to perform such measurements is the determination of the $b$- and $c$-hadron production asymmetries. These proceedings summarise the present status at LHCb for $D_s^+$, $D^+$, $B^0$, $B_s^0$ and $\Lambda_b^0$ hadrons. LHCb will continue to produce results in this sector as more data are being collected.

## References

[1] Chaichian M and Fridman A, *On a possibility for measuring effects of CP violation at pp colliders*, *PLB*, 298, 1, 1993.

[2] Norrbin E and Vogt R, *Bottom Production Asymmetries at the LHC*, `arXiv:hep-ph/0003056`.

[3] Norrbin E and Sjöstrand T, *Production and Hadronization of Heavy Quarks*, *European Physics Journal C*, 17, 1, 2000.

[4] Aaij R *et al.* (LHCb Collaboration), *Measurement of the $D_s^+ - D_s^-$ production asymmetry in 7 TeV pp collisions*, Physics Letters B 713 (2012), pp. 186-195.

[5] Grossman Y, Kagan A L, and Nir Y, *New physics and CP violation in singly Cabibbo suppressed D decays*, *Phys. Rev.* D 75 (2007) 036008, `arXiv:hep-ph/0609178`.

[6] Bergmann S and Nir Y, *New physics effects in doubly Cabibbo suppressed D decays*, *JHEP* 09 (1999) 031, `arXiv:hep-ph/9909391`.

[7] Aaij R *et al.* (LHCb Collaboration), *Measurement of the $D^{\pm}$ production asymmetry in 7 TeV pp collisions*, Phys. Lett. B. 718 (2013) pp. 902-907.

[8] del Amo Sanchez P *et al.* (BaBar Collaboration), *Study of $B \to X\gamma$ decays and determination of $|V_{td}/V_{ts}|$*, *Phys. Rev.* D82 (2010) 051101, arXiv:1005.4087.

[9] Bigi I I and Yamamoto H, *Interference between Cabibbo allowed and doubly forbidden transitions in $D^+ \to K_S^0$, $K_L^0 + \pi$ decays*, *Phys. Lett.* B349 (1995) 363, arXiv:hep-ph/9502238.

[10] Aaij R *et al.* (LHCb Collaboration), *Measurement of the $\overline{B}^0 - B^0$ and $\overline{B}_s^0 - B_s^0$ production asymmetries in pp collisions at $\sqrt{(s)} = 7$ TeV*, Phys. Lett. B739 (2014) 218.

[11] Cranmer K S, *Kernel estimation in high-energy physics*, *Comput. Phys. Commun.* 136 (2001) 198, arXiv:hep-ex/0011057.

[12] Aaij R *et al.* (LHCb Collaboration), *Study of the production of $\Lambda_b^0$ and $\overline{B}^0$ hadrons in pp collisions and first measurement of the $\Lambda_b^0 \to J/\psi p K^-$ branching fraction*, Chinese Physics C Vol. 40, No. 1 (2016) 011001.

[13] Aaij R *et al.* (LHCb Collaboration), *Measurement of CP asymmetry in $D^0 \to K^- K^+$ and $D^0 \to \pi^- \pi^+$ decays*, *JHEP* 07 (2014) 041, arXiv:1405.2797.

# Measurements of Open heavy-flavour Production with ALICE at the LHC

**S De[1,2] for the ALICE collaboration**

1. Departamento de Física Nuclear, Instituto de Física, Universidade de São Paulo, CEP: 05508-090, São Paulo, Brasil
2. Discipline of Physics, School of Basic Sciences, Indian Institute of Technology Indore, Simrol, M.P.- 453552, India

E-mail: `Sudipan.De@cern.ch`

**Abstract.**
In ALICE, open heavy-flavour production is studied through the measurements of the leptons (electrons and muons) from heavy-flavour hadron decays at central and forward rapidity and via the reconstruction of D-meson hadronic decays at mid-rapidity. An overview of the open heavy-flavour production with ALICE in pp ($\sqrt{s}$ = 2.76 TeV and 7 TeV), p–Pb ($\sqrt{s_{\rm NN}}$ = 5.02 TeV) and Pb–Pb ($\sqrt{s_{\rm NN}}$ = 2.76 TeV) collisions will be presented. We will discuss the production cross sections, modifications of the transverse momentum distributions, azimuthal anisotropic emissions and correlations with hadrons in comparison with various theoretical predictions.

## 1. Introduction

A deconfined state of quarks and gluons, generally known as the Quark–Gluon Plasma (QGP), is expected to be formed in ultrarelativistic heavy-ion collisions. Due to their large mass, heavy-quarks, i.e. charm and beauty quarks, are produced at the early stages of the collisions via hard scattering processes. Therefore, they can experience the complete evolution of the system and interact with the created hot medium. Theoretical model calculations based on Quantum Chromodynamics (QCD) predict a mass-dependent energy loss due to the mass-dependent restriction of the phase space of the gluon radiation. Therefore, heavy quarks are expected to lose less energy compared to the light quarks [1, 2]. In addition, gluons are considered to lose more energy than quarks due to their stronger colour coupling [3]. Experimentally this can be studied by measuring the nuclear modification factor of hadrons containing heavy quarks, which is defined as the ratio between the differential yield measured in Pb–Pb (or p–Pb) collisions and the cross section measured in pp collisions multiplied by the average nuclear overlap function. The degree of the thermalisation of the heavy quarks in the medium can be studied by measuring the anisotropic flow parameter $v_2$, second harmonic of the Fourier expansion of the particle azimuthal distribution, of the heavy-flavour particles at low transverse momentum ($p_{\rm T}$). The anisotropic flow is caused by the initial asymmetries in the geometry of the system produced in non-central collisions [4]. The elliptic flow measurement at high $p_{\rm T}$ gives information about the path-length dependence of energy loss of the heavy quarks. Measurements of heavy-flavour particles in pp collisions are studied to test the perturbative QCD (pQCD) calculations and to set a baseline for both Pb–Pb and p–Pb collisions. Studies of heavy-quark production as a function of charged-particle multiplicity in pp collisions are expected to be sensitive to the interplay between hard and soft QCD processes and therefore they give insight into the multi-parton interactions (MPI) [5]. Measurements of open heavy-flavour production in p–Pb collisions can be used to disentangle Cold Nuclear Matter (CNM) effects, such as the modification of nuclear parton distribution functions (PDF) [6, 7] and $k_T$ broadening, from final-state effects occurring in the QGP. In this article we report the open heavy-flavour production via hadronic decays of D mesons ($D^0$, $D^+$ and $D^{*+}$) and the semileptonic decays of charm and beauty hadrons in pp collisions at $\sqrt{s} = 2.76$ TeV and 7 TeV, in p–Pb collisions at $\sqrt{s_{NN}} = 5.02$ TeV and in Pb–Pb collisions at $\sqrt{s_{NN}} = 2.76$ TeV.

The details of the ALICE detector are described in [8, 9]. Charged particles were reconstructed and identified in the central barrel, which consists of the Inner Tracking System (ITS), Time Projection Chamber (TPC) and Time-Of-Flight (TOF) detector at mid-rapidity ($|\eta| < 0.9$). Muons were identified using the Muon Spectrometer covering $-4 < \eta < -2.5$. Triggering and multiplicity determination are performed by two arrays of scintillator detectors V0A ($2.8 < \eta < 5.1$) and V0C ($-3.7 < \eta < -1.7$) placed on both sides of the interaction point. ElectroMagnetic Calorimeter (EMCal) is used to select the high $p_T$ electrons candidates. D-meson candidates are selected based on particle identification of the decay tracks and the decay topology using ITS, TPC and TOF detectors. The details of the reconstruction procedure can be found in [10]. Electron candidates are reconstructed using ITS, TPC, TOF and EMCal detector and muons are reconstructed with the Muon Spectrometer. The detailed procedures to measure the heavy-flavour hadrons decay electrons and muons are described in [11] and [12], respectively.

## 2. Results

### *2.1. Results in pp collisions*

The differential cross sections of prompt D mesons ($D^0$, $D^+$, $D^{*+}$) [13, 14], heavy-flavour hadron decay leptons [15, 12] and beauty-hadron decay electrons [16, 17] were measured in pp collisions at $\sqrt{s} = 2.76$ TeV and 7 TeV. The results are well described by the pQCD calculations [18, 19] within the systematic uncertainties. Differential cross section of prompt $D^0$ meson was measured down to $p_T = 0$ at $\sqrt{s} = 7$ TeV and results are well described by the pQCD calculations [20]. The left panel of Fig. 1 shows the self-normalised D-meson yield as a function of charged-particle multiplicity at $\sqrt{s} = 7$ TeV at mid-rapidity ($|y| < 0.5$). It shows a stronger than the linear increase as a function of the multiplicity for all the $p_T$ intervals. The results in the different $p_T$ ranges are found to be in agreement within the uncertainties. The results are found in qualitative agreement with the model predictions, which contain MPI and hydrodynamic effects (right panel of Fig. 1). It suggests that MPIs play an important role in heavy-flavour production in case of high-multiplicity events in pp collisions [10].

### *2.2. Results in p–Pb collisions*

The left panel of the Fig. 2 shows the combined measurement of the average nuclear modification factor ($R_{pPb}$) of prompt $D^0$, $D^+$ and $D^{*+}$ mesons as a function of $p_T$ for $1 < p_T < 20$ GeV/$c$ and $D^0$ mesons for $0 < p_T < 1$ GeV/$c$ at mid-rapidity in p–Pb collisions at $\sqrt{s_{NN}} = 5.02$ TeV [20]. Results are compared with the theoretical results containing CNM effects [20]. All the model predictions describe the data well for the entire $p_T$ range within the uncertainties. The right panel of Fig. 2 shows the $R_{pPb}$ of heavy-flavour hadron decay electrons as a function of $p_T$ for $0.5 < p_T < 20$ GeV/$c$ at mid-rapidity [11]. The $R_{pPb}$ of prompt D mesons and heavy-flavour hadron decay electrons is consistent with unity within uncertainties. Fig. 3 shows the $R_{pPb}$ of the heavy-flavour hadron decay muons at both forward rapidity ($2.5 < y_{cm} < 3.54$, proton-beam direction) and backward rapidity ($-4 < y_{cm} < -2.96$, Pb-beam direction). The $R_{pPb}$ is consistent with unity at both at forward and backward rapidity. However there is an enhancement above unity for $2 < p_T < 4$ GeV/$c$ for the backward rapidity region. In both

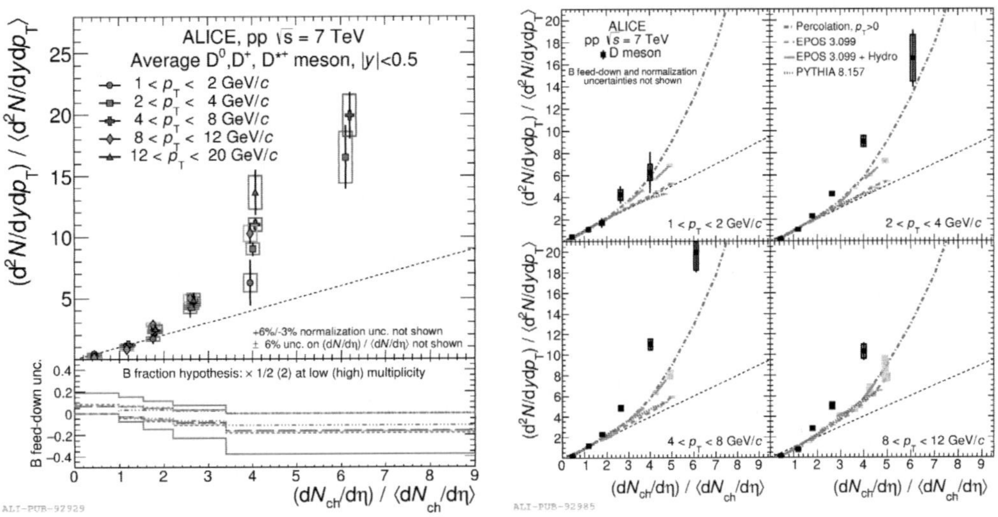


**Figure 1.** Left: Self-normalised yield of D mesons as a function of charged particle multiplicity in pp collisions at $\sqrt{s} = 7$ TeV [10]. Different symbols are for different $p_{\mathrm{T}}$ intervals. Right: Self-normalised yield of D mesons as a function of charged-particle multiplicity in different $p_{\mathrm{T}}$ intervals with different model predictions [10]. Different lines show the predictions from different theoretical calculations. Uncertainties on the data points are the statistical uncertainties and boxes represent the systematic uncertainties. The dotted lines show the linear increase to guide the eye.

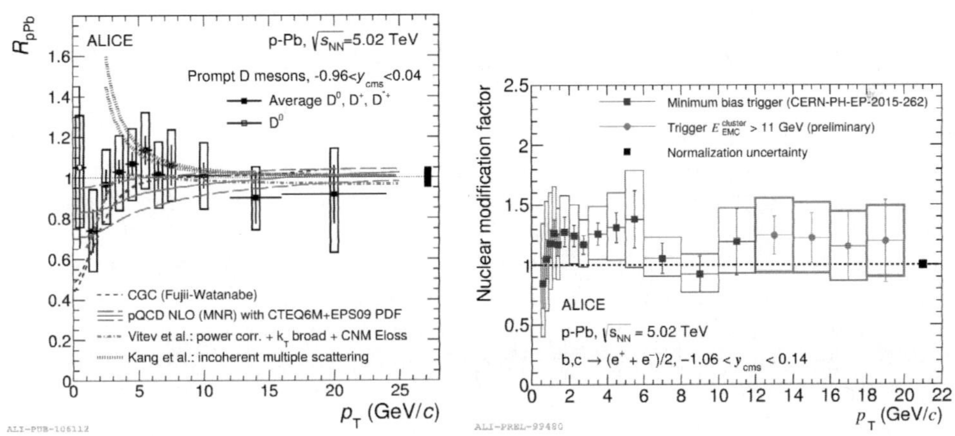


**Figure 2.** Left: Nuclear modification factor $R_{\mathrm{pPb}}$ of D mesons in p–Pb collisions at $\sqrt{s_{\mathrm{NN}}} = 5.02$ TeV [20]. Different lines represent different model predictions. Right: The $R_{\mathrm{pPb}}$ of heavy-flavour hadron decay electrons. Uncertainties on the data points are the statistical uncertainties and boxes represent the systematic uncertainties.

cases the results are well described by the model predictions within uncertainties. The $R_{\mathrm{pPb}}$ of electrons from beauty-hadron decays is consistent with unity within systematic uncertainties at high $p_{\mathrm{T}}$ and is well described by theoretical calculations that include CNM effects [21].

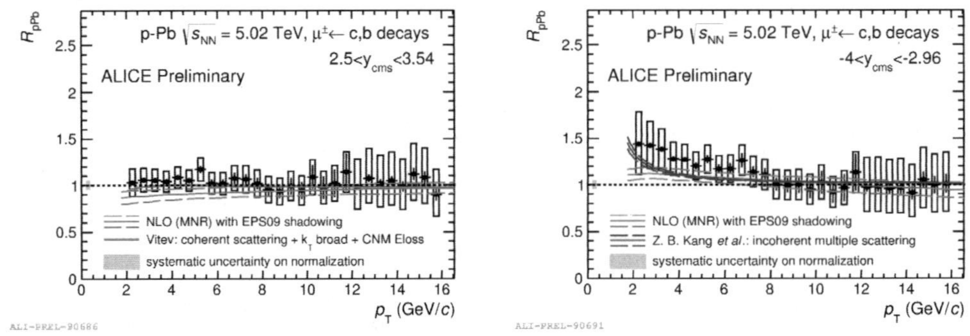


**Figure 3.** Nuclear modification factor $R_{\rm pPb}$ of heavy-flavour hadrons decay muons in p–Pb collisions at $\sqrt{s_{\rm NN}}$ = 5.02 TeV at forward rapidity ($2.5 < y_{\rm cm} < 3.54$) (left) and at backward rapidity ($-4 < y_{\rm cm} < -2.96$) (right). Uncertainties on the data points are the statistical uncertainties and boxes represent the systematic uncertainties. The lines show the different model predictions.

### *2.3. Results in Pb–Pb collisions*

The left panel of Fig. 4 shows the nuclear modification factor ($R_{\rm AA}$) as a function of $p_{\rm T}$ of D mesons for central Pb–Pb collisions at $\sqrt{s_{\rm NN}} = 2.76$ TeV at mid-rapidity. A strong suppression of the D-meson yield is observed at intermediate and high $p_{\rm T}$. The D-meson $R_{\rm AA}$ is compatible with the charged particle and pion $R_{\rm AA}$ within the uncertainties [22]. The results can be described by the model calculations including mass and colour charged dependent energy loss and different fragmentation functions for light quarks, heavy quarks and gluons [26]. The right panel of Fig. 4 shows the $R_{\rm AA}$ of prompt D mesons and non-prompt J/ψ from beauty-hadron decays from the CMS collaboration [23, 24] as function of the Pb–Pb collision centrality. A larger suppression for D mesons than for non-prompt J/ψ is observed in central collisions [25]. The data are well reproduced by the model calculations including mass dependent energy loss [26].

The left panel of Fig. 5 shows the $R_{\rm AA}$ of D mesons in central Pb–Pb collisions (0–10%) and right panel shows the elliptic flow ($v_2$) measurements of D mesons for mid-central collisions (30–50%) with the different model predictions [22, 27]. The positive $v_2$ of D mesons at low $p_{\rm T}$ indicates that charm quarks participate in the collective expansion of the medium. The models which explain the strong suppression seen in the $R_{\rm AA}$ tend to underestimate the $v_2$. The simultaneous model description of the heavy-flavour $R_{\rm AA}$ and $v_2$ is still challenging and can therefore help to constrain the theoretical understanding of the interaction of heavy quarks in the medium.

## 3. Summary

We presented the open heavy-flavour production in pp collisions at $\sqrt{s} = 2.76$ TeV and 7 TeV, in p–Pb collisions at $\sqrt{s_{\rm NN}} = 5.02$ TeV and in Pb–Pb collisions at $\sqrt{s_{\rm NN}} = 2.76$ TeV with ALICE at the LHC. The results from pp collisions are well described by the pQCD calculations within the uncertainties. Measurements of the multiplicity dependent self-normalised yield of D mesons in pp collisions indicate the MPI contribute to high-multiplicity events and are affecting heavy-flavour production. Strong suppression of heavy-flavour yield is observed in most central (0–10%) Pb–Pb collisions at intermediate and high $p_{\rm T}$. In p–Pb collisions the $R_{\rm pPb}$ is measured at mid-rapidity for D mesons and heavy-flavour hadron decay electrons as well as at forward and backward rapidity for heavy-flavour hadrons decay muons. For all the cases, the results are

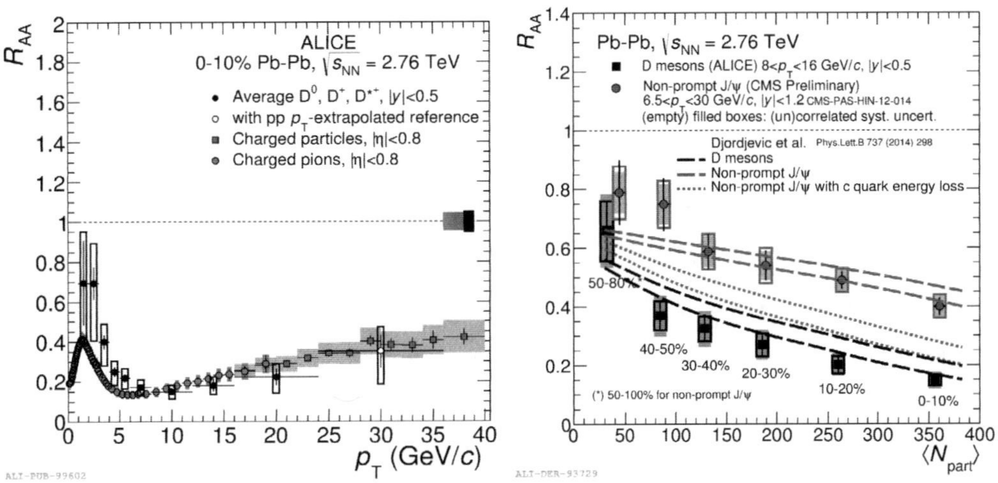


**Figure 4.** Left: Nuclear modification factor $R_{AA}$ of D mesons in Pb–Pb collisions with the $R_{AA}$ of charged-particles and pions at $\sqrt{s_{NN}} = 2.76$ TeV for 0–10% centrality [22]. Right: The $R_{AA}$ of D mesons in comparison with non-prompt J/ψ [25]. The dotted and dashed lines are from model predictions. Uncertainties on the data points are the statistical uncertainties and boxes represent the systematic uncertainties.

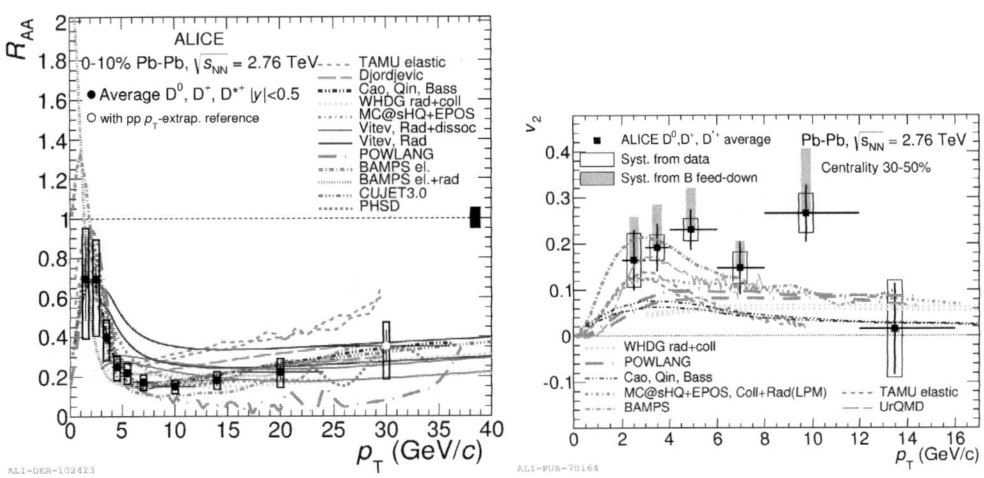


**Figure 5.** Nuclear modification factor $R_{AA}$ (left) and elliptic flow $v_2$ (right) of D mesons in Pb–Pb collisions at $\sqrt{s_{NN}} = 2.76$ TeV [22, 27]. The different lines represent various model predictions. Uncertainties on the data points are the statistical uncertainties and boxes represent the systematic uncertainties.

Journal of Physics: Conference Series **770** (2016) 012006 doi:10.1088/1742-6596/770/1/012006

consistent with unity indicating that the initial state effects are small and the strong suppression of the heavy-flavour yield observed in Pb–Pb collisions is due the presence of the hot and dense QCD matter. The nuclear modification factor of prompt D mesons is smaller than that of non-prompt J/ψ suggesting a mass dependent energy loss of the heavy quarks. The positive $v_2$ of the D mesons indicates collective motion of the charm quarks in the medium at low $p_{\rm T}$. The theoretical models are challenged by simultaneously reproducing the $R_{\rm AA}$ and $v_2$.

## References

[1] Yu. L. Dokshitzer and D. E. Kharzeev 2001 *Phys. Lett.* B **519** 199.
[2] N. Armesto *et al.* 2004 *Phys. Rev.* D **69** 114003.
[3] M. Gyulassy *et al.* 1994 *Nucl. Phys.* B **420** 583-614.
[4] K. Aamodt *et al.* [ALICE Collaboration] 2010 *Phys. Rev. Lett.* **105** 252302.
[5] T. Sjöstrand *et al.* 2006 *JHEP* **05** 026.
[6] K. Eskola *et al.* 2009 *JHEP* **065** 0904.
[7] D. de Florian *et al.* 2004 *Phys.Rev.* D **69** 074028.
[8] K. Aamodt *et al.* [ALICE Collaboration] 2008 *JINST* **3** S08002.
[9] B. Abelev *et al.* [ALICE Collaboration] 2014 *Int.J.Mod.Phys.* **A29** 1430044.
[10] J. Adam *et al.* [ALICE Collaboration] 2015 *JHEP* **09** 148.
[11] J. Adam *et al.* [ALICE Collaboration] 2016 *Phys. Lett.* B **754** 81-93.
[12] B. Abelev *et al.* [ALICE Collaboration] 2012 *Phys. Lett.* B **708** 265.
[13] B. Abelev *et al.* [ALICE Collaboration] 2012 *JHEP* **1201** 128.
[14] B. Abelev *et al.* [ALICE Collaboration] 2012 *JHEP* **1207** 191.
[15] B. Abelev *et al.* [ALICE Collaboration] 2012 *Phys.Rev.* D **86** 112007.
[16] B. Abelev *et al.* [ALICE Collaboration] 2014 *Phys. Lett.* B **738** 97.
[17] B. Abelev *et al.* [ALICE Collaboration] 2013 *Phys. Lett.* B **721** 13.
[18] M. Cacciari *et al.* 2012 *JHEP* **1210** 137.
[19] B. Kniehl *et al.* 2012 *EPJ* C **72** 2082.
[20] J. Adam *et al.* [ALICE Collaboration] arXiv: 1605.07569v1 (submitted to JHEP).
[21] J. Adam *et al.* [ALICE Collaboration] arXiv: 1609.03898v1 (submitted to JHEP).
[22] J. Adam *et al.* [ALICE Collaboration] 2016 *JHEP* **03** 081.
[23] S. Chatrchyan *et al.* [CMS Collaboration] 2012 *JHEP* **05** 063
[24] S. Chatrchyan *et al.* [CMS Collaboration] CMS-PAS-HIN-12-014.
[25] J. Adam *et al.* [ALICE Collaboration] 2015 *JHEP* **1511** 205.
[26] M. Djordjevic *et al.* 2014 *Phys. Rev. Lett.* **112** 042302, 2014 *Phys. Lett.* B **734** 286.
[27] B. Abelev *et al.* [ALICE Collaboration] 2014 *Phys. Rev.* C **90** 034904.

# Recent heavy flavor measurements from PHENIX at RHIC

**T O S Haseler for the PHENIX collaboration**

Department of Physics and Astronomy, Georgia State University, Atlanta, GA 30303, USA

E-mail: thaseler1@student.gsu.edu

**Abstract.** Heavy flavor quarks are an important probe of the initial state of the Quark Gluon Plasma formed in heavy-ion collisions. Bottom and charm quarks are primarily produced through hard interactions, early in the collision and experience the full time evolution of the medium. Measuring their production in $p+p$ collisions can also give a baseline reference to study larger collision systems, including asymmetric systems and can directly test pQCD calculations. At PHENIX open heavy flavor states can be measured through leptonic decay channels. Some measurements have utilized silicon vertex detectors to determine meson decay lengths in order to separate $D$ mesons from $B$ mesons. Recent measurements have been made at $\sqrt{s_{NN}} = 200$ and 500 GeV, with a variety of collision species, in both forward/backward and central rapidities.A review of the recent heavy flavor measurements from PHENIX will be presented in this proceeding.

## 1. Introduction

The Quark Gluon Plasma (QGP) is a strongly coupled state of matter composed of "free" quarks and gluons that is produced through heavy-ion collisions at the Relativistic Heavy Ion Collider (RHIC). The Pioneering High Energy Nuclear Interaction eXperiment (PHENIX) is positioned to study QGP through a variety of observables including open heavy flavor. Charm and bottom quarks are produced early in the collision system due to their large masses, $\Lambda^{QCD} << mass_{c,b}$, through the initial hard scattering phase of the collision process. They are a good probe of the QGP matter since they interact with it through scattering, leading to energy losses and ultimately experiencing the full space-time evolution of the medium. Open heavy flavor is studied in heavy ion collisions in which the QGP forms and it is also measured in small asymmetric collision systems in which no QGP is expected to form in order to understand Cold Nuclear Matter (CNM) effects. These effects are those associated with the presence of normal nuclear matter in a collision. Baseline measurements need to be made in $p+p$ collisions to serve as a reference to both Hot Nuclear Matter (HNM) and CNM effects and can be used to directly test pQCD theory. One such way of quantifying the HNM and CNM effects is through a parameter known as the nuclear modification factor

$$R_{AA} = \frac{dN^{AA}/dp_T}{< N_{coll} > dN^{pp}/dp_T}. \tag{1}$$

Recent results of open heavy flavor measurements made at PHENIX will be presented. The first result relies on the secondary displaced vertex from a silicon vertex detector in order to separate charm and bottom contributions due to their hadronic decay lengths in $Au + Au$ collisions at $\sqrt{s_{NN}} = 200$ GeV. The second result will detail production via a di-electron measurement in $p+p$ collisions as well as explore CNM in $d + Au$ collisions both measured at $\sqrt{s_{NN}} = 200$ GeV. Finally the last measurement discussed is that of $\sigma_{b\bar{b}}$ in $p + p$ collisions, at $\sqrt{s_{NN}} = 500$ GeV, via the like sign di-muon signal that will use the unique properties of $B^0$ oscillation.

## 2. Electrons from Semi-Leptonic $D$ and $B$ Meson Decay in $Au + Au$ Collisions

This measurement utilizes a new silicon vertex (VTX) detector installed for the RHIC 2011 run, in order to measure displaced vertices of heavy flavor contributions in the single electron yield. $D$ mesons have known lifetimes of $c\tau_{D^{\pm}} = 318\ \mu m$, $c\tau_{D^0} = 123\ \mu m$ and $B$ mesons lifetimes are $c\tau_{B^{\pm}} = 491\ \mu m$, $c\tau_{B^0} = 456\ \mu m$ which are then used to help separate the contributions through a distance of closest approach (DCA) analysis [1]. The analysis procedure breaks down the DCA distribution in the transverse plane (DCA$_T$) into five $p_T$ bins and accounts for known backgrounds using a variety of data driven and Monte Carlo methods. It then samples the inclusive heavy flavor invariant $p_T$ distribution from a previous PHENIX measurement [2], shown in Fig. 1 (a), and uses a Bayesian inference (a.k.a. unfolding) method to separate the charm and bottom contributions. One such $p_T$ bin is shown in Fig. 1 (b) that also includes the results of the unfolding procedure.

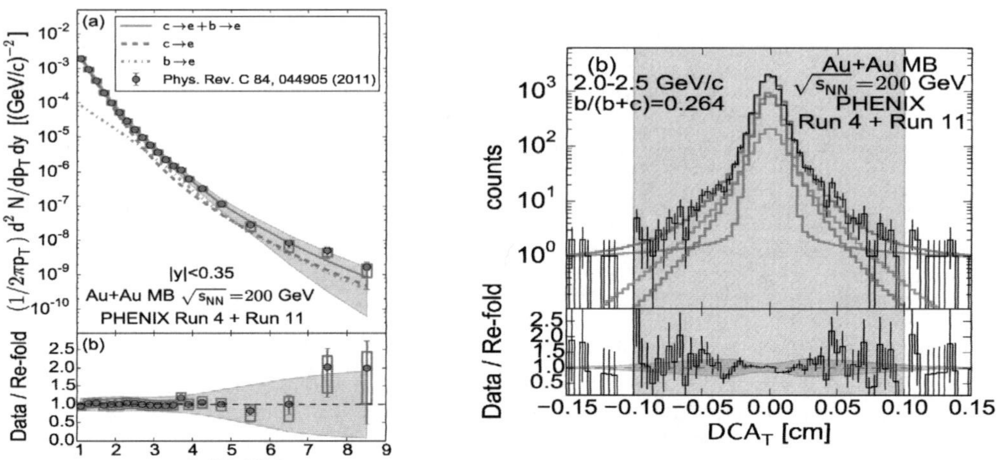


**Figure 1.** (a) The sampled inclusive invariant yield from heavy flavor single electrons as measured in [2]. (b) A sample DCA$_T$ distribution in the 2.0-2.5 GeV/$c$ $p_T$ kinematic window. The distribution is broken into charm, bottom and background contributions and summed together. The shaded region is the region of interest for the analysis. On the bottom of both plots are the ratios of the data and final results from the unfolding method.

Once the two contributions are separated, the bottom fraction, shown in Fig. 2 (a), is calculated as the ratio of heavy flavor from bottom decays to the inclusive heavy flavor contribution of the single electron yield. The result is compared to FONLL calculations and previous STAR [3] and PHENIX [4] results. There is good agreement in the ratio to the data in the higher $p_T$ region (5-8 GeV/$c$) which will have implications in the $R_{AA}$ calculation. This fraction is then used to determine the individual charm and bottom $R_{AA}$ distributions as shown in Fig. 2 (b). It was found that charm is more suppressed than the bottom in the 3-5 GeV/$c$ region and that they have similar suppression in the 5-8 GeV/$c$ region [5].

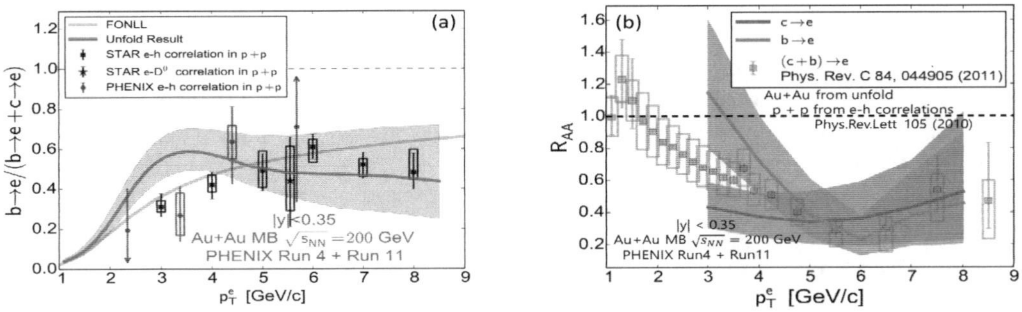


**Figure 2.** (a) The bottom fraction determined from the unfolding method, compared to FONLL calculations and previous $p + p$ measurements. (b) The $R_{AA}$ calculation for charm and bottom contributions compared to previous inclusive heavy flavor $R_{AA}$ results from single electrons.

## 3. Heavy Flavor Production via $e^+e^-$ in $d + Au$ and $p + p$ Collisions

Measurements of electron pairs are made in the PHENIX central arm spectrometers, $|y| < 0.35$, without the capabilities of the VTX detector, and the heavy flavor contributions are separated through a phase space analysis. The unlike sign pair mass spectrum consists of a variety of contributions that need to be accounted for including the inclusive heavy flavor signal, combinatorial pairs, and correlated background. After like sign pairs are used to estimate the combinatorial background and subtracted off, a hadronic cocktail is used to determine and remove correlated background signals such as vector mesons, pseudo scalar mesons and Drell-Yan pairs. See Fig. 3 for the correlated invariant mass distribution and the hadronic cocktail.

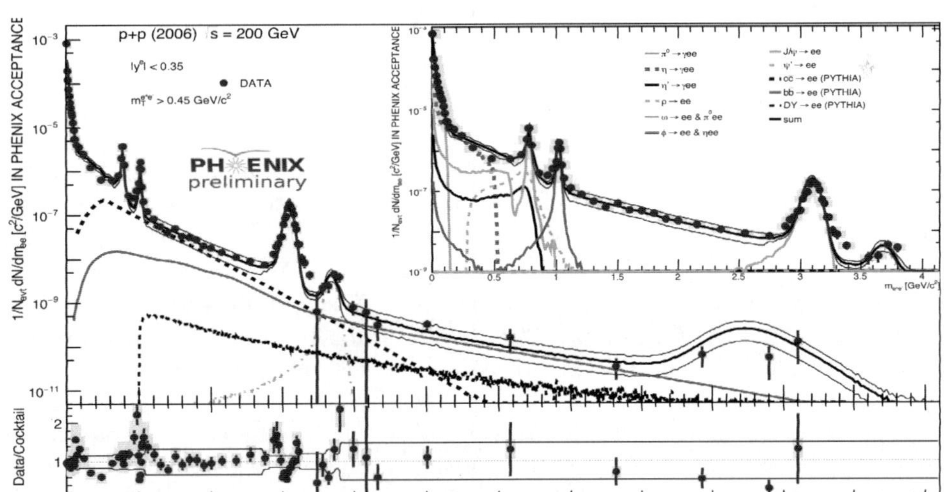


**Figure 3.** Correlated invariant $e^+e^-$ mass distribution. The individual components of the hadronic cocktail and the total cocktail are shown alongside the data. The insert in the upper right corner shows the low mass region and the lower plot is the ratio of data to the total cocktail.

The inclusive heavy flavor signal is then simultaneously fit, using PYTHIA and MC@NLO models, to mass distributions in $p_T$ bins. An example of this process can be seen in Fig. 4 and the method is detailed in [6]. As expected, charm contributions dominate the low mass and low $p_T$ regions and the bottom contributions dominate the high $p_T$, low mass and low $p_T$, high mass regions of phase space. A total fit to mass and $p_T$ is then determined. The fit is shown to match data and from it a total cross section can be determined for both the charm and bottom quarks.

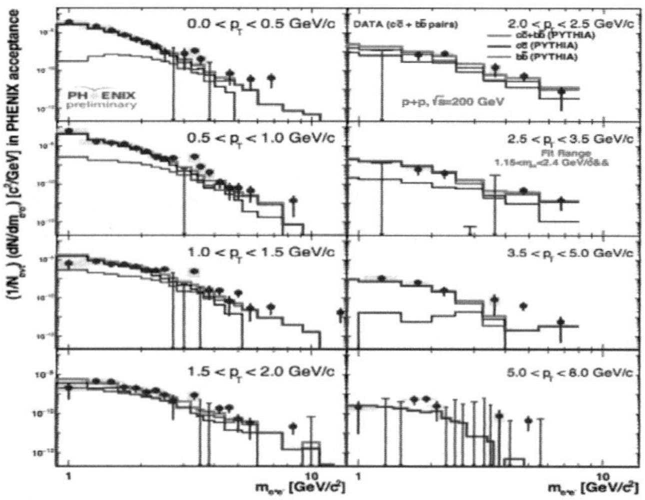


**Figure 4.** Inclusive heavy flavor invariant $e^+e^-$ mass shown in eight $p_T$ bins. An example of the simultaneous fitting procedure is shown with the PYTHIA model where the separate charm and bottom model contributions are shown and the total model is fit to the data in the individual $p_T$ bins.

Shown in Fig. 5 (a), the models are in agreement in the bottom cross section determination but not in the charm cross sections. This is due to the handling of the decay kinematics of the $D$ and $B$ mesons and the differences in their masses relative to the decay electron. The bottom quark mass is much greater then that of the electron and therefore has a larger uncertainty associated with the electrons decay angle relative to the bottom quark. The charm quark mass is smaller and there is less uncertainty associated with the kinematics of the electrons relative to the charm quark. The differences in mass smear the opening angles of the electrons relative to the opening angle of the heavy flavor quarks differently. These cross sections are then used as the baseline measurement for the $d+Au$ measurement. The $R_{dAu}$ for the heavy flavor di-electron pairs, shown in Fig. 5 (b) is consistent with 1 which indicates there is no modification in the inclusive heavy flavor production.

## 4. Production of $b\bar{b}$ via $\mu^\pm\mu^\pm$ in $p+p$ Collisions $\sqrt{s_{NN}} = 500$ GeV

This measurement takes advantage of $B^0$ oscillation in the 5-10 GeV/$c^2$ di-muon mass window which are measured in the muon arm spectrometer rapidity acceptance of $1.2 < |y| < 2.2$. $B^0$ mesons oscillate with known frequencies, $B^0_s \sim 17\%$ and $B^0_d \sim 50\%$ [1], and $b\bar{b}$ pairs in which a $B$ meson oscillates will produce like sign di-muons. In the case of $b\bar{b}$ pairs not oscillating only unlike sign pairs are produced. In the like sign di-muon high mass region the signal is dominated by primary-primary decay $B$ mesons and pairs that include a $D$ meson decay chain, $B \to D \to \mu$. Charm contributions are negligible in the mass region greater then 5 GeV/$c^2$ and within the PHENIX forward and backward muon arms acceptance. Like sign pairs also have an additional benefit of few other background contributions that would otherwise be found in unlike sign pairs such as Quarkonia and Drell-Yan pairs.

To separate out the open $B$ signal from the $\mu^\pm\mu^\pm$ pairs in the invariant mass distribution, a mixed event background subtraction technique is used and the remaining correlated signal includes two components which are the open $B$ meson signal and hadronic jets. Simulation is used to account for this jet background and a fit to both the total correlated signal and hadronic

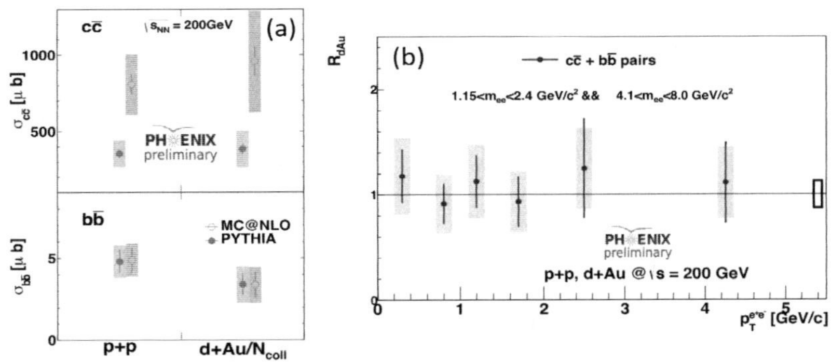


**Figure 5.** (a) The total charm and bottom cross sections as determined through each model. Shown along side is the cross section as found in $d + Au$ collisions normalized by the average number of nucleons involved in a collision. (b) The $R_{dAu}$ $vs$ $p_T$ of inclusive heavy flavor $e^+e^-$.

background is performed. The correlated like sign mass distributions are shown in Fig. 6 (a) and (b). The difference between the correlated signal and the hadronic background is the signal from open bottom mesons.

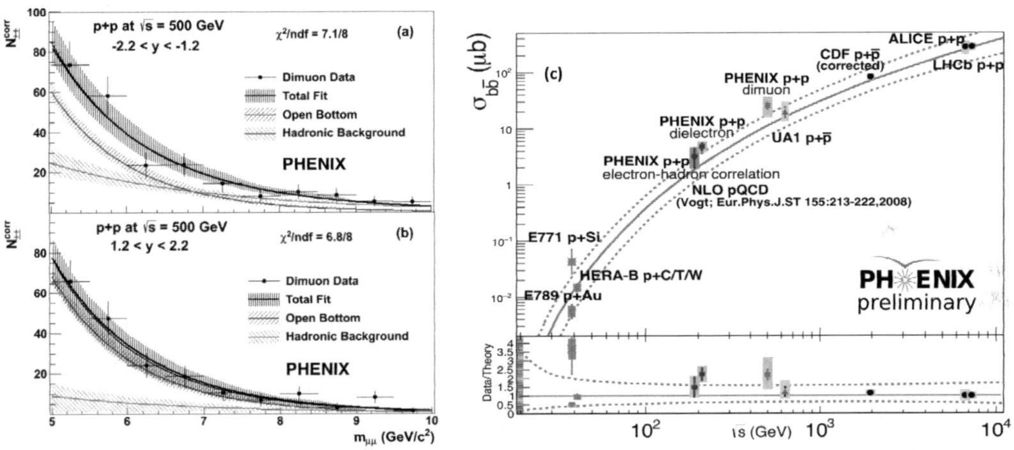


**Figure 6.** (a) Like sign mass distribution in the the 5-10 GeV/$c^2$ mass region for $-2.2 < y < -1.2$ (a) and $1.2 < y < 2.2$ (b). For both (a) and (b) the correlated data are the points and the jet background is subtracted to get the open bottom signal. There is some difference between the two panels and that is due to the differences of absorber material in the muon detectors. (c) The total measured cross section $vs$ center of mass energies is plotted along side pQCD theory [7] with other world data [4, 7–15]. The bottom panel is the ratio of the data to the pQCD theory.

Once the open bottom signal is determined from the like sign pairs and through the use of MC@NLO and PYTHIA models, the signal is further separated into like sign pair muons in which at least one $\mu$ comes from a neutral $B$ meson that has oscillated. Then, from the fraction of like sign pairs from oscillation to total $\mu\mu$ pairs, a total yield is extracted, the cross section is calculated and then extrapolated to full phase space. The total $\sigma_{b\bar{b}}$ cross section is shown to be in agreement with theory within uncertainties and is plotted alongside the pQCD theory curve as well as with other world data in Fig. 6 (c) [4, 7–15].

## 5. Summary

In this proceeding recent heavy flavor measurements from PHENIX have been presented. Using a variety of collision species at $\sqrt{s_{NN}} = 200$ and 500 GeV, an understanding of HNM effects and CNM effects was obtained by comparison with baseline measurements in $p + p$ collisions.

In 2011 PHENIX began running with the VTX silicon vertex detector in order to separate contributions from their secondary displaced vertex measurements. In the first such measurement to do so, individual charm and bottom $R_{AA}$ measurements were made from the single electron signal. It was found that the bottom quark was less suppressed than the charm in the lower $p_T$ region but they both had similar suppression levels in the higher $p_T$ region.

To understand CNM effects $p$+$p$ baseline cross section measurements were made as a reference to $d + Au$ measurements through the $e^+e^-$ decay channel. It was found that there is no modification within uncertainties across the 0-5 GeV/$c$ $p_T$ range. This analysis focused on extracting the signal through a phase space separation.

$b\bar{b}$ production was studied at $\sqrt{s_{NN}} = 500$ GeV through the di-muon decay channel. This analysis technique separated out contributions in the 5-10 GeV/$c^2$ mass region and relied on the $B^0$ oscillation properties of the mesons. The measured cross section was found to be consistent with pQCD calculations [7] within the known uncertainties.

## Acknowledgments

The author would like to thank the organizers of the BEACH 2016 conference for giving him the opportunity to present these results and for their financial support.

## 6. References

[1] Olive K A *et al.* (Particle Data Group) 2014 and 2015 *Chin. Phys.*C **38** 090001.
[2] Adare A *et al.* (PHENIX Collaboration) 2011 Heavy-quark production in $p + p$ and energy loss and flow of heavy quarks in Au + Au collisions at $\sqrt{s_{NN}} = 200$ GeV *Phys. Rev.* C **84** 044905-47.
[3] Aggarwal M M *et al.* (STAR Collaboration) 2010 Measurement of the bottom quark contribution to nonphotonic electron production in $p + p$ collisions at $\sqrt{s} = 200$ GeV *Phys. Rev. Lett.* **105** 202301-7.
[4] Adare A *et al.* (PHENIX Collaboration) 2009 Measurement of bottom versus charm as a function of transverse momentum with electron-hadron correlations in $p + p$ collisions at $\sqrt{s} = 200$ GeV *Phys. Rev. Lett.* **103** 082002-9.
[5] Adare A *et al.* (PHENIX Collaboration) 2016 Single electron yields from semileptonic charm and bottom hadron decays in Au + Au collisions at $\sqrt{s_{NN}} = 200$ GeV *Phys. Rev.* C **93** 034904-33.
[6] Adare A *et al.* (PHENIX Collaboration) 2015 Cross section for $b\bar{b}$ production via dielectrons in $d$ + Au collisions at $\sqrt{s_{NN}} = 200$ GeV *Phys. Rev.* C **91** 014907-25.
[7] Vogt R 2008 The total charm cross section *Eur. Phys. J. Spec. Top.* **155** 213-22.
[8] Jansen D M *et al.* 1995 Measurement of the bottom-quark production cross section in 800 GeV/$c$ proton-gold collisions *Phys. Rev. Lett.* **74** 3118-21.
[9] Alexopoulos T *et al.* (E771 Collaboration) 1999 Measurement of the $b\bar{b}$ cross section in 800 GeV/$c$ proton-silicon interactions *Phys. Rev. Lett.* **82** 41-4.
[10] Abt I *et al.* (HERA-B Collaboration) 2006 Improved measurement of the $b\bar{b}$ production cross section in 920 GeV fixed-target proton-nucleus collisions *Phys. Rev.* D **73** 052005-22.
[11] Adare A *et al.* (PHENIX Collaboration) 2009 Dilepton mass spectra in $p + p$ collisions at $\sqrt{s} = 200$ GeV and the contribution from open charm *Phys. Lett.* B **670** 313-20.
[12] Albajar C *et al.* (UA1 Collaboration) 1991 Beauty production at the CERN $p\bar{p}$ collider *Phys. Lett.* B **256** 121-8.
[13] Acosta D *et al.* (CDF Collaboration) 2005 Measurement of the J/$\psi$ meson and $b$-hadron production cross sections in $p\bar{p}$ collisions at $\sqrt{s} = 1960$ GeV *Phys. Rev.* D **71** 032001-27.
[14] Abelev B *et al.* (ALICE Collaboration) 2013 Measurement of electrons from beauty hadron decays in $pp$ collisions at $\sqrt{s} = 7$ TeV *Phys. Lett.* B **721** 13-23.
[15] Aaij R *et al.* (LHCb Collaboration) 2013 Measurement of $B$ meson production cross-sections in proton-proton collisions at $\sqrt{s} = 7$ TeV *J. High Energ. Phys.* 8 **2013** 117.

Journal of Physics: Conference Series **770** (2016) 012008 doi:10.1088/1742-6596/770/1/012008

# Searches for Heavy Quark States at ATLAS

**Hok-Chuen (Tom) Cheng (on behalf of the ATLAS Collaboration)**
University of Michigan, Ann Arbor, USA
E-mail: hccheng@umich.edu

**Abstract.** This talk highlights the latest results of heavy quark searches from the ATLAS collaboration, mainly on resonance searches and vector-like quarks (VLQs) searches. Searches for $t\bar{t}$ resonances using lepton-plus-jets events in proton-proton collisions at center-of-mass energy of 8 and 13 TeV are presented. Limits are set for BSM particles such as topcolor-assisted technicolor $Z'_{TC}$, Kaluza-Klein (K-K) gluons $g_{KK}$ and K-K excitations of graviton $G_{KK}$ in the Randall-Sundrum model of extra dimensions. VLQs arise naturally in many models such as Little Higgs and Composite Higgs and typically couple preferably to the third generation SM quarks and weak bosons. Limits are set for vector-like bottom (B) and top (T) quarks decay to lepton-plus-jets final states via Hb+X and Ht+X channels in two analyses using 8 and 13 TeV datasets from ATLAS.

## 1. Introduction

There are two main categories of heavy quark searches at ATLAS [1]. Firstly, there are model-agnostic searches in which we look for resonances such as $X \rightarrow t\bar{t}$, $t\bar{b}$ or single $t+E_T^{miss}$. Secondly, there are vector-like quark (VLQ) searches in which we look for either a pair of VLQs or a single VLQ production. This talk intends to give an overview and describe several examples of these two categories of searches.

## 2. Resonance Searches

### *2.1. Overview of Resonance Searches at ATLAS*

ATLAS has several searches for resonances with different final states. At $\sqrt{s} = 8$ TeV, there is a $t\bar{t}$ resonance search using the lepton-plus-jet final state [2], a $t\bar{b}$ resonance search using the lepton-plus-jet final state [3] and a $t\bar{b}$ resonance search using the $qqbb$ final state [4]. At $\sqrt{s} = 13$ TeV, there is a $t\bar{t}$ resonance search using the lepton-plus-jet final state looking for $Z'$ [5]. The following sections focus on the two $t\bar{t}$ resonance searches using the lepton-plus-jet final state.

### *2.2. $t\bar{t}$ Resonance Searches: Motivations*

Several benchmark models predict new resonances decaying into $t\bar{t}$. For example, the topcolor-assisted technicolor $Z'$, which is a spin-1 color singlet, decays into $t\bar{t}$. In the Kaluza-Klein Randall-Sundrum model, the K-K excitation of gluon $g_{KK}$ decays to $t\bar{t}$ and is a spin-1 color octet. Also, there is the excited graviton $G_{KK}$ which is a spin-2 color singlet and decays to $t\bar{t}$. There are other color singlet scalar particles which also decay into $t\bar{t}$.

The general strategy here is to select $t\bar{t}$ events, then reconstruct the $t\bar{t}$ invariant mass $m_{t\bar{t}}$ and look for an excess or deviation from the Standard Model (SM) predicted $t\bar{t}$ background.

### *2.3. $t\bar{t}$ Resonance Searches at $\sqrt{s} = 8$ TeV*

This analysis uses the 20.3 fb$^{-1}$ of data collected at the ATLAS detector at a center-of-mass energy of 8 TeV in 2012. The events are categorized into two types depending on the topology.

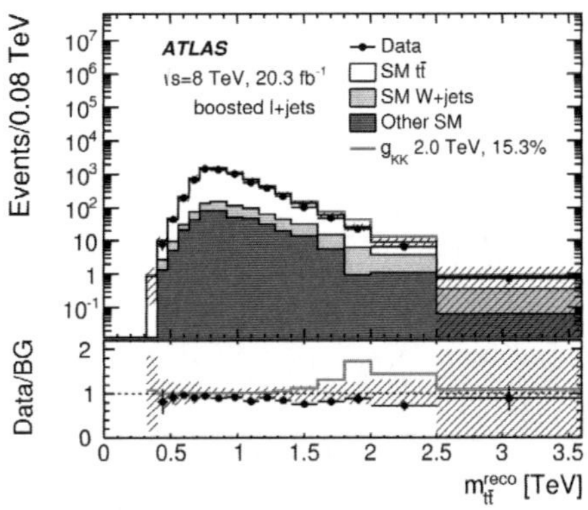


**Figure 1.** Reconstructed mass spectrum for the boosted topology overlaid with $g_{KK}$ signal at 2.0 TeV. The ratio of data to SM background and signal is shown in the bottom. [2]

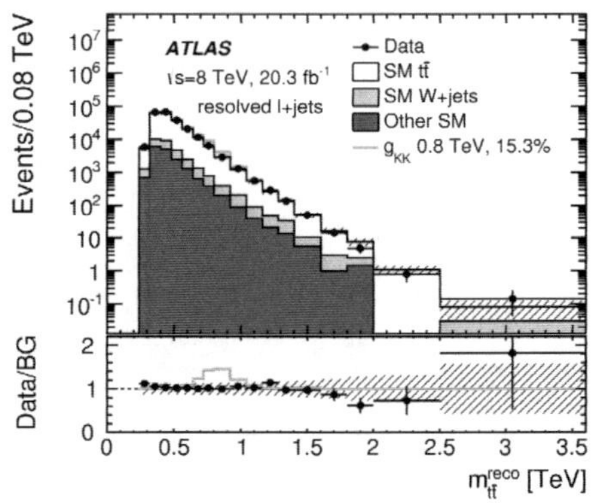


**Figure 2.** Reconstructed mass spectrum for the resolved topology overlaid with $g_{KK}$ signal at 0.8 TeV. The ratio of data to SM background and signal is shown in the bottom. [2]

The first category is the boosted topology where the hadronic top is reconstructed using the largest $p_T$ large-R jet. At least one small-R jet must be b-tagged and matched to at least one of the top candidates. The semi-leptonic top is reconstructed using $E_T^{miss}$, lepton, and the largest $p_T$ small-R jet near jet. Events that failed to meet the requirement for the boosted topology will pass to the selection for the resolved topology where at least 4 small-R jets are required and at least one of them has to be b-tagged. In both categories, the b-tagged jet has to match to either the hadronic-top jet, the leptonic-top jet or both, otherwise it is discarded. The reconstructed mass $m_{t\bar{t}}^{reco}$ spectrum is scanned for deviations from the SM expectation as shown in figures 1 and 2. The search is performed on 6 boosted, 6 resolved and 6 combined channels and a deviation is required in all channels at the same place.

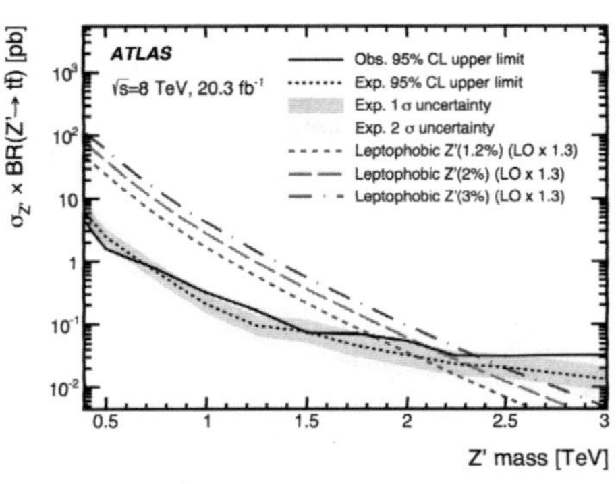


**Figure 3.** Limit plot for cross-section times branching fraction as a function of the $Z'$ mass. [2]

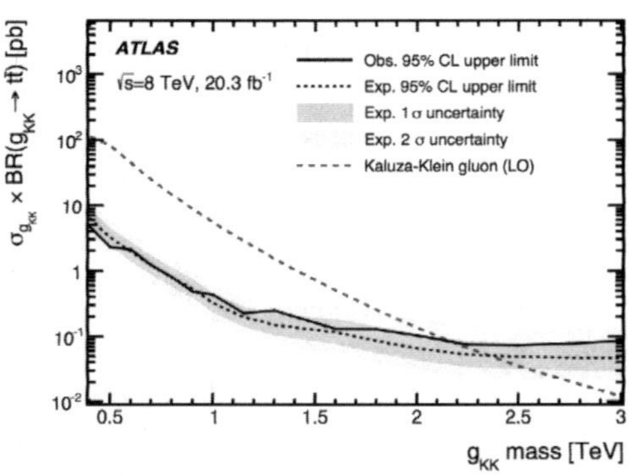


**Figure 4.** Limit plot for cross-section times branching fraction as a function of the $g_{KK}$ mass. [2]

No excess above the SM prediction is observed. Cross-section limits are set for different benchmark models. In particular, the topcolor-assisted technicolor $Z'$ has been excluded between 0.4 TeV and 1.8 TeV at 95% confidence level (CL). The K-K gluon in the R-S model $g_{KK}$ has been excluded between 0.4 TeV and 2.2 TeV at 95% CL. The corresponding limit plots are shown in figures 3 and 4. This analysis is not sensitive enough to set any limit to the K-K graviton or the color singlet scalar particle.

### *2.4. $t\bar{t}$ Resonance Searches at $\sqrt{s} = 13$ TeV*

This analysis uses 3.2 $\text{fb}^{-1}$ data collected at the ATLAS detector at $\sqrt{s} = 13$ TeV in 2015. In each event, exactly one lepton is reconstructed and must match candidates that triggered the event. For the leptonic-top b-jet, at least one small-R jet is reconstructed in the $\Delta R(l, \text{jet}) < 1.5$ cone. For the hadronic-top jet, at least one top-tagged large-R jet is reconstructed in the $\Delta\phi(l,$ large-R jet$) > 2.3$ and $\Delta\phi$(leptonic-top jet, large-R jet) $> 1.5$ region. In addition, at least one b-tagged small-R jet is reconstructed.

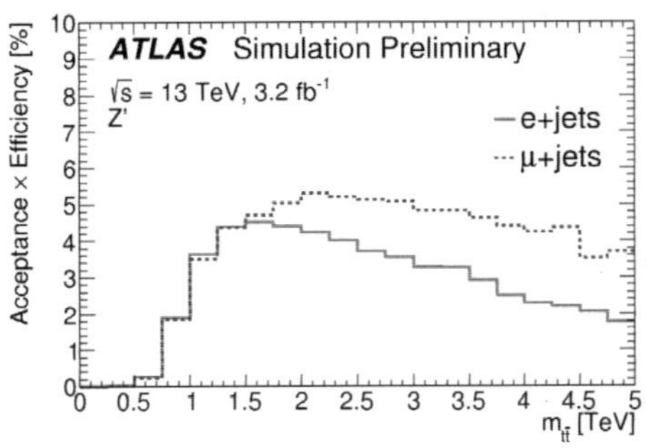


**Figure 5.** Acceptance times efficiency for the $e$+jets and $\mu$+jets channels as a function of the $Z'$ mass. [5]

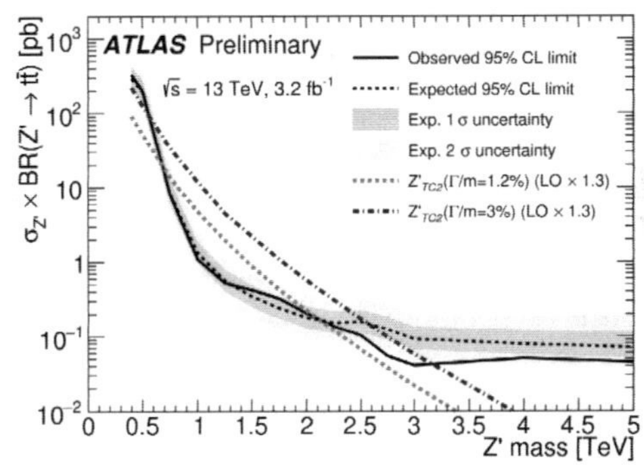


**Figure 6.** Limit plot for cross-section times branching fraction as a function of the reconstructed $Z'$ mass. [5]

The acceptance times efficiency for $e$+jets and $\mu$+jets channels as a function of reconstructed mass $m_{t\bar{t}}$ is shown in figure 5. The $e$+jets channel has lower acceptance times efficiency for $m_{t\bar{t}}$ above 1.5 TeV because of the inefficiency of $e^-$ ID and overlap removal in highly boosted top quarks. The major sources of systematic uncertainties are the large-R jet energy scale, the light flavor and charm jet mistag rate and luminosity. No excess above the SM prediction is observed. The topcolor-assisted technicolor $Z'$ has been excluded from 0.7 TeV to 2.7 TeV at 95% CL and the corresponding limit plot is shown in figure 6.

## 3. Vector-like Quark (VLQ) Searches

### *3.1. VLQ Searches: Motivations*

VLQs that arise naturally in models alternative to supersymmetry such as the Little Higgs [6, 7] and the Composite Higgs [8, 9] have no Yukawa coupling to the newly discovered Higgs boson and are therefore not constrained by the recent Higgs measurements. They typically couple preferentially to the third generation quarks and can have flavor-changing neutral current decays. They can be pair-produced via QCD below 1 TeV and can be dominated by single production at high mass, depending on the weak couplings to the gauge bosons. Depending on the models, 4 types of VLQs exist. They include the vector-like T ($Q = +2/3$), B ($Q = -1/3$), X ($Q = +5/3$) and Y ($Q = -4/3$) quarks. The branching ratio of B and T quarks for different models are shown in figures 7 and 8.

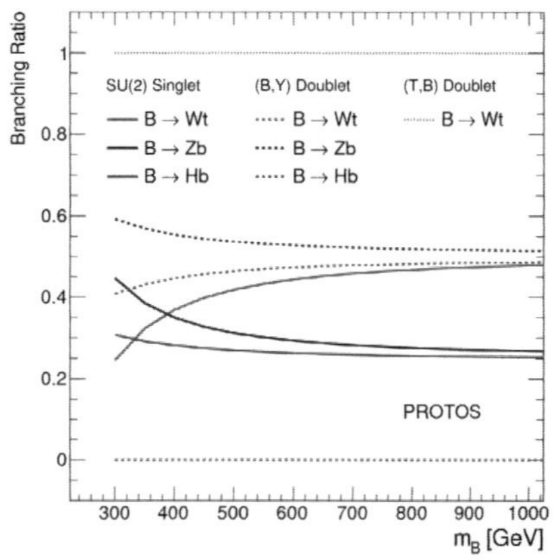


**Figure 7.** Branching ratio of $SU(2)$ singlet, $(B, Y)$ and $(T, B)$ doublet models as a function of the B quark mass. [10]

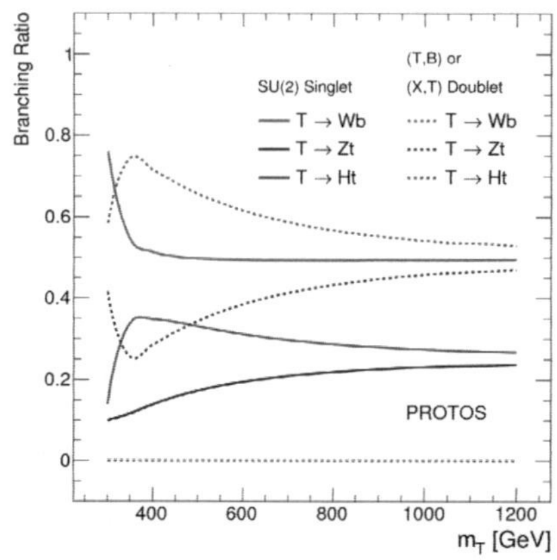


**Figure 8.** Branching ratio of $SU(2)$ singlet, $(T, B)$ and $(X, T)$ doublet models as a function of the T quark mass. [10]

*3.2. Vector-like Bottom (B) and Top (T) Quark Searches at ATLAS*

There are several searches for VLQs at ATLAS, each sensitive to different decay modes. At $\sqrt{s} = 8$ TeV, there is one for VLQs that decay into a $Z$ boson and a third-generation quark [11] and another one for the final state of b-jets and a pair of leptons of the same charge [12]. In the following sections, searches for pair production of VLQ and of four top quarks in the lepton-plus-jets final state at 8 TeV [10] and 13 TeV [13] will be described.

*3.3. B Quark Search using $Hb \to X$ Decay Mode at 8 TeV*

The sample dominated by $t\bar{t}$ is selected and is sub-divided into 8 categories using jet multiplicity $n_{jets}$, b-multiplicity and the mass of the closest two jets. Background rich regions have low b-multiplicity and signal rich regions are dominated by high b-multiplicity events. The search is done in the distribution of the scalar sum of the transverse momentum of jets, $l^{\pm}$ and $E_T^{miss}$ $(H_T)$. The $H_T$ distribution for the signal rich regions are shown in figure 9.

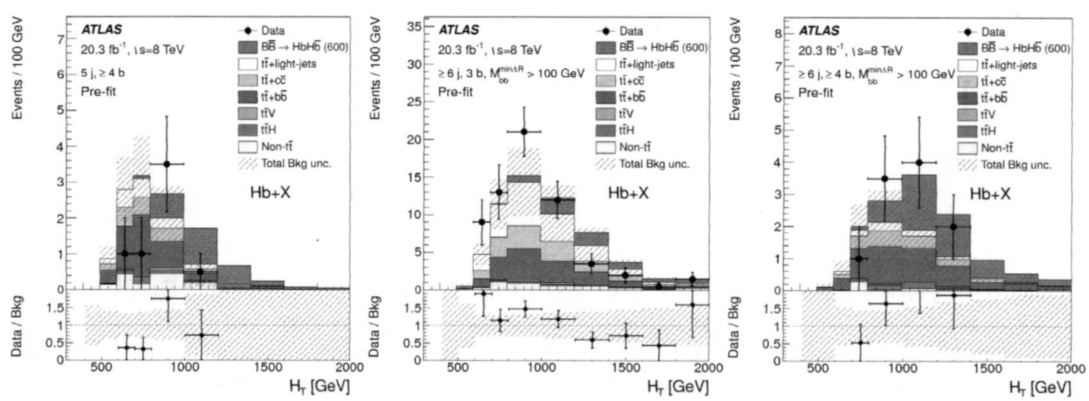


**Figure 9.** $H_T$ distribution of the signal rich regions. [10]

The major systematic uncertainties are the jet energy scale and the theoretical cross-section. No excess above the SM prediction is observed. Assuming $\mathcal{BR}(B \to Hb) = 1$, $m_B$ is excluded from 350 GeV to 580 GeV and from 635 GeV to 700 GeV at 95% CL with expected exclusion region below 625 GeV. In the SU(2) singlet model, $m_B$ is excluded below 735 GeV at 95% CL with expected exclusion region below 635 GeV. The results are summarized in the exclusion limit plots as shown in figure 10.

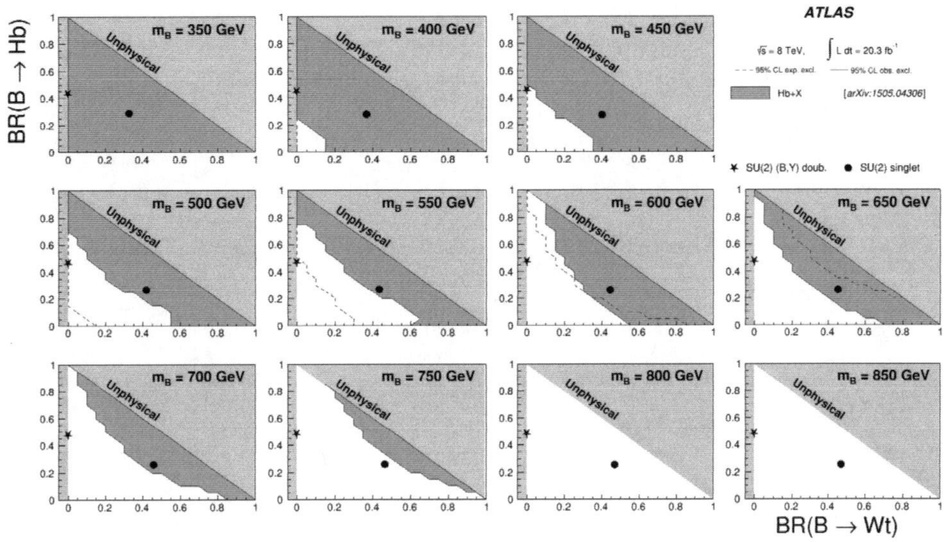


**Figure 10.** 2D exclusion limit plots of $\mathcal{BR}(B \to Hb)$ vs $\mathcal{BR}(B \to Wt)$ for the B quark mass $m_B$ ranging from 350 GeV to 850 GeV. [10]

### 3.4. T Quark Search using $Ht \to X$ Decay Mode at 13 TeV

The sample is sub-divided into 20 categories using jet multiplicity $n_{jets}$, b-multiplicity, the mass of the two closest jets and the number of large-R jets with mass greater than 100 GeV. The background rich regions have low b-multiplicity while the signal rich regions have high b-multiplicity. The search is done using the distribution of the scalar sum of the transverse momentum of jets, $l^{\pm}$ and $E_T^{miss}$ ($m_{eff}$) using a binned likelihood fit under the background-only hypothesis. The $m_{eff}$ distribution of the signal-sensitive regions are shown in figure 11.

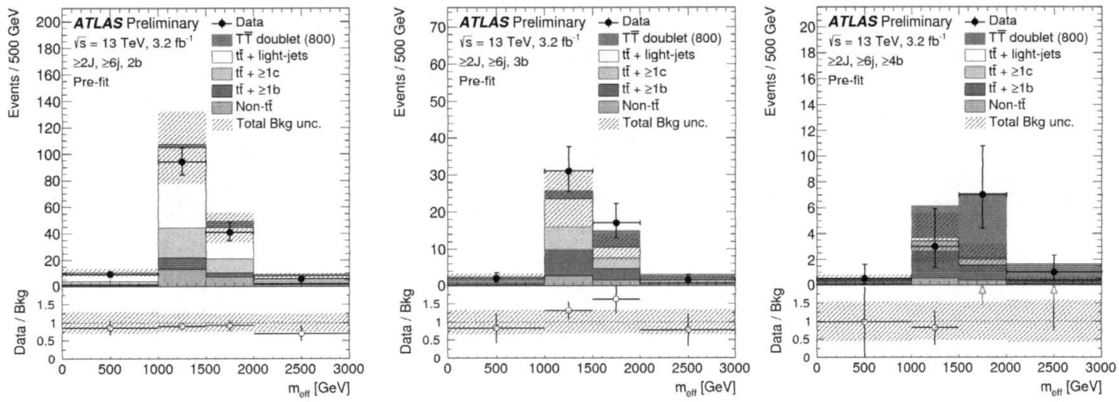


**Figure 11.** $m_{eff}$ distribution of the signal-sensitive regions. [13]

The main systematic uncertainties include background production cross-section, which includes factorization and renormalization scales and parton distribution function uncertainty both at NLO and NNLO, luminosity and the jet energy scale. No excess above the SM prediction is observed. Under the $\mathcal{BR}(T \to Ht) = 1$ hypothesis, $m_T$ is excluded below 900 GeV with an expected exclusion region below 980 GeV at 95% CL. In the SU(2) singlet model, $m_T$ is excluded below 750 GeV with an expected exclusion region below 780 GeV at 95% CL. In the SU(2) doublet model, $m_T$ is excluded below 800 GeV with an expected exclusion region below 900 GeV at 95% CL. The cross-section limits as a function of the T quark mass are shown in figure 12. The 13 TeV analysis sensitivity reaches 950 GeV, exceeding that of 8 TeV analysis. However, the observed limits are weaker than expected above 700 GeV due to statistical fluctuations in data.

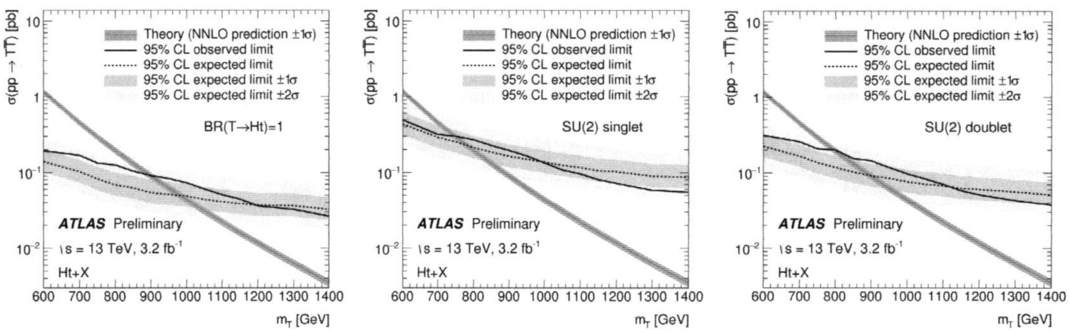


**Figure 12.** Cross-section limit plots as a function of the T quark mass $m_T$ under $\mathcal{BR}(T \to Ht) = 1$, SU(2) singlet and SU(2) doublet hypotheses. [13]

## 4. Acknowledgement

I am grateful to the ATLAS speakers committee for providing me the opportunity to give this talk at the BEACH 2016 conference. I would also like to express my gratitude to the ATLAS conveners Joseph Haley, Danilo Enoque Ferreira De Lima and Koji Terashi for their invaluable advice and suggestions during the preparation of this talk.

## References

[1] ATLAS Collaboration 2008 *JINST* **3** S08003
[2] ATLAS Collaboration 2015 *Journal of High Energy Physics* **2015** 1–54 ISSN 1029-8479
[3] ATLAS Collaboration 2015 *Physics Letters B* **743** 235 – 255 ISSN 0370-2693
[4] ATLAS Collaboration 2015 *The European Physical Journal C* **75** 1–23 ISSN 1434-6052
[5] ATLAS Collaboration 2016 Search for heavy particles decaying to pairs of highly-boosted top quarks using lepton-plus-jets events in proton–proton collisions at $\sqrt{s} = 13$ TeV with the ATLAS detector Tech. Rep. ATLAS-CONF-2016-014 CERN Geneva
[6] Arkani-Hamed N, Cohen A G, Katz E and Nelson A E 2002 *Journal of High Energy Physics* **2002** 034
[7] Schmaltz M and Tucker-Smith D 2005 *Annual Review of Nuclear and Particle Science* **55** 229–270 (*Preprint* `http://dx.doi.org/10.1146/annurev.nucl.55.090704.151502`)
[8] Kaplan D B, Georgi H and Dimopoulos S 1984 *Physics Letters B* **136** 187 – 190 ISSN 0370-2693
[9] Agashe K, Contino R and Pomarol A 2005 *Nuclear Physics B* **719** 165 – 187 ISSN 0550-3213
[10] ATLAS Collaboration 2015 *Journal of High Energy Physics* **2015** 1–86 ISSN 1029-8479
[11] ATLAS Collaboration 2014 *Journal of High Energy Physics* **2014** 1–54 ISSN 1029-8479
[12] ATLAS Collaboration 2015 *Journal of High Energy Physics* **2015** 1–51 ISSN 1029-8479
[13] ATLAS Collaboration 2016 Search for production of vector-like top quark pairs and of four top quarks in the lepton-plus-jets final state in *pp* collisions at $\sqrt{s} = 13$ TeV with the ATLAS detector Tech. Rep. ATLAS-CONF-2016-013 CERN Geneva

# Search for Scalar Top-Quark Production in the all Hadronic Channel at 13 TeV

**S Norberg On behalf of the CMS Collaboration**
University of Puerto Rico Mayaguez, Peurto Rico, USA
E-mail: snorberg@cern.ch

**Abstract.** A search for supersymmetry in all-hadronic events with missing transverse momentum is presented. The data were collected in proton-proton collisions at a center-of-mass energy of 13 TeV with the CMS detector at the LHC and correspond to an integrated luminosity of 2.3 $fb^{-1}$. Search regions are defined using the properties of reconstructed jets, the presence of b quark and t quark candidates, and missing transverse momentum. No statistically significant excess of events above the expected contribution from standard model processes is observed. Exclusion limits are set on the masses of potential new particles in the context of simplified models of direct and gluino-mediated top squark production.

## 1. Introduction

A direct top squark production search is described in this document and refers to the high efficiency top tagger (HETT) portion of Ref. [1] . The 'T2tt' and 'T2tb' models explored here are illustrated in Fig. 1. In the T2tt model the stop decays to a top quark (t quark) and neutralino. In the T2tb model the stop decays to a bottom quark (b quark) and a neutralino.

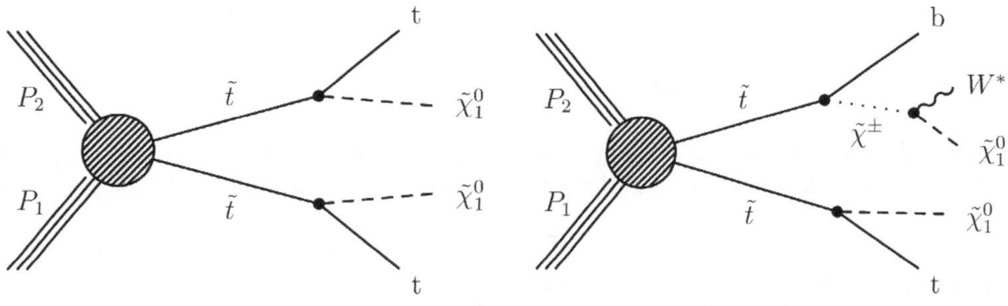


**Figure 1.** The top squark pair production with the top squark decaying into a t quark and neutralino or into a b quark and chargino.

## 2. CMS Detector

The detector used is the Compact Muon Solenoid (CMS) and is a multi-purpose detector of cylindrical design. More details of the main features of the detector can be found in Ref. [2]. For a jet to be considered a b-quark jet (b-tagged) it has to pass the medium working point requirements of the 'Combined Secondary Vertex' (CSV) method [3] and have transverse momentum ($p_T$) greater than 30 GeV and be within $|\eta| < 2.4$. The efficiency to identify the b-quarks is 67% overall. The probability of a jet originating from a light quark or gluon to be mis-identified as a b-quark jet is 1.4%, averaged over $p_T$ in $t\bar{t}$ events [3].

## 3. Search Strategy

Top tagging in this analysis is done using all jets with a cone of radius 0.4 in $(\eta, \phi)$ space that satisfy $p_T > 30$ GeV and $|\eta| < 5$. There are three categories in this algorithm for these jets: trijet, dijet, and monojet candidates. In the trijet category we require the jets to lie within a cone of radius 1.5 in $(\eta, \phi)$ space. Within this cone they are subject to a set of conditions that are kinematically consistent with a t quark decay, more details are available in Refs. [1, 4]. Next the dijets category where one jet originates from the decay products of a W boson ($W$ jet) that is merged, the mass is required to be between 70–110 GeV. Also the ratio of the Wjet and dijet masses are set to be between the ratio of the W boson and the t quark masses. Last but not least is the monojet category, for this we have a single jet which we require to have a jet mass between 110–220 GeV. After all the categories are looked at and all the top candidates are found the final step is to require that there be one b jet and that the mass of the candidates be within this range 100–250 GeV. If these candidates pass these requirements they are considered a top jet Ref. [1].

In this analysis the 37 search regions are defined in terms of the magnitude of the missing transverse momentum($\not{E}_{\mathrm{T}}$), 'stransverse' mass $M_{T2}$ [5, 6], and the number of b-tagged jets and top-tagged objects which is shown in Fig. 2. $M_{T2}$ is defined using a pair of top candidates and $\not{E}_{\mathrm{T}}$. When there are more than two top candidates the combination of all possible tops are found and the smallest $M_{T2}$ is used. When only one top candidate is found the second top candidate is taken from the remanent of the event using the bottom-tagged jet. If this is the case than we find the jet closest to the bottom-tagged jet and if that jet has a value between 50 GeV and the top quark mass than we use that jet as the top candidate. If no combination is found that satisfies this requirement than the bottom tagged jet is considered the only remanent and used as the second top candidate. Fig. 3 demonstrates the background composition following the pre-selection cuts for $\mathrm{N_t}$, and $\not{E}_{\mathrm{T}}$. There are three main backgrounds in this analysis. There is the $\tau$ background, the QCD multijet background, and the $\mathrm{Z} \rightarrow \nu\nu$ background, in the next section techniques to understand these backgrounds are discussed.

## 4. Backgrounds

### *4.1. Estimation of the $\tau$ background*

The biggest contribution to the SM background comes from $t\bar{t}$ and W+jets events with leptonic W decays. In one case where the W boson decays to an electron or muon, then the veto on leptons are satisfied unless the lepton is 'lost'. By 'lost' we mean not isolated, not identified/reconstructed, or out of the acceptance region. In the other case a $\tau$ lepton decays hadronically from the W boson which is reconstructed as a jet. In this case the lepton would pass all of our veto requirements, this we call the hadronic $\tau$ background.

To model these lost leptons (LL) a weighted data sample that consists mainly of $t\bar{t}$ events is used. The search trigger and the pre-selection are the same in this region but the muon veto is replaced requiring there be exactly one isolated muon with $p_T > 10$ GeV and $|\eta| < 2.4$, where the isolated track veto is removed. Also required is that the $M_T < 100$ GeV this is used to reduce any signal contamination. The $M_T$ is reconstructed from the muon $p_T$ and is defined as $M_T = \sqrt{2p_{\mathrm{T}}^{\mu}\not{E}_{\mathrm{T}}(1 - \cos(\Delta\phi))}$, where $\Delta\phi$ is the distance in $\phi$ between the muon and the $\not{E}_{\mathrm{T}}$ vector.

The weight factor is defined as taking into account acceptance, reconstruction and

Journal of Physics: Conference Series **770** (2016) 012009 doi:10.1088/1742-6596/770/1/012009

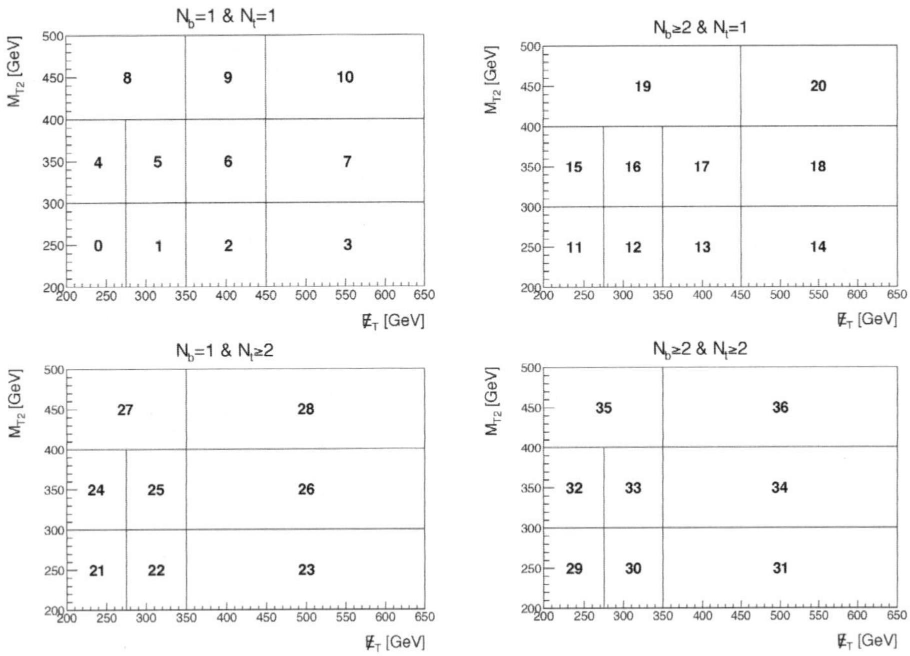


**Figure 2.** Search bin definitions. The highest $\not{E}_T$ and $M_{T2}$ bins are open-ended. An example of this is in bin 10 which requires $\not{E}_T$> 450 GeV and $M_{T2}$> 400 GeV.

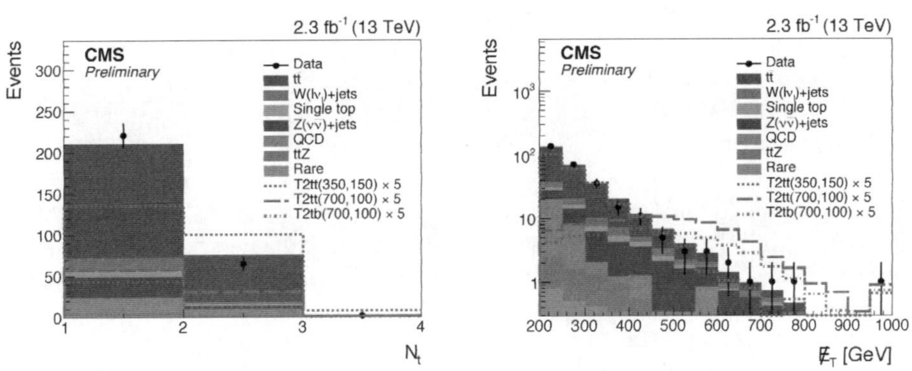


**Figure 3.** Background composition as a function of $N_t$, and $\not{E}_T$ between SM backgrounds (filled histograms) and tow mass points of the T2tt signal model (dashed lines) and one mass point of the T2tb model.

identification, and isolation efficiencies multiplied by a factor that accounts for dilepton events where both leptons are lost is applied to the sum over the events in the single-muon control region, where this includes events with lost leptons from $t\bar{t}$, $W$+jets, and single-top processes. A further correction is applied in order to take into account the efficiency of the $M_T < 100$ GeV requirement. The final correction that is applied is to take into account the isolated track veto efficiency.

The estimate of the remaining hadronic $\tau$'s background is calculated by taking the signal region after applying the isolated track veto and is based on a control sample of $\mu+$jets events selected from data using a muon and $\mathrm{H_T}$-based trigger, and requiring exactly one muon with $p_T^{\mu} > 20$ GeV and $|\eta| < 2.4$. A cut on the transverse mass of the W boson, $M_T < 100$ GeV, is required to select events containing a $\mathrm{W} \to \mu\nu$ decay and to suppress potential signal events from being present in the $\mu+$jets sample.

The muon $p_T$ is smeared by response template distributions derived for a hadronically-decaying $\tau$ lepton to correct the leptonic part of the event. The response templates are derived using $t\bar{t}$ and $W$+jets MC by comparing the true $\tau$ lepton $p_T$ with the reconstructed hadronic $\tau$'s jet $p_T$. The kinematic variables of the event are recalculated with this hadronic $\tau$'s jet, and the search selections are applied. The probability to mistag a hadronic $\tau$'s jet as a b jet is significant and affects the $\mathrm{N_b}$ distribution of hadronic $\tau$'s background events. This effect is taken into account in the same way as in Ref. [1].

The hadronic $\tau$'s background prediction is calculated as a sum over all events in the $\mu+$jets control sample weighted by the $\tau_h$ response. Additional corrections are applied. A few corrections are the muon acceptance, the $M_T$ selection efficiency, the isolated track veto efficiency, and others.

### *4.2. Estimation of the Z→ νν background*

For the $\mathrm{Z} \to \nu\nu$ background the $\mathrm{Z} \to \nu\nu$ simulation is used to have more statistics in the interpretation. This background is calculated in a way that any difference seen between data and simulation is taken into account through a correction factor that is applied. Two scale factors are applied to simulated events and are called $R_{\mathrm{norm}}$ and $S_{\mathrm{DY}}(N_{\mathrm{j}})$, and they correct the normalization of the simulation and the shape of the simulated $N_{\mathrm{j}}$ distribution. To calculate the factors a dimuon control region is used.

The $R_{\mathrm{norm}}$ scale factor, is derived in data using the same selection as the search region pre-selection, apart from the muon requirement and the requirement on the b-tagged jets. The expected yield in the DY simulation in this region is compared to the observed event yield in data after subtraction of the other SM processes for the scale factor calculation.

The factor, $S_{\mathrm{DY}}$, depends on the number of jets ($N_{\mathrm{j}}$) in the event and is derived in a loose dimuon control region in which the signal region requirement on $\mathrm{H_T}$ is relaxed to $\mathrm{H_T} > 200$ GeV and the requirements on $\not{E}_{\mathrm{T}}$, $\mathrm{N_t}$ and $M_{T2}$ are removed. This scale factor is derived for each $N_{\mathrm{j}}$ bin as the ratio between the data, with non-DY backgrounds subtracted, and the DY simulation.

### *4.3. Estimation of the QCD multijet background*

In order to predict the QCD multijet background a QCD multijet-rich data control region with minimum signal is achieved by inverting the pre-selection requirements on $\Delta\phi(\not{E}_{\mathrm{T}}, j_{1,2,3})$. A little of the other SM backgrounds remain, such as $t\bar{t}$, $W$+jets, and $Z$+jets so they are subtracted out. The method used for lost leptons and hadronic $\tau$'s in this analysis are also used to subtract out those backgrounds in this region. For $\mathrm{Z} \to \nu\nu$ simulation is used because the contribution is so small. The number of QCD multijet events is measured in the data control region and a translation factor is applied to predict the amount of the QCD multijet background in each search bin. The translation factor is calculated in data and partly by simulation. It is computed as a simulated ratio between the signal region and the inverted-$\Delta\phi$ control region, in bins of $\not{E}_{\mathrm{T}}$ and $M_{T2}$ where the bin boundaries follow those of the signal bins. The overall shape of this translation factor is kept from simulation. The value is normalized to data. Where the data measurement is in a sideband of the pre-selection region and is defined by the requirement $175 < \not{E}_{\mathrm{T}} < 200$ GeV, where the amount of data is sufficiently large to make an accurate measurement.

## 5. Results

In Fig. 4 the number of events observed in data and the SM background predictions for the search region defined above are shown. In general the most significant background across the search regions comes from the SM $\mathrm{t\bar{t}}$ production or W-boson production where the lepton is not detected or where the W boson decays hadronically. The next largest contribution is $\mathrm{Z} \rightarrow \nu\nu$ production in association with jets, which also including heavy-flavor jets, where the neutrino pair gives large $\not{E}_{\mathrm{T}}$ and the t quark conditions are satisfied by a combination of jets accidentally. The rest of the backgrounds are small contributions.

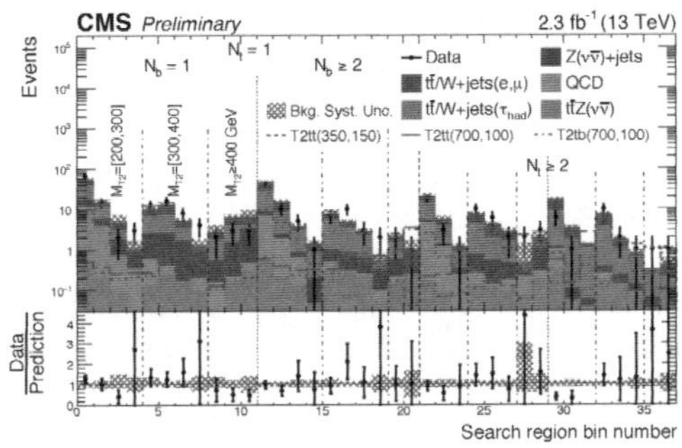


**Figure 4.** The black points are observed event yields in data and the filled solid areas are the predicted SM backgrounds for the 37 search bins of this analysis. The ratio of data over total background prediction in each search bin is shown in the latter half of the plot. In the ration only statistical uncertainties are propagated.

The binned likelihood fit to the observed data is the statistical interpretation of the results for the exclusion limits for both T2tt and T2tb models. This takes into account the predicted background and expected signal yields with their uncertainties in each bin. Exclusion limits are extracted based on a modified frequentist approach [1] using a profile likelihood ratio as test statistic. If the 95% upper limit on the production cross section falls below the theoretical cross section the signal models are considered to be excluded by the analysis.

Figures 5 show the 95% CL exclusion limits obtained, for both the pure T2tt scenario, and in the mixed T2tb scenario assuming a 50% branching fraction for each of the two decay modes ($\tilde{\mathrm{t}} \rightarrow \mathrm{t}\tilde{\chi}_1^0 / \tilde{\mathrm{t}} \rightarrow \mathrm{b}\, \tilde{\chi}_1^{\pm}$). In the T2tb model the $\tilde{\chi}_1^{\pm}$ and $\tilde{\chi}_1^0$ have a 5 GeV mass difference this is because there masses are assumed to be nearly degenerate. Using the 2.3 fb$^{-1}$ dataset, for the T2tt model stop masses are up to 780 GeV and LSP masses up to 260 GeV are probed. In the T2tb model, stop masses up to 750 GeV and LSP masses up to 200 GeV are probed. Observed weaker limits than expected for the LSP mass in the T2tb model because the most sensitive bins have a small excess.

## 6. Conclusions

An over view of the search results for direct top squark production in the all-hadronic final state have been presented here. In this analysis a top tagger is used and events with $\not{E}_{\mathrm{T}}$, and $M_{T2}$ are

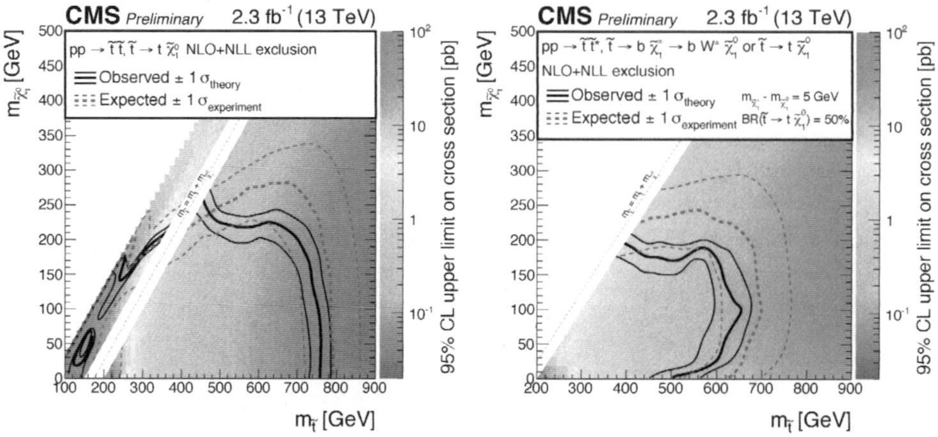


**Figure 5.** The solid black curves represent the observed exclusion contours and the corresponding ±1 standard deviations. The expected exclusion contour is the dashed red curves and includes a ±1 standard deviations with experimental uncertainties.

selected from a data sample corresponding to an integrated luminosity of 2.3 $\mathrm{fb}^{-1}$ collected in proton-proton collisions at a center-of-mass energy of 13 TeV with the CMS detector. No statistically significant excess of events above the expected standard model background is observed. Exclusion limits are set at the 95% confidence level in the context of two simplified models of direct top squark pair production. For simplified models in which both top squarks decay to a top quark and a neutralino (T2tt), top squark masses up to 780 GeV and neutralino masses up to 250 GeV are probed. In the case of models that assume 50% branching fractions for top squark decays to a top quark and a neutralino, or to a bottom quark and a chargino that is nearly degenerate in mass with respect to the neutralino (T2tb), top squark masses up to 625 GeV and neutralino masses up to 200 GeV are probed.

## References

[1] CMS Collaboration. Search for direct production of top squark pairs decaying to all-hadronic final states in pp collisions at sqrt(s) = 13 TeV. Technical Report CMS-PAS-SUS-16-007, CERN, Geneva, 2016.

[2] CMS Collaboration. CMS technical design report, volume II: Physics performance. *J. Phys. G*, 34:995, 2007.

[3] CMS Collaboration. Performance of b tagging at $\sqrt{s} = 8$ tev in multijet, $t\bar{t}$and boosted topology events. CMS Physics Analysis Summary CMS-PAS-BTV-13-001, CERN, 2013.

[4] T Plehn, M Spannowsky, M Takeuchi, and D Zerwas. Stop Reconstruction with Tagged Tops. *JHEP*, 1010:078, 2010.

[5] C G Lester, and D J Summers. Measuring masses of semiinvisibly decaying particles pair produced at hadron colliders. Phys.Lett., B463, 99-103 , 10.1016/S0370-2693(99)00945-4, 1999.

[6] A Barr, C Lester, and P Stephens. m(T2): The Truth behind the glamour. J.Phys., G29, 2343-2363 , 10.1088/0954-3899/29/10/304, 2003.

Journal of Physics: Conference Series **770** (2016) 012010 doi:10.1088/1742-6596/770/1/012010

# The KLOE-2 Experiment at DAΦNE

**A De Santis**[1]

INFN - Laboratori Nazionali di Frascati, v. Enrico Fermi, 50, 00044

E-mail: antonio.desantis@lnf.infn.it

**Abstract.** The KLOE-2 experiment is currently collecting data at DAΦNE the INFN $e^+e^-$ collider located in the Frascati National Laboratories.

The experiment has a wide physics program ranging from: discrete symmetries test, study of light unflavored mesons, searches for light mass for dark matter candidates.

In this contribution the upgrade of the detector will be briefly discussed before starting a more detailed presentation on some results concerning: CPT and Lorentz Invariance tests with neutral kaons, dark forces massive boson mediator searches, hadron structure and low energy mesons interaction.

## 1. KLOE-2 Experiment

The KLOE-2 experiment [1] is currently taking data at DAΦNE collider [2][3], the $e^+e^-$ accelerator of the Frascati National Laboratory running at the $\phi$ resonance peak. DAΦNE began operations for KLOE-2 experiment in November 2014 with a new Crab-Waist collision scheme and customized permanent magnet optics in the interaction region [4] to increase the specific luminosity. Several consolidation intervention [5] allowed to deliver 3.0 fb$^{-1}$ integrated luminosity: 2.4 fb$^{-1}$acquired that corresponds to 7.4 billion of $\phi$ decays. A record peak-luminosity of $2.21 \times 10^{32}$ cm$^{-2}$s$^{-1}$ has been achieved. The best daily integrated luminosity has been 13.4 pb$^{-1}$ and the best weekly integrated luminosity has been 76.3 pb$^{-1}$. Full time evolution of data delivery is shown in figure 1.

The KLOE-2 experiment is an upgraded version of the previous KLOE apparatus with the inclusion of new sub-detectors allowing for larger physics program with increased reconstruction performance. The original KLOE detector consists of a large cylindrical drift chamber (DC) [6] surrounded by a lead-scintillating fiber electromagnetic calorimeter (EMC) [7]. A superconducting coil around the EMC provides a 0.52 T axial field.

Two pairs of electron-positron taggers have been installed for $\gamma\gamma$ physics: a small LYSO crystal calorimeter matrix, the Low Energy Tagger [8] inside KLOE-2 apparatus and a plastic scintillator hodoscope, the High Energy Tagger (HET) [9], along the beam lines outside KLOE-2 detector.

To increase the acceptance two new calorimeters have been developed. A pair of LYSO crystal calorimeters (CCALT) [10] have been installed near the interaction region to cover the low-$\theta$ range. This calorimeter will be also useful to provide fast signals for luminosity measurement and beam instability feedback to help DAFNE tune-up [11].

[1] on behalf of KLOE-2 Collaboration

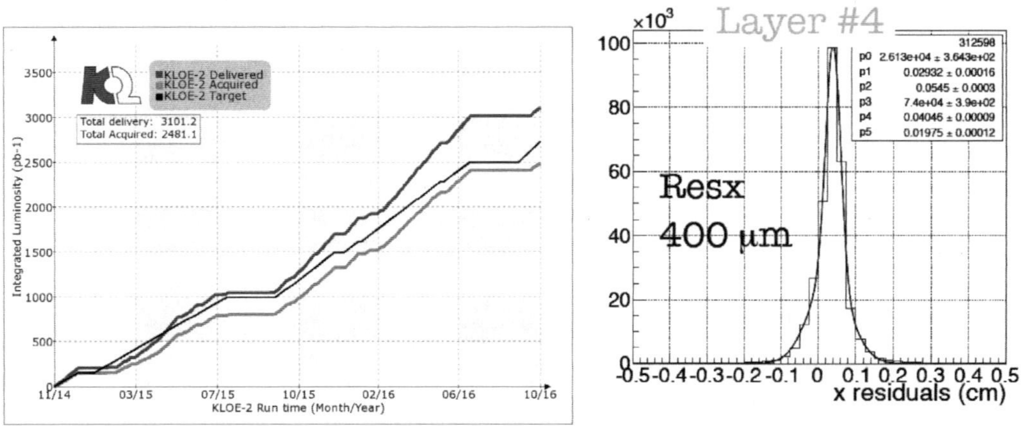


**Figure 1.** Left: time evolution of the DAΦNE data delivery since November 2014. The blue line represent the integrated luminosity delivered, while the red curve is the corresponding amount of integrated luminosity acquired. The black line represent the target luminosity fixed to achieve KLOE-2 goal of 5 $\mathrm{fb}^{-1}$ by the end of KLOE-2 physics run. The black line is horizontal during DAΦNE planned shut-downs (maintenance/laboratory closure). Riht: preliminary results on the residuals between observed and projected impact point with Bhabha scattering events on the outermost plane of the IT. The width of this distribution is the convolution of the IT and DCH resolutions.

A pair of tile calorimeters (QCALT) [12], covers the quadrupoles inside KLOE-2 detector and along the beam pipe. These calorimeters are made of tungsten slabs and singly read-out scintillator tiles to improve the angular coverage for particles coming from the active volume of the DC (e.g. $\mathrm{K_L}$ decay).

The most important and challenging upgrade is a multilayer cylindrical GEM [13] tracker for the reconstruction of decay in the vicinity of the primary interaction point (IP). Inner Tracker (IT) is made of four planes covering the tranverse radius interval between 10 and 25 cm from KLOE-2 center. It is expected to double decay vertex resolution at the IP. The current status of alignment of this detector is shown in figure 1.

## 2. Kaon interferometry

Entangled neutral kaon pair can be observed at KLOE-2 because $\phi$ decay via strong interaction preserves the $\phi$ quantum numbers: $J^{PC} = 1^{--}$. For an exhaustive review on this topic refer to [14]. The initial state $|\phi\rangle \propto |\mathrm{K_S}, \vec{p}\rangle|\mathrm{K_L}, -\vec{p}\rangle - |\mathrm{K_S}, -\vec{p}\rangle|\mathrm{K_L}, \vec{p}\rangle$ evolves preserving the correlation. Observing the two kaons decaying in the same final state ($\pi^+\pi^-$) as a function of the difference of proper decay times ($\Delta t$) the following distribution is expected:

$$I_{f_1 f_2}(\Delta t) \propto e^{-\Gamma_\Sigma |\Delta t|}\left[|\eta_1|^2 e^{\frac{\Delta\Gamma}{2}\Delta t} + |\eta_{f_2}|^2 e^{-\frac{\Delta\Gamma}{2}\Delta t} - 2|\eta_1||\eta_2|\cos(\Delta m \Delta t + \Delta\phi_{12})\right] \quad (1)$$

where $\Gamma_\Sigma = \Gamma_\mathrm{S} + \Gamma_\mathrm{L}$, $\Delta\Gamma = \Gamma_\mathrm{S} - \Gamma_\mathrm{L}$, $\Delta m = m_{\mathrm{K_L}} - m_{\mathrm{K_S}}$ and $\eta_j = \langle\pi^+\pi^-|\mathrm{K_L}(\vec{P_j})\rangle / \langle\pi^+\pi^-|\mathrm{K_S}\rangle = \varepsilon_K - \delta_K$, where $\varepsilon_K$ and $\delta_K$ are the CP and CPT violation parameters in the kaon system, respectively. According to the Standard Model Extension (SME) framework [15] and Greenberg theorem [16], the $\delta_K$ parameter is expressed as:

$$\delta_K \equiv \delta_K(P_\mu) \approx i \sin\phi_{SW} e^{i\phi_{SW}} \frac{E_K}{m_K}(\Delta a_0 - \frac{\vec{P}_K}{E_K} \cdot \Delta\vec{a})/\Delta m, \quad (2)$$

where $P_\mu$ is the kaon 4-momentum, $\phi_{SW}$ is the super-weak phase and $\Delta a_\mu$ are the SME parameters for the kaon system that explicitly violates Lorentz invariance.

At KLOE-2 the two kaons are produced almost back-to-back in the $\phi$ decay and therefore evolve with two different $\delta_K$ ($\delta_K(\tilde{P}_1) \neq \delta_K(\tilde{P}_2)$), this induces small asymmetries in the equation 1 that are expected to be modulated by the Earth rotation if the kaon momentum is observed in a reference frame defined by fixed stars. The data distributions have been fitted including SME effects to extract $\Delta a_\mu$ as show in figure 2

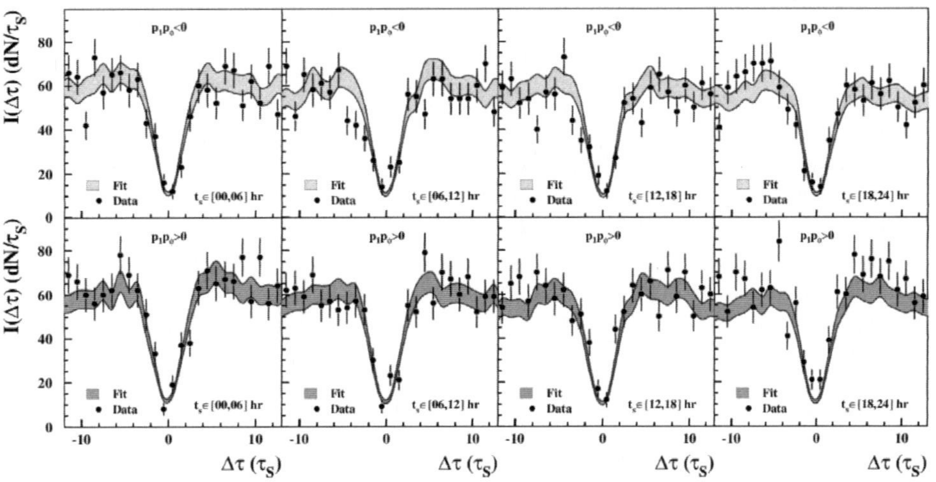


**Figure 2.** Fit results for $\Delta a_\mu$ parameters. left and right columns are relative to different angular selection in the laboratory frame while rows refers to different sidereal time interval. Black points are data while colored bands are the fit output. The fit $\chi^2/N_{\mathrm{DoF}}$ is 211.7/187 corresponding to a probability of 10%.

The results for the $\Delta a_\mu$ parameters, in $10^{-18}$ GeV units, are [17]: $\Delta a_0 = (-6.0 \pm 7.7_{stat} \pm 3.1_{syst})$, $\Delta a_X = (0.9 \pm 1.5_{stat} \pm 0.6_{syst})$, $\Delta a_Y = (-2.0 \pm 1.5_{stat} \pm 0.5_{syst})$, $\Delta a_Z = (3.1 \pm 1.7_{stat} \pm 0.5_{syst})$. That are the most precise result in the quark sector of the SME. The total error is fully dominated by the statistical uncertainty.

### 3. Dark forces searches

Several astrophysical observations have obtained results not fitting the Standard Model (SM) framework [18]. To explain these anomalies a new gauge U(1) interaction mediated by a massive vector gauge boson, the U boson (dark photon), with mass below two proton masses and kinetically mixed with the SM U(1) hypercharge gauge field has been proposed. The small coupling between dark matter and the SM is described by a single kinetic mixing parameter: $\epsilon(= \alpha_D/\alpha_{EW})$.

Depending on the interaction type, the U bosons should appear as a sharp resonance at $m_U$ in the invariant-mass distributions of final-state particles in reactions of the type $e^+e^- \to U\gamma$ with $U \to X^+X^-(X = e/\mu/\pi)$ or in meson Dalitz decays.

At KLOE two different searches [19] were performed using the decay chain $\phi \to \eta U$ with $U \to e^+e^-$ and $\eta \to \pi^+\pi^-\pi^0$ and $\eta \to \pi^0\pi^0\pi^0$ using 1.7 fb$^{-1}$of data. Irreducible background from Dalitz decay $\phi \to \eta e^+e^-$ was studied separately, as discussed in the next section. A resonant peak was not observed and the CLS technique [20] was used to set an upper limit on the strength of kinetic mixing parameter as a function of the U boson mass. The 90% confidence level limit is shown in figure 3.

KLOE data have been used to search for all the two body decays: $U \to X^+X^-(X = e/\mu/\pi)$. For the $U \to \mu^+\mu^-$ process [21] and $U \to \pi^+\pi^-$ [24] the two charged tracks are required to be emitted at large-angle (detected in the EMC barrel) while the initial-state radiation (ISR) photon is not detected. For the $e^+e^-$ final state [23] all the particles are observed at large-angle.

In all the three analysis the final invariant mass spectrum was obtained after subtracting residual background obtained form Monte-Carlo (MC) simulation normalized with dedicated data control samples. No resonant peak was observed so the CLS technique was used to estimate the maximum number of U boson signal events that can be excluded at 90% confidence level, $N_{CLS}$ as a function of the U invariant mass. This upper limit translate in the corresponding value for the $\epsilon$ parameter as shown in figure 3.

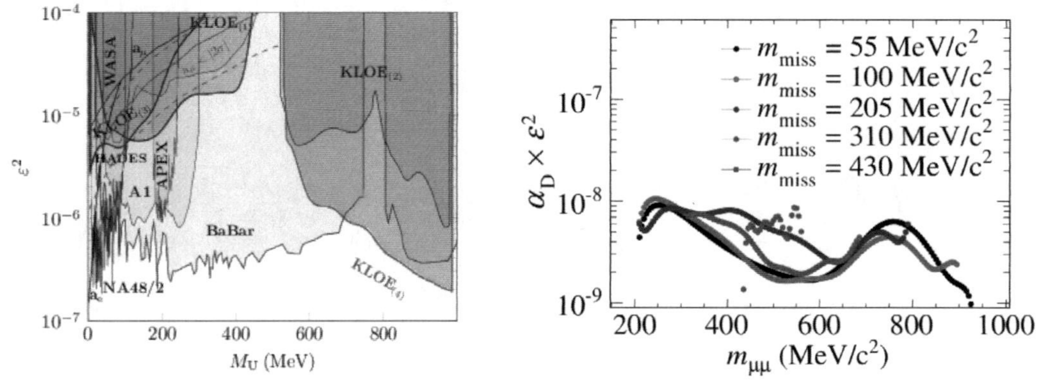


**Figure 3.** Left: summary of 90% CL exclusion limits on the dark sector coupling as a function of the U boson invariant mass. In the plot label KLOE(1) is for the combination of the results relative to $\phi \to \eta\gamma$ [19], KLOE(2) and KLOE(3) is relative to $\mu^+\mu^-\gamma$ [21] and $e^+e^-\gamma$ [23] final state, respectively. KLOE(4) is for the $\pi^+\pi^-\gamma$ final state [24]. KLOE-2 results are also compared with the exclusion plots obtained by other experiments [22]. Right: Limits from dark Higgs-strahlung process. The limit is presented as a function of the dilepton invariant mass ($m_U$) for different values of the event missing mass (dark Higgs mass). Lines represents the exclusion plots at the 90% CL.

A different scenario has been considered: the U(1) gauge symmetry is spontaneously broken by a Higgs-like mechanism, thus implying the existence of at least one other scalar particle, $h'$ [25]. The hypothetical dark Higgs-strahlung process $e^+e^- \to Uh'$ with $U \to \mu\mu$ has been investigated using KLOE data [26].

Assuming $m_{h'} < m_U$ the dark Higgs boson would have a large lifetime and escape KLOE detector without interacting. 1.65 fb$^{-1}$ of data collected at the $\phi$-peak energy, and 0.2 fb$^{-1}$ of data at 1 GeV energy have been used. A sharp peak in the two-dimensional distribution missing mass versus $m_{\mu\mu}$ is expected for the signal. No evidence of the dark Higgs-strahlung process was found. Using uniform prior distributions, 90% confidence level Bayesian upper limits on the number of events, $N_{90\%}$, were derived separately for the two samples used and then combined. The combined 90% confidence level limits are shown figure 3.

## 4. Hadron physics

The low-energy structure of hadrons is strongly connected with precise knowledge of meson transition form factors (TFF) that are also connected in the determination of the "light-by-light" contribution to the anomalous magnetic moment of the muon, one the largest source of uncertainty in the $(g-2)_\mu$ calculation [27].

KLOE-2 collaboration has measured the $\phi \to \eta$ and $\phi \to \pi^0$ TFFs studying the decays $\phi \to \eta(3\pi^0)\mathrm{e^+e^-}$ [28] and $\phi \to \pi^0\mathrm{e^+e^-}$ [29]. In figure 4 the TFFs are shown as a function of $\sqrt{q^2} = m_{ee}$, the invariant mass of $e^+e^-$ system. The observed mass spectrum has been described according to [30] where the TFF is included with a single pole parametrization, $|F_{\phi\eta}(q^2)|^2 = (1 - q^2/\Lambda^2)^{-1}$. The $\Lambda$ parameter is extracted from the $m_{ee}$ invariant mass spectrum and the corresponding slope parameter ($\mathrm{b}_{\phi\eta} = \frac{\mathrm{dF(q^2)}}{\mathrm{dq^2}}|_{\mathrm{q^2=0}} = \Lambda^{-2}$) is evaluated. The value $\mathrm{b}_{\phi\eta} = 1.17 \pm 0.10^{+0.07}_{-0.11}\,\mathrm{GeV}^{-2}$ has been obtained while $1\mathrm{GeV}^{-2}$ is expected by pure VMD description. In the analysis has been measured also the branching fraction $\mathrm{B}(\phi \to \eta\mathrm{e^+e^-}) = 1.075 \pm 0.007 \pm 0.038) \times 10^{-4}$.

For the decay $\phi{\to}\pi^0e^+e^-$, using the same parametrization [30], the invariant mass spectrum has been fitted and a slope of $\mathrm{b}_{\phi\pi^0} = 2.02 \pm 0.11\,\mathrm{GeV}^{-2}$ has been obtained. Naive VMD extimate would give $\mathrm{b}_{\phi\pi^0} \sim 1\,\mathrm{GeV}^{-2}$, while prediction based on disperve calculation [31] are $\mathrm{b}_{\phi\pi^0} \sim 2.52 ... 2.68\,\mathrm{GeV}^{-2}$.

The branching fraction $B(\phi \to \pi^0e^+e^-) = (1.035 \pm 0.005^{+0.05}_{-0.10}) \times 10^{-5}$ has been measured.

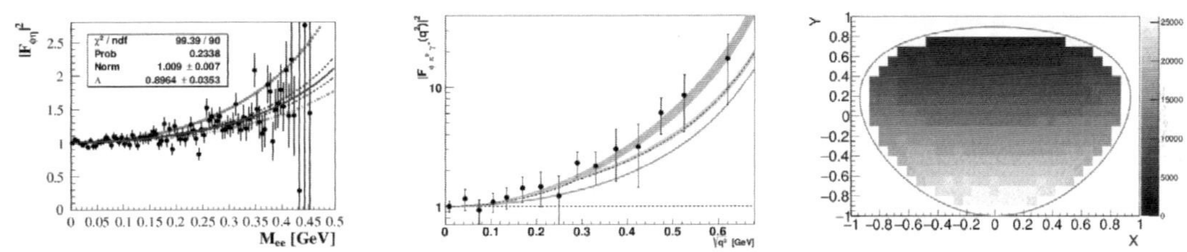


**Figure 4.** TFF $|F_{\phi P}(q^2)|^2$ as a function of $\sqrt{q^2}$ (invariant mass of $e^+e^-$) for $\phi \to \eta\mathrm{e^+e^-}$ (left) and $\phi \to \pi^0\mathrm{e^+e^-}$ (center). Black dots are data. In the top panel the blue curve is the single pole approximation of the TFF with the slope measured on the $m_{ee}$ mass spectrum. Red and purple are different theoretical model: Terschluensen/Leupold [32] and VMD, respectively. In the bottom panel the green band is the Ivashyn model [31], the red curve is the VMD expectation, while the dashed line is KLOE-2 fit result. Right: Dalitz density distribution as a function of $X = \sqrt{3}\frac{T_{\pi^+} - T_{\pi^-}}{Q_\eta}$ and $Y = \frac{3T_{\pi^0}}{Q_\eta} - 1$ and $Q_\eta = m_\eta - 2m_{\pi^+} - m_{\pi^0}$.

Recently a new result on the $\eta \to \pi^+\pi^-\pi^0$ Dalitz plot density with the highest statistics in the world ($\sim 4.5 \cdot 10^6$ events) [33] has been released. This analysis allows to extract the light quark mass difference and to perform a test of the charge conjugation violation. The $\eta$ decay is selected among the $\phi \to \eta\gamma$ radiative decay events having two tracks and two extra gamma's. The density of the Dalitz plot as a function of the kinetic energies of the pions is then fitted with a polynomial:

$$|A(X,Y)|^2 = \mathcal{N}(1 + aY + bY^2 + cX + dX^2 + eXY + fY^3 + gX^2Y + ...) \qquad (3)$$

where X and Y are the normalized Dalitz plot variables. The equation 3 should be symmetric under X parity (charge conjugation) thus resulting in $c = e = 0$, confirmed by data. Fitting the distribution show in figure 4 KLOE-2 obtained the following results for the non-vanishing coefficients:

$$a = -1.095 \pm 3 \pm 2;\ b = 0.145 \pm 3 \pm 5;\ \mathrm{d} = 0.081 \pm 3^{+5}_{-6};$$

$$\mathrm{f} = 0.141 \pm 7^{+7}_{-8};\ \mathrm{g} = -0.044 \pm 9^{+12}_{-13}$$

The $g$ parameter has been observed different from zero for the first time because of the high statistics collected and high purity of the signal selected.

## 5. Conclusions

The data delivery evolution, shown in figure 1, demonstrates DAΦNE capabilities to fulfill the physics run goal within the allocated time: 5 $fb^{-1}$ within 2017. The KLOE-2 detector is fully functional and all subdetectors are constantly monitored in order to keep the detector fully calibrate and perfectly operational during the whole data taking period as well as the DAΦNE performance.

Waiting for the full understanding of the new collected data, KLOE-2 collaboration progress in the study of the original KLOE dataset producing interesting physics results with the aim of use the new analysis on the old data as benchmark for the new data especially on the $\gamma\gamma$ physics and kaon interferometry.

## 6. References

[1] G. Amelino-Camelia *et al.*, Eur. Phys. J. C **68** (2010) 619
[2] G. Vignola et al., Frascati Phys. Ser. 4:19-30, 1996.
[3] P. Raimondi et al., arXiV:physics/0702033; C. Milardi et al., Int.J.Mod.Phys.A24:360-368, 2009; M.
[4] C. Milardi et al 2012 JINST 7 T03002.
[5] C. Milardi et al, THPRI002, IPAC14 (2014); A. De Santis et al, THPRI015, IPAC14 (2014); C. Milardi et al, WEOCA03, IPAC14 (2014).
[6] M. Adinolfi *et al.*, [KLOE],NIM A 488 (2002) 51.
[7] M. Adinolfi *et al.*, [KLOE],NIM A 482 (2002) 363.
[8] D. Babusci *et al.*, NIM A **617**, 81 (2010).
[9] F. Archilli, *et al.*, NIM A **617** (2010) 266.
[10] F. Happacher *et al.*, Nucl. Phys. Proc. Suppl. **197**, 215.
[11] A. De Santis DAΦNE Tech. Note BM-18 (2016).
[12] M. Cordelli *et al.*, NIM A **617**, 105 (2010).
[13] A. Balla, *et al.*, NIM A **628** (2011) 194
[14] A. Di Domenico, Frascati Physics Series, Vol. 43 (2007)
[15] V. A. Kostelecky, Phys. Rev. D **64** (2001) 076001
[16] O. W. Greenberg, Phys. Rev. Lett. **89** (2002) 231602.
[17] D. Babusci *et al.* [KLOE-2 Collaboration], Phys. Lett. B **730**, 89 (2014)
[18] O. Adriani, et al., Nature 458 (2009) 607. P. Jean, et al., Astron. Astrophys. 407 (2003) L55. J. Chang, et al., Nature 456 (2008) 362. HESS Coll., Phys. Rev. Lett. 101 (2008) 261104. HESS Coll., Astron. Astrophys. 508 (2009) 561. A. A. Abdo, et al., Phys. Rev. Lett. 102 (2009) 181101. R. Bernabei, et al., Int. J. Mod. Phys. D 13 (2004) 2127. R. Bernabei, et al., Eur. Phys. J. C 56 (2008) 333. CoGeNT Coll., Phys. Rev. Lett. 106 (2011) 131301. CoGeNT Coll., Phys. Rev. Lett. 107 (2011) 141301. AMS Coll., Phys. Rev. Lett. 110 (2013) 141102.
[19] KLOE-2 Coll., Phys. Lett. B 706 (2012) 251255. KLOE-2 Coll., Phys. Lett. B 720 (2013) 111115.
[20] G. C. Feldman, R. D. Cousins, Phys. Rev. D 57 (1998) 3873.
[21] KLOE-2 Coll., Phys. Lett. B 736 (2014) 459464.
[22] E141 Coll., Phys. Rev. Lett. 59 (1987) 755. E774 Coll., Phys. Rev. Lett. 67 (1991) 2942. MAMI/A1 Coll., Phys. Rev. Lett. 106 (2011) 251802. APEX Coll., Phys. Rev. Lett. 107 (2011) 191804. WASA Coll., Phys. Lett. B 726 (2013) 187. HADES Coll., Phys. Lett. B 731 (2014) 265271. A1 Coll., Phys. Rev. Lett. 112 (2014) 221802. BaBar Coll., arXiv:1406.2980.
[23] A. Anastasi *et al.* [KLOE-2], Phys. Lett. B **750** (2015) 633
[24] A. Anastasi *et al.* [KLOE-2], Phys. Lett. B **757** (2016) 356
[25] A. R. B. Batell, M. Pospelov, Phys. Rev. D 79 (2009) 115008.
[26] A. Anastasi *et al.* [KLOE-2], Phys. Lett. B **747** (2015) 365
[27] F. Jegerlehner, A. Nyffeler, Phys.Rept.477(2009)1
[28] D. Babusci *et al.* [KLOE-2], Phys. Lett. B **742** (2015) 1
[29] A. Anastasi *et al.* [KLOE-2], Phys. Lett. B **757** (2016) 362
[30] L.G. Landsberg, Phys. Rep.**128**(1985)301.
[31] S. Ivashyn, Prob. Atomic Sci. Technol. **2012N1**(2012)179.
[32] C. Terschlusen, S. Leupold, Phys. Lett. B**691**(2010)191.
[33] A. Anastasi *et al.* [KLOE-2 Coll.], JHEP **1605** (2016) 019

# Status of Pion Decay Experiments

**T Numao**
TRIUMF, 4004 Wesbrook Mall, Vancouver, British Columbia, Canada V6T 2A3
E-mail: toshio@triumf.ca

**Abstract.** The branching ratio of pion decays, $R_{e/\mu} = \Gamma(\pi^+ \to e^+\nu + e^+\nu\gamma)/\Gamma(\pi^+ \to \mu^+\nu + \mu^+\nu\gamma)$, has provided a sensitive test of electron-muon universality in weak interactions. The uncertainty of the Standard Model prediction is at a 0.01 % level. Although a recent measurement, $R_{e/\mu} = (1.2344 \pm 0.0023(stat) \pm 0.0019(syst)) \times 10^{-4}$, reduced the experimental uncertainty by a factor of two, there is room for improvement by more than an order of magnitude. The status of two $\pi^+ \to e^+\nu$ experiments at TRIUMF and PSI as well as related pion decay experiments is presented.

## 1. Introduction

The focus of the present paper is on the measurements of the branching ratio $R_{e/\mu} = \Gamma(\pi^+ \to e^+\nu + e^+\nu\gamma)/\Gamma(\pi^+ \to \mu^+\nu + \mu^+\nu\gamma)$ , which provides the best test of $e - \mu$ universality in charged-current weak interactions. As indicated in the expression of $R_{e/\mu}$, the experimental result includes radiative decays with virtual and real photons. A recent theoretical prediction, $R_{e/\mu}= (1.2352 \pm 0.0002) \times 10^{-4}$ [1, 2], included uncertainties from uncalculated two-loop effects. Model dependent effects were calculated based on the Chiral Perturbation Theory (ChPT) up to the $O(m^2/p^4)$ terms.

The experimental status till 2014 was: $R_{e/\mu}= (1.2265 \pm 0.0050) \times 10^{-4}$ (TRIUMF) [3] and $(1.235 \pm 0.005) \times 10^{-4}$ (PSI) [4]. Presently, there are two experiments, PIENU at TRIUMF [5] and PEN at PSI [6], to improve the uncertainties to a level of <0.1%. The PIENU group released a result of a partial analysis in 2015, $R_{e/\mu} = (1.2344 \pm 0.0023(stat) \pm 0.0019(syst)) \times 10^{-4}$ [7]. The new result improved the test of $e - \mu$ universality to be $g_e/g_\mu = 0.9996 \pm 0.0012$. This still leaves room for experimental improvement by more than an order of magnitude in the uncertainty.

Since the uncertainty of the standard model (SM) calculation for $R_{e/\mu}$ is very small and the decay $\pi^+ \to e^+\nu$ is helicity-suppressed by the V–A structure of charged currents, the branching ratio $R_{e/\mu}$ is extremely sensitive to helicity-unsuppressed couplings such as the pseudo-scalar coupling. At the level of 0.1 % measurements, new physics beyond the SM up to the mass scale of 1000 TeV/c$^2$ may be accessed by a deviation from the precise SM expectation [2]. Possible sources of deviation include Majorons [8], charged Higgs particles and leptoquarks [9]. Also, SUSY models with and without R-parity violation [10]. or with lepton flavor violating terms [11] could cause a deviation from the SM prediction. Another example of the new physics is sterile massive neutrinos [12] which were sought in the PIENU experiment [13].

In this paper, the present status of two on-going experiments [5, 6], which aim at a precision level of < 0.1 % for the measurement of $R_{e/\mu}$, will be discussed, and also precision measurements of other pion decays will be reviewed in the context of $\pi^+ \to e^+\nu$ measurements.

## 2. Experimental method

Both PIENU and PEN experiments measure pion decays at rest. A 70-MeV positron is emitted in the $\pi^+ \to e^+\nu$ decay. Since muons from $\pi^+ \to \mu^+\nu$ decays, having a short range (~1.4 mm) in plastic scintillator, remain in the target, the solid angle for positron detection from the subsequent muon decay $\mu^+ \to e^+\nu\overline{\nu}$ ($E_{e+} = 0 - 52$ MeV) is almost the same as that from $\pi^+ \to e^+\nu$ decays. On the other hand, muon detection for $\pi^+ \to \mu^+\nu$ decays involving entirely different solid angles and corrections is anticipated to have larger experimental uncertainties. In both experiments, the ratio of the positron yields from $\pi^+ \to e^+\nu$ and $\mu^+ \to e^+\nu\overline{\nu}$ following $\pi^+ \to \mu^+\nu$ ($\pi^+ \to \mu^+ \to e^+$) was measured with an inorganic crystal calorimeter. No magnetic field was applied.

Events from $\pi^+ \to e^+\nu$ (decay with the pion lifetime of 26 ns) and $\pi^+ \to \mu^+ \to e^+$ (a sharp increase with the pion lifetime and a slow decrease with the muon lifetime of 2.2 $\mu$s) are separated at the upper end of $\pi^+ \to \mu^+ \to e^+$ spectrum (nominally, at 52 MeV), and the time spectra for the low- and high-energy regions are fitted together with background terms. Low-energy $\pi^+ \to e^+\nu$ events below 52 MeV due to the response function of the calorimeter and radiative pion decays will be corrected. Other small corrections for energy-dependent effects are also applied. The main background for $\pi^+ \to \mu^+ \to e^+$ decay comes from the presence of muons from previously stopped pions ("old muons"). Other backgrounds arising from various mechanisms will be discussed later.

## 3. PIENU experiment

The concept of the PIENU experiment [5, 14] was based on the previous TRIUMF experiment [3] with improvements in the acceptance by an order of magnitude and in the systematic uncertainties. Expected uncertainties are summarized in Table 1 together with those from the previous experiments.

A 75 MeV/c $\pi^+$ beam was degraded by two thin plastic scintillators (6 mm and 3 mm thick) and stopped in an 8-mm thick active target at a rate of $5 \times 10^4$ pions/s. Pion tracking was provided by six planes of wire chambers at the exit of the beam line, and two sets of X-Y Si-strip counters immediately upstream of the target.

In order to gain a large solid angle (~25 %) while reducing the variation of the amount of material along the positron path, the positron telescope consisting of two thin plastic scintillator counters and a calorimeter system was placed on the beam axis. The extension of the beam line with a capability of particle separation [15] allowed this geometry by suppressing positrons in the beam to < 2 % of pions. It also reduced (together with more shielding) the neutron rate from the pion-production target by an order of magnitude. The primary calorimeter was a 48-cm (diam.) × 48-cm (length) single-crystal NaI(T$\ell$) surrounded by two rings of 97 pure CsI crystals (9 radiation lengths) to reduce the shower leakage. The improvement factor of 30 in statistics was therefore expected to come from a larger solid angle and a longer running period. Positron tracking came from one set of X-Y Si-strip counters immediately downstream of the target, and three layers of wire chambers in front of the NaI crystal.

In the analysis, events with a single pion in the beam detector and a single positron in the calorimeter were selected. By requiring no preceding beam particles within a time window up to -6 $\mu$s, the presence of "old" muons in the target region was reduced by an order of magnitude to achieve a low beam-rate condition for a clean measurement.

The "old" muon was the primary source of the background in the low-energy time spectrum. In the high-energy time spectrum, one type of the backgrounds which had the identical distribution to the low-energy one was due to the calorimeter resolution and $\mu^+ \to e^+\nu\overline{\nu}\gamma$ decays (the $\gamma$-ray also hit the calorimeter to raise the observed energy). Neutrons from the production target and calibration sources also contributed to this type of background.

Radiative pion decays $\pi^+ \to \mu^+\nu\gamma$ (the branching ratio of $2 \times 10^{-4}$) followed by a

$\mu^+ \rightarrow e^+\nu\overline{\nu}$ decay could be a background in the high-energy region when the $\gamma$-ray hit the calorimeter. Since the calorimeter energy used to separate the two regions was measured with respect to the positron time, the contribution of the extra $\gamma$-ray energy to the observed energy varied with the time difference between the two decays.

Another type of background occurred when a positron from an "old" muon deposited extra energy into the calorimeter in addition to the true event. The one-hit requirement in the positron telescope usually protected against this type of background, but if one of the positrons was emitted to a large angle to avoid detection the protection logic did not function. Also, if two decays occurred during the two-pulse resolution of the telescope, the events were kept. These types of the backgrounds were suppressed by reducing the "old" muon component. All the backgrounds described above were included in the fitting function.

When the muon from $\pi^+ \rightarrow \mu^+\nu$ decayed in flight, the Lorentz boosting might raise the positron energy above 52 MeV. This contribution has a time distribution of the pion decay. The effect was estimated by simulation to be 0.2 % of the branching ratio and included in the branching ratio obtained by the time-spectrum fit.

The validity of the background shape and amplitude was confirmed by enhancing each type of the backgrounds by changing the related selection criteria.

**Table 1.** Comparison of uncertainties (%) in $R_{e/\mu}$ for PIENU [5], 2015 analysis [7] and the previous TRIUMF result [3].

| Sources | PIENU | 2015 analysis | Previous exp. |
|---|---|---|---|
| Statistical error | 0.05 | 0.19 | 0.28 |
| Tail correction | 0.03 | 0.12 | 0.25 |
| Acceptance difference | 0.03 | 0.03 | 0.11 |
| Others | 0.03 | 0.08 | 0.14 |
| Total | 0.06 | 0.24 | 0.45 |

The correction for the fraction of low-energy $\pi^+ \rightarrow e^+\nu$ events was estimated with two methods. The response function of the calorimeter to 70-MeV/c positrons was measured with a positron beam directly injected into the calorimeter. The effects of hadronic interactions, which were not reproduced accurately enough in the simulation for the present condition, were parametrized based on the measured spectra at various angles, and the response function for positrons from the $\pi^+ \rightarrow e^+\nu$ decay including the radiative mode was generated by simulation. Since the amount of the contamination of low-energy positrons in the beam was not estimated, the present tail correction was considered to be an upper bound.

In order to further constrain the amount of the low-energy tail, $\pi^+ \rightarrow e^+\nu$ events were enhanced by suppressing the dominant $\pi^+ \rightarrow \mu^+ \rightarrow e^+$ background using the decay time, the total charge deposited in the target ($\pi^+ \rightarrow \mu^+ \rightarrow e^+$ decays involve extra 4 MeV) and tracking information (to suppress background from decays-in-flight of pions). Assuming that the lowest energy area of the suppressed spectrum did not contain the low-energy tail of $\pi^+ \rightarrow e^+\nu$ events, the number of events in the area was scaled to the whole background region using the shape of the known $\pi^+ \rightarrow \mu^+ \rightarrow e^+$ background spectrum and the background was subtracted from the background-suppressed spectrum. This method might result in over-subtraction of the background due to the presence of actual signal in the lowest energy area, and the correction obtained by this method was considered to be a lower bound.

The PIENU experiment completed the data taking stage with about $1 \times 10^7$ clean $\pi^+ \rightarrow e^+\nu$ events at the end of 2012. The result of the partial analysis of 2010 data was published in 2015 [7]. A full analysis based on ten times more event statistics with reduced background

Journal of Physics: Conference Series **770** (2016) 012011 doi:10.1088/1742-6596/770/1/012011

and additional improvements in systematics is in progress. In the full analysis of the low-energy tail estimation, additional positron-beam data taken with various extreme slit and magnetic-field combinations will be analyzed, and hopefully the measured response function could be taken as the tail correction itself. Also, the uncertainty of the lower bound limited by statistics in the partial analysis will be reduced. The lower bound could be further improved by including the well-known electro-magnetic shower contribution instead of the assumption of no $\pi^+ \to e^+\nu$ events in the lowest energy area.

## 4. PSI experiment

In the new PSI experiment (PEN) [6], ~70 MeV/c pions were tagged by a counter in the beam line, slowed down by a 5-mm thick active plastic-scintillator degrader, tracked by a 40-mm × 40-mm × 50-mm mini-TPC (for pion tracking), and stopped in a 15-mm thick fast-scintillator target at a rate of 20 k/s.

Positrons from the $\pi^+ \to e^+\nu$ and $\pi^+ \to \mu^+ \to e^+$ decays were measured by two sets of cylindrical wire chambers, a thin 20-element plastic-scintillator hodoscope, and 240 12-radiation-length thick pure CsI detectors that covered a solid angle of $\sim 3\pi$ sr.

Fast wave-form digitizers (2 GHz) were employed for the beam and target counters, which were effective in suppressing the background by identifying two-pulse events ($\pi^+ \to e^+\nu$) from three-pulse events ($\pi^+ \to \mu^+ \to e^+$).

**Table 2.** Expected uncertainties (%) in $R_{e/\mu}$ for PEN [6].

| Sources | Uncertainties |
|---|---|
| Statistical | 0.02 |
| Systematic | 0.03 |
| $\pi/\mu$ discri. | 0.01 |
| $E_e < E_{threshold}$ | 0.01 |
| Acceptance | |
| Radiative decay | 0.01> |
| Photo-nuclear int. | 0.01 |
| T0 variation | 0.02> |
| Total | 0.05 |

The data taking stage was over in 2010. About $2 \times 10^7$ $\pi^+ \to e^+\nu$ events (before cuts) were accumulated. Because of the large solid angle for positron detection and the method for the low-energy tail correction, the PEN experiment has different systematic uncertainties from the PIENU experiment. Expected uncertainties are summarized in Table 2, which include corrections for pion and muon event discrimination, positrons below the detection threshold for $\mu^+ \to e^+\nu\overline{\nu}$ decays, acceptance difference between $\pi^+ \to e^+\nu$ and $\pi^+ \to \mu^+ \to e^+$ events, effects of radiative decays, photo-nuclear interactions, and the knowledge of the time zero for both decay modes. The analysis is in progress.

## 5. Related pion decay experiments

In this section, other pion decay experiments especially which are related to the $R_{e/\mu}$ measurement through radiative corrections are briefly summarized.

Precision measurements of other pion decay modes were used for inputs of ChPT calculations as well as for the confirmation of the ChPT predictions. The amplitudes of structure-dependent radiative decays in the $\pi^+ \to e^+\nu\gamma$ decay, which are not helicity-suppressed, are parametrized

Journal of Physics: Conference Series **770** (2016) 012011 doi:10.1088/1742-6596/770/1/012011

by the vector and axial-vector form factors, $F_V$ and $F_A$, respectively. The structure-dependent radiative correction for $R_{e/\mu}$ came from the ratio $F_A/F_V$. The PIBETA group [16], using the CsI crystal calorimeter measured $F_V = 0.0258(17)$ and $F_A = 0.0117(17)$—they also set a tight bound for the presence of the tensor form factor, $F_T = (-0.6 \pm 2.8) \times 10^{-4}$. While in the PIBETA experiment the data was collected at a high beam rate to precisely measure the rare branching ratio of the $\pi^+ \to \pi^0 e^+ \nu$ decay (BR$\sim 10^{-8}$), the PEN experiment [6] collected data at a low beam rate together with the $\pi^+ \to e^+\nu$ decay data. This allowed an access to the energy combination of the positron and $\gamma$-ray which was sensitive to $F_V - F_A$. The data analysis is in progress.

The vector form factor $F_V$ in the $\pi^+ \to e^+\nu\gamma$ decay is related to the lifetime of the $\pi^0$ through the Conserved-Vector-Current theory. A recent result $(8.32 \pm 0.15 \pm 0.18) \times 10^{-17}$ s by the PrimEx group at JLAB [18], being consistent with the previous measurements [17] though slightly off from the direct measurement, improved the uncertainty by a factor of two. Another approach to the $\pi^0$ lifetime is expected through the measurement of $e^-e^+ \to \pi^0$. The KLOE-2 group is planning to improve the measurement in the near future [19].

There are other exotic rare $\pi^0$ decay modes being sought. Since the NA62 experiment [20] is a "tagged $\pi^0$ factory" via copious $K^+ \to \pi^+\pi^0$ decays, it is expected to improve related measurements, such as $\pi^0 \to \nu\bar{\nu}$ [21], together with a better photon vetoing capability.

## 6. Conclusion

Although the decay $\pi^+ \to e^+\nu$ was discovered 50 years ago, it still remains one of the topical subjects. Precision measurements of $R_{e/\mu}$ provide the best test of e-$\mu$ universality in weak interactions. It is also sensitive to the presence of pseudo-scalar interactions arising from physics beyond the SM.

The two on-going experiments at TRIUMF and PSI are expected to finish the analysis at a level of <0.1 % uncertainty in a few years. At this level of precision, new physics up to 1000 TeV/c$^2$ could be surveyed.

## References

[1] Cirigliano V and Rosell I 2007 *Phys. Rev. Lett.* **99** 231801; Cirigliano V and Rosell I 2007 *JHEP* **0710** 005
[2] Bryman D A, Marciano W and Yamanaka T 2011 *Ann. Rev. Nucl. Part. Sci.* **61** 331-354
[3] Britton D I *et al* 1992 *Phys. Rev. Lett.* **68** 3000; Britton D I *et al* 1994 *Phys. Rev.* D **49** 28
[4] Czapek G *et al* 1993 *Phys. Rev. Lett.* **70** 17
[5] PIENU collaboration 2005 PIENU experiment *TRIUMF Proposal S1072*
[6] PEN collaboration 2006 PEN experiment *PSI Proposal R-05-01*
[7] Aguilar-Arevalo A *et al* 2015 *Phys. Rev. Lett.* **115** 071801
[8] Lessa A P and Peres O L G 2007 *Phys. Rev.* D **75** 094001
[9] Campbell B A and Ismail A 2008 *arXiv:0810.4918*
[10] Ramsey-Musolf M J, Su S and Tulin S 2007 *Phys. Rev.* D **76** 095017.
[11] Masiero A, Paradisi P and Petronzio R 2006 *Phys. Rev.* D **74** 011701
[12] Kusenko A 2009 *Phys. Rep.* A **481** 1; and references therein.
[13] Aguilar-Arevalo A *et al* 2011 *Phys. Rev.* D **84** 052002
[14] Aguilar-Arevalo A *et al* 2015 *Nucl. Instrum. Method* A **791** 38
[15] Aguilar-Arevalo A *et al* 2009 *Nucl. Instrum. Method* A **609** 102
[16] Bychkov M *et al* 2009 *Phys. Rev. Lett.* **103** 051802
[17] Particle Data Group, Beringer J *et al* 2012 *Phys. Rev.* D **86** 1
[18] Larin I *et al* 2011 *Phys. Rev. Lett.* **106** 162303
[19] De Santis A 2016 The KLOE-2 experiment at DAFNE *Presentation at this conference*
[20] NA62 collaboration 2005 Proposal to measure the rare $K^+ \to \pi^+\nu\bar{\nu}$ decay at the CERN SPS *CERN proposal P326*; Cenci P 2016 Neutral pion form factor measurement by the NA62 experiment *Presentation at this conference*
[21] Artamonov A V *et al* 2002 *Phys. Rev.* D **91** 052001

Journal of Physics: Conference Series **770** (2016) 012012 doi:10.1088/1742-6596/770/1/012012

# Charm physics at BESIII

**P Weidenkaff for the BESIII collaboration**

Institut für Kernphysik, Johann-Joachim-Becher-Weg 45, 55128 Mainz, Germany

E-mail: weidenka@uni-mainz.de

**Abstract.** The study of mesons and baryons which contain at least one charm quark is referred to as open charm physics. It offers the possibility to study up-type quark transitions. Since the $c$ quark can not be treated in any mass limit, theoretical predictions are difficult and experimental input is crucial. BESIII collected large data samples of $e^+e^-$ collisions at several charm thresholds. The at-threshold decay topology offers special opportunities to study open charm decays.

We present a selection of recent BESIII results. The $D_s^+$ decay constant is measured using the leptonic decays to $\mu^+\nu$ and $\tau^+\nu$. Using the semi-leptonic decays of $D^0$ and $D^\pm$ to $Ke^+\nu_e$ and $\pi e^+\nu_e$, a measurement of the form factors $f_+^K(q^2)$ and $f_+^\pi(q^2)$ is performed and furthermore, we show preliminary results of a model independent measurement of the strong phase difference between $D^0$ and $\overline{D}^0$ in the channel $D^0 \to K_S^0\pi^+\pi^-$ which is an experimental input to the measurement of the CKM angle $\gamma/\phi_3$.

## 1. Introduction

The BESIII experiment is located at the Beijing Electron-Positron Collider. The accelerator is a $e^+e^-$ storage ring located at the Institute of High Energy Physics in Beijing. It provides symmetric collisions in the energy range between 2.0 GeV and 4.6 GeV. The maximum luminosity of BEPCII is achieved at $\sqrt{s}$ =3.773 GeV. In April 2016 a luminosity of $1 \times 10^{33}\,\mathrm{cm}^{-2}\,\mathrm{s}^{-1}$ was surpassed. The detector measures charged track momenta with a relative precision of 0.5 % (@1.0 GeV) using a multi-wire drift chamber in a 1 T magnetic field. Electromagnetic showers are measured in a caesium iodide calorimeter with a relative precision of 2.5 % (@1.0 GeV) and a good particle identification is achieved by combining information from energy loss in the drift chamber, from the time-of-flight system and from the calorimeter. Muons can be identified using 9 layers of resistive plate chambers integrated in the magnet return yoke. The BESIII experiment is described in detail elsewhere [1]. BESIII has collected large data samples in the tau-charm region. The interesting samples for the study of charmed hadrons are usually at a center-of-mass energy close to a threshold. The largest such sample was recorded at $\sqrt{s} = 3.773$ GeV with an integrated luminosity of 2.81 $\mathrm{fb}^{-1}$. The corresponding threshold is the $D^0\overline{D}^0/D^+D^-$ threshold. At $\sqrt{s} = 4.009$ GeV, close to the $D_s^+\overline{D}_s^+$ threshold, a sample of 0.482 $\mathrm{fb}^{-1}$ was recorded. $\Lambda_c$ decays can be studied using a sample of 0.5669 $\mathrm{fb}^{-1}$ at the $\Lambda_c\overline{\Lambda_c}$ threshold ($\sqrt{s} = 4.5995$ GeV).

The implications of those at-threshold decay topologies are discussed in section 2. In the following we present the measurements of the $D_s^+$ decay constant (section 3), form factor measurements of neutral and charged $D$ decays to $Ke^+\nu$ and $\pi e^+\nu$ (section 4) as well as a model independent measurement of the strong phase difference between $D^0$ and $\overline{D}^0 \to K_S^0\pi^+\pi^-$ (section 5). Recent results on $\Lambda_c$ decays are presented in [2].

Journal of Physics: Conference Series **770** (2016) 012012 doi:10.1088/1742-6596/770/1/012012

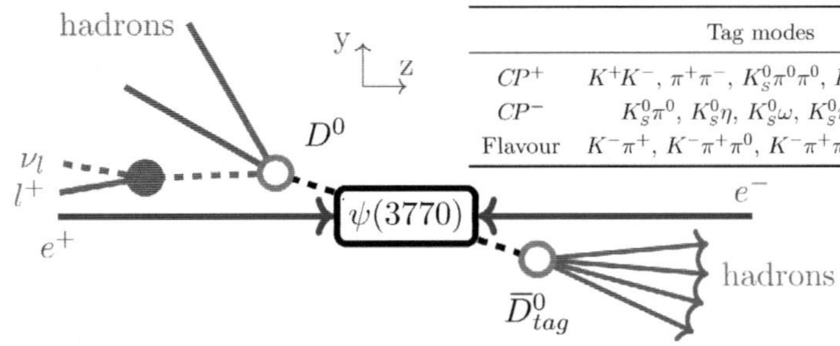


| | Tag modes |
|---|---|
| $CP^+$ | $K^+K^-$, $\pi^+\pi^-$, $K^0_S\pi^0\pi^0$, $K^0_L\pi^0$, ... |
| $CP^-$ | $K^0_S\pi^0$, $K^0_S\eta$, $K^0_S\omega$, $K^0_S\eta'$, ... |
| Flavour | $K^-\pi^+$, $K^-\pi^+\pi^0$, $K^-\pi^+\pi^+\pi^-$, ... |

**Figure 1.** $\psi(3770)$ decay topology in the $\psi(3770)$ rest frame. An undetected particle track can be reconstructed using the constrained kinematics of the decay. Typical tag modes for *CP* and flavour eigenstates are listed.

## 2. Open charm decays at threshold

The at-threshold decay topology at a center-of-mass energy of 3.773 GeV is illustrated in figure 1. A pair of particles is produced and it is possible to conclude from the decay of one meson (so-called tag meson) properties of the second decay. For instance in case of neutral $D$ decays the flavour or the *CP* quantum numbers of the signal decay can be measured, even if the signal final state does not provide this information. In case of charged $D$ decays the reconstruction of both decays is used to reduce the background and furthermore if undetected particles are involved in the signal decay the four momenta of those can be calculated. In particular the study of leptonic and semi-leptonic decays benefits from this. The reconstruction of both decays in each event is referred to as double tag technique.

## 3. $D_s^+$ decay constant

The simplest and cleanest modes of $D_s^+$ decays are pure leptonic modes. We analyse the final state $\tau^+\nu_\tau$ and $\mu^+\nu_\mu$ and measure their branching fractions. Detailed information can be found in [3].

The leptonic decay of $D_s^+$ proceeds via the annihilation of $c$ and $\overline{s}$ to a virtual $W^\pm$ boson and its decay to $l^+\nu_l$. The decay rate can be parametrized as:

$$\Gamma(D_s^+ \to l^+\nu_l) = \frac{G_F^2}{8\pi} f_{D_s^+}^2 m_l^2 m_{D_s^+} \left(1 - \frac{m_l^2}{m_{D_s^+}^2}\right)^2 |V_{cs}|^2, \tag{1}$$

with the Fermi constant $G_F$, the lepton mass $m_l$, the CKM matrix element $|V_{cs}|^2$, the $D_s^+$ mass $m_{D_s^+}$ and the decay constant $f_{D_s^+}^2$. The decay constant parametrizes the QCD effect on the decay. From the measurement of the decay width $\Gamma(D_s^+ \to l^+\nu_l)$ the decay constant $f_{D_s^+}^2$ can be extracted. The CKM matrix element $|V_{cs}|^2$ is an external input [4].

The branching fraction can be measured via the previously described double tag technique. In each event the tag decay is reconstructed via numerous decay channels. The number of events that contain a tag candidate is denoted by $N_{\text{tag}}$. Among those events the signal decay is reconstructed and the number of events that contain a tag decay and a signal decay is denoted by $N_{\text{sig,tag}}$. The branching fraction is then given by:

$$\mathcal{B}(D_s^+ \to l^+\nu_l) = \frac{N_{\text{sig,tag}}}{\epsilon_{\text{sig,tag}}} \times \frac{\epsilon_{\text{tag}}}{N_{\text{tag}}}. \tag{2}$$

The efficiencies for reconstruction and selection $\epsilon_i$ are obtained from simulation.

Since the final state contains a neutrino which is not detected the signal yield is determined using the missing mass:

$$MM^2 = \frac{(E_{\text{beam}} - E_\mu)^2}{c^4} - \frac{\left(-\vec{p}_{D_s^+} - \vec{p}_{\mu^+}\right)^2}{c^2}. \quad (3)$$

The beam energy is denoted by $E_{\text{beam}}$ and the reconstructed momentum of the tag $D_s^+$ decay candidate by $\vec{p}_{D_s^+}$.

The distribution is shown in figure 2. The yield is determined via a simultaneous fit to signal and sideband regions, whereas the sideband regions are defined in the $D_s^+$ mass spectrum. The $\mu^+\nu_\mu$ signal is shown as red dotted curve and the $\tau^+\nu_\tau$ signal as black dot-dashed curve. Background from misreconstructed tag $D_s^+$ decays and background from non-$D_s^+\overline{D}_s^+$ events is shown as green short dashed and violet long dashed curve, respectively. Within a sample of $15\,127 \pm 312$ event which contain a tag candidate we find $69.3 \pm 9.3$ $D_s^+ \to \mu^+\nu_\mu$ decays and $32.5 \pm 4.3$ $D_s^+ \to \tau^+\nu_\tau$ decays. In the fitting procedure the ratio of $D_s^+ \to \mu^+\nu_\mu$ to $D_s^+ \to \tau^+\nu_\tau$ was constraint to its Standard model prediction. Those yields are corrected for radiative effects and we obtain:

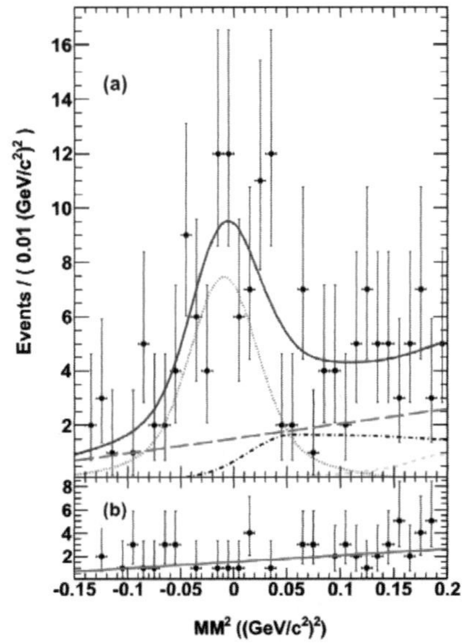


**Figure 2.** $MM^2$ distribution of the signal are (a) and from the sideband region in the $D_s^+$ mass of the tag candiate (b).

$$\mathcal{B}(D_s^+ \to \mu^+\nu_\mu) = (0.495\ \pm 0.067\ (\text{stat.}) \pm 0.026\ (\text{sys.}))\,\%$$
$$\mathcal{B}(D_s^+ \to \tau^+\nu_\tau) = (4.83\ \pm 0.65\ (\text{stat.}) \pm 0.26\ (\text{sys.}))\,\%. \quad (4)$$

The branching fractions $\mathcal{B}$ $(D_s^+ \to \mu^+\nu_\mu)$ and $\mathcal{B}$ $(D_s^+ \to \tau^+\nu_\tau)$ is consistent with the world average within 1 and 1.5 standard deviations, respectively. Furhtermore, the branching fractions are consistently determined using a fitting method which does not rely on the ratio of $D_s^+ \to \mu^+\nu_\mu$ to $D_s^+ \to \tau^+\nu_\tau$. For further details we refer to [3].

Using $\mathcal{B}$ $(D_s^+ \to \mu^+\nu_\mu)$ the decay constant $f_{D_s^+}$ is determined using eq. (1):

$$f_{D_s^+} = (241.0\ \pm 16.3\ (\text{stat.}) \pm 6.6\ (\text{sys.}))\,\text{MeV}. \quad (5)$$

The CKM matrix element $|V_cd| = 0.974\,25 \pm 0.000\,22$ [5] and the $D_s^+$ lifetime [5] is used. The result is consistent with LQCD calculations.

## 4. Form factor measurement

In pure leptonic decays, as described previously, the coupling strength is parametrized using a constant. In case of semi-leptonic decays the coupling strength depends on the momentum of the leptonic system and therefore, has a $q^2$ dependence. The coupling strength is then referred to as form factor. We analyse the decays of charged and neutral $D$ mesons to the semi-leptonic final states $Ke^+\nu_e$ and $\pi e^+\nu_e$. The decay rate can be parameterized as:

$$\frac{d\Gamma}{dq^2} = \frac{G_F^2}{24\pi^3}\left|V_{cs(d)}\right|^2\left|\vec{p}_{K^-(\pi^-)}\right|^3\left|f_+^{K(\pi)}(q^2)\right|^2, \quad (6)$$

with the Fermi constant $G_F^2$, the three momentum of the hadronic system $\left|\vec{p}_{K^-(\pi^-)}\right|$, the corresponding CKM matrix element $\left|V_{cs(d)}\right|^2$ and the form factor $\left|f_+^{K(\pi)}(q^2)\right|$. From the

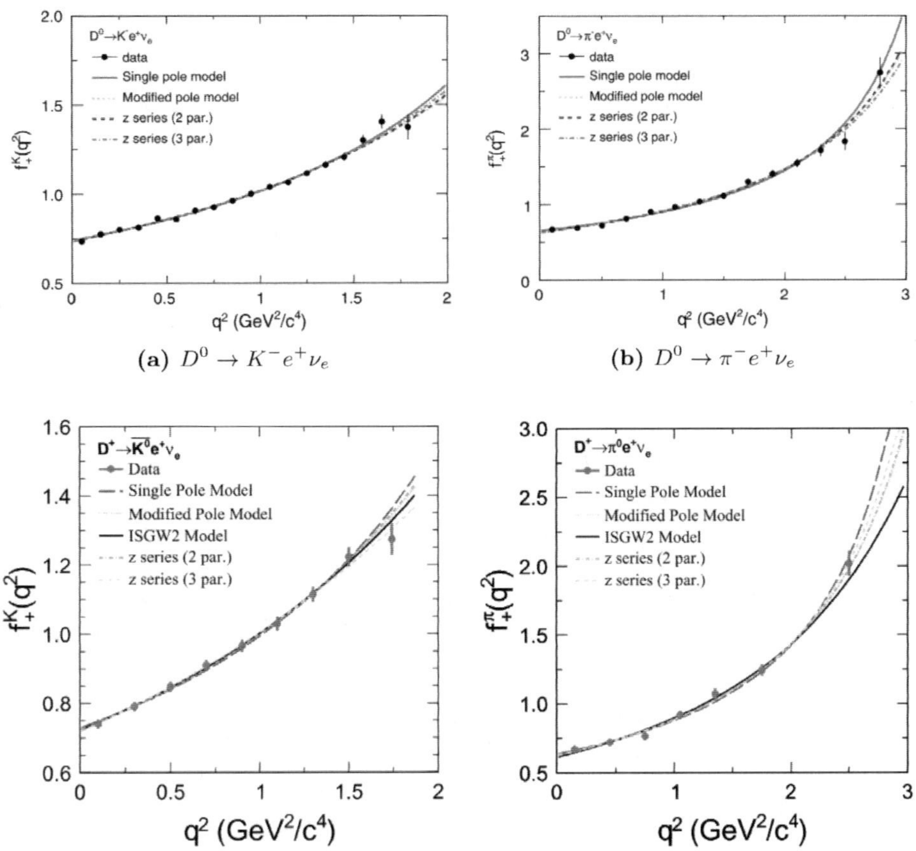


(a) $D^0 \to K^- e^+ \nu_e$ (b) $D^0 \to \pi^- e^+ \nu_e$

(c) $D^+ \to K^0 e^+ \nu_e$ (d) $D^+ \to \pi^0 e^+ \nu_e$

**Figure 3.** Dependence of the form factor $f_+^{K(\pi)}$ on the momentum transfer $q^2$ to the leptonic system.

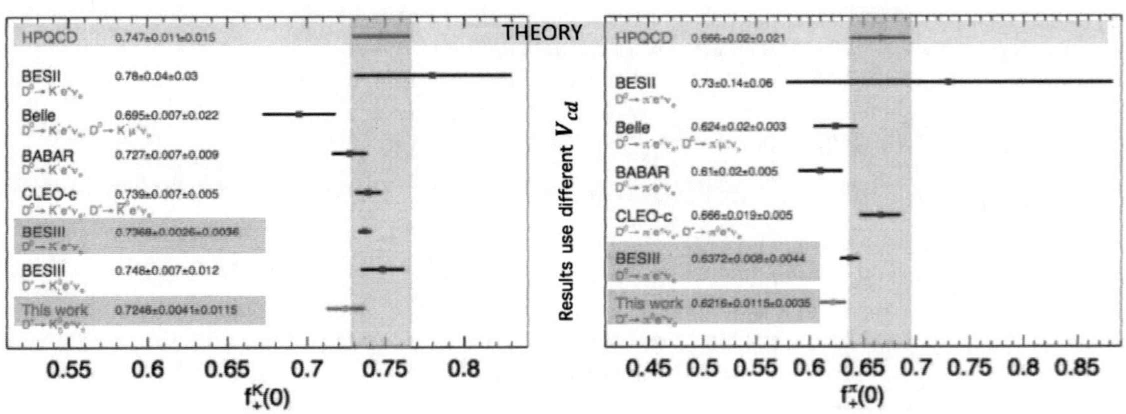


**Figure 4.** Comparison of form factor measurements with previous results and theory.

**Table 1.** Single and double tag yields for the form factor measurement. The branching fraction is calculated according to eq. (1). Uncertainties are statistical followed by systematical uncertainty.

| | $D^0$ | $D^+$ |
|---|---|---|
| $N_{tag}$ | $2\,793\,317 \pm 3684$ | $1\,703\,054 \pm 3405$ |
| $N_{sig}(Ke^+\nu_e)$ | $70\,727 \pm 278$ | $26\,008 \pm 168$ |
| $N_{sig}(\pi e^+\nu_e)$ | $6297 \pm 87$ | $3402 \pm 70$ |
| $\mathcal{B}\ (Ke^+\nu_e)$ [%] | $3.505 \pm 0.014 \pm 0.033$ | $8.60 \pm 0.06 \pm 0.15$ |
| $\mathcal{B}\ (\pi e^+\nu_e)$ [%] | $0.295 \pm 0.004 \pm 0.003$ | $0.363 \pm 0.008 \pm 0.005$ |

measurement of the branching fractions in bins of $q^2$, the product of matrix elements and form factor can be extracted. Using external measurements as input for the CKM matrix element, the form factor can be extracted or vice-versa, from a lattice QCD calculation of the form factor, the matrix element can be calculated. Since the final state contains a neutrino which is not detected, we use the constrained kinematics of the at-threshold production of $D^0\overline{D}^0/D^+D^-$ at $\sqrt{s} = 3.773\,\text{GeV}$. Similarly to the previously described measurement the double tag technique is used and the branching fraction is given by eq. (2). Yields and branching fractions are listed in table 1. The $q^2$ dependence of the form factor is shown in figure 3. It is then extrapolated to $q^2 = 0$ using various models. A comparison of measurements of $f_+^h(0)$ with theory, given in figure 4, shows that BESIII is able the reduce the uncertainty significantly.

The analysis of $D^0 \to K^-/\pi^- e^+\nu_e$ is published in [6] and the analysis of $D^+ \to \overline{K}^0/\pi^0 e^+\nu_e$ is a BESIII preliminary result, but is expected to be published soon.

## 5. Strong phase measurement in $D^0 \to K_{S/L}\pi^+\pi^-$

The CKM angle $\gamma/\phi_3$ can be measured in $B^-$ decays to the final state $D^0K^-/\overline{D}^0K^-$, as illustrated in figure 5. Since the $D^0$ final state $f(D)$ needs to be accessible from $D^0$ as well as from $\overline{D}^0$ it has to be a charge conjugate state. The strong phase difference $\delta_D$ between $D^0$ and $\overline{D}^0$ to the same final state is an external input to this type of measurement. $\delta_D$ can be extracted from a amplitude analysis of the final state but since those analyses have large uncertainties on the amplitude model, a model independent measurement is of advantage. We present preliminary results for such a model independent measurement of $\delta_D$ using the final state $K_{S/L}\pi^+\pi^-$. For simplicity we only discuss the final state $K_S^0\pi^+\pi^-$ here.

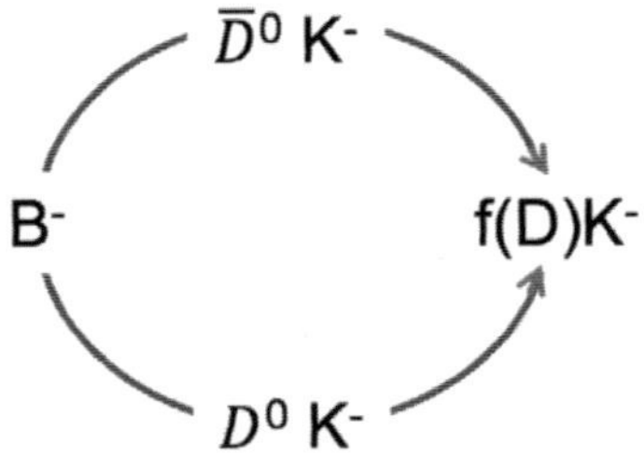


**Figure 5.** Measurement of the CKM angle $\gamma/\phi_3$ in $B$ decays via the GGSZ method [7].

The measurement is performed on a binned phase space. The binning in figure 6(a) relies on an amplitude model and is chosen such that the sensitivity on $\delta_D$ is maximal. In each event both $D^0$ mesons are reconstructed, one in the final state $K_S^0\pi^+\pi^-$ and the second decay in either a $CP^\pm$ eigenstate or in a flavour eigenstate. Furthermore, a sample with both $D^0$ mesons decaying to $K_S^0\pi^+\pi^-$ is used. The number of events in each bin of the phase space of the flavour tagged sample is denoted by $K_i$. Then, for the $CP$ tagged samples and for the $K_S^0\pi^+\pi^-$ versus $K_S^0\pi^+\pi^-$ sample the expected event rate in each bin $i$ is given in table 2 and depends on the (average) value of $\delta_D$ in this bin, which is denotes by $c_i = \cos\delta_i$ and $s_i = \sin\delta_i$. $c_i$ and $s_i$ are

**Table 2.** Expected number of events per bin for three different combinations of tag signal decays. $K_i$ is the event rate in bin $i$ measured using the flavour tagged sample. The cosine and sine of $\delta_D$ are denoted by $c_i$ and $s_i$, respectively.

| Tag mode | $K_S^0\,\pi^+\pi^-$ |
|---|---|
| Flavour | $K_i$ |
| $CP\,\pm$ | $M_i^{\pm} \propto \left(K_i \pm 2c_i\sqrt{K_iK_{-i}} + K_{-i}\right)$ |
| $K_S^0\,\pi^+\pi^-$ | $M_{i,j}^{\pm} \propto \left(K_iK_{-j} + K_{-i}K_j - 2\sqrt{K_iK_{-j}K_{-i}K_j}\,(c_ic_j + s_is_j)\right)$ |

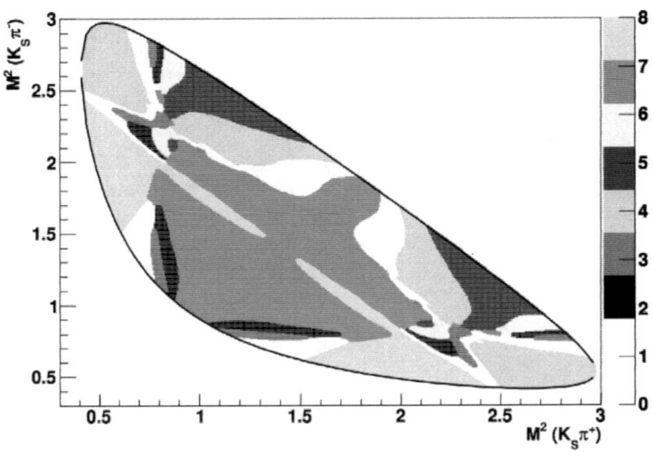


(a) Phase space binning scheme [8]. The color denotes the bin number.

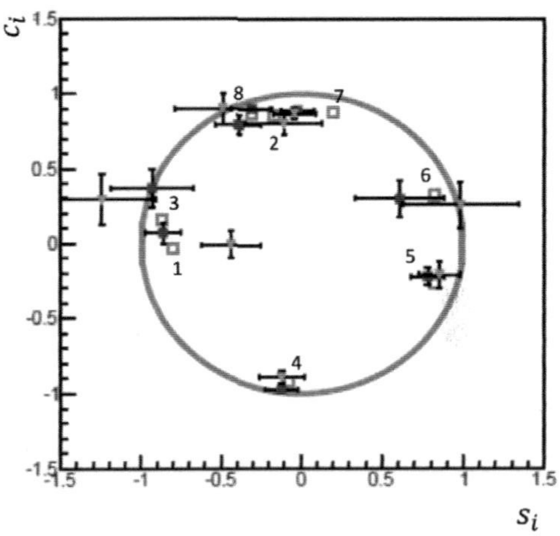


(b) $c_i$ versus $s_i$. The bin number is given for each data point. The corresponding bin to each number is given on the left.

**Figure 6.** Strong phase between $D^0$ and $\overline{D}^0$ in phase space bins. The measurement (blue) is compatible with the model dependent prediction (red open squares). BESIII is able to improve the existing result from CLEO-c (pink) significantly.

then determined in a simultaneous fit to all bins in both samples. The preliminary result is shown in figure 6(b) (blue) and compared to the previous result of CLEO-c [8]. We are able to significantly reduce the uncertainty on the measurement. Since the sample which contains two $K_S^0\pi^+\pi^-$ decays is very limited we also use the channel $K_L^0\pi^+\pi^-$ for this result.

## References

1. M. Ablikim *et al.*, *Nucl. Instrum. Meth. A* **614**, 345–399 (2010).
2. B.-C. Ke, presented at the same Conf.
3. M. Ablikim *et al.*, arXiv: 1608.06732 (2016).
4. J. C. Hardy, I. S. Towner, *Phys. Rev. C* **79**, 055502 (2009).
5. K. Olive, *Chin. Phys. C* **38**, 090001 (2014).
6. M. Ablikim *et al.*, *Phys. Rev. D* **92**, 072012 (2015).
7. A. Giri *et al.*, *Phys. Rev. D* **68** (2003).
8. J. Libby *et al.*, *Phys. Rev. D* **82**, 112006 (2010).

Journal of Physics: Conference Series **770** (2016) 012013 doi:10.1088/1742-6596/770/1/012013

# Recent Results on $\Lambda_c$ Decays at BESIII

**Bai-Cian Ke on behalf of the BESIII Collaboration**
Carnegie Mellon University
5000 Forbes Ave, Pittsburgh, PA 15213

E-mail: baiciank@andrew.cmu.edu

**Abstract.** As the lightest and most common charmed baryon, the $\Lambda_c^+$ plays a key role in our understanding of particles of this type. The BESIII detector has collected a 567 pb$^{-1}$ sample of $e^+e^-$ annihilation data near the $\Lambda_c^+\bar{\Lambda}_c^-$ threshold. Using a double-tag technique, we make absolute measurements of twelve Cabibbo-favored $\Lambda_c^+$ hadronic decay modes, including the golden reference mode $pK^-\pi^+$, for which we find $\mathcal{B}(\Lambda_c^+ \to pK^-\pi^+) = (5.84 \pm 0.27(\text{stat}) \pm 0.23(\text{syst}))\%$. We also determine $\mathcal{B}(\Lambda_c^+ \to \Lambda e^+\nu_e) = (3.63 \pm 0.38(\text{stat}) \pm 0.20(\text{syst}))\%$. Preliminary results for other final states, including $nK_S\pi^+$ and $\Lambda\mu\nu_\mu$, are also presented, along with future prospects.

## 1. Introduction

The $\Lambda_c^+$ baryon is the ground state of charmed baryons. Most excited $\Lambda_c$ and $\Sigma_c$ baryons eventually decay into a $\Lambda_c^+$ (with the exception of the decay to the $pD$ final state), and the major decays of $\Lambda_b$ baryons include a $\Lambda_c^+$ [1]. Thus, the study of the $\Lambda_c^+$ provides an important normalization to the measurements of $\Lambda_b$, $\Lambda_c$ and $\Sigma_c$ baryons, reducing the uncertainties in measurements of these heavier baryons. Since there is no lighter baryon containing a charm quark, $\Lambda_c^+$ can only decay through the weak interaction, and is the most common of the four weakly-decaying charmed baryons.

It has been more than 30 years since the Mark II experiment discovered the $\Lambda_c^+$ baryon in 1979 [2]. However, many hadronic decays have not been measured. Among the reported measurements of branching fractions, most are relative to the decay mode $\Lambda_c^+ \to pK^-\pi^+$. Recently, the Belle experiment reported $\mathcal{B}(\Lambda_c^+ \to pK^-\pi^+) = (6.84 \pm 0.24^{+0.21}_{-0.27})\%$ [3]. The absolute branching fraction of this decay mode has not yet been measured using threshold data and many other hadronic branching fractions still have poor precision [1]. The high statistics $\Lambda_c^+$ data collected at the BESIII experiment near the $\Lambda_c^+\bar{\Lambda}_c^-$ threshold therefore provide an excellent opportunity to perform precise measurement of $\Lambda_c^+$ decays.

## 2. Measurements near $\Lambda_c^+\bar{\Lambda}_c^-$ threshold

In 2014, the BESIII experiment collected the largest $\Lambda_c^+$ data sample to date near the $\Lambda_c^+\bar{\Lambda}_c^-$ threshold. Utilizing $e^+e^-$ annihilations, BESIII collected an integrated luminosity of 567 pb$^{-1}$ of data at $\sqrt{s} = 4.599$ GeV, which is 26 MeV above the $\Lambda_c^+\bar{\Lambda}_c^-$ pair mass. This energy is not enough for the production of even one additional pion. The data taken near the $\Lambda_c^+\bar{\Lambda}_c^-$ threshold is therefore very clean and can take advantage of "tagging" techniques.

There are two types of samples used in the tagging technique: single tag (ST) and double tag (DT) samples. In the ST sample, only one $\Lambda_c$ candidate is reconstructed through a chosen

Journal of Physics: Conference Series **770** (2016) 012013 doi:10.1088/1742-6596/770/1/012013

hadronic decay without any requirement on the remaining tracks and showers. In the DT sample, both a $\Lambda_c^+$ and a $\bar{\Lambda}_c^-$ are reconstructed, where the $\Lambda_c^+$ is reconstructed through the hadronic decay of interest and is called "the signal side" and the $\bar{\Lambda}_c^-$, called "the tag side", is usually reconstructed through well-known and clean hadronic decay modes. Charge-conjugate states are implied throughout this paper. The signal yield of the ST sample is given by $N_j^{\rm ST} = 2N_{\Lambda_c^+\bar{\Lambda}_c^-}\mathcal{B}_j\varepsilon_j$, and the signal yield of the DT sample is given by $N_{i,j}^{\rm DT} = 2N_{\Lambda_c^+\bar{\Lambda}_c^-}\mathcal{B}_i\mathcal{B}_j\varepsilon_{i,j}$, where $i$ and $j$ indicate a certain decay mode, $N_{\Lambda_c^+\bar{\Lambda}_c^-}$ is the total number of $\Lambda_c^+\bar{\Lambda}_c^-$ pairs, $\mathcal{B}$ is the branching fraction, and $\varepsilon$ is the corresponding efficiency. The branching fraction of the signal side is determined by isolating $\mathcal{B}_i$ such that $\mathcal{B}_i = N_{i,j}^{\rm DT}\varepsilon_j/N_j^{\rm ST}\varepsilon_{i,j}$. The total number of produced $\Lambda_c^+\bar{\Lambda}_c^-$ events cancels, and many systematic uncertainties associated with the tag side also cancel.

## 3. Absolute hadronic branching fractions of $\Lambda_c^+$ baryon

Most branching fractions of $\Lambda_c^+$ decays are studied by their ratios to "the golden reference mode", $\Lambda_c^+ \to pK^-\pi^+$. Hence, an absolute measurement of the golden reference mode is crucial. Using the largest $\Lambda_c^+\bar{\Lambda}_c^-$ threshold sample, the BESIII Collaboration recently reported absolute branching fractions of twelve Cabibbo-favored $\Lambda_c^+$ hadronic decay modes, including $\Lambda_c^+ \to pK^-\pi^+$ [4].

To obtain signal yields, we define the beam-constrained mass $M_{\rm BC}$ of the $\Lambda_c$ candidates calculated by substituting the beam energy $E_{\rm beam}$ for the measured $\Lambda_c$ energy. The fits to ST $M_{\rm BC}$ distributions are shown in Figure 1. BESIII obtains the ST and DT signal yields of the twelve Cabibbo-favored decay modes and further implements a global least-squares fit by considering the correlations to improve precision and obtain proper uncertainties. The branching fraction of $\Lambda_c^+ \to pK^-\pi^+$ is determined to be $\mathcal{B}(\Lambda_c^+ \to pK^-\pi^+) = (5.84\pm0.27({\rm stat})\pm 0.23({\rm syst}))\%$. The precision of the other eleven Cabibbo-favored hadronic decay modes is also improved significantly, compared to 2014 PDG values, shown in Table 1.

**Table 1.** Comparison of the measured branching fractions in this work with previous results from 2014 PDG [1]. For our results, the first uncertainties are statistical and the second are systematic.

| Mode | This work (%) | PDG (%) |
|---|---|---|
| $pK_S^0$ | $1.52 \pm 0.08 \pm 0.03$ | $1.15 \pm 0.30$ |
| $pK^-\pi^+$ | $5.84 \pm 0.27 \pm 0.23$ | $5.0 \pm 1.3$ |
| $pK_S^0\pi^0$ | $1.87 \pm 0.13 \pm 0.05$ | $1.65 \pm 0.50$ |
| $pK_S^0\pi^+\pi^-$ | $1.53 \pm 0.11 \pm 0.09$ | $1.30 \pm 0.35$ |
| $pK^-\pi^+\pi^0$ | $4.53 \pm 0.23 \pm 0.30$ | $3.4 \pm 1.0$ |
| $\Lambda\pi^+$ | $1.24 \pm 0.07 \pm 0.03$ | $1.07 \pm 0.28$ |
| $\Lambda\pi^+\pi^0$ | $7.01 \pm 0.37 \pm 0.19$ | $3.6 \pm 1.3$ |
| $\Lambda\pi^+\pi^-\pi^+$ | $3.81 \pm 0.24 \pm 0.18$ | $2.6 \pm 0.7$ |
| $\Sigma^0\pi^+$ | $1.27 \pm 0.08 \pm 0.03$ | $1.05 \pm 0.28$ |
| $\Sigma^+\pi^0$ | $1.18 \pm 0.10 \pm 0.03$ | $1.00 \pm 0.34$ |
| $\Sigma^+\pi^+\pi^-$ | $4.25 \pm 0.24 \pm 0.20$ | $3.6 \pm 1.0$ |
| $\Sigma^+\omega$ | $1.56 \pm 0.20 \pm 0.07$ | $2.7 \pm 1.0$ |

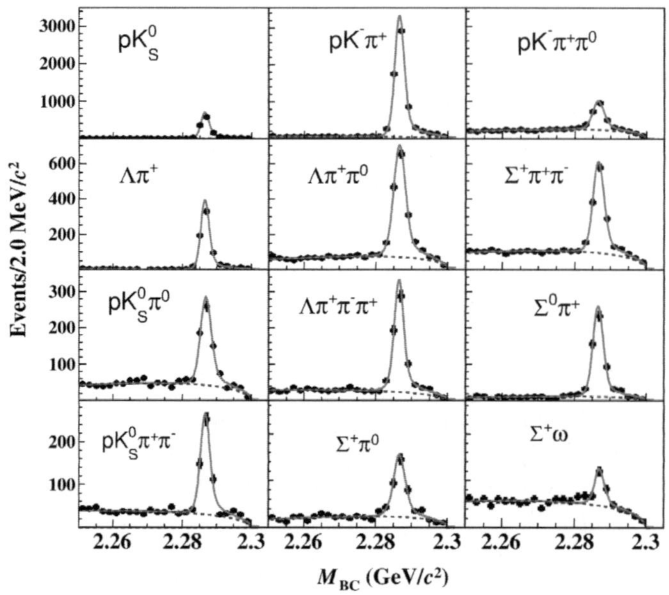


**Figure 1.** Fits to the ST $M_{\rm BC}$ distributions for the different decay modes. Points with error bars are data, solid lines are the sum of the fit functions, and dashed lines are the background shape. (Taken from [4])

## 4. Measurement of the absolute branching fraction for $\Lambda_c^+ \to \Lambda e^+\nu_e$

The BESIII Collaboration recently published the first absolute measurement of the branching fraction for the semi-leptonic decay $\Lambda_c^+ \to \Lambda e^+\nu_e$ [5]. This semi-leptonic decay of $\Lambda_c^+$, coming from the dominant Cabibbo-favored $c \to se^+\nu_e$ transition, is a benchmark for all other $\Lambda_c^+$ semi-leptonic decays. However, various $\Lambda_c^+$ form-factor models result in a wide range of theoretical predictions, from 1.4% to 9.2%, for the branching fraction.

We first single-tag the $\bar{\Lambda}_c^-$ through the eleven of the hadronic decay modes in Table 1, excluding $\Sigma^+\omega$. Then, we reconstruct the $\Lambda$ and $e^+$, leaving the neutrino undetected. To obtain the neutrino information, the variable $U_{\rm miss} = E_{\rm miss} - c|\vec{p}_{\rm miss}|$ is used, where $E_{\rm miss}$ and $\vec{p}_{\rm miss}$ are the missing energy and momentum carried by the neutrino, respectively. They can be calculated as $E_{\rm miss} = E_{\rm beam} - E_\Lambda - E_{e^+}$ and $\vec{p}_{\rm miss} = \vec{p}_{\Lambda_c^+} - \vec{p}_\Lambda - \vec{p}_{e^+}$, where $\vec{p}_{\Lambda_c^+}$, $\vec{p}_\Lambda$, $\vec{p}_{e^+}$ are the momenta of the $\Lambda_c^+$, the $\lambda$ and the $e^+$, respectively, while $E_\Lambda$ and $E_{e^+}$ are the energies of the $\Lambda$ and the $e^+$, respectively. Here, $\vec{p}_{\Lambda_c^+}$ is obtained by using the direction opposite the measured momentum of the $\bar{\Lambda}_c^-$ with the magnitude constrained to be $|\vec{p}_{\Lambda_c^+}| = \sqrt{E^2_{\rm beam} - m^2_{\Lambda_c^+}}$, where $m_{\Lambda_c^+}$ is the mass of $\Lambda_c^+$ from the PDG.

The fit to $U_{\rm miss}$ is shown in Figure 2. The branching fraction for $\Lambda_c^+$ decays to $\Lambda e^+\nu_e$ is measured to be $\mathcal{B}(\Lambda_c^+ \to \Lambda e^+\nu_e) = (3.63\pm0.38({\rm stat})\pm0.20({\rm syst}))\%$. This result is a significant improvement on the PDG value of $(2.9\pm0.5)\%$, and provides a powerful constraint on theoretical models.

## 5. Measurement of the absolute branching fraction for $\Lambda_c^+ \to \Lambda\mu^+\nu_\mu$

We also studied $\Lambda\mu^+\nu_\mu$ decays of the $\Lambda_c^+$. With the result of $\mathcal{B}(\Lambda_c^+ \to \Lambda e^+\nu_e)$ in hand, the ratio $\mathcal{B}(\Lambda_c^+ \to \Lambda e^+\nu_e)/\mathcal{B}(\Lambda_c^+ \to \Lambda\mu^+\nu_\mu)$ tests lepton universality in baryonic decays. We use a technique analogous to that of $\Lambda_c^+ \to \Lambda e^+\nu_e$, with $U_{\rm miss}$ used as the final signal

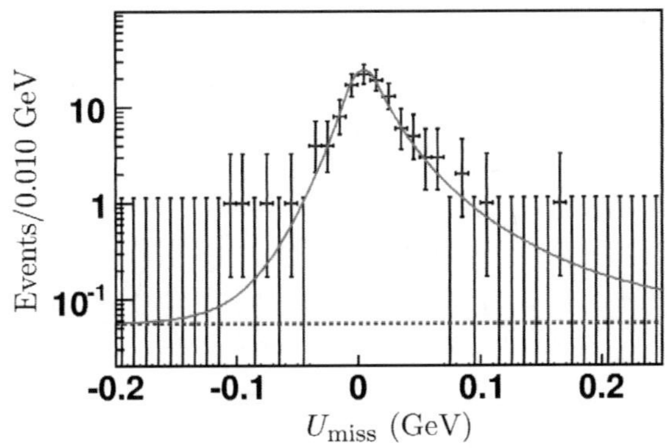


**Figure 2.** Fit to the $U_{\text{miss}}$ distribution. The points with error bars are data, the solid curve shows the total fit, and the dashed curve is the background shape. (Taken from [5])

variable. However, the background is higher in the case of $\Lambda_c^+ \to \Lambda\mu^+\nu_\mu$, due to the ambiguity in separating $\mu$ and $\pi$. Thus, a cut on the mass of the $\Lambda\mu^+$ system is used to suppress $\Lambda_c^+ \to \Lambda\pi^+$ and $\Lambda_c^+ \to \Sigma^0\pi^+$ processes. A cut on extra energy deposits in the electro magnetic calorimeter is applied to suppress the $\Lambda_c^+ \to \Lambda\pi^+\pi^0$ background. The fitting model includes MC-drive background shapes to simulate the remaining background, shown in Figure 3. The preliminary result for the branching fraction is determined to be $\mathcal{B}(\Lambda_c^+ \to \Lambda\mu^+\nu_\mu) = (3.49 \pm 0.46(\text{stat}) \pm 0.26(\text{syst}))\%$, and the ratio is determined to be $\mathcal{B}(\Lambda_c^+ \to \Lambda e^+\nu_e)/\mathcal{B}(\Lambda_c^+ \to \Lambda\mu^+\nu_\mu) = (0.96 \pm 0.16(\text{stat}) \pm 0.04(\text{syst}))\%$.

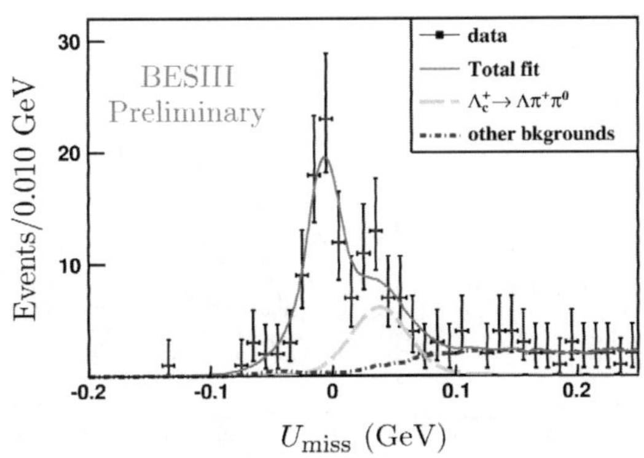


**Figure 3.** Fit to the $U_{\text{miss}}$ distribution. Data are shown as the dots with error bars. The long-dashed curve shows the $\Lambda_c^+ \to \Lambda\pi^+\pi^0$ background while the dot-dashed curve shows other $\Lambda_c^+$ decay backgrounds. The solid line shows the total fit.

## 6. Observation of $\Lambda_c^+ \to nK_S^0\pi^+$

Comparing branching fractions for $\Lambda_c^+ \to p(K\pi)^0$ and $\Lambda_c^+ \to nK_S^0\pi^+$ provides an excellent opportunity to test final state interactions and isospin symmetry in the charmed baryon sector

Journal of Physics: Conference Series **770** (2016) 012013 doi:10.1088/1742-6596/770/1/012013

[6]. Hence, the BESIII Collaboration performed the first direct measurement of $\Lambda_c^+ \to nK_S^0\pi^+$, which is also the first direct measurement of any $\Lambda_c^+$ decay involving a neutron in the final state.

The eleven hadronic decay modes in Table 1 (excluding $\Sigma^+\omega$) are used to reconstruct the $\bar{\Lambda}_c^-$ baryons as the tag side. Since the neutron is not detected, we define missing mass squared, $M^2_{\rm miss} = E^2_{\rm miss} - c^2|\vec{p}_{\rm miss}|^2$, to access to the missing neutron, where $E_{\rm miss}$ and $\vec{p}_{\rm miss}$ are the missing energy and momentum carried by the neutron, respectively. They are calculated as $E_{\rm miss} = E_{\rm beam} - E_{K_S^0} - E_{\pi^+}$ and $\vec{p}_{\rm miss} = \vec{p}_{\Lambda_c^+} - \vec{p}_{K_S^0} - \vec{p}_{\pi^+}$, where $\vec{p}_{\Lambda_c^+}$, $\vec{p}_{K_S^0}$, and $\vec{p}_{\pi^+}$ are the momenta of the $\Lambda_c^+$, the $K_S^0$ and the $\pi^+$, respectively, while $E_{K_S^0}$ and $E_{\pi^+}$ are the energies of the $K_S^0$ and the $\pi^+$, respectively. The momentum, $\vec{p}_{\Lambda_c^+}$, is obtained from the tag by the same method as in the previous analysis.

In order to obtain signal yields, we perform a simultaneous fit of the two-dimensional $M^2_{\rm miss}$ vs. $M_{\pi^+\pi^-}$ distributions in both $M_{\rm BC}$ signal and sideband regions of the tag side, shown in Figure 4. The purpose of fitting the two-dimensional $M^2_{\rm miss}$ vs. $M_{\pi^+\pi^-}$ distributions is to simulate the background from $\Lambda_c^+ \to \Sigma^\pm\pi^+\pi^\mp$ with $\Sigma^\pm \to n\pi^\pm$, which has same final state as the signal process when the $K_S^0$ is reconstructed through $K_S^0 \to \pi^+\pi^-$. These background events form a peaking background in $M^2_{\rm miss}$ but, unlike the $K_S^0$, are distributed flat along $M_{\pi^+\pi^-}$. The purpose of performing simultaneous fits in both the $M_{\rm BC}$ signal and sideband regions of the tag side is to constrain the non-$\Lambda_c^+$ decay background under the $M_{\rm BC}$ peak to that of the $M_{\rm BC}$ sideband. The preliminary result for the absolute branching fraction is determined to be $\mathcal{B}(\Lambda_c^+ \to nK_S^0\pi^+) = (1.82 \pm 0.23(\rm stat) \pm 0.11(\rm syst))\%$. With the measurement of $\mathcal{B}(\Lambda_c^+ \to pK^-\pi^+)$ and $\mathcal{B}(\Lambda_c^+ \to p\bar{K}^0)$ [4], we determine $\mathcal{B}(\Lambda_c^+ \to n\bar{K}^0\pi^+)/\mathcal{B}(\Lambda_c^+ \to pK^-\pi^+) = 0.62\pm0.09$ and $\mathcal{B}(\Lambda_c^+ \to n\bar{K}^0\pi^+)/\mathcal{B}(\Lambda_c^+ \to p\bar{K}^0\pi^0) = 0.97 \pm 0.16$. These ratios are a key input to test isospin symmetry and extract strong phases of final states in the charmed baryon sector.

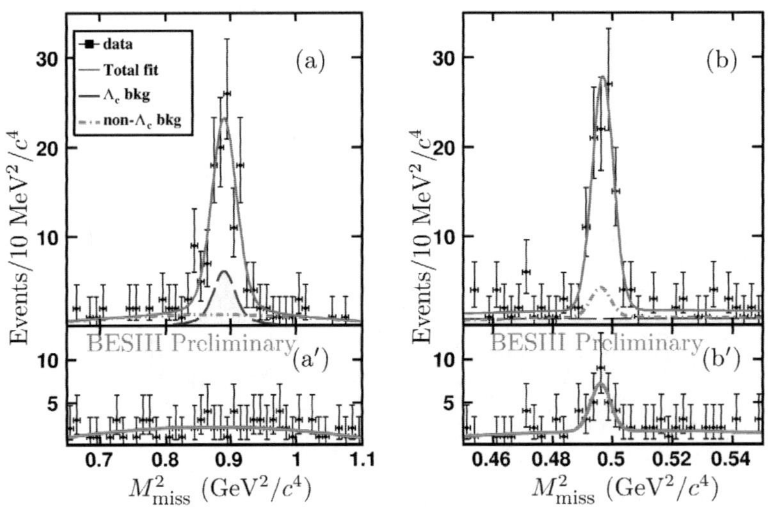


**Figure 4.** Simultaneous fit to $M^2_{\rm miss}$ and $M\pi^+\pi$ of events in (a, b) the $\bar{\Lambda}_c^-$ signal region and (a′, b′) sideband regions. Data are shown as the dots with error bars. The long-dashed lines show the $\Lambda_c^+$ backgrounds while the dot-dashed curves show the non-$\Lambda_c^+$ backgrounds. The solid curves show the total fit. The shaded area show the MC simulated backgrounds from $\Lambda_c^+$ decay.

Journal of Physics: Conference Series **770** (2016) 012013 doi:10.1088/1742-6596/770/1/012013

## 7. Summary

The $\Lambda_c^+\bar{\Lambda}_c^-$ near-threshold data at BESIII have proven to be an effective sample to study the $\Lambda_c^+$ baryon. BESIII has published the first absolute measurement of twelve Cabibbo-favored $\Lambda_c^+$ hadronic decay modes, including $\Lambda_c^+ \to pK^-\pi^+$. BESIII has also published the first absolute measurement of $\Lambda_c^+ \to \Lambda e\nu_e$. Preliminary results are also given for the first absolute measurement of $\Lambda_c^+ \to \Lambda\mu\nu_\mu$ and $\Lambda_c^+$ decays involving a neutron, $\Lambda_c^+ \to nK_S^0\pi^+$. Many other analyses of $\Lambda_c^+$, including more hadronic modes, more modes with neutrons, more semi-leptonic modes, and inclusive studies, are in progress.

## References


[1] Olive K A *et al.* (Particle Data Group) 2014 *Chin. Phys. C* **38** 090001.
[2] Abrams G S *et al.* (MARKII Collaboration) 1980 *Phys. Rev. Lett.* **44**(1) 10.
[3] Zupanc A *et al.* (Belle Collaboration) 2014 *Phys. Rev. Lett.* **113**(4) 042002.
[4] Ablikim M *et al.* (BESIII Collaboration) 2016 *Phys. Rev. Lett.* **116**(5) 052001.
[5] Ablikim M *et al.* (BESIII Collaboration) 2015 *Phys. Rev. Lett.* **115**(22) 221805.
[6] Lü C D, Wang W and Yu F S 2016 *Phys. Rev. D* **93**(5) 056008.

Journal of Physics: Conference Series **770** (2016) 012014 doi:10.1088/1742-6596/770/1/012014

# Bayronic $b$ Decays at LHCb

**D Hutchcroft, on behalf of the LHCb collaboration**

Department of Physics, University of Liverpool, Oxford Street, Liverpool, L69 7ZE, UK

E-mail: David.Hutchcroft@cern.ch

**Abstract.** The properties and decays of baryonic $b$ hadrons have been studied using $3\,\mathrm{fb}^{-1}$ of LHCb data. A determination of $|V_{\mathrm{ub}}|$ is made using the ratio of $\frac{\mathcal{B}(\Lambda_b^0\to p\mu^-\overline{\nu}_\mu)}{\mathcal{B}(\Lambda_b^0\to\Lambda_c^+\mu^-\overline{\nu}_\mu)}$, the mass and lifetime of the $\Omega_b^-$ and $\Xi_b^*$ are measured, the mass of the $\Lambda_b^0$ is precisely determined using $\Lambda_b^0 \to (c\overline{c})pK^-$ and the angular distributions and the forward-backward asymmetries in the dimuon and hadron systems are measured in the decay $\Lambda_b^0 \to \Lambda\mu^+\mu^-$.

## 1. Introduction

The study of baryons containing a b quark provides a window into the structure of hadrons and allows sensitive tests of the Standard Model. Bound states of three quarks allow complementary measurement of parameters also accessible from B meson decays. They also allow tests of QCD models, where heavy quark effective theory (HQET) [1] can predict the quark structure in baryons. Branching fractions, masses and lifetimes measurements for b baryons were performed with LHCb data, as well as the CKM parameter $|V_{\mathrm{ub}}|$ and distrubutions and asymmetries of the decays of $\Lambda_b^0 \to \Lambda\mu^+\mu^-$. Charge conjugate processes are implied throughout.

## 2. LHCb Detector and data sample

The LHCb detector [2] is a single-arm forward spectrometer covering the pseudorapidity range $2 < \eta < 5$, designed for the study of particles containing b or c quarks. The detector elements that are particularly relevant to this analysis are: a silicon-strip vertex detector surrounding the pp interaction region that allows c- and b-hadrons to be identified from their characteristically long flight distance; a tracking system that provides a measurement of momentum, $p$, of charged particles; and two ring-imaging Cherenkov detectors that are able to discriminate between different species of charged hadrons. The analyses mentioned in this proceeding use pp collision data, corresponding to an integrated luminosity of $3.0\,\mathrm{fb}^{-1}$, collected during 2011 and 2012 at centre-of-mass energies of 7 and 8 TeV, respectively.

## 3. $|V_{\mathrm{ub}}|$ measured with decays of the $\Lambda_b^0$

The most precise measurements of the CKM parameter $|V_{\mathrm{ub}}|$ come from measurements of the semileptonic quark-level transition $b \to u\ell^-\overline{\nu}_\ell$. The world average from Ref. [3] for this method, using the decays $B^0 \to \pi^+\ell^-\overline{\nu}_\ell$ and $B^- \to \pi^0\ell^-\overline{\nu}_\ell$, is $|V_{\mathrm{ub}}| = (3.28 \pm 0.29) \times 10^{-3}$, where the most precise experimental inputs come from the BaBar [4,5] and Belle [6,7] experiments. The results are also senstive to lattice QCD (LQCD) calculations [8,9]. These have a tension with the measurements of $|V_{\mathrm{ub}}|$ from the differential decay rate in an inclusive way over all possible B meson decays containing the $\mathrm{b} \to \mathrm{u}\ell^-\overline{\nu}_\ell$ quark level transition. This result is $|V_{\mathrm{ub}}| = (4.41 \pm 0.15^{+0.15}_{-0.17}) \times 10^{-3}$ [10], which is approximately three standard deviations from the other measurement.

$|V_{\mathrm{ub}}|$ can also be measured using the ratio of branching fractions of the $\Lambda_b^0$ to $\mathrm{p}\mu^-\overline{\nu}_\mu$ and $\Lambda_c^+\mu^-\overline{\nu}_\mu$ final states, using the ratio

$$\frac{|V_{\mathrm{ub}}|^2}{|V_{\mathrm{cb}}|^2} = \frac{\mathcal{B}(\Lambda_b^0 \to \mathrm{p}\mu^-\overline{\nu}_\mu)}{\mathcal{B}(\Lambda_b^0 \to \Lambda_c^+\mu^-\overline{\nu}_\mu)} R_{\mathrm{FF}}, \tag{1}$$

where $\mathcal{B}$ denotes the branching fraction and $R_{\mathrm{FF}}$ is a ratio of the relevant form factors, calculated using LQCD. Many of the systematic uncertainties cancel in the ratio and the relatively large yields of these decays reconstructed in the LHCb detector allow a precise measurement to resolve the discrepancy.

The LHCb event selection [11] is deliberately biased toward high values of $q^2$, the square of the invariant mass of the leptons, as the systematic uncertainties on $R_{\mathrm{FF}}$ are smaller there.

The measurement is

$$\frac{\mathcal{B}(\Lambda_b^0 \to \mathrm{p}\mu^-\overline{\nu}_\mu)_{q^2>15\,\mathrm{GeV}^2/c^4}}{\mathcal{B}(\Lambda_b^0 \to \Lambda_c^+\mu^-\overline{\nu}_\mu)_{q^2>7\,\mathrm{GeV}^2/c^4}} = (1.00 \pm 0.04 \pm 0.08) \times 10^{-2}$$

which can be combined with equation 1 and $R_{\mathrm{FF}} = 0.68 \pm 0.07$ from [12] for the restricted $q^2$ regions, the measurement

$$\frac{|V_{\mathrm{ub}}|^2}{|V_{\mathrm{cb}}|^2} = 0.083 \pm 0.004 \pm 0.004$$

is obtained, where the uncertainties are experimental and from the LQCD prediction. Finally, using the world average $|V_{\mathrm{cb}}| = (39.5 \pm 0.8) \times 10^{-3}$ measured using exclusive decays [10], giving $|V_{\mathrm{ub}}| = (3.27 \pm 0.15 \pm 0.16 \pm 0.06) \times 10^{-3}$, where the uncertainties are experimental, from LQCD and the normalisation to $|V_{\mathrm{cb}}|$. This disfavours models with a right-handed component to the weak coupling, which were proposed to explain the tension in the meson measurements [13].

## 4. Measurements of $\Omega_b^- \to \Omega_c^0(\to \mathrm{pK^+K^-}\pi^+)\pi^-$

The HQET predictions for the lifetimes of heavy b baryons suggest a hierarchy of lifetimes [14–16] of $\tau(\Omega_b^-) \simeq \tau(\Xi_b^-) > \tau(\Xi_b^0) \approx \tau(\Lambda_b^0)$. The signal of $\Omega_b^- \to \Omega_c^0(\to \mathrm{pK^-K^-}\pi^+)\pi^-$ were searched for in the LHCb data with the signal $\Xi_b^- \to \Xi_c^0(\to \mathrm{pK^-K^-}\pi^+)\pi^-$ as the control channel [17]. In total $62.9 \pm 9.0$ $\Omega_b^-$ decays and $1384 \pm 39$ $\Xi_b^-$ decays were reconstructed, which were then split into four decay lifetime bins, fit to get the baryon lifetimes and corrected for the relative efficiencies for the two decay chains. The lifetime ratio and absolute lifetime of the $\Omega_b^-$ baryon are measured to be

$$\frac{\tau(\Omega_b^-)}{\tau(\Xi_b^-)} = 1.11 \pm 0.16 \pm 0.03 \text{ and}$$

$$\tau(\Omega_b^-) = 1.78 \pm 0.26 \pm 0.05 \pm 0.06\,\mathrm{ps},$$

where the uncertainties are statistical, systematic and from the calibration mode for the absolute measurement. A measurement is also made of the mass differences which yields

$$m(\Omega_b^-) - m(\Xi_b^-) = 247.4 \pm 3.2 \pm 0.5\,\mathrm{MeV}/c^2 \text{ and}$$

$$m(\Omega_b^-) = 6045.1 \pm 3.2 \pm 0.5 \pm 0.6\,\mathrm{MeV}/c^2.$$

These results are consistent with previous measurements and HQET predictions for the lifetime ratios.

## 5. Properties of the $\Xi_b^*$

The $\Xi_b^*$ resonance was first reported by the CMS collaboration [18], close to the threshold for $\Xi_b^* \to \Xi_b^0\pi^+$ decays and is reconstructed with that decay followed by $\Xi_b^0 \to \Xi_c^0(\to \mathrm{pK^+K^-}\pi^+)\pi^-$. The measurement at LHCb shows only one significant resonance in the $X \to \Xi_b^0\pi^+$ distribution [19], which is compatible with the $J^P = \frac{3}{2}^+$ state expected in the quark model [20]. The measured properties of the $\Xi_b^*$ resonance are

$$\begin{aligned} m(\Xi_b^*) - m(\Xi_b^-) - m(\pi^+) &= 15.727 \pm 0.068 \pm 0.023\ \mathrm{MeV}/c^2, \\ m(\Xi_b^*) &= 5953.02 \pm 0.07 \pm 0.02 \pm 0.55\ \mathrm{MeV}/c^2 \text{ and} \\ \Gamma(\Xi_b^*) &= 0.90 \pm 0.16 \pm 0.08\ \mathrm{MeV} \end{aligned}$$

where the uncertainties are first experimental then systematic, with the third error on the mass being the uncertainty on $m(\Xi_b^-)$ [21]. The inclusive ratio of production cross-sections is $\frac{\sigma(\mathrm{pp}\to\Xi_b^* X)\mathcal{B}(\Xi_b^*\to\Xi_b^-\pi^+)}{\sigma(\mathrm{pp}\to\Xi_b^- X)} = 0.27 \pm 0.03 \pm 0.01$ showing that a large fraction of $\Xi_b^-$ are produced in feed-down decays from higher mass states.

## 6. Evidence for strangeness-changing weak decays in $\Xi_b^- \to \Lambda_b^0\pi^-$

HQET predicts that the decay of the b quark dominates the decay width of b baryons, with the s quark contributing about 1% of the total decay width. There are models [22] where the light quarks in the b baryon are in a $J^P = 0^+$ state, which could enhance the contribution from the weak decay of the s in the $\Xi_b^-$ baryon to a level that ranges from 2% to 8%. This is tested with LHCb data by evaluating

$$r_s \equiv \frac{f(\Xi_b^-)}{f(\Lambda_b^0)}\mathcal{B}(\Xi_b^- \to \Lambda_b^0\pi^-) = \frac{N(\Xi_b^- \to \Lambda_b^0\pi^-)}{N(\Lambda_b^0)}\epsilon_{\mathrm{rel}}, \tag{2}$$

where $f(X)$ are the fragmentation functions, $N(X)$ is the signal yield and $\epsilon_{\mathrm{rel}}$ is the relative efficiency. The results [23] is that

$$\frac{f(\Xi_b^-)}{f(\Lambda_b^0)}\mathcal{B}(\Xi_b^- \to \Lambda_b^0\pi^-) = (5.7 \pm 1.8^{+0.8}_{-0.9}) \times 10^{-4}.$$

Assuming $\frac{f(\Xi_b^-)}{f(\Lambda_b^0)}$ is bounded between 0.1 and 0.3, the branching fraction $\mathcal{B}(\Xi_b^- \to \Lambda_b^0\pi^-)$ would be in the range from 0.19% – 0.76% as predicted in Ref. [24] assuming the diquark transitions have roughly the same weak amplitude as in $B$, $D$ and $K$ meson decays. This disfavours a large enhancement to the decay rate of $\Xi_b^-$ baryons from the $s \to u\bar{u}d$ transition, which could occur if the short-distance correlations within the $J^P = 0^+$ diquark system are enhanced.

## 7. Observation of the decays $\Lambda_b^0 \to \psi(2S)\mathrm{pK}^-$, $\Lambda_b^0 \to \mathrm{J}/\psi\,\pi^+\pi^-\mathrm{pK}^-$ and a measurement of $m(\Lambda_b^0)$

The decays of $\Lambda_b^0$ to $\psi(2S)\mathrm{pK}^-$ and $\mathrm{J}/\psi\,\pi^+\pi^-\mathrm{pK}^-$ have been observed, where the final state $\mathrm{J}/\psi\,\pi^+\pi^-\mathrm{pK}^-$ implicitly includes all intermediate resonances, such as $\psi(2S)\mathrm{pK}^-$. The branching fraction measurements are made in the ratio to the control channel $\Lambda_b^0 \to \mathrm{J}/\psi\,\mathrm{pK}^-$.

The measurements [25] are

$$\frac{\mathcal{B}(\Lambda_b^0 \to \psi(2S)\mathrm{pK}^-)}{\mathcal{B}(\Lambda_b^0 \to \mathrm{J}/\psi\,\mathrm{pK}^-)} = (20.70 \pm 0.76 \pm 0.56 \pm 0.37) \times 10^{-2} \text{ and}$$

$$\frac{\mathcal{B}(\Lambda_b^0 \to \mathrm{J}/\psi\,\pi^+\pi^-\mathrm{pK}^-)}{\mathcal{B}(\Lambda_b^0 \to \mathrm{J}/\psi\,\mathrm{pK}^-)} = (20.86 \pm 0.96 \pm 1.34) \times 10^{-2},$$

**Table 1.** Measured $\Lambda_b^0$ mass in different decay channels and for the combination of all channels. The first uncertainty is statistical and the second is systematic.

| Channel | $m(\Lambda_b^0)[\text{MeV}/c^2]$ |
|---|---|
| $\Lambda_b^0 \to J/\psi\, pK^-$ | $5619.62 \pm 0.04 \pm 0.34$ |
| $\Lambda_b^0 \to \psi(2S)(\to \mu^+\mu^-)pK^-$ | $5619.84 \pm 0.18 \pm 0.19$ |
| $\Lambda_b^0 \to \psi(2S)(\to J/\psi\, \pi^+\pi^-)pK^-$ | $5619.38 \pm 0.33 \pm 0.18$ |
| $\Lambda_b^0 \to J/\psi\, \pi^+\pi^- pK^-$ excluding $\psi(2S)$ | $5619.08 \pm 0.30 \pm 0.27$ |
| Combined | $5619.65 \pm 0.17 \pm 0.17$ |

where the first uncertainties are statistical, the second are systematic and the third is related to knowledge of $J/\psi$ and $\psi(2S)$ branching fractions.

The mass of the $\Lambda_b^0$ can be determined from each of these modes plus the control channel, while $\mathcal{B}(\Lambda_b^0 \to J/\psi\, pK^-)$ is the largest of the branching fractions, it also has the largest momentum scale uncertainty. This is a consequence of the detector calibration in which the momentum scale is set to optimise the reconstructed $m(J/\psi)$. The other channels have fewer events and a larger statistical uncertainty, in combination the most precise measurement of the mass of the $\Lambda_b^0$ can be made, shown in table 1.

## 8. Differential branching fraction and angular analysis of $\Lambda_b^0 \to \Lambda\mu^+\mu^-$ decays

The decay channel $\Lambda_b^0 \to \Lambda\mu^+\mu^-$ is evaluated using $\Lambda_b^0 \to J/\psi\,(\to \mu^+\mu^-)\Lambda$ as a control channel. The results are observed as a function of $q^2$, the square of the invariant mass of the $\mu^+\mu^-$ system. The reconstruction of the $\Lambda \to p\pi^-$ is done for decays inside the VELO using *long* tracks and after the VELO using *downstream* tracks which have less precise directional information.

The signal yield is evaluated as

$$N_S(\Lambda\mu^+\mu^-)_k = \left[\frac{d\mathcal{B}(\Lambda\mu^+\mu^-)/dq^2}{\mathcal{B}(J/\psi\,\Lambda)}\right] \cdot N_s(J/\psi\,\Lambda)_k \cdot \epsilon_k^{\text{rel}} \cdot \frac{\Delta q^2}{\mathcal{B}(J/\psi \to \mu^+\mu^-)}$$

where $k$ is the candidate category (long or downstream), $\Delta q^2$ is the width of the $q^2$ interval considered and $\epsilon_k^{\text{rel}}$ is the relative efficiency.

The distribution of $d\mathcal{B}(\Lambda\mu^+\mu^-)/dq^2$ is shown in figure 1 [26], compared to predictions from the Standard Model. Figure 1 also shows the first measurements of the forward-backward asymmetries, in the dimuon and $p\mu$ systems, $A_{\text{FB}}^l$ and $A_{\text{FB}}^h$ respectively. The measurements of the $A_{\text{FB}}^h$ are in good agreement with the predictions of the Standard Model, while for the $A_{\text{FB}}^l$ measurements are consistently above the prediction.

## 9. Conclusion

Many measurements of hadrons with a b quark have been made with LHCb data. The results improve our knowledge on the CKM parameters, as well as the properties and decays of heavy baryons. These constrain models extending the Standard Model and allow the study of the internal structure of hadrons. LHCb is continuing to take data and more results on properties of b baryons will follow and extend our understanding of both QCD and new physics models.

We express our gratitude to our colleagues in the CERN accelerator departments for the excellent performance of the LHC. We thank the technical and administrative staff at the LHCb institutes. We acknowledge support from CERN and from the national agencies: CAPES,

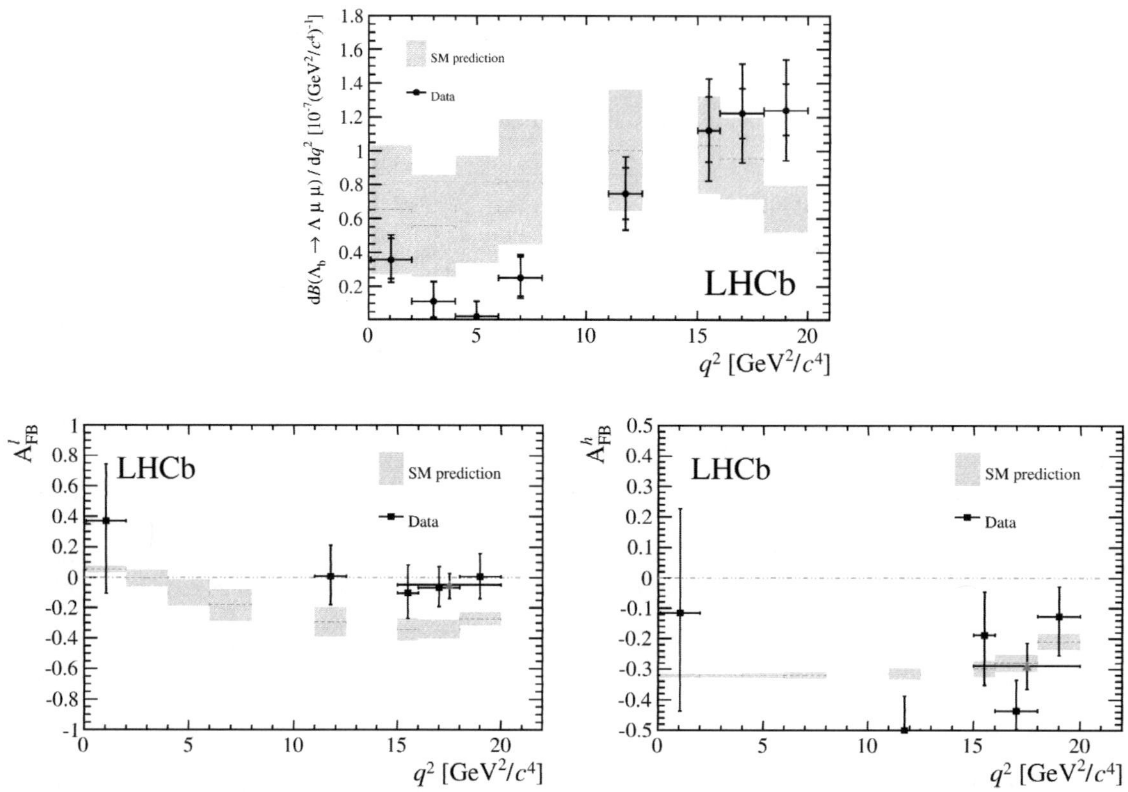


**Figure 1.** Top row: $d\mathcal{B}(\Lambda\mu^+\mu^-)/dq^2$ with the predictions of the SM [27] superimposed. The inner error bars on data points represent the total uncertainty on the relative branching fraction (statistical and systematic); the outer error bar also includes the uncertainties from the branching fraction of the normalisation mode.
Bottom row: Measured values of (left) the leptonic and (right) the hadronic forward-backward asymmetries in bins of $q^2$. Data points are only shown for $q^2$ intervals where a statistically significant signal yield is found. The (red) triangle represents the values for the $15 < q^2 < 20$ GeV$^2/c^4$ interval. Standard Model predictions are obtained from [28].

CNPq, FAPERJ and FINEP (Brazil); NSFC (China); CNRS/IN2P3 (France); BMBF, DFG and MPG (Germany); INFN (Italy); FOM and NWO (The Netherlands); MNiSW and NCN (Poland); MEN/IFA (Romania); MinES and FANO (Russia); MinECo (Spain); SNSF and SER (Switzerland); NASU (Ukraine); STFC (United Kingdom); NSF (USA). We acknowledge the computing resources that are provided by CERN, IN2P3 (France), KIT and DESY (Germany), INFN (Italy), SURF (The Netherlands), PIC (Spain), GridPP (United Kingdom), RRCKI and Yandex LLC (Russia), CSCS (Switzerland), IFIN-HH (Romania), CBPF (Brazil), PL-GRID (Poland) and OSC (USA). We are indebted to the communities behind the multiple open source software packages on which we depend. Individual groups or members have received support from AvH Foundation (Germany), EPLANET, Marie Skłodowska-Curie Actions and ERC (European Union), Conseil Général de Haute-Savoie, Labex ENIGMASS and OCEVU, Région Auvergne (France), RFBR and Yandex LLC (Russia), GVA, XuntaGal and GENCAT (Spain), Herchel Smith Fund, The Royal Society, Royal Commission for the Exhibition of 1851 and the Leverhulme Trust (United Kingdom).

## References

[1] Neubert M 1994 *Phys. Rept.* **245** 259–396 (*Preprint* hep-ph/9306320)
[2] Alves Jr A A *et al.* (LHCb collaboration) 2008 *JINST* **3** S08005
[3] Amhis Y *et al.* (Heavy Flavor Averaging Group (HFAG)) 2014 (*Preprint* 1412.7515)
[4] Lees J P *et al.* (BaBar) 2012 *Phys. Rev.* **D86** 092004 (*Preprint* 1208.1253)
[5] del Amo Sanchez P *et al.* (BaBar) 2011 *Phys. Rev.* **D83** 032007 (*Preprint* 1005.3288)
[6] Ha H *et al.* (Belle) 2011 *Phys. Rev.* **D83** 071101 (*Preprint* 1012.0090)
[7] Sibidanov A *et al.* (Belle) 2013 *Phys. Rev.* **D88** 032005 (*Preprint* 1306.2781)
[8] Flynn J M, Izubuchi T, Kawanai T, Lehner C, Soni A, Van de Water R S and Witzel O 2015 *Phys. Rev.* **D91** 074510 (*Preprint* 1501.05373)
[9] Bailey J A *et al.* (Fermilab Lattice, MILC) 2015 *Phys. Rev.* **D92** 014024 (*Preprint* 1503.07839)
[10] Olive K A *et al.* (Particle Data Group) 2014 *Chin. Phys.* **C38** 090001 and 2015 update
[11] Aaij R *et al.* (LHCb collaboration) 2015 *Nature Physics* **11** 743 (*Preprint* 1504.01568)
[12] Detmold W, Lehner C and Meinel S 2015 *Phys. Rev.* **D92** 034503 (*Preprint* 1503.01421)
[13] Bernlochner F U, Ligeti Z and Turczyk S 2014 *Phys. Rev.* **D90** 094003 (*Preprint* 1408.2516)
[14] Bigi I I Y 1995 (*Preprint* hep-ph/9508408)
[15] Cheng H Y 1997 *Phys. Rev.* **D56** 2783–2798 (*Preprint* hep-ph/9704260)
[16] Ito T, Matsuda M and Matsui Y 1998 *Prog. Theor. Phys.* **99** 271–280 (*Preprint* hep-ph/9705402)
[17] Aaij R *et al.* (LHCb collaboration) 2016 *Phys. Rev.* **D93** 092007 (*Preprint* 1604.01412)
[18] Chatrchyan S *et al.* (CMS) 2012 *Phys. Rev. Lett.* **108** 252002 (*Preprint* 1204.5955)
[19] Aaij R *et al.* (LHCb collaboration) 2016 *JHEP* **05** 161 (*Preprint* 1604.03896)
[20] Klempt E and Richard J M 2010 *Rev. Mod. Phys.* **82** 1095–1153 (*Preprint* 0901.2055)
[21] Aaij R *et al.* (LHCb collaboration) 2014 *Phys. Rev. Lett.* **113** 242002 (*Preprint* 1409.8568)
[22] Li X and Voloshin M B 2014 *Phys. Rev.* **D90** 033016 (*Preprint* 1407.2556)
[23] Aaij R *et al.* (LHCb) 2015 *Phys. Rev. Lett.* **115** 241801 (*Preprint* 1510.03829)
[24] Faller S and Mannel T 2015 *Phys. Lett.* **B750** 653–659 (*Preprint* 1503.06088)
[25] Aaij R *et al.* (LHCb collaboration) 2016 *JHEP* **05** 132 (*Preprint* 1603.06961)
[26] Aaij R *et al.* (LHCb collaboration) 2015 *JHEP* **06** 115 (*Preprint* 1503.07138)
[27] Detmold W, Lin C J D, Meinel S and Wingate M 2013 *Phys. Rev.* **D87** 074502 (*Preprint* 1212.4827)
[28] Meinel S 2014 *PoS* **LATTICE2013** 024 (*Preprint* 1401.2685)

# Recent Results from the Belle Experiment

**Z King and A J Schwartz**
**(on behalf of the Belle Collaboration)**

Physics Department, University of Cincinnati, P.O. Box 210011, Cincinnati, Ohio 45221 USA

E-mail: kingze@mail.uc.edu, alan.j.schwartz@uc.edu

**Abstract.** We review recent results from the Belle experiment, which took data at the KEKB asymmetric-energy $e^+e^-$ collider in Japan. The experiment recorded about 1000 $\text{fb}^{-1}$ of data running mainly at the $\Upsilon(4S)$ and $\Upsilon(5S)$ resonances. The results presented here are obtained from the full data set.

## 1. Introduction

The Belle experiment successfully operated for more than a decade until 2010 at the asymmetric-energy $e^+e^-$ collider KEKB [1]. The experiment took data at center-of-mass energies corresponding to several $\Upsilon(n\text{S})$ resonances; the total data sample recorded exceeds $1\,\text{ab}^{-1}$. Here we present recent results based on the full data sample.

The Belle detector is a large-solid-angle magnetic spectrometer that consists of a silicon vertex detector (SVD), a 50-layer central drift chamber (CDC), an array of aerogel threshold Cherenkov counters (ACC), a barrel-like arrangement of time-of-flight scintillation counters (TOF), and an electromagnetic calorimeter comprised of CsI(Tl) crystals (ECL) located inside a superconducting solenoid coil that provides a 1.5 T magnetic field. An iron flux-return located outside of the coil is instrumented to detect $K_L^0$ mesons and to identify muons (KLM). The detector is described in detail elsewhere [2].

## 2. Search for the Decay $B^0 \to \phi\gamma$

In the Standard Model (SM), the decay $B^0 \to \phi\gamma$ proceeds through electroweak and gluonic $b \to d$ penguin annihilation processes. These amplitudes are proportional to the small Cabibbo-Kobayashi-Maskawa [3] matrix element $V_{td}$ and thus are highly suppressed. The branching fraction has been estimated based on naive QCD factorization [4] and perturbative QCD [5] and found to be in the range $10^{-12}$ to $10^{-11}$. However, the internal loop can also be mediated by non-SM particles such as a charged Higgs boson or supersymmetric squarks, and thus the decay is sensitive to new physics (NP). It is estimated that such NP could enhance the branching fraction to the level of $10^{-9}$ to $10^{-8}$ [4]. Previously, no evidence for this decay has been found. Here we describe a recent Belle search for this decay with much higher sensitivity than previously that uses the full data set of 711 $\text{fb}^{-1}$ recorded on the $\Upsilon(4\text{S})$ resonance [6].

Candidate $\phi$ mesons are reconstructed via $\phi \to K^+K^-$ decays. The $K^+K^-$ invariant mass is required to be in the range $[1.000, 1.039]$ GeV/$c^2$, which corresponds to $4.5\sigma$ in resolution around the $\phi$ mass. Candidate $B$ mesons are identified using a modified beam-energy-constrained mass $M_{\text{bc}} = \sqrt{E_{\text{beam}}^2 - |\vec{p}_B|c^2/c^2}$, and the energy difference $\Delta E = E_B - E_{\text{beam}}$, where $E_{\text{beam}}$

is the beam energy and $\vec{p}_B$ and $E_B$ are the momentum and energy, respectively, of the $B^0$ candidate. All quantities are evaluated in the center-of-mass frame. To improve the $M_{\rm bc}$ resolution, the momentum $\vec{p}_B$ is calculated as $\vec{p}_\phi + (\vec{p}_\gamma/|p_\gamma|)\sqrt{(E_{\rm beam} - E_\phi)^2}/c$, where $\vec{p}_\gamma$ is the photon momentum, and $E_\phi$ and $\vec{p}_\phi$ are the energy and momentum, respectively, of the $\phi$ candidate. We require that events satisfy $5.25\ {\rm GeV}/c^2 < M_{\rm bc} < 5.29\ {\rm GeV}/c^2$ and $-0.30\ {\rm GeV} < \Delta E < 0.15\ {\rm GeV}$. The signal yield is calculated in a smaller "signal region" $5.27\ {\rm GeV}/c^2 < M_{\rm bc} < 5.29\ {\rm GeV}/c^2$ and $-0.20\ {\rm GeV} < \Delta E < 0.10\ {\rm GeV}$.

Charmless hadronic decays suffer from large backgrounds arising from continuum $e^+e^- \to q\bar{q}$ $(q = u, d, s, c)$ production. To suppress this background, we use a multivariate analyzer based on a neural network (NN) [7]. The NN generates an output variable $C_{\rm NN}$, which ranges from $-1$ for background-like events to $+1$ for signal-like events. We make a loose requirement $C_{\rm NN} > 0.3$ and then translate $C_{\rm NN}$ to $C'_{\rm NN}$, defined as $C'_{\rm NN} = \ln\left[(C_{\rm NN} - C_{\rm min})/(C_{\rm max} - C_{\rm NN})\right]$ where $C_{\rm min} = 0.3$ and $C_{\rm max} = 1.0$. The variable $C'_{\rm NN}$ is well-described by a sum of Gaussian functions. After the above selections, 961 events remain. The remaining background consists of continuum events and rare charmless $b$-decay processes.

We obtain the signal yield using an unbinned extended maximum likelihood fit to the observables $M_{\rm bc}$, $\Delta E$, $C'_{\rm NN}$, and $\cos\theta_\phi$. The helicity angle $\theta_\phi$ is the angle between the $K^+$ momentum and the opposite of the $B$ flight direction in the $\phi$ rest frame. The resulting branching fraction is calculated as

$$\mathcal{B}\left(B^0 \to \phi\gamma\right) = \frac{N_{\rm sig}}{N_{B\overline{B}} \cdot \varepsilon \cdot \mathcal{B}(\phi \to K^+K^-)} \tag{1}$$

where $N_{\rm sig} = 3.4\,^{+4.6}_{-3.8}$ is the signal yield in the signal region, $N_{B\overline{B}} = (772 \pm 11) \times 10^6$ is the number of $B\overline{B}$ events, $\varepsilon = 0.296 \pm 0.001$ is the signal efficiency as calculated from Monte Carlo (MC) simulation, and $\mathcal{B}(\phi \to K^+K^-) = (48.9 \pm 0.5)\%$ [8]. We find no evidence for this decay and set an upper limit on the branching fraction of $\mathcal{B}(B^0 \to \phi\gamma) < 1.0 \times 10^{-7}$ at 90% confidence level (C.L.). This limit is almost an order of magnitude lower than the previous most stringent result [9].

## 3. Angular Analysis of $B^0 \to K^*(892)^0\ell^+\ell^-$

The rare decay $B^0 \to K^*(892)^0\ell^+\ell^-$ involves the quark transition $b \to s\ell^+\ell^-$, a flavor changing neutral current that is forbidden at tree level in the SM. Higher order SM processes such as penguin or $W^+W^-$ box diagrams allow for such transitions, leading to branching fractions below $10^{-6}$. Various extensions to the SM predict NP amplitudes that can interfere with the SM amplitudes and lead to enhanced or suppressed branching fractions and modified angular distributions.

Belle performed an angular analysis using the decay modes $B^0 \to K^*(892)^0e^+e^-$ and $B^0 \to K^*(892)^0\mu^+\mu^-$ [10]. $K^{*0}$ candidates are reconstructed in the channel $K^{*0} \to K^+\pi^-$ [11]. Large combinatoric background is suppressed by applying requirements on kinematic variables. Large backgrounds also arise from charmonium decays $B \to K^{(*)}J/\psi$ and $B \to K^{(*)}\psi(2S)$, in which $\psi \to \ell^+\ell^-$. To maximize signal efficiency and purity, neural networks are applied sequentially beginning from the end of the decay chain. Signal and background yields are determined from an unbinned extended maximum likelihood fit to $M_{\rm bc}$ in bins of $q^2$. The bin ranges used and the resulting fitted yields are listed in Table 1. In total, $117.6 \pm 12.4$ signal candidates are obtained for $B^0 \to K^*(892)^0\mu^+\mu^-$, and $69.4 \pm 12.0$ candidates for $B^0 \to K^*(892)^0e^+e^-$.

We subsequently perform the angular analysis. The decay is kinematically described by three angles ($\theta_\ell$, $\theta_K$, and $\phi$) and the invariant mass squared of the lepton pair ($q^2$). The definitions of the angles and the full angular distribution are described in Ref. [12]. For this fit we require $M_{\rm bc} > 5.27\ {\rm GeV}/c^2$, and the number of signal ($N_{\rm sig}$) and background events ($N_{\rm bkg}$) are fixed to the values in Table 1.

Journal of Physics: Conference Series **770** (2016) 012015 doi:10.1088/1742-6596/770/1/012015

**Table 1.** Fitted yields in bins of $q^2$ for signal ($N_{\rm sig}$) and background ($N_{\rm bkg}$), for electron and muon channels combined.

| Bin | $q^2$ range in $\mathrm{GeV}^2/c^4$ | $N_{\rm sig}$ | $N_{\rm bkg}$ |
|---|---|---|---|
| 0 | $1.00 - 6.00$ | $49.5 \pm 8.4$ | $30.3 \pm 5.5$ |
| 1 | $0.10 - 4.00$ | $30.9 \pm 7.4$ | $26.4 \pm 5.1$ |
| 2 | $4.00 - 8.00$ | $49.8 \pm 9.3$ | $35.6 \pm 6.0$ |
| 3 | $10.09 - 12.90$ | $39.6 \pm 8.0$ | $19.3 \pm 4.4$ |
| 4 | $14.18 - 19.00$ | $56.5 \pm 8.7$ | $16.0 \pm 4.0$ |

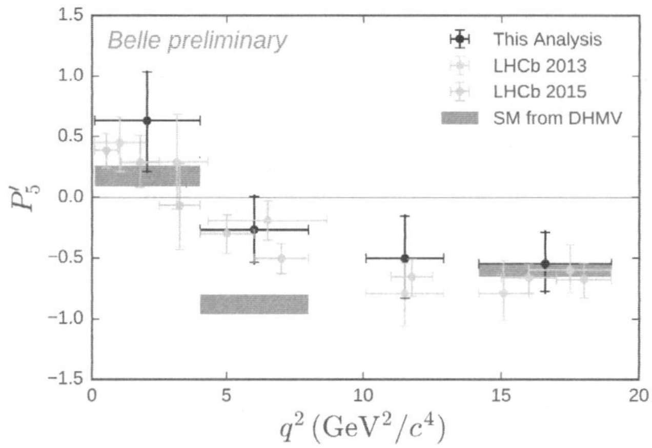


**Figure 1.** Results for $P_5'$ as compared to SM predictions and to previous measurements by LHCb [15][16].

The angular observables $P'_{i=4,5,6,8}$ introduced in Ref. [13] contain information about short-distance effects. These observables are considered to be largely free from form-factor uncertainties [14], and thus they are a promising place to search for NP. As the statistics in this analysis are insufficient to perform a full eight-dimensional fit, a "folding technique" is used as described in Ref. [15]. We determine $P'_{4,5,6,8}$ by performing a three-dimensional unbinned maximum likelihood fit in four bins of $q^2$ using the folded signal probability density function and fixed background yields and shapes. We also fit for the longitudinal polarization $F_L$ and the transverse polarization asymmetry $A_T^{(2)}$. The fit results are shown in Fig. 1. These results are consistent with SM predictions, although one value of $P_5'$ differs by $2.1\sigma$ from the SM value. It is notable that this deviation is for the same $q^2$ bin and in the same direction as a similar deviation reported by LHCb [15][16].

## 4. Measurement of the branching fraction of $\overline{B}^0 \to D^{*+}\tau^-\overline{\nu}_\tau$ relative to that of $\overline{B}^0 \to D^{*+}\ell^-\overline{\nu}_\ell$

Semi-tauonic $B$ decays of the type $b \to c\tau\nu_\tau$ are sensitive probes to search for NP. Charged Higgs bosons, which appear in supersymmetry and other models with at least two Higgs doublets, may induce measurable effects in the branching fraction due to the large mass of the $\tau$ lepton. Similarly, leptoquarks, which carry both baryon number and lepton number, can also contribute to this process. The ratio of branching fractions

$$\mathcal{R}(D^{(*)}) = \frac{\mathcal{B}(\overline{B} \to D^{(*)}\tau^-\overline{\nu}_\tau)}{\mathcal{B}(\overline{B} \to D^{(*)}\ell^-\overline{\nu}_\ell)} \quad (\ell = e, \mu) \tag{2}$$

is typically measured instead of the absolute branching fraction $\mathcal{B}(\overline{B} \to D^{(*)}\tau^-\overline{\nu}_\tau)$ to reduce systematic uncertainties arising from reconstruction efficiencies, form factors, and the CKM matrix element $|V_{cb}|$. Standard Model calculations predict $\mathcal{R}(D^*) = 0.252 \pm 0.003$ [17] and $\mathcal{R}(D) = 0.297 \pm 0.017$ [18][19]. Semi-tauonic $B$ decays were first observed by Belle [20], with subsequent studies reported by Belle [19][21], *BABAR* [22], and LHCb [23]. The world average values [24] are $\mathcal{R}(D^*) = 0.322 \pm 0.018 \pm 0.012$ and $\mathcal{R}(D) = 0.391 \pm 0.041 \pm 0.028$, which exceed the SM predictions by $3.0\sigma$ and $1.7\sigma$, respectively.

Thus far, measurements of $\mathcal{R}(D^{(*)})$ at Belle and *BABAR* have been performed using either a hadronic [19][22] or an inclusive [20][21] tagging method. Here we report the first measurement of $\mathcal{R}(D^*)$ using a semileptonic tagging method. We reconstruct signal events in modes in which one $B$ decays as $\overline{B}^0 \to D^{*+}\ell^-\overline{\nu}_\ell$ and the the other $B$ decays as $\overline{B}^0 \to D^{*+}\tau^-\overline{\nu}_\tau$, $\tau^- \to \ell^-\overline{\nu}_\ell\nu_\tau$. To reconstruct normalization events corresponding to the denominator in Eq. (2), we require that both $B$ mesons decay to $D^{*+}\ell^-\overline{\nu}_\ell$.

Signal and normalization events are identified using a neural network. The dominant background contribution arises from events with falsely reconstructed $D^{(*)}$ mesons. To separate signal and normalization events from backgrounds, we use the energy variable $E_{\rm ECL}$, which is defined as the sum of the energies of neutral clusters detected in the electromagnetic calorimeter (ECL) that are not associated with reconstructed particles.

We extract the signal and normalization yields using a two-dimensional extended maximum-likelihood fit to the neural network output NN and $E_{\rm ECL}$. The resulting yields of signal and normalization events are $231 \pm 23$(stat) and $2800 \pm 57$(stat), respectively. The ratio $\mathcal{R}(D^*)$ is $0.302 \pm 0.030 \pm 0.011$, where the first and second errors correspond to statistical and systematic uncertainties, respectively. Our measurement is $1.6\sigma$ higher than the SM prediction.

We investigate the compatibility of the data with a Type II two-Higgs doublet model (2HDM) [25] and with $R_2$-type and $S_1$-type leptoquark models [26]. We find that the data allows for additional contributions from 2HDM scalar operators and also vector operators, but contributions from a tensor operator with $0.34 < C_T < 0.39$, an $R_2$-type leptoquark model with $0.34 < C_T < 0.38$, or an $S_1$-type leptoquark model with $0.22 < C_T < 0.28$ are disfavored.

## 5. Observation of the decay $B_s^0 \to K^0\overline{K}^0$

The two-body decays $B_s^0 \to hh'$, where $h^{(\prime)}$ is either a charged $\pi^\pm$ or $K^\pm$, have all been observed [8]. However, the neutral modes $\pi^0\pi^0$, $\pi^0K^0$, and $K^0\overline{K}^0$ have not. The decay $B_s^0 \to K^0\overline{K}^0$ is of particular interest because the branching fraction is predicted to be large: $(16 - 27) \times 10^{-6}$ [27]. The presence of non-SM particles or couplings could measurably affect this value [28]. The current upper limit, $\mathcal{B}(B_s^0 \to K^0\overline{K}^0) < 6.6 \times 10^{-5}$ at 90% C.L., was set by Belle using only 23.6 fb$^{-1}$ of data recorded at the $\Upsilon(5S)$ [29]. Here we update that result using the full data set of 121.4 fb$^{-1}$ [30]. At the $\Upsilon(5S)$, $B_s^0\overline{B}_s^0$ pairs are produced in three decay channels: $B_s^0\overline{B}_s^0$, $B_s^{*0}\overline{B}_s^0$ or $B_s^0\overline{B}_s^{*0}$, and $B_0^{*0}\overline{B}_s^{*0}$. The latter two channels dominate, with production fractions of $f_{B_s^{*0}\overline{B}_s^0} = (7.3 \pm 1.4)\%$ and $f_{B_s^{*0}\overline{B}_s^{*0}} = (87.0 \pm 1.7)\%$ [31].

Candidate $K^0$ mesons are reconstructed via the decay $K_S^0 \to \pi^+\pi^-$ using a neural network technique. To suppress background arising from $e^+e^- \to q\overline{q}$ ($q = u, d, s, c$) production, a second NN is used. The output of the latter ($C_{\rm NN}$) ranges from $-1$ for background-like events to $+1$ for signal-like events. We make a loose requirement $C_{\rm NN} > -0.1$ and then translate $C_{\rm NN}$ to the variable $C'_{\rm NN}$ as described in Section 2.

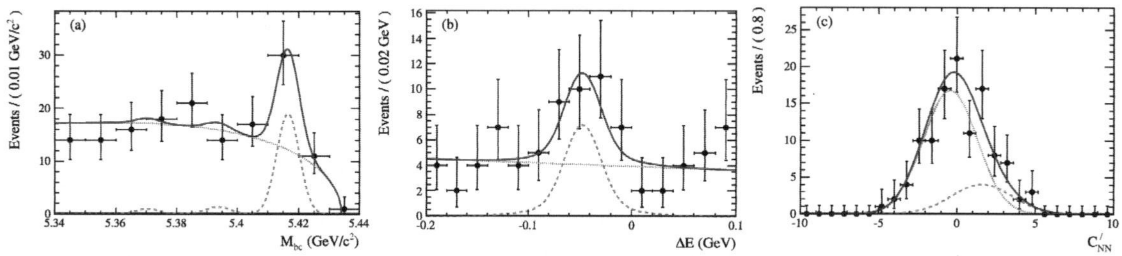


**Figure 2.** Projections of the fit for $B_s^0 \to K^0\overline{K}^0$: (a) $M_{\rm bc}$ for $-0.11$ GeV $< \Delta E <$ 0.02 GeV and $C'_{\rm NN} > 0.5$; (b) $\Delta E$ for 5.405 GeV/$c^2 < M_{\rm bc} <$ 5.427 GeV/$c^2$ and $C'_{\rm NN} > 0.5$; and (c) $C'_{\rm NN}$ for 5.405 GeV/$c^2 < M_{\rm bc} <$ 5.427 GeV/$c^2$ and $-0.11$ GeV $< \Delta E <$ 0.02 GeV. The points with error bars are data, the (green) dashed curves show the signal, (magenta) dotted curves show the continuum background, and (blue) solid curves show the total.

We measure the signal yield by performing an unbinned extended maximum likelihood fit to variables $M_{\rm bc}$, $\Delta E$, and $C'_{\rm NN}$. The results are $29.0\,^{+8.5}_{-7.6}$ signal events and $1095.0\,^{+33.9}_{-33.4}$ background events. Projections of the fit result are shown in Fig. 2. The branching fraction is calculated as

$$\mathcal{B}(B_s^0 \to K^0\overline{K}^0) \quad = \quad \frac{N_{\rm sig}}{2N_{B_s^0\overline{B}_s^0}(0.50)\mathcal{B}^2_{K^0}\varepsilon} \tag{3}$$

where $N_{\rm sig}$ is the fitted signal yield; $N_{B_s^0\overline{B}_s^0} = (6.53{\pm}0.66){\times}10^6$ is the number of $B_s^0\overline{B}_s^0$ events [32]; $\mathcal{B}_{K^0} = (69.20 \pm 0.05)\%$ is the branching fraction for $K_S^0 \to \pi^+\pi^-$ [8]; and $\varepsilon = (46.3 \pm 0.1)\%$ is the reconstruction efficiency as determined from MC simulation. The factor 0.50 accounts for the 50% probability for $K^0\overline{K}^0 \to K_S^0K_S^0$ (since $K^0\overline{K}^0$ is $CP$ even). Inserting these values gives $\mathcal{B}(B_s^0 \to K^0\overline{K}^0) = (19.6\,^{+5.8}_{-5.1} \pm 1.0 \pm 2.0) \times 10^{-6}$, where the first uncertainty is statistical, the second is systematic, and the third reflects the uncertainty due to the total number of $B_s^0\overline{B}_s^0$ pairs. This result is in good agreement with the SM prediction [27].

The signal significance is calculated as $\sqrt{-2\ln(\mathcal{L}_0/\mathcal{L}_{\rm max})}$, where $\mathcal{L}_0$ is the likelihood value when the signal yield is fixed to zero, and $\mathcal{L}_{\rm max}$ is the likelihood value of the nominal fit. Systematic uncertainties are included in the significance by convolving the likelihood function with a Gaussian function whose width is equal to the systematic uncertainty associated with the signal yield. We obtain a signal significance of $5.1\sigma$, and thus our measurement constitutes the first observation of this decay.

## 6. Measurement of the branching fraction and $CP$ asymmetry for $D^0 \to V\gamma$

$CP$ asymmetries in the charm sector are a promising area in which to search for new physics. The $CP$ asymmetry ($\mathcal{A}_{CP}$) of many two-body $D$ decays has been measured [24], but all results are consistent with zero. Here we present a first search for a $CP$ asymmetry in radiative $D^0 \to V\gamma$ decays, where $V$ is a neutral vector meson $\phi$, $\overline{K}^{*0}$, or $\rho^0$. Within the SM, such asymmetries are predicted to be small: $\mathcal{A}_{CP} \sim 10^{-3}$. However, theoretical studies [33][34] indicate that extensions to the SM with chromomagnetic dipole operators can increase $\mathcal{A}_{CP}$ up to a few percent. Several $D^0 \to V\gamma$ decay modes have been measured by Belle and *BABAR* [35][36]. The current world average branching fractions are $\mathcal{B}(D^0 \to \phi\gamma) = (2.70 \pm 0.35) \times 10^{-5}$ and $\mathcal{B}(D^0 \to \overline{K}^{*0}\gamma) = (32.7 \pm 3.4) \times 10^{-5}$ [8]. The $D^0 \to \rho^0\gamma$ mode has not yet been observed, and the current upper limit on the branching fraction is $\mathcal{B}(D^0 \to \rho^0\gamma) < 24 \times 10^{-5}$ at 90% C.L. [8].

Here we present new measurements of all three modes using the full Belle data set. Our results include the first measurement of the $CP$ asymmetries. For our analysis, the $D^0$ is required to originate from $D^{*+} \to D^0\pi^+$ decays; this tags the $D^0$ flavor and suppresses combinatoric background. The vector mesons in the final state are reconstructed via $\phi \to K^+K^-$, $\overline{K}^{*0} \to K^-\pi^+$, and $\rho^0 \to \pi^+\pi^-$.

All branching fractions and $\mathcal{A}_{CP}$ are determined by normalizing to decay channels $D^0 \to K^+K^-$ for $D^0 \to \phi\gamma$, $D^0 \to K^-\pi^+$ for $D^0 \to \overline{K}^{*0}\gamma$, and $D^0 \to \pi^+\pi^-$ for $D^0 \to \rho^0\gamma$. The branching fractions are calculated as

$$\mathcal{B}_{\rm sig} = \mathcal{B}_{\rm norm} \times \frac{N_{\rm sig}}{N_{\rm norm}} \times \frac{\varepsilon_{\rm norm}}{\varepsilon_{\rm sig}} \tag{4}$$

where $\mathcal{B}$ and $N$ are the branching fraction and fitted yield, respectively, of signal or normalization modes, and $\varepsilon$ is the reconstruction efficiency. For $\mathcal{B}_{\rm norm}$, the world-average value [8] is used.

The extracted raw asymmetry

$$A_{\rm raw} = \frac{N(D^0) - N(\overline{D}^0)}{N(D^0) + N(\overline{D}^0)} \tag{5}$$

has several contributions: $A_{\rm raw} = \mathcal{A}_{CP} + A_{\rm FB} + A_{\varepsilon}^{\pm}$, where $A_{\rm FB}$ is the forward-backward production asymmetry and $A_{\varepsilon}^{\pm}$ is the detection asymmetry between positively and negatively charged particles. Both of these asymmetries are eliminated by measuring $A_{CP}$ of a signal mode relative to that of its normalization mode, which has the same final-state charged particles. The $CP$ asymmetry of a signal mode is then $\mathcal{A}_{CP}^{\rm sig} = A_{\rm raw}^{\rm sig} - A_{\rm raw}^{\rm norm} + \mathcal{A}_{CP}^{\rm norm}$, where $\mathcal{A}_{CP}^{\rm norm}$ is the PDG value of the $CP$ asymmetry for the normalization mode [8].

To extract the signal yields and $CP$ asymmetries, we perform a simultaneous two-dimensional unbinned extended maximum likelihood fit of the $D^0$ and $\overline{D}^0$ samples. The fit variables are the invariant mass of the $D^0$ and the cosine of the helicity angle $\theta_H$, defined as the angle between the $D^0$ and the positively or negatively charged hadron in the rest frame of the $V$ meson. For $D^0$ candidates we use the $K^+/K^-/\pi^+$ for $\phi/\overline{K}^{*0}/\rho^0$ decays, and we use the oppositely charged particles for $\overline{D}^0$ candidates. The fitted signal yields are $N_{\phi\gamma} = 524 \pm 35$, $N_{\overline{K}^{*0}\gamma} = 9104 \pm 396$, and $N_{\rho^0\gamma} = 500 \pm 85$. The resulting branching fractions are

$$\begin{aligned}
\mathcal{B}(D^0 \to \phi\gamma) &= (2.76 \pm 0.20 \pm 0.08) \times 10^{-5} \\
\mathcal{B}(D^0 \to \overline{K}^{*0}\gamma) &= (4.66 \pm 0.21 \pm 0.18) \times 10^{-4} \\
\mathcal{B}(D^0 \to \rho^0\gamma) &= (1.77 \pm 0.30 \pm 0.08) \times 10^{-5},
\end{aligned}$$

where the first uncertainty is statistical and the second is systematic. The result for $\phi\gamma$ is improved with respect to the previous result [35]. The result for $\overline{K}^{*0}\gamma$ is $3.3\sigma$ higher than the BABAR result [36]. The result for $\rho^0\gamma$ is close to that for $\phi\gamma$, which is consistent with theoretical expectations. The significance is calculated as $\sqrt{-2\ln(\mathcal{L}_0/\mathcal{L}_{\rm max})}$ as described in Section 5. Systematic uncertainties are included in the significance by convolving the likelihood function with a Gaussian whose width is equal to the systematic uncertainty associated with the signal yield. The significance for $\rho^0\gamma$ is $5.5\sigma$, and thus our measurement constitutes the first observation of this decay.

The results for $\mathcal{A}_{CP}$ are

$$\begin{aligned}
\mathcal{A}_{CP}(D^0 \to \phi\gamma) &= -0.094 \pm 0.066 \pm 0.001 \\
\mathcal{A}_{CP}(D^0 \to \overline{K}^{*0}\gamma) &= -0.003 \pm 0.020 \pm 0.000 \\
\mathcal{A}_{CP}(D^0 \to \rho^0\gamma) &= 0.056 \pm 0.151 \pm 0.006\,.
\end{aligned}$$

No $CP$ asymmetry is seen in any of these radiative decays.

## Acknowledgements

The authors thank the BEACH 2016 organizers for a well-run workshop and excellent hospitality. This research is supported by the U.S. Department of Energy.

## References

[1] Kurokawa S and Kikutani E 2003 *Nucl. Instrum. Methods Phys. Res., Sect.* A **499** 1 and other papers included in this volume
Abe T *et al* 2013 *Prog. Theor. Exp. Phys.* 03A001 and following articles up to 03A011
[2] Abashian A *et al* (Belle Collaboration) 2002 *Nucl. Instr. and Meth.* A **479** 117
[3] Cabibbo N 1963 *Phys. Rev. Lett.* **10** 531
Kobayashi M and Maskawa T 1973 *Prog. Theor. Phys.* **49** 652
[4] Li X-Q, Lu G-R, Wang R-M and Yang Y 2004 *Eur. Phys. J.* C **36** 97
Hua J, Kim C and Li Y 2010 *Eur. Phys. J.* C **69** 139
[5] Lu C-D, Shen Y-L and Wang W 2006 *Chin. Phys. Lett.* **23** 2684
[6] King Z *et al* (Belle Collaboration) 2016 *Phys. Rev.* D **93** 111101
[7] Feindt M and Kerzel U 2006 *Nucl. Instrum. Methods Phys. Res. Sect.* A **559** 190
[8] Olive K A *et al* (Particle Data Group) 2014 *Chin. Phys.* C **38** 090001
[9] Aubert B *et al* (*BABAR* Collaboration) 2005 *Phys. Rev.* D **72** 091103
[10] Abdesselam A *et al* (Belle Collaboration) 2016 arXiv:1604.04042
[11] Charge-conjugate modes are implicitly included unless noted otherwise
[12] Aaij R *et al* (LHCb Collaboration) 2013 *JHEP* **08** 131
[13] Descotes-Genon S, Matias J, Ramon M and Virto J 2013 *JHEP* **01** 048
[14] Descotes-Genon S, Hurth T, Matias J and Virto J 2013 *JHEP* **1305** 137
[15] Aaij R *et al* (LHCb Collaboration) 2013 *Phys. Rev. Lett.* **111** 191801
[16] Aaij R *et al* (LHCb Collaboration) 2016 *JHEP* **02** 104
[17] Fajfer S, Kamenik J F and Nisandzic I 2012 *Phys. Rev.* D **85** 094025
[18] Kamenik J F and Mescia F 2008 *Phys. Rev.* D **78** 014003
[19] Huschle M *et al* (Belle Collaboration) 2015 *Phys. Rev.* D **92** 072014
[20] Matyja A *et al* (Belle Collaboration) 2007 *Phys. Rev. Lett.* **99** 191807
[21] Bozek A *et al* (Belle Collaboration) 2010 *Phys. Rev.* D **82** 072005
[22] Lees J P *et al* (*BABAR* Collaboration) 2012 *Phys. Rev. Lett.* **109** 101802
Lees J P *et al* (*BABAR* Collaboration) 2013 *Phys. Rev.* D **88** 072012
[23] Aaij R *et al* (LHCb Collaboration) 2015 *Phys. Rev. Lett.* **115** 111803
[24] Amhis Y *et al* (Heavy Flavor Averaging Group) 2014 arXiv:1412.7515 and online update at www.slac.stanford.edu/xorg/hfag/
[25] Tanaka M 1995 *Zeit. Phys.* C **67** 321
[26] Sakaki Y, Watanabe R, Tanaka M and Tayduganov A 2013 *Phys. Rev.* D **88** 094012
[27] Chen C H 2001 *Phys. Lett.* B **520** 33
Williamson A R and Zupan J 2006 *Phys. Rev.* D **74** 014003
Ali A, Kramer G, Li Y, Lu C-D, Shen Y-L, Wang W and Wang Y-M 2007 *Phys. Rev.* D **76** 074018
Chua C K 2008 *Phys. Rev.* D **78** 076002
Wang K and Zhu G 2013 *Phys. Rev.* D **88** 014043
Wang J J, Lin D T, Sun W, Ji Z J, Cheng S and Xiao Z J 2014 *Phys. Rev.* D **89** 074046
Chang Q, Sun J, Yang Y and Li X 2015 *Phys. Lett.* B **740** 56
Cheng H Y, Chiang C W and Kuo A L 2015 *Phys. Rev.* D **91** 014011
[28] Chang Q, Li X Q and Yang Y D 2014 *J. Phys.* G **41** 105002
[29] Peng C C *et al* (Belle Collaboration) 2010 *Phys. Rev.* D **82** 072007
[30] Pal B *et al* (Belle Collaboration) 2016 *Phys. Rev. Lett.* **116** 161801
[31] Esen S *et al* (Belle Collaboration) 2013 *Phys. Rev.* D **87** 031101(R)
[32] Oswald C *et al* (Belle Collaboration) 2015 *Phys. Rev.* D **92** 072013
[33] Isidori G and Kamenik J F 2012 *Phys. Rev. Lett.* **109** 171801
[34] Lyon J and Zwicky R 2012 arXiv:1210.6546
[35] Tajima O *et al* (Belle Collaboration) 2014 *Phys. Rev. Lett.* **92** 101803
[36] Aubert B *et al* (*BABAR* Collaboration) 2008 *Phys. Rev.* D **78** 071101

Journal of Physics: Conference Series **770** (2016) 012016 doi:10.1088/1742-6596/770/1/012016

# Charmless $b$-meson and $b$-baryon decays at LHCb

**Irina Nasteva on behalf of the LHCb collaboration**

Instituto de Física, Universidade Federal do Rio de Janeiro
Av. Athos da Silveira Ramos 149, CT bloco A, Cidade Universitária
Rio de Janeiro, RJ 21941–972, Brazil

E-mail: irina.nasteva@cern.ch

**Abstract.** The latest LHCb results for charmless $b$-meson and $b$-baryon decays, using Run I data collected by the LHCb experiment at centre-of-mass energies of 7 and 8 TeV, are presented. The search for charmless three-body decays of the $\Lambda_b^0$ and $\Xi_b^0$ baryons to the final states $\Lambda h^+ h'^-$ ($h^{(\prime)} = \pi$ or $K$) results in the first observation of $\Lambda_b^0 \to \Lambda K^+ \pi^-$ and $\Lambda_b^0 \to \Lambda K^+ K^-$ decays and measurements of their branching fractions and $CP$ asymmetries, as well as evidence of the $\Lambda_b^0 \to \Lambda \pi^+ \pi^-$ decay and limits on the branching fractions of $\Xi_b^0$ baryon decays. The first observation of $\Lambda_b^0 \to \Lambda\phi$ decay is also reported and the branching fraction and triple-product asymmetries are measured. The search for $\Lambda_b^0 \to \Lambda\eta'$ and $\Lambda_b^0 \to \Lambda\eta$ decays yields evidence for the latter and upper limits on the former. A measurement of the branching fraction of $B_s^0 \to \phi\phi$ decay and a limit on the $B^0 \to \phi\phi$ mode are also presented. The decay $B_s^0 \to K_S^0 K^*(892)^0$ is observed for the first time and its branching fraction measured.

## 1. Introduction

Decays of $b$-mesons and $b$-baryons to charmless final states proceed via $b \to u$ tree and flavour-changing neutral current $b \to s, d$ loop (penguin) transitions, both of which are suppressed in the Standard Model. The tree and penguin transition amplitudes are of similar size and have a relative weak phase, which can lead to $CP$ violation in decay. The possibility of new physics particles contributing to the loop makes these decay modes sensitive probes of new physics and motivates their branching fraction and $CP$ asymmetry measurements. The $CP$ asymmetries can be compared in corresponding $b$-meson and $b$-baryon decays, such as $B_s^0 \to \phi\phi$ ($B^0 \to K_S^0\phi$, $B^0 \to K^{*0}\phi$) where no $CP$ violation has been found so far [1], [2], [3], [4], and $\Lambda_b^0 \to \Lambda\phi$ reported here. The sector of charmless $b$-baryon decays is almost unexplored and particular to LHCb.

## 2. Datasets and analysis strategies

The results presented in these proceedings are recent measurements, performed by the LHCb experiment in its Run I with a dataset corresponding to 1 fb$^{-1}$ of integrated luminosity collected at centre-of-mass energy of 7 TeV in 2011 and 2 fb$^{-1}$ at 8 TeV in 2012. The LHCb detector [5], [6] is a single-arm forward spectrometer designed for the study of particles containing $b$ or $c$ quarks.

The selection criteria for signal candidates share some common characteristics in the analyses of charmless $b$-hadron decays presented here. The selections are based on multivariate classifiers that exploit the kinematic properties of final-state particles and the topologies of three-body and quasi-two-body decays. In addition, particle identification requirements on charged hadrons, based primarily on information from ring-imaging Cherenkov detectors, are applied to reduce

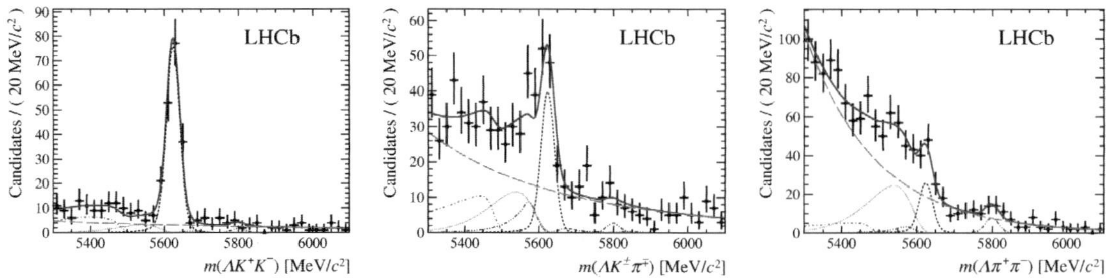


**Figure 1.** Results of the invariant mass fits for (left) $\Lambda K^+K^-$, (middle) $\Lambda K^{\pm}\pi^{\mp}$, and (right) $\Lambda\pi^+\pi^-$ final states. The total result of the fit is superimposed as solid lines, the $\Lambda_b^0$ ($\Xi_b^0$) decay as short-dashed (double dot-dashed) lines, cross-feed as triple dot-dashed lines, the combinatorial background as long-dashed lines, and partially reconstructed background components with either a missing neutral pion as dot-dashed lines or a missing soft photon as dotted lines. From [8].

background from misidentified particles. Possible contributions from $b$-hadron decays with intermediate $\Lambda_c^+$, $\Xi_c^+$ or $D^0$ particles are removed by excluding the corresponding regions in phase space. In some of the analyses, these charm decays are retained separately as control channels for the normalisation of branching fraction measurements, while in others different modes with high yields and similar final states to the signal are used as control channels.

In order to extract the signal, control channel and background yields, fits are performed to the invariant masses of the selected candidates. The relative branching fractions with respect to a control channel are obtained as the product of the ratio of the yields and the ratio of the efficiencies for the two modes. The absolute branching fractions of the signal channels are then calculated by multiplying with the branching fractions of the control channel, taken from previous measurements or the PDG [7].

Long-lived particles such as $\Lambda$ and $K_S^0$ can travel considerable distances and sometimes decay outside the acceptance of the LHCb vertex detector (VELO). They are therefore reconstructed in two categories: the first, referred to as "long", involving $K_S^0/\Lambda$ particles that decay early enough for their decay products to be reconstructed in the VELO; and the second, referred to as "downstream", containing those that decay later, such that track segments cannot be formed in the VELO. The two data categories have different vertex, momentum and mass resolutions, and are therefore treated separately.

## 3. Observations of $\Lambda_b^0 \to \Lambda K^+\pi^-$ and $\Lambda_b^0 \to \Lambda K^+K^-$ decays and searches for other decays of $\Lambda_b^0$ and $\Xi_b^0$ to $\Lambda h^+h'^-$ final states

Few charmless decays of $b$-baryons have been studied to date. In this study [8], a search is performed for the charmless three-body decays of the $\Lambda_b^0$ and $\Xi_b^0$ baryons to the final states $\Lambda h^+h'^-$, with $h^{(\prime)} = \pi$ or $K$. The $\Lambda$ baryon is reconstructed in its decay $\Lambda \to p\pi^-$. The charm decay $\Lambda_b^0 \to (\Lambda\pi^+)_{\Lambda_c^+}\pi^-$ is used as a control mode.

A simultaneous fit is performed to all final states and the control mode. Figure 1 shows the invariant mass distributions of the $\Lambda K^+K^-$, $\Lambda K^{\pm}\pi^{\mp}$, and $\Lambda\pi^+\pi^-$ final states with the fit results overlaid. The decay modes $\Lambda_b^0 \to \Lambda K^+\pi^-$, and $\Lambda_b^0 \to \Lambda K^+K^-$ are observed for the first time with significances of $8.1\,\sigma$ and $15.8\,\sigma$, respectively, including the effects of systematic uncertainties on the yields. Evidence is seen for the $\Lambda_b^0 \to \Lambda\pi^+\pi^-$ mode with a total significance of $4.7\,\sigma$. The dominant systematic uncertainties are due to the fit model, efficiency estimation and the yield of the normalisation channel.

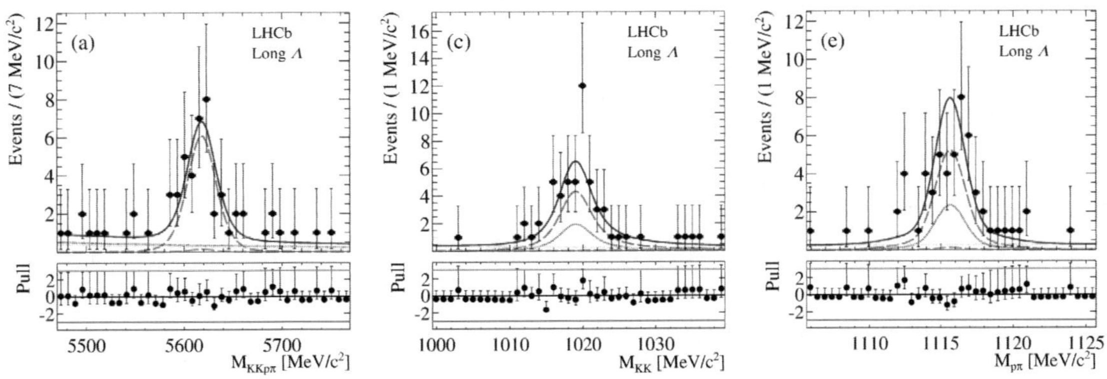


**Figure 2.** Fit projections to the (left) $p\pi^-K^+K^-$, (middle) $K^+K^-$and (right) $p\pi^-$invariant masses of the long dataset of $\Lambda_b^0 \to \Lambda\phi$ candidates. The total fit is given by the solid lines, the $\Lambda_b^0 \to \Lambda\phi$ signal by the long-dashed line peaks, the $\Lambda_b^0 \to \Lambda K^+K^-$ non-resonant component by the long-dashed line lower peaks, and the $\phi + \Lambda$ and pure combinatorial fit components by the dotted lines. From [10].

The absolute branching fractions are measured using the normalisation channel to be

$$\begin{aligned}
\mathcal{B}(\Lambda_b^0 \to \Lambda\pi^+\pi^-) &= (4.6 \pm 1.2\,(\text{stat}) \pm 1.4\,(\text{syst}) \pm 0.6\,(\text{norm})) \times 10^{-6}\,, \\
\mathcal{B}(\Lambda_b^0 \to \Lambda K^+\pi^-) &= (5.6 \pm 0.8\,(\text{stat}) \pm 0.8\,(\text{syst}) \pm 0.7\,(\text{norm})) \times 10^{-6}\,, \\
\mathcal{B}(\Lambda_b^0 \to \Lambda K^+K^-) &= (15.9 \pm 1.2\,(\text{stat}) \pm 1.2\,(\text{syst}) \pm 2.0\,(\text{norm})) \times 10^{-6}\,,
\end{aligned}$$

where the first quoted uncertainty is statistical, the second systematic, and the third is due to the normalisation mode branching fraction.

The measured yields of the corresponding $\Xi_b^0$ modes are compatible with zero. The results are translated into limits on the product of their branching fractions and the ratio of fragmentation fractions $f_{\Xi_b^0}$ and $f_{\Lambda_b^0}$ that give the probability for a $b$ quark to hadronise into either a $\Xi_b^0$ or $\Lambda_b^0$,

$$\begin{aligned}
f_{\Xi_b^0}/f_{\Lambda_b^0} \times \mathcal{B}(\Xi_b^0 \to \Lambda\pi^+\pi^-) &< 1.7\ (2.1) \times 10^{-6} \text{ at } 90\ (95)\,\% \text{ confidence level}\,, \\
f_{\Xi_b^0}/f_{\Lambda_b^0} \times \mathcal{B}(\Xi_b^0 \to \Lambda K^-\pi^+) &< 0.8\ (1.0) \times 10^{-6} \text{ at } 90\ (95)\,\% \text{ confidence level}\,, \\
f_{\Xi_b^0}/f_{\Lambda_b^0} \times \mathcal{B}(\Xi_b^0 \to \Lambda K^+K^-) &< 0.3\ (0.4) \times 10^{-6} \text{ at } 90\ (95)\,\% \text{ confidence level}\,.
\end{aligned}$$

The two $\Lambda_b^0$ modes with highest yields offer the possibility to search for *CP* violation, as has been seen in the corresponding three-body $B^\pm$ decays [9]. The raw asymmetries are obtained from efficiency-corrected yields of $\Lambda_b^0$ and $\overline{\Lambda}_b^0$ fitted separately. The *CP* asymmetries are then calculated by correcting for the production and detection asymmetries using the control mode which has negligible *CP* violation. The measured *CP* asymmetries are

$$\begin{aligned}
\mathcal{A}_{CP}(\Lambda_b^0 \to \Lambda K^+\pi^-) &= -0.53 \pm 0.23\,(\text{stat}) \pm 0.11\,(\text{syst})\,, \\
\mathcal{A}_{CP}(\Lambda_b^0 \to \Lambda K^+K^-) &= -0.28 \pm 0.10\,(\text{stat}) \pm 0.07\,(\text{syst})\,,
\end{aligned}$$

consistent with the conservation of *CP* symmetry.

## 4. Observation of the $\Lambda_b^0 \to \Lambda\phi$ decay

This analysis presents the search for the quasi-two-body $\Lambda_b^0 \to \Lambda\phi$ decay [10]. The $\phi$ mesons are reconstructed as $\phi \to K^+K^-$, and the $\Lambda$ baryons as $\Lambda \to p\pi^-$. The $B^0 \to K_S^0\phi$ decay mode is used as a normalisation channel, with $K_S^0 \to \pi^+\pi^-$.

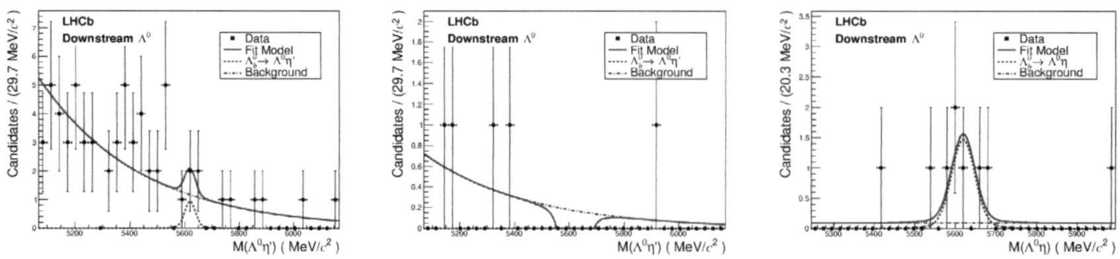


**Figure 3.** Invariant mass distributions of the (left) $\Lambda_b^0 \to \Lambda\eta'$ ($\eta' \to \pi^+\pi^-\gamma$), (middle) $\Lambda_b^0 \to \Lambda\eta'$ ($\eta' \to \pi^+\pi^-\eta$) and (right) $\Lambda_b^0 \to \Lambda\eta$ ($\eta \to \pi^+\pi^-\pi^0$) downstream candidates. The solid lines show the total fit, the dashed lines the signals, and the dash-dotted lines the background. From [11].

A three-dimensional fit to the $p\pi^-K^+K^-$ ($\pi^+\pi^-K^+K^-$), $p\pi^-$ ($\pi^+\pi^-$) and $K^+K^-$ invariant masses is performed, as shown in Figure 2 for the signal mode. The resulting signal yield of $89 \pm 13$ events represents the first observation of $\Lambda_b^0 \to \Lambda\phi$ decay, with a total significance of $5.9\sigma$. The branching fraction is measured to be

$$\mathcal{B}(\Lambda_b^0 \to \Lambda\phi) = \left(5.18 \pm 1.04\,(\text{stat}) \pm 0.35\,(\text{syst})\,^{+0.50}_{-0.43}\,(\text{norm}) \pm 0.44\,(f_d/f_{\Lambda_b^0})\right) \times 10^{-6}.$$

Here the third and fourth uncertainties are due to the normalisation mode and the ratio of fragmentation fractions, respectively. The systematic uncertainty is dominated by the efficiency estimation and the fit model.

The $\Lambda_b^0 \to \Lambda\phi$ decay is a spin-$\frac{1}{2}$ to spin-$\frac{1}{2}$ plus vector meson transition, and five angles are needed to describe the decay. $CP$ violation in the decay can be probed without the need for a control channel, through $T$-odd triple products. The angles $\Phi_\Lambda$ ($\Phi_\phi$) are defined as the azimuthal angles of the final-state $K^+$ (proton) in the rest frame of the $\phi$ ($\Lambda$). The sines and cosines of these angles are odd under time reversal, and their corresponding asymmetries $\mathcal{A}^{c,s}_{\Lambda,\phi}$ are measured from fits to the subsets of the data where the observables are positive and negative. The $T$-odd triple-product asymmetries are obtained to be

$$\begin{aligned} \mathcal{A}^c_\Lambda &= -0.22 \pm 0.12\,(\text{stat}) \pm 0.06\,(\text{syst}), & \mathcal{A}^s_\Lambda &= \phantom{-}0.13 \pm 0.12\,(\text{stat}) \pm 0.05\,(\text{syst}), \\ \mathcal{A}^c_\phi &= -0.01 \pm 0.12\,(\text{stat}) \pm 0.03\,(\text{syst}), & \mathcal{A}^s_\phi &= -0.07 \pm 0.12\,(\text{stat}) \pm 0.01\,(\text{syst}), \end{aligned}$$

and are consistent with zero.

## 5. Search for $\Lambda_b^0 \to \Lambda\eta'$ and $\Lambda_b^0 \to \Lambda\eta$ decays

The search for the as yet unobserved $\Lambda_b^0 \to \Lambda\eta'$ and $\Lambda_b^0 \to \Lambda\eta$ decays is presented [11]. Decays to states with $\eta$ and $\eta'$ are sensitive to the mixing angle between light and strange quarks [12]. In addition to the dominant $b \to s$ loop diagram for both $\Lambda_b^0 \to \Lambda\eta'$ and $\Lambda_b^0 \to \Lambda\eta$ decays, $\Lambda_b^0 \to \Lambda\eta'$ decays have further contributions from non-spectator and anomalous diagrams, which can alter its branching fraction due to gluon mixing. Thus, branching fraction measurements are important to constrain the $\eta - \eta'$ mixing angles.

The $B^0 \to K^0_S\eta'$ mode is used as a normalisation channel for the measurement of the branching fractions. The decays are reconstructed as $\Lambda \to p\pi^-$, $K^0_S \to \pi^+\pi^-$, $\eta'$ candidates as $\eta' \to \pi^+\pi^-\gamma$ and $\eta' \to \pi^+\pi^-\eta$ (with $\eta \to \gamma\gamma$), and $\eta$ candidates as $\eta \to \pi^+\pi^-\pi^0$ (with $\pi^0 \to \gamma\gamma$). The signal invariant mass distributions are shown in Figure 3. No significant signal is observed for $\Lambda_b^0 \to \Lambda\eta'$ decays, while the $\Lambda_b^0 \to \Lambda\eta$ yield is $5.3 \pm 3.8$ events which represents evidence with a significance of $3.0\sigma$.

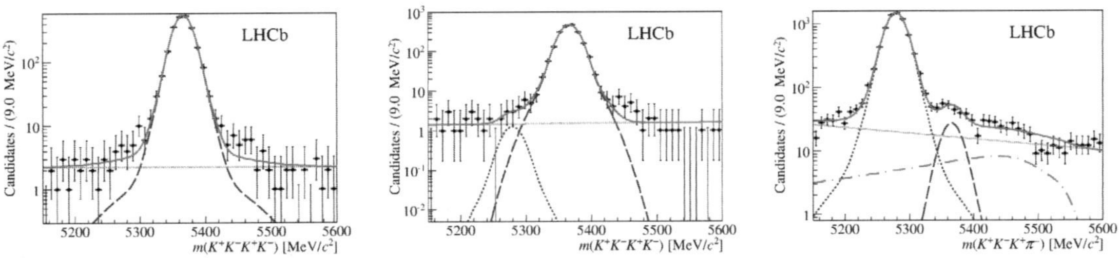


**Figure 4.** Distribution of (left) $K^+K^-K^+K^-$ mass, (middle) $K^+K^-K^+K^-$ mass with a tighter selection applied, and (right) $K^+K^-K^+\pi^-$ mass of $B^0_{(s)} \to \phi\phi$ candidates. The total fitted function is shown by the solid lines, the $B^0_s \to \phi\phi$ component by the long-dashed lines, $B^0 \to \phi\phi$ by the short-dashed lines, and the combinatorial background by the dotted lines. In the right plot, the short-dashed line shows the $B^0 \to \phi K^*$, the long-dashed line $B^0_s \to \phi\overline{K}^*(892)^0$, and the dash-dotted line $\Lambda^0_b \to \phi p\pi^-$ and $\Lambda^0_b \to \phi pK^-$ contributions. From [13].

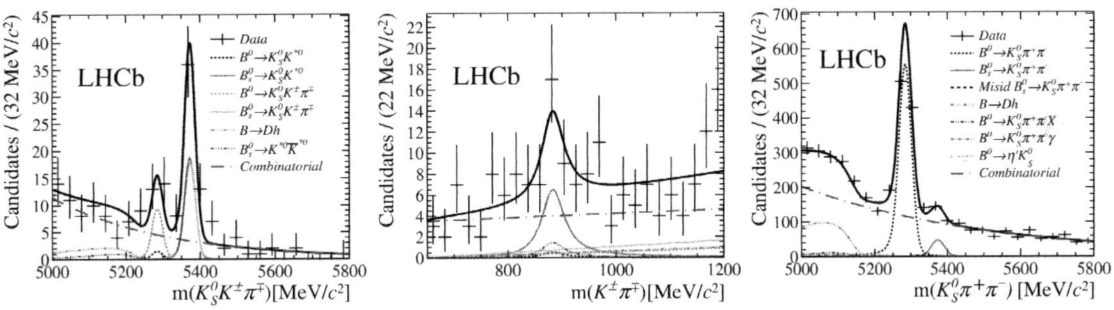


**Figure 5.** Distributions of (left) $K^0_SK^\pm\pi^\mp$ mass, (middle) $K^\pm\pi^\mp$ mass and (right) $K^0_S\pi^+\pi^-$ mass for downstream $B^0_{(s)} \to K^0_SK^*(892)^0$ candidates, with the overall fit represented by the solid black lines, the $B^0$ ($B^0_s$) signal components by the black short-dashed (dotted) lines, the combinatorial background by the long-dash-dotted lines, and the rest of the fit components as shown in the legends. From [15].

A branching fraction measurement is performed using the normalisation channel, where a limit is placed for $\Lambda^0_b \to \Lambda\eta'$ decays and 68% CL intervals are calculated for $\Lambda^0_b \to \Lambda\eta$ decays,

$$\mathcal{B}(\Lambda^0_b \to \Lambda\eta') < 3.1 \times 10^{-6} \text{ at } 90\% \text{ CL}, \qquad \mathcal{B}(\Lambda^0_b \to \Lambda\eta) = (9.3^{+7.3}_{-5.3}) \times 10^{-6}.$$

The results suggest the possible presence of gluonic mixing of $\eta'$ mesons.

## 6. Measurement of the $B^0_s \to \phi\phi$ branching fraction and search for $B^0 \to \phi\phi$ decay

In this study [13], a branching fraction measurement of $B^0_s \to \phi\phi$ decay and a search for the suppressed $B^0 \to \phi\phi$ decay are performed, using $B^0 \to \phi K^*(892)^0$ as a control channel.

The invariant mass fits are shown in Figure 4, where the selection used in the $B^0 \to \phi\phi$ search is optimised for sensitivity to that decay, but its yield is negligible. The upper limit on the branching ratio is obtained as $\mathcal{B}(B^0 \to \phi\phi) < 2.8\,(3.4) \times 10^{-8}$ at 90 (95)% CL, and provides a constraint on possible new physics contributions [14].

The yields of the $B^0_s \to \phi\phi$ and control modes are corrected for the S-wave fractions of candidates, whose uncertainty is the largest systematic contribution. The branching fraction is

$$\mathcal{B}(B^0_s \to \phi\phi) = (1.84 \pm 0.05\,(\text{stat}) \pm 0.07\,(\text{syst}) \pm 0.11\,(f_s/f_d) \pm 0.12\,(\text{norm})) \times 10^{-5},$$

representing the most precise measurement of this quantity to date.

## 7. Observation of $B_s^0 \to K_S^0 K^*(892)^0$ decay

The branching fraction measurement of $B_s^0 \to K_S^0 K^*(892)^0$ and search for $B^0 \to K_S^0 K^*(892)^0$ decays [15] is performed using 2011 data and the $B^0 \to K_S^0 \pi^+\pi^-$ mode for normalisation. The LHCb experiment has already measured inclusive $B_{(s)}^0 \to K_S^0 h^{\pm} h'^{\mp}$ decays, including first observation of the decay $B_s^0 \to K_S^0 K^{\pm}\pi^{\mp}$ [16]. In this analysis the resonant structure in the $K^+\pi^-$ invariant mass region around $1\,\mathrm{GeV}/c^2$ is analysed to determine the number of decays that proceed through an intermediate $K^{*0}$ resonance.

A two-dimensional fit is performed to the $K_S^0 K^+\pi^-$ and $K^+\pi^-$ invariant masses in order to select the $K^*(892)^0$ resonant signal mode, as shown in Figure 5. The $B_s^0 \to K_S^0 K^*(892)^0$ decay is observed for the first time, with a total significance of 7.1 standard deviations. For the $B^0$ decay, the yields are negligible and an upper limit is determined. The results for the branching fractions are obtained in terms of the sum of final states with either $K^0\overline{K}^{*0}$ or $\overline{K}^0 K^{*0}$,

$$\begin{aligned}
\mathcal{B}(B_s^0 \to \bar{K}^0 K^*(892)^0) + \mathcal{B}(B_s^0 \to K^0 \bar{K}^*(892)^0) &= (16.4 \pm 3.4 \pm 1.9 \pm 1.0 \pm 0.7) \times 10^{-6}\,, \\
\mathcal{B}(B^0 \to \bar{K}^0 K^*(892)^0) + \mathcal{B}(B^0 \to K^0 \bar{K}^*(892)^0) &< 0.96\ (1.04) \times 10^{-6} \text{ at } 90\%\ (95\%)\ \mathrm{CL}\,,
\end{aligned}$$

where the first uncertainty is statistical, the second systematic, the third due to the ratio of the fragmentation fractions and the fourth due to the branching fraction of the normalisation mode.

## 8. Summary

Several recent searches for $b$-meson and $b$-baryon decays with the LHCb detector were presented, including the first observations of $\Lambda_b^0$ decays to $\Lambda K^+\pi^-$, $\Lambda K^+K^-$ and $\Lambda\phi$ states, and of $B_s^0 \to K_S^0 K^*(892)^0$ decays. Branching ratio measurements were performed for these modes, and limits were set for the unobserved modes. The *CP* asymmetries and $T$-odd triple-product asymmetries of some modes were measured to be consistent with zero. With the larger datasets being collected by LHCb during its Run II, the branching fraction and *CP* violation measurements will be improved statistically, and new charmless $b$-hadron decay modes will become accessible.

## Acknowledgments

Financial support from CNPq and George Mason University is gratefully acknowledged.

## References

[1] Aaij R *et al.* (LHCb collaboration) 2014 *Phys. Rev.* **D90** 052011 (*Preprint* 1407.2222)
[2] Aaij R *et al.* (LHCb collaboration) 2013 *Phys. Rev. Lett.* **110** 241802 (*Preprint* 1303.7125)
[3] Abe K *et al.* (Belle) 2003 *Phys. Rev. Lett.* **91** 261602 (*Preprint* hep-ex/0308035)
[4] Aaij R *et al.* (LHCb collaboration) 2014 *JHEP* **05** 069 (*Preprint* 1403.2888)
[5] Alves Jr A A *et al.* (LHCb) 2008 *JINST* **3** S08005
[6] Aaij R *et al.* (LHCb collaboration) 2015 *Int. J. Mod. Phys.* **A30** 1530022 (*Preprint* 1412.6352)
[7] Olive K A *et al.* (Particle Data Group) 2014 *Chin. Phys.* **C38** 090001
[8] Aaij R *et al.* (LHCb collaboration) 2016 *JHEP* **05** 081 (*Preprint* 1603.00413)
[9] Aaij R *et al.* (LHCb collaboration) 2014 *Phys. Rev.* **D90** 112004 (*Preprint* 1408.5373)
[10] Aaij R *et al.* (LHCb collaboration) 2016 *Phys. Lett.* **B759** 282 (*Preprint* 1603.02870)
[11] Aaij R *et al.* (LHCb collaboration) 2015 *JHEP* **09** 006 (*Preprint* 1505.03295)
[12] Di Donato C, Ricciardi G and Bigi I 2012 *Phys. Rev.* **D85** 013016 (*Preprint* 1105.3557)
[13] Aaij R *et al.* (LHCb collaboration) 2015 *JHEP* **10** 053 (*Preprint* 1508.00788)
[14] Bar-Shalom S, Eilam G and Yang Y D 2003 *Phys. Rev.* **D67** 014007 (*Preprint* hep-ph/0201244)
[15] Aaij R *et al.* (LHCb collaboration) 2016 *JHEP* **01** 012 (*Preprint* 1506.08634)
[16] Aaij R *et al.* (LHCb collaboration) 2013 *JHEP* **10** 143 (*Preprint* 1307.7648)

Journal of Physics: Conference Series **770** (2016) 012017 doi:10.1088/1742-6596/770/1/012017

# Radiative B decays at LHCb

**Violaine Bellée**

École Polytechnique Fédérale de Lausanne, Switzerland

E-mail: `violaine.bellee@epfl.ch`

**Abstract.** Radiative B decays are sensitive probes for new physics. In particular, the measurement parameters sensitive to the photon polarisation could help constrain various models of new physics. We present three methods used by the LHCb collaboration to measure the photon polarisation using $3\,\mathrm{fb}^{-1}$ of integrated luminosity.

## 1. Introduction

Radiative decays of B mesons are flavour-changing neutral current processes which are forbidden at tree level in the Standard Model (SM). These decays predominantly occur via $b \to s\gamma$ loop diagrams and are therefore sensitive to new physics (NP) effects.

Numerous studies of radiative decays have been performed at LHCb since the beginning of the data taking in 2010, two of which provided the first constraints on the photon polarisation parameter. The observation of a non-zero photon polarisation in $B^+ \to K^+\pi^-\pi^+\gamma$ decays[1] at LHCb is presented here [1], along with the measurement of angular observables related to the photon polarisation in $B^0 \to K^{0*}e^+e^-$ decays at low dielectron invariant masses, due to the contribution of the virtual photon [2]. The decay time analysis of $B_0^s \to \phi\gamma$ decays, which also constrains this parameter, will be introduced [3].

## 2. Photon polarisation in radiative B decays

Photons produced in $b \to s\gamma$ transitions are expected to be predominantly left handed in the SM. However, some NP contributions may introduce a significant right handed component to the photon polarisation. Although the measured inclusive $b \to s\gamma$ rate [4] has already put strong constraints on NP scenarios, there is still room for observation of non-SM effects in the photon polarisation.

The decay amplitude for the $b \to s\gamma$ transition is proportional to [5]:

$$< f|H_{\text{eff}}|i >= -4\frac{G_F}{\sqrt{2}}V_{tb}V_{ts}^*[\mathcal{C}_7^{\text{eff}}(m_b) < f|O_7(m_b)|i > + \mathcal{C}_7^{\prime\text{eff}}(m_b) < f|O_7'(m_b)|i >] \quad (1)$$

where $\mathcal{C}_7^{\text{eff}}$ and $\mathcal{C}_7^{\prime\text{eff}}$ are the Wilson coefficients coupled to the left and right handed photon respectively.

In the SM, those coefficients are such that:

[1] The charge conjugation is implied throughout the text.

$$\frac{C_7^{\prime\text{eff}}}{C_7^{\text{eff}}} \simeq \frac{m_s}{m_b} \simeq 0.02 \tag{2}$$

The photon polarisation parameter $\lambda_\gamma$ is defined as the asymmetry between the right and left components:

$$\lambda_\gamma = \frac{|C_7^{\prime\text{eff}}|^2 - |C_7^{\text{eff}}|^2}{|C_7^{\prime\text{eff}}|^2 + |C_7^{\text{eff}}|^2} \tag{3}$$

In the SM, the value of this parameter is -1 up to corrections of the order of $(m_s^2/m_b^2)$ for decays of a $b$ quark.

## 3. Studies of Radiative B decays at LHCb

The measurement of radiative B decays at LHCb [6] is allowed by the good quality of the tracking systems with a momentum resolution $\delta$p/p of 0.5-1% in the full momentum range, a time resolution of 45 fs and a resolution on the impact parameter of 20 $\mu$m for high $p_T$ tracks. The Ring Imaging CHerenkov detector (RICH) provides a good particle identification, allowing the separation of kaons and protons from pions. Finally, the calorimeter system, which is essential to the detection of radiative decays, has a resolution of $\sigma_E/E \simeq 10\%/\sqrt{E} + 1\%$.

Studies of radiative decays at LHCb face three main challenges: the use of $pp$ collisions which generate high levels of background, the misidentification of $\pi^0$s due to unresolved photons at transverse energies above 4 GeV, and the mass resolution dominated by the photon reconstruction.

Despite these difficulties, LHCb has managed to perform high precision measurements of radiative B decays. Using the 0.37 fb$^{-1}$ of LHCb data obtained in the first half of 2011, the ratio of branching ratios of $B_0 \to K^*\gamma$ over $B_0^s \to \phi\gamma$ decays has been measured to be $1.12 \pm 0.08(stat)^{0.06}_{0.04}(syst)^{0.09}_{0.08}(f_s/f_d)$ [7], which is in agreement with the theory prediction of $1.0 \pm 0.2$ [8]. In particular, searches for $B^0 \to J/\psi\gamma$ and $B_s^0 \to J/\psi\gamma$ have been conducted on the 3 fb$^{-1}$ of data collected by LHCb in 2011 and 2012 using converted photons to suppress background from decays like $B^0 \to J/\psi\pi^0$ [9]. For the $B^0 \to J/\psi\gamma$ mode, branching ratios higher than $1.5 \times 10^{-6}$ are excluded at 90% confidence level, which is in agreement with the BABAR limit [10]. Branching ratio values over $7.3 \times 10^{-6}$ have been excluded at 90% confidence level for $B_s^0 \to J/\psi\gamma$ decays, which is the first limit ever set on this decay.

## 4. Measuring the photon polarisation

Several methods have been proposed to measure the photon polarisation in $b \to s\gamma$ transitions. The four most promising methods respectively use:

- ◇ Angular distribution of radiative decays with three charged particles in the final state (e.g. $B^+ \to K^+\pi^-\pi^+\gamma$)
- ◇ The transverse asymmetry in $B^0 \to K^{0*}e^+e^-$ decays in the low dielectron invariant mass region
- ◇ Time-dependent analyses of $B^0$ or $B_s^0$ decaying into a photon and a $CP$-eigenstate (e.g. $B^0 \to K_S\pi^0\gamma$, $B_s^0 \to \phi\gamma$)
- ◇ Angular distributions in $\Lambda_b \to \Lambda\gamma$ and other $b$ baryon decays.

The first two methods have been exploited at LHCb and their results are presented below. The third one is ongoing.

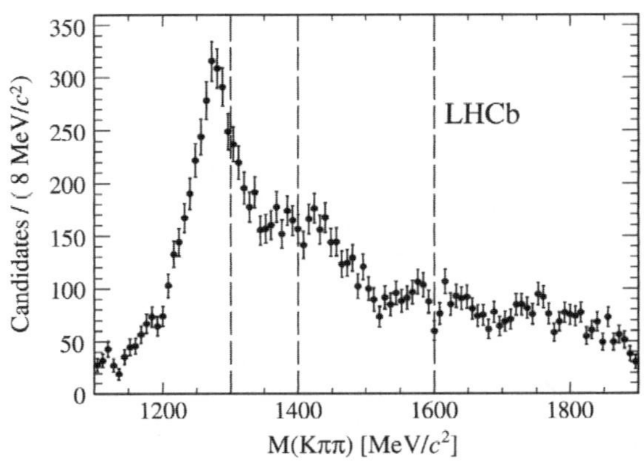


Figure 1: Background-subtracted $K\pi\pi$ mass distribution of the $B^+ \to K^+\pi^-\pi^+\gamma$ signal. The four intervals of interest are shown separated by dashed lines [1].

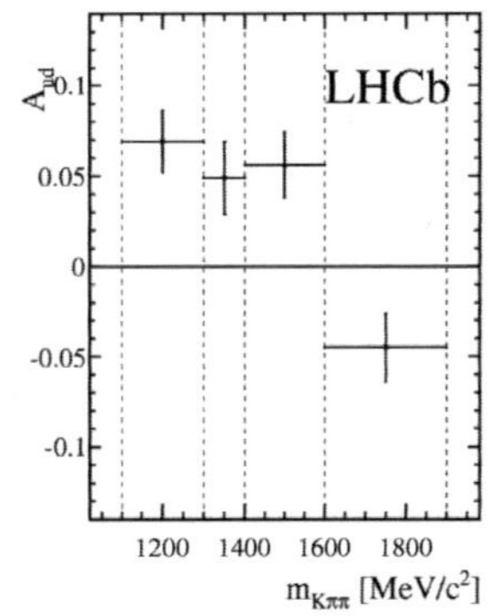


Figure 2: Values obtained for $\mathcal{A}_{\rm ud}$ from fits to the normalized background substracted $\cos\widehat{\theta}$ distribution in each bin of $K\pi\pi$ mass [1].

*4.1. Observation of a non-zero photon polarisation in $B^+ \to K^+\pi^-\pi^+\gamma$ decays*

The $B^+ \to K^+\pi^-\pi^+\gamma$ decay is expected to proceed via different $K^*$ resonances so the decay can be written as $B \to K_{\rm res}(\to P_1P_2P_3)\gamma$ where $P_1$, $P_2$ and $P_3$ refer to the final state hadrons. Using the final-state momenta $\overrightarrow{{\rm p}_i}$ (i = 1, 2, 3) and ${\rm p}_\gamma$ in the $K_{\rm res}$ rest frame, $\cos\theta$ is defined as [11]:

$$\cos\theta = -\frac{\overrightarrow{{\rm p}_\gamma}}{|\overrightarrow{{\rm p}_\gamma}|} \cdot \widehat{n}, \text{ with } \widehat{n} \equiv \frac{\overrightarrow{{\rm p}_1} \times \overrightarrow{{\rm p}_2}}{|\overrightarrow{{\rm p}_1} \times \overrightarrow{{\rm p}_2}|} \tag{4}$$

As $\cos\theta$ changes sign under the exchange of ${\rm s}_{13}$ and ${\rm s}_{23}$, a new variable $\cos\tilde{\theta} = {\rm sign}({\rm s}_{12}-{\rm s}_{23})\cos\theta$ is introduced. Then the up-down asymmetry is defined as:

$$\mathcal{A}_{\rm ud} = \frac{\int_0^1 {\rm d}\cos\tilde{\theta} \times \frac{{\rm d}\Gamma}{{\rm d}\cos\tilde{\theta}} - \int_{-1}^0 {\rm d}\cos\tilde{\theta} \times \frac{{\rm d}\Gamma}{{\rm d}\cos\tilde{\theta}}}{\int_{-1}^1 {\rm d}\cos\tilde{\theta} \times \frac{{\rm d}\Gamma}{{\rm d}\cos\tilde{\theta}}} = C \times \lambda_\gamma \tag{5}$$

where the proportionality coefficient $C$ depends on the content of the $K\pi\pi$ system. Due to missing knowledge of the $K\pi\pi$ system and lack of theoretical predictions, this proportionality coefficient can not be determined yet.

This analysis used a data sample of $pp$ collisions recorded by LHCb in 2011 and 2012, corresponding to a total luminosity of 3 fb$^{-1}$ at $\sqrt{s} = 7$ and 8 TeV [1]. The signal is reconstructed by selecting events with a high ${\rm p}_{\rm t}$ photon and three charged tracks, one of which is associated to a charged kaon and the other two to oppositely-charged pions. High transverse-energy photons are reconstructed from energy deposits in the calorimeter. The signal events are separated from the background from $B$ decays with $\pi^0$s using the shape of the electromagnetic shower in the calorimeter.

A fit to the mass distribution of $B^+ \to K^+\pi^-\pi^+\gamma$ candidates is performed and yields 13876 $\pm$ 153 events. The background-subtracted $K\pi\pi$ spectrum is obtained and shown on Fig. 1. The peak of the $K_1(1270)$ is visible and the rest of the spectrum is populated by many $K$ resonances of different masses and spins. The data set is divided in 4 bins following recommendations from theory [11]. In each of those four bins, the background subtracted angular distribution is fitted with a fourth-order Legendre polynomial normalized to unit area:

$$\mathrm{f}(\cos\widehat{\theta};\, c_0 = 0.5,\, c_1,\, c_2,\, c_3,\, c_4) = \sum_{i=0}^{4} c_i \mathrm{L}_i(\cos\widehat{\theta}) \tag{6}$$

where $\mathrm{L}_i(\mathrm{x})$ is the Legendre Polynomial of order $i$ and $\cos\widehat{\theta}$ is equal to $\cos\tilde{\theta}$ corrected for the charge of the $B$ meson. The up-down asymmetry can be expressed in terms of the $c_i$ coefficients by:

$$\mathcal{A}_{\mathrm{ud}} = c_1 - \frac{c_3}{4} \tag{7}$$

The results obtained from a $\chi^2$ fit of the normalized binned angular distribution are shown in Fig. 2. The combined observed values translate into a 5.2 $\sigma$ significance for non-zero up-down asymmetry. As the up-down asymmetry is proportional to the photon polarisation, this study also lead to the observation of a non-zero photon polarisation with a significance of 5.2 $\sigma$. However, as the proportionality coefficient between these observables is unknown, a partial wave analysis is needed and in progress at LHCb to disentangle the contributions from the different resonances.

### *4.2. $B^0 \to K^{0*}e^+e^-$ at low dielectron invariant mass*

$B^0 \to K^{0*}e^+e^-$ decays occur through penguin diagrams or electroweak box diagrams. The contribution of each of these diagrams varies with the dilepton mass. In the region of low dilepton invariant mass squared ($q^2$) below 6 GeV$^2/c^4$, couplings from a virtual photon to the dilepton pair dominate, allowing for sensitivity to the helicity of the photon in $b \to s\gamma$ transitions.

The partial decay rate of the $B^0 \to K^{0*}e^+e^-$ decays can be described with four variables [12]: $q^2$ and three angular variables, $\theta_l$, $\theta_K$ and $\phi$. $\theta_l$ is defined as the angle between the electron and the opposite direction of the $B$ in the dilepton rest frame. $\theta_K$ is the angle of the $K$ meson with respect to the opposite direction of the $B$ in the rest frame of the $K\pi$ system. Finally, $\phi$ is the angle between the plane containing the $e^+$ and the $e^-$ and the plane containing the kaon and the pion, in the rest frame of the $B$ meson. The $\phi$ angle is redefined as $\tilde{\phi} = \phi + \pi$ if $\phi < 0$, which cancels out terms proportional to $\cos\phi$ or $\sin\phi$. Neglecting the $K\pi$ S-wave contribution, the angular distributions for $B^0 \to K^{0*}e^+e^-$ decays in the limit of massless leptons reads:

$$\begin{aligned} \frac{1}{d(\Gamma+\overline{\Gamma})/\mathrm{d}q^2}\frac{\mathrm{d}^4(\Gamma+\overline{\Gamma})}{\mathrm{d}q^2\mathrm{dcos}\theta_l\mathrm{dcos}\theta_K\mathrm{d}\tilde{\phi}} &= \frac{9}{16\pi}[\frac{3}{4}(1-\mathrm{F}_L)\sin2\theta_K + \mathrm{F}_L\cos2\theta_K + \\ (\frac{1}{4}(1-\mathrm{F}_L)\sin^2\theta_K - \mathrm{F}_L\cos^2\theta_K)\cos2\theta_l &+ \frac{1}{2}(1-\mathrm{F}_L)A_{\mathrm{T}}^{(2)}\sin^2\theta_K\sin^2\theta_l\cos2\tilde{\phi} + \\ (1-\mathrm{F}_L)A_{\mathrm{T}}^{\mathrm{Re}}\sin^2\theta_K\cos\theta_l &+ \frac{1}{2}(1-\mathrm{F}_L)A_{\mathrm{T}}^{\mathrm{Im}}\sin^2\theta_K\sin^2\theta_l\sin2\tilde{\phi} \end{aligned} \tag{8}$$

where $A_{\mathrm{T}}^{\mathrm{Im}}$ is a $CP$ asymmetry, $\mathrm{F_L}$ the longitudinal polarisation of the $K^{0*}$ (which is expected to be small at low $q^2$), and $A_{\mathrm{T}}^{\mathrm{Re}}$ is related to the forward-backward asymmetry. In the limit $q^2 \to 0$, the variables $A_{\mathrm{T}}^{(2)}$ and $A_{\mathrm{T}}^{\mathrm{Im}}$ are sensitive to the photon polarisation. They depend on the Wilson coefficients $C_7$ and $C_7'$ as:

$$A_{\mathrm{T}}^{(2)}(q^2 \to 0) \to \frac{\mathcal{R}e(C_7C_7')}{|C_7|^2+|C_7'|^2} \text{ and } A_{\mathrm{T}}^{\mathrm{Im}}(q^2 \to 0) \to \frac{\mathcal{I}m(C_7C_7')}{|C_7|^2+|C_7'|^2}. \tag{9}$$

Using the 3 fb$^{-1}$ of data collected by LHCb in 2011 and 2012 at 7 and 8 TeV respectively, the $B^0 \to K^{0*}e^+e^-$ signal candidates are reconstructed from candidate dilepton pairs and K* mesons [2]. Oppositely charged tracks with high $\mathrm{p_T}$ and a good quality vertex are used to select signal candidate dilepton pairs. Bremsstralung radiation is accounted for by adding to

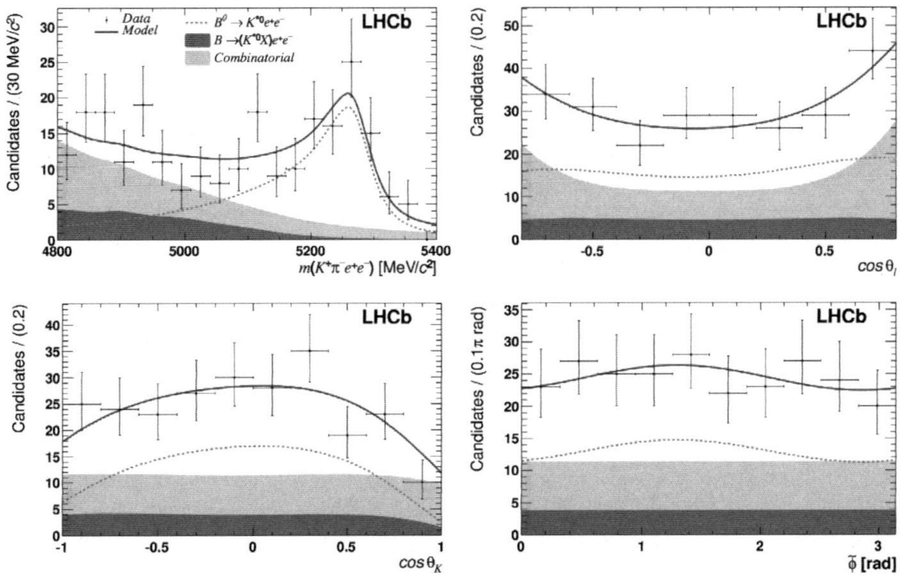


Figure 3: Distributions of the $K^+\pi^-e^+e^-$ invariant mass, cos$\theta_l$, cos$\theta_K$ and $\tilde{\phi}$ variables for the $B^0 \to K^{0*}e^+e^-$ decay mode. The dashed line is the signal PDF, the light grey area corresponds to the combinatorial background, the dark grey area is the partially reconstructed background. The solid line is the total PDF [2].

the measured electron momentum the contributions from neutral calorimeter clusters that are close to the extrapolation of the electron track. The reconstructed $e^+e^-$ pair is required to have an invariant mass squared in the range [0.0004, 1]GeV/$c^2$. The choice of the lower bound is a compromise between the gain in sensitivity from measurements at lowest possible $q^2$ values and the decrease in resolution due to multiple scattering when $q^2$ decreases.

First, a wide mass window of $K^+\pi^-e^+e^-$ invariant mass between 4300 and 6300 GeV/$c^2$ is used to obtain the fitted value of the $B^0 \to K^{0*}e^+e^-$ yields. An unbinned maximum likelihood fit is performed. The fit yields 150 ± 17 $B^0 \to K^{0*}e^+e^-$ signal events.

In order to reduce the contribution from combinatorial background, whose angular distribution is difficult to model, the angular fit is performed on a reduced mass window of 4800 – 5400 MeV/$c^2$, which contains 124 $B^0 \to K^{0*}e^+e^-$ signal events.

The measurement of $A_{\mathrm{T}}^{\mathrm{Im}}$, $A_{\mathrm{T}}^{\mathrm{Re}}$, $A_{\mathrm{T}}^{(2)}$ and $\mathrm{F_L}$ is performed by fitting simultaneously the m($K^+\pi^-e^+e^-$), cos$\theta_l$, cos$\theta_K$ and $\tilde{\phi}$ distributions shown in Fig. 3. The fitted values for these parameters are corrected for the (3.8 ± 1.9)% contamination from $B^0 \to K^{0*}\gamma_{e^+e^-}$ decays assuming the physics parameter values are zero for these decays, giving the following results:

$$\begin{aligned} \mathrm{F_L} &= 0.16 \pm 0.06 \pm 0.03 \quad & A_{\mathrm{T}}^{(2)} &= -0.23 \pm 0.23 \pm 0.05 \\ A_{\mathrm{T}}^{\mathrm{Im}} &= 0.14 \pm 0.22 \pm 0.05 \quad & A_{\mathrm{T}}^{\mathrm{Re}} &= 0.10 \pm 0.18 \pm 0.05 \end{aligned} \tag{10}$$

Taking into account the bin migration due to reconstruction effects, the true $q^2$ effective range for the reported results on $A_{\mathrm{T}}^{\mathrm{Im}}$, $A_{\mathrm{T}}^{\mathrm{Re}}$, $A_{\mathrm{T}}^{(2)}$ and $\mathrm{F_L}$ ranges from 0.0020 ± 0.0008 to 1.120 ± 0.060 GeV$^2$/$c^4$.

For this range of $q^2$ values, the equations relating $A_{\mathrm{T}}^{(2)}$ and $A_{\mathrm{T}}^{\mathrm{Im}}$ to $\mathcal{C}_7$ and $\mathcal{C}_7'$ are accurate at the 5% level, leading to a $\mathcal{C}_7'/\mathcal{C}_7$ ratio compatible with zero, in agreement with the SM. The value of $\mathrm{F_L}$ is also consistent with the SM prediction [13].

### *4.3. Photon polarisation in $B_s^0 \to \phi\gamma$ decays*

The time-dependent decay rates of $B_s^0$ and $\overline{B_s^0}$ decaying to $\phi\gamma$ can be written [14]:

$$\begin{aligned}\Gamma(t)(B_s^0 \to \phi\gamma) &\propto \exp^{-\Gamma_s t}[\cosh(\frac{\Delta\Gamma_s}{2}t) - A^{\Delta}\sinh(\frac{\Delta\Gamma_s}{2}t) + C\cos\Delta m_s t - S\sin\Delta m_s t] \\ \Gamma(t)(\overline{B_s^0} \to \phi\gamma) &\propto \exp^{-\Gamma_s t}[\cosh(\frac{\Delta\Gamma_s}{2}t) - A^{\Delta}\sinh(\frac{\Delta\Gamma_s}{2}t) - C\cos\Delta m_s t + S\sin\Delta m_s t]\end{aligned} \tag{11}$$

where $C$ is the direct CP asymmetry, $S$ is the asymmetry associated with $B_0^s - \overline{B_0^s}$ mixing, $\Delta\Gamma_s$ and $\Delta m_s$ are the decay width and mass differences between the $B_0^s$ $CP$ eigenstates.

Hence, if an untagged analysis is performed, *i.e.* if the $B_0^s$ and $\overline{B_0^s}$ states can not be disentangled, the total decay rate reads:

$$\Gamma(t) \propto \exp^{-\Gamma_s t}[\cosh(\frac{\Delta\Gamma_s}{2}t) - A^{\Delta}\sinh(\frac{\Delta\Gamma_s}{2}t)] \tag{12}$$

where $A^{\Delta}$ depends on the Wilson coefficients $\mathcal{C}_7$ and $\mathcal{C}_7'$ as:

$$A^{\Delta} \sim \frac{2\mathcal{R}e(e^{-i\phi_s}\mathcal{C}_7\mathcal{C}_7')}{(|\mathcal{C}_7|^2 + |\mathcal{C}_7'|^2)} \tag{13}$$

Therefore the decay-time analysis of $B_0^s \to \phi\gamma$ is sensitive to the photon polarisation. This analysis has been conducted in LHCb using $3\,\mathrm{fb}^{-1}$ of 2011 and 2012 data and the results have been made public after the conference.

## 5. Conclusion

Despite a very challenging environment, the LHCb Collaboration has performed competitive measurements with Run 1 data. A first limit has been set on the branching ratio of $B_s^0 \to J/\psi\gamma$ decays and a competitive limit has been found for $B^0 \to J/\psi\gamma$ decays. The first observation of a non-zero photon polarisation in $B^+ \to K^+\pi^-\pi^+\gamma$ decays has been obtained using the full $3\,\mathrm{fb}^{-1}$ of Run 1. The first measurement of angular observables sensitive to the photon polarisation in $B^0 \to K^{0*}e^+e^-$ decays (in an effective $q^2$ range [0.0020, 1.120] $\mathrm{GeV}^2/c^4$) has been performed. The results of the decay-time analysis of $B_0^s \to \phi\gamma$ have just been published.

From 2015 to 2018, LHCb Run 2 will significantly increase statistics with an integrated luminosity of the order of $5\,\mathrm{fb}^{-1}$, along with an improved trigger implementation, offering the possibility of an improvement in statistically limited analyses, as well as new searches for baryonic radiative decays such as $\Lambda_b \to \Lambda\gamma$. New methods are also being developed to enable a first measurement of the photon polarisation using a partial wave analysis of $B^+ \to K^+\pi^-\pi^+\gamma$ decays.

## References

[1] Aaij R *et al.* (LHCb) 2014 *Phys. Rev. Lett.* **112** 161801 (*Preprint* 1402.6852)
[2] Aaij R *et al.* (LHCb) 2015 *JHEP* **04** 064 (*Preprint* 1501.03038)
[3] Sanchez Mayordomo C 2016 Radiative *b*-hadron decays at lhcb paper presented at ICHEP, Chicago
[4] Beringer J *et al.* (Particle Data Group) 2012 *Phys. Rev.* **D86** 010001
[5] Becirevic D, Kou E, Le Yaouanc A and Tayduganov A 2012 *JHEP* **08** 090 (*Preprint* 1206.1502)
[6] Alves Jr A A *et al.* (LHCb) 2008 *JINST* **3** S08005
[7] Aaij R *et al.* (LHCb) 2012 *Phys. Rev.* **D85** 112013 (*Preprint* 1202.6267)
[8] Ali A, Pecjak B D and Greub C 2008 *Eur. Phys. J.* **C55** 577–595 (*Preprint* 0709.4422)
[9] Aaij R *et al.* (LHCb) 2015 *Phys. Rev.* **D92** 112002 (*Preprint* 1510.04866)
[10] Aubert B *et al.* (BaBar) 2004 *Phys. Rev.* **D70** 091104 (*Preprint* hep-ex/0408018)
[11] Gronau M and Pirjol D 2002 *Phys. Rev.* **D66** 054008 (*Preprint* hep-ph/0205065)
[12] Aaij R *et al.* (LHCb) 2013 *JHEP* **08** 131 (*Preprint* 1304.6325)
[13] Jager S and Martin Camalich J 2016 *Phys. Rev.* **D93** 014028 (*Preprint* 1412.3183)
[14] Muheim F, Xie Y and Zwicky R 2008 *Phys. Lett.* **B664** 174–179 (*Preprint* 0802.0876)

Journal of Physics: Conference Series **770** (2016) 012018 doi:10.1088/1742-6596/770/1/012018

# Rare and semi-rare decays at ATLAS

**U De Sanctis** *on behalf of the ATLAS Collaboration*
University of Sussex, Brighton BN1 9RH, United Kingdom
E-mail: umberto.de.sanctis@cern.ch

**Abstract.** The measurements of the rare $B^0$-meson decay processes performed by the ATLAS experiment at the LHC are reviewed. Particular attention is given to the measurement of the branching ratios of the $B_s^0$ and $B_d^0$ meson decays into a pair of muons with the full Run 1 dataset corresponding to an integrated luminosity of 25 fb$^{-1}$.

## 1. Introduction

The measurements of rare and semi-rare decays of $B^0$ mesons are extremely interesting as tests of the predictions of the Standard Model (SM). In fact, they are forbidden at lowest perturbative order in the SM due to flavour-changing neutral-current (FCNC) suppression and can only proceed through loop diagrams. New heavy particles in extensions of the SM can appear in competing Feynman diagrams and significantly affect both branching fractions and the angular distributions of final state particles. Therefore these processes are sensitive to a wide range of new physics phenomena which are not predicted by the SM.

Among these processes, decays of a $B_s$ and $B_d$ meson into a pair of oppositely charged muons (referred to in the following as $B_{(s)}^0 \rightarrow \mu\mu$) are of particular interest. In fact, the $B_{(s)}^0 \rightarrow \mu\mu$ branching fractions are accurately predicted in the SM: $BR(B_s^0 \rightarrow \mu\mu) = (3.65 \pm 0.23) \times 10^{-9}$ and $BR(B_d^0 \rightarrow \mu\mu) = (1.06 \pm 0.09) \times 10^{-10}$ [1]. Hence, any significant deviation from these predictions would be a hint of New Physics.

Recently, both CMS and LHCb collaborations have reported an observation of $B_s^0 \rightarrow \mu\mu$ and an evidence of $B_d^0 \rightarrow \mu\mu$ decays, leading to a combined measurement of $BR(B_s^0 \rightarrow \mu\mu) = 2.8^{+0.7}_{-0.8} \times 10^{-9}$ and $BR(B_d^0 \rightarrow \mu\mu) = 3.9^{+1.6}_{-1.4} \times 10^{-10}$ [2].

In this report, the measurement of the branching ratios of the $B_s^0 \rightarrow \mu\mu$ and $B_d^0 \rightarrow \mu\mu$ decays performed by the ATLAS Collaboration is described. The results are extracted from the entire dataset collected by the experiment during the LHC Run 1 campaign, corresponding to an integrated luminosity of 25 fb$^{-1}$ collected at a center-of-mass energy of 7 TeV (4.7 fb$^{-1}$) and 8 TeV (20.3 fb$^{-1}$) [3].

## 2. Analysis outline

A blind analysis is performed in the di-muon invariant mass region between 5166 and 5526 MeV. All the events belonging to this region are removed until the procedures for event selection and the details of signal yield extraction are completely defined. Data used in this analysis have been collected by the ATLAS detector during stable beams conditions. Di-muon triggers have been used to select the events from the *pp* collisions. The data sample has been split into

Journal of Physics: Conference Series **770** (2016) 012018 doi:10.1088/1742-6596/770/1/012018

four mutually exclusive trigger categories $T_i$ depending on the requirements on the transverse momentum and the pseudorapidity of the two muons:

$T_1$: "higher threshold" trigger with $p_T > 6$ GeV and $> 4$ GeV for the two muons in the 2012 data sample;

$T_2$: "barrel" trigger with $p_T > 4$ GeV for both muon candidates and at least one of them with $|\eta| < 1.05$ in the 2012 data sample (and $T_1$ requirement not satisfied);

$T_3$: basic di-muon trigger with $p_T > 4$ GeV for both muon candidates in the 2012 data sample (and $T_1$, $T_2$ requirements not satisfied).

$T_4$: basic di-muon trigger with $p_T > 4$ GeV for both muon candidates in the 2011 data sample.

In order to minimise the systematic uncertainties, both branching ratios are extracted relative to the $B^{\pm} \to J/\psi(\to \mu\mu)K^{\pm}$ reference channel, which has a well-measured branching ratio. Therefore, the procedure to extract the BR($B^0_{(s)} \to \mu\mu$) has to be corrected taking into account the difference in the hadronisation probability $f_{u,s,d}$ to form, respectively, a $B^+$, $B_s$ or $B_d$ meson giving a $b$ quark. Also the different acceptances and efficiencies between the signal and the reference channel need to be taken into account. Summarising, the BR($B^0_{(s)} \to \mu\mu$), can be expressed as:

$$BR(B^0_{(s)} \to \mu^+\mu^-) = N_{d(s)} \times [BR(B^{\pm} \to J/\psi(\to \mu\mu)K^{\pm}) \times BR(J/\psi \to \mu\mu)] \times \frac{f_u}{f_{d(s)}} \times \frac{1}{\mathcal{D}_{\text{norm}}}, \tag{1}$$

with

$$\mathcal{D}_{\text{norm}} = \sum_k N^k_{J/\psi K^{\pm}} \alpha_k \left(\frac{\varepsilon_{\mu^+\mu^-}}{\varepsilon_{J/\psi K^{\pm}}}\right)_k. \tag{2}$$

Here the index $k$ runs over the four different trigger categories used in the analysis, $\alpha_k$ takes into account the different trigger prescale factors and integrated luminosities in the signal and reference channels. $N_{d(s)}$ and $N^k_{J/\psi K^{\pm}}$ are the event yields, while $\varepsilon_{\mu^+\mu^-}$ and $\varepsilon_{J/\psi K^{\pm}}$ represent the detector acceptance and analysis efficiencies of the signal and the reference channels, respectively. $B_s \to J/\psi(\to \mu\mu)\phi(\to KK)$ events are also used in the analysis as control sample for the signal kinematic variables used in the analysis.

Muons are required to have $p_T > 4$ GeV and $|\eta| < 2.5$ and to be reconstructed in both the Muon Spectrometer and the Inner Detector (ID). Kaons are reconstructed in the ID system and are required to have $p_T > 1$ GeV and $|\eta| < 2.5$. $B$ mesons kinematic observables are reconstructed imposing quality requirements on the di-muon vertex for signal, or on the di-muon combined with one track vertex for the reference channel. The matching of a B candidate to a primary vertex (PV) is made by projecting the direction of the candidate to the point of closest approach to the collision axis (z) and choosing the PV with the smallest separation in z. This association procedure has an efficiency of 99% evaluated using simulated signal samples.

The reconstructed $B$ candidates are finally required to satisfy the following criteria: $p_T(B) > 8$ GeV and $|\eta(B)| < 2.5$. The invariant mass regions used in the analysis for the signal, the reference and the control channels are reported in Table 1.

## 3. Backgrounds

The main backgrounds for this analysis can be classified into three categories: combinatorial background, partially reconstructed $B$-decays (PRD) and peaking background.

The combinatorial background consists mainly of real muons produced in the $b$ or $\bar{b}$ decay chains. It is the dominant background for the analysis and it is several orders of magnitude larger than the signal. Therefore, a multivariate classifier, namely a Boosted Decision Tree [4] (c-BDT),

Journal of Physics: Conference Series **770** (2016) 012018 doi:10.1088/1742-6596/770/1/012018

is trained in order to reduce its contribution. The c-BDT is based on 15 kinematics variables discriminating the signal from this background. These variables describe: the kinematics of the $B$ meson candidate, the secondary vertex displacement, the properties of the muons and the rest of the event (such as the $B$-isolation $I$ [1]). The c-BDT is trained on a Monte-Carlo (MC) sample simulating decays of $b$ and $c$ quarks (and anti-quarks correspondingly) containing muons. This simulated sample is validated against data candidates from the mass sideband regions. A cut on the c-BDT output at 0.24 is applied, corresponding to an efficiency of 54% on simulated signal and a rejection of about $10^3$ of the combinatorial background.

The PRD background consists of decays where the two muons in the final state come from one of the following topologies: ($a$) originating from a common ancestor in 'cascade' transitions (e.g. $b \to c\mu\nu \to s\mu\mu\nu\nu$) referred to as *same-side* muons (SS); ($b$) the same decay diagram (e.g. $B \to J/\psi X$ or $B_d \to \mu\mu K^*$) referred to as *same vertex* muons (SV); ($c$) the $B_c \to J/\psi\mu\nu$ decay; ($d$) semileptonic $B$ decays where a hadron $h$ ($\pi$ or K) is misidentified as a muon (e.g. $B \to \mu h\nu$). All these backgrounds populate the low-mass sideband and in some cases may have tails extending into the signal region.

The *peaking background* consists of events where the hadronic decays $B^0_{(s)} \to hh'$ ($h$ being $\pi$ or K) are reconstructed as signal events due to hadron misidentification. This background component peaks in the signal region and needs to be reduced since it overcomes the signal by several orders of magnitude. In addition to the reduction given by the muon identification algorithm (which gives a rejection factor of 250 and 500 for pions and kaons, respectively), a different BDT (called f-BDT) is setup based on calorimetric and tracking variables that are able to distinguish real muons from hadrons. A cut on f-BDT output is tuned to have a 95% efficiency on the signal while reducing the peaking background to $1.0 \pm 0.4$ events in the signal region. The f-BDT selection also effectively reduces any other background containing hadrons misidentified as muons (such as the semileptonic decays $B \to \mu h\nu$). Applying this f-BDT cut, an improvement of a factor of seven in the peaking background rejection is achieved with respect to the previous result [5].

## 4. Reference channel yield extraction and efficiencies computation

In order to extract the BR($B^0_{(s)} \to \mu\mu$) using Eq. 1 the yield of the reference channel and the relative efficiency and acceptance ratios between the two channels need to be computed.

The $B^\pm \to J/\psi(\to \mu\mu)K^\pm$ and the $B^\pm \to J/\psi(\to \mu\mu)\pi^\pm$ yields are simultaneously extracted, using an unbinned extended maximum likelihood (ML) fit to the $\mu\mu K^\pm$ invariant mass distribution. The functional forms used in the fit are extracted from simulation, while the shape parameters are fitted simultaneously to data and to simulation. As a byproduct, the ratio between $B^\pm \to J/\psi(\to \mu\mu)\pi^\pm$ and $B^\pm \to J/\psi(\to \mu\mu)K^\pm$ was also measured to be 0.035 $\pm$ 0.003 $\pm$ 0.012, in agreement with the PDG average [6].

[1] $I = \frac{|p_T^{\to}(B)|}{|p_T^{\to}(B)| + \sum_k p_{T,k}}$ where $k$ runs over all additional tracks contained within a cone of size $\Delta R$ that are matched to the same PV as the B candidate

| Channel | Low-mass sideband (MeV) | Signal region (MeV) | High-mass sideband (MeV) |
|---|---|---|---|
| $B^0_{(s)} \to \mu\mu$ | 4766–5166 | 5166–5526 | 5526–5966 |
| $B^\pm \to J/\psi K^\pm$ | 4930–5180 | 5180–5380 | 5380–5630 |
| $B^0_s \to J/\psi\phi$ | 5050–5297 | 5297–5437 | 5437– 5650 |

**Table 1.** List of the invariant mass regions used in the analysis for signal, reference and control channels.

The efficiency ratio $\frac{\varepsilon_{\mu^+\mu^-}}{\varepsilon_{J/\psi K^\pm}}$ between signal and reference channels is computed using simulated samples within the fiducial volume of $p_{\mathrm{T}}(B) > 8$ GeV and $|\eta(B)| <2.5$. The simulated samples are reweighted in order to match the distributions of number of primary vertices (and therefore pile-up), $p_{\mathrm{T}}(B)$, $|\eta(B)|$ and trigger efficiencies (as a function of $p_{\mathrm{T}}(\mu)$ and $|\eta(\mu)|$) in data. In addition, a correction to the $B_s$ lifetime in the simulated signal sample is applied to take into account that the $B_s \to \mu\mu$ decay proceeds almost exclusively through the heavy $B_s$ mass eigenstate [7]. A comparison between the spectra of the 15 variables used in the c-BDT in simulated samples and in data is made to look for possible discrepancies that could affect $\frac{\varepsilon_{\mu^+\mu^-}}{\varepsilon_{J/\psi K^\pm}}$. For all variables, but the $B$ isolation $I$, the agreement is found to be good. The simulated spectra are reweighted, one at a time, to match those in data. For each variable, the variation in $\frac{\varepsilon_{\mu^+\mu^-}}{\varepsilon_{J/\psi K^\pm}}$, after applying the reweighting procedure, is found to be below 2% and considered as a systematic uncertainty on $\frac{\varepsilon_{\mu^+\mu^-}}{\varepsilon_{J/\psi K^\pm}}$. Signal and reference MC samples are instead reweighted to match the $B$ isolation in data ($B_s \to J/\psi\phi$ data for the signal and $B^\pm \to J/\psi K^\pm$ data for the reference channel) and used throughout the analysis. The total systematic uncertainty on $\frac{\varepsilon_{\mu^+\mu^-}}{\varepsilon_{J/\psi K^\pm}}$ is found to be 5.9%.

## 5. Signal yield extraction

The numbers of signal events $N_{B_s}$ and $N_{B_d}$ are extracted simultaneously using an unbinned extended ML fit to the di-muon invariant mass $m_{\mu\mu}$ in the three invariant mass regions shown in Table 1. In order to enhance the discriminating power of the analysis, three bins in the c-BDT output are defined: 0.24–0.346, 0.346–0.446 and 0.446–1. The signal efficiency, as evaluated in simulations, is 18% in each bin. The fit is performed simultaneously in the three c-BDT bins.

The signals are parameterised by a double Gaussian function to take into account the different resolution of the di-muon invariant mass pair in the four trigger categories. The shape and the relative signal efficiencies are constrained to be the same in all three c-BDT bins. The combinatorial background is parameterised with a first order polynomial, with mass slope and relative fractions in the three c-BDT bins as independent parameters. The background coming from SS and SV events (see Section 3) is fitted with an exponential function. The background invariant mass shape is constrained to be the same in all three c-BDT bins with independent normalisations. The peaking background is fitted with the same model as the signal and normalised to a total of $1.0 \pm 0.4$ events (see Section 3) equally distributed in the three c-BDT bins. The resulting fit is shown in Fig. 1 for the second and the third c-BDT bin. All backgrounds are normalised independently in each c-BDT bin.

The main systematic uncertainties affecting $N_s$ and $N_d$ as extracted from the fit are evaluated using a toy MC. In particular, different fit PDFs are chosen for the various backgrounds. Systematic uncertainties related to mass resolution and scale and to the constraint on the signal efficiencies to be the same across the three c-BDT bins are also considered. The total systematic uncertainties on the fitted signal yields are: $\sigma_{syst}(N_{B_d}) = 3$ and $\sigma_{syst}(N_{B_s}) = \sqrt{4 \pm 0.06 \times N_{B_s}}$ events, respectively. These systematic uncertainties are included in the likelihood as Gaussian constraints before the fit to data.

The expected number of signal events, based on the SM calculations and taking the relevant input parameters from PDG [6] and recent ATLAS results [8], is: $N_{B_s} = 41$ and $N_{B_d} = 5$. The number of fitted signal events is found to be: $N_{B_s} = 16 \pm 12$ and $N_{B_d} = -11 \pm 9$.

## 6. BR($B^0_{(s)} \to \mu\mu$) extraction

The BR($B^0_{(s)} \to \mu\mu$) are extracted from data through a profile-likelihood fit using the terms written in Eq. 1. The $BR(B^\pm \to J/\psi(\to \mu\mu)K^\pm) \times BR(J/\psi \to \mu\mu)$ term is taken from PDG [6],

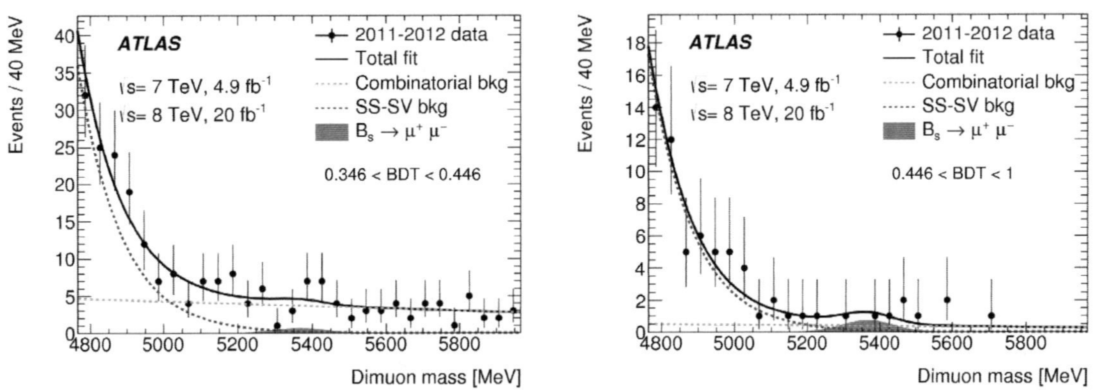


**Figure 1.** Di-muon invariant mass distributions in the unblinded data, for the second and the third c-BDT bin. Superimposed is the result of the maximum-likelihood fit, obtained imposing the boundary of non-negative signal contributions. The total fit is shown as a black continuous line, the filled area corresponds to the observed signal component, the blue dashed line to the SS+SV background, and the green dashed line to the combinatorial background.

while the hadronisation probability ratio $\frac{f_u}{f_{d(s)}}$ from a measurement made by ATLAS [8].

The single-event-sensitivities are found to be $(8.9 \pm 1.0) \times 10^{-11}$ and $(2.21 \pm 0.15) \times 10^{-11}$ for $B_s$ and $B_d$, respectively. The values maximising the likelihood with the constraints on $N_{B_s}$ and $N_{B_d}$ to be positive are found to be: $\text{BR}(B_s^0) = 0.9^{+1.1}_{-0.8} \times 10^{-9}$ and $\text{BR}(B_d^0) = 0$. All the results in the following are obtained applying those constraints if not otherwise stated. The compatibility of the results with the SM prediction is evaluated with toy-MC experiments using a hypothesis test based on the likelihood ratio -ln[L(SM)/L(max)] for the simultaneous fit to $\text{BR}(B_s^0)$ and $\text{BR}(B_d^0)$. The result is a $p$-value of 0.048, corresponding to 2.0 standard deviations.

Upper limits on both BR are also computed at the 95%CL, leading to the following results: $\text{BR}(B_s^0) < 3.0 \times 10^{-9}$ and $\text{BR}(B_d^0) < 4.2 \times 10^{-10}$. The $\text{BR}(B_s^0)$ upper limit is obtained with toy MC experiments under the background only hypothesis leaving the $\text{BR}(B_d^0)$ free to be determined by the fit. Similarly, the $\text{BR}(B_d^0)$ upper limit is set leaving the $\text{BR}(B_s^0)$ free to be determined by the fit.

Figure 2 summarises the outcomes of the measurement in comparison with the SM predictions as well as previous combined results published by the CMS and LHCb collaborations [2]. Blue curves represent intervals of $-2\,\Delta\ln(L)$ equal to 2.3, 6.2 and 11.8 that are obtained with respect to the absolute maximum of the likelihood (i.e. without constraining both BR to be positive), while gray curves represent the CL intervals obtained by CMS and LHCb experiments.

## 7. Conclusions

This report summarised the measurement of the $\text{BR}(B^0_{(s)} \to \mu\mu)$ performed by the ATLAS Collaboration analysing the full Run 1 dataset corresponding to an integrated luminosity of 25 $\text{fb}^{-1}$. The results found are compatible with the SM prediction within 2.0 standard deviations and are in agreement with the measurements performed by the CMS and LHCb collaborations. Other measurements of semi-rare decay processes, such as $B_d \to \mu\mu K^*$ are still under scrutiny by the ATLAS Collaboration. These decays, together with other rare and semi-rare processes, will be an essential part of the B-physics programme of the ATLAS Collaboration for the ongoing Run 2 campaign, where the ATLAS experiment will benefit from improved detector and trigger performance and from the increased integrated luminosity that will be available.

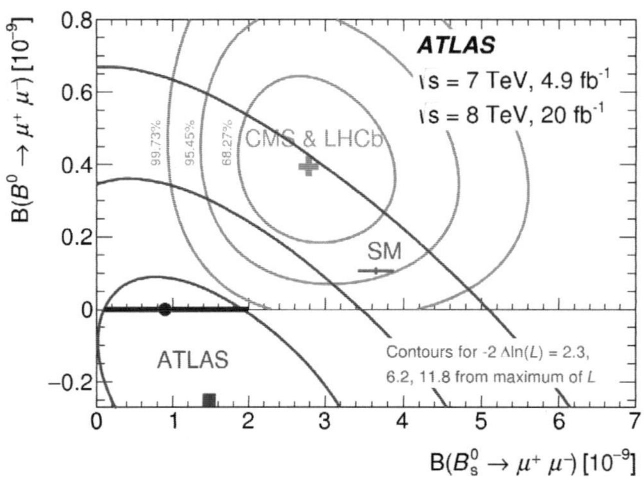


**Figure 2.** Contours in the plane $\mathrm{BR}(B^0_s \to \mu\mu)$, $\mathrm{BR}(B^0_d \to \mu\mu)$ for intervals of $-2\,\Delta\ln(L)$ equal to 2.3, 6.2 and 11.8 relative to the absolute maximum of the likelihood (blue square), without imposing the constraint of non-negative branching fractions. Also shown are the corresponding contours for the combined result of the CMS and LHCb experiments, the SM prediction, and the maximum of the likelihood within the boundary of non-negative branching fractions (blue dot), with the error bars covering the 68.3% confidence range for $\mathrm{BR}(B^0_s \to \mu\mu)$ under the conditions explained in the text.

## References

[1] Bobeth C et al, 2014 $B_{s,d} \to l^+l^-$ in the Standard Model with reduced theoretical uncertainty, *Phys. Rev. Lett.*, **112**, 101801

[2] CMS and LHCb Collaborations, 2015 Observation of the rare $B^0_s \to \mu^+\mu^-$ decay from the combined analysis of CMS and LHCb data, *Nature*, **522**

[3] ATLAS Collaboration, Study of the rare decays of $B^0_s$ and $B^0$ into muon pairs from data collected during the LHC Run 1 with the ATLAS detector, 2016 *Preprint* arXiv:1604.04263 [hep-ex] *Submitted to Eur.Phys. J. C.*

[4] Hoecker A et al, 2007 TMVA 4 Toolkit for Multivariate Data Analysis with ROOT users guide, *Preprint* arXiv:physics/0703039

[5] ATLAS Collaboration, 2012 Search for the decay $B^0_s \to \mu^+\mu^-$ with the ATLAS detector, *Phys. Lett. B* **713** 387

[6] Particle Data Group, 2014 The review of particle physics, *Chin.Phys.C*, **38**, 090001

[7] De Bruyn K et al, 2012 Probing New Physics via the $B^0_s \to \mu\mu$ effective lifetime, *Phys. Rev. Lett.* **109** 041801

[8] ATLAS Collaboration, 2015 Determination of the ratio of $b$-quark fragmentation fractions $f_s/f_d$ in $pp$ collisions at $\sqrt{s} = 7$ TeV with the ATLAS detector, *Phys. Rev. Lett.*, **115**, 262001

# Effects of Scalar Leptoquarks in $b \to s$ Transitions

**R Mohanta***, **S Sahoo**

School of Physics, University of Hyderabad, Hyderabad - 500046, India

E-mail: *rmsp@uohyd.ac.in

**Abstract.** We investigate the effect of scalar leptoquarks on the recent anomalies observed in rare semileptonic $B$ meson decays involving the quark level transition $b \to s$. The leptoquark parameter space is constrained by using the measured branching ratio of $B_s \to \mu^+\mu^-$ process. We estimate the branching ratios of $B \to K^{(*)}\mu^+\mu^-(\nu\bar{\nu})$ processes, using the constrained leptoquark couplings. We also compute forward-backward asymmetry, polarization fractions of $K^*$ and $P_5'$ observable in the $B \to K^*\mu^+\mu^-$ process. The $R_K$ anomaly in the $B \to Kl^+l^-$ process is also studied. Furthermore, we predict the branching ratios of the lepton flavour violating decays, such as $B_s \to l_i^+ l_j^-$, $B \to K^{(*)} l_i^+ l_j^-$ and $B_s \to \phi l_i^+ l_j^-$, which are found to be within the experimental reach of LHCb and the upcoming Belle II experiments.

## 1. Introduction

The study of rare semileptonic decays of $B$ mesons involving the flavour-changing neutral current (FCNC) transition $b \to s$, plays an important role to critically test the standard model (SM) and to look for the possible existence of new physics (NP). Such rare processes are highly suppressed in the SM as they occur at one-loop level (penguin and box diagrams). Recently LHCb has reported a $3\sigma$ discrepancy in $B \to K^*\mu^+\mu^-$ angular observables, mainly in the decay rate [1] and $P_5'$ [2]. Furthermore, the lepton non-universality parameter in the $B \to Kl^+l^-$ decay is found to be $R_K^{\rm LHCb} = {\rm Br}(B^+ \to K^+\mu^+\mu^-)/{\rm Br}(B^+ \to K^+e^+e^-) = 0.745^{+0.09}_{-0.074} \pm 0.036$ [3] in the low $q^2$ bin $(1 \le q^2 \le 6)$ GeV$^2$, which has a $2.6\sigma$ deviation from the SM prediction $R_K^{\rm SM} = 1.0003 \pm 0.0001$.

In this article, we intend to study the effect of scalar leptoquark (LQ) on the branching ratios as well as other asymmetry parameters of the $B \to K^{(*)}\mu^+\mu^-(\nu\bar{\nu})$ processes. We also investigate the lepton flavour violating (LFV) decays, such as $B_s \to l_i^+ l_j^-$, $B_{(s)}^+ \to K^{(*)+}(\phi) l_i^+ l_j^-$ mediated via the scalar LQ. Leptoquarks are color triplet bosonic particles which can couple to a quark and a lepton simultaneously and can occur in various extensions of the SM, e.g., grand unified theory, Pati Salam model, composite scenarios, etc. We consider the simple renormalizable scalar LQ model, for which the bounds from proton decays may not be relevant, and LQ may give signatures in other low-energy processes. In this work, we would like to see whether the scalar LQ model can accommodate some of the recent anomalies observed at LHCb.

The outline of the paper is follows. In section 2 we discuss the new physics contributions to the SM values due to the exchange of scalar LQ and the constraint on the leptoquark couplings from $B_s \to \mu^+\mu^-$ process. The branching ratios and other recent anomalies in the $B \to K^{(*)}\mu^+\mu^-$ process are computed in section 3. In section 4 we estimate the branching ratio of the $B \to K^{(*)}\nu\bar{\nu}$ process. The rare LFV decays are studied in section 5 and section 6 contains the summary and conclusion.

## 2. New physics contributions due to the leptoquark exchange

In the SM, the effective Hamiltonian for processes involving $b \to s l^+ l^-$ quark level transition is given by [4]

$$\mathcal{H}_{eff} = -\frac{4G_F}{\sqrt{2}} V_{tb} V_{ts}^* \left[ \sum_{i=1}^{6} C_i(\mu) \mathcal{O}_i + C_7 \frac{e}{16\pi^2} \Big( \bar{s} \sigma_{\mu\nu} (m_s P_L + m_b P_R) b \Big) F^{\mu\nu} + C_9^{eff} \frac{\alpha}{4\pi} (\bar{s} \gamma^\mu P_L b) \bar{l} \gamma_\mu l + C_{10} \frac{\alpha}{4\pi} (\bar{s} \gamma^\mu P_L b) \bar{l} \gamma_\mu \gamma_5 l \right], \quad (1)$$

where $G_F$ is the Fermi constant, $V_{qq'}$ are the CKM matrix elements, $\alpha$ denotes the fine structure constant, $P_{L,R}$ are the projection operators and $C_i$'s are the Wilson coefficients.

The SM effective Hamiltonian (1) can be modified in the scalar leptoquark model due to the exchange of LQ. Here we will consider two minimal renormalizable scalar leptoquark multiplets $X = (3, 2, 7/6)$ and $(3, 2, 1/6)$, which are invariant under the SM gauge group $SU(3)_c \times SU(2)_L \times U(1)_Y$ and do not allow proton decay. The interaction Lagrangian of $X = (3, 2, 7/6)$ leptoquark with the SM fermion bilinear is given by [4]

$$\mathcal{L} = -\lambda_u^{ij} \, \bar{u}_{\alpha R}^i (V_\alpha e_L^j - Y_\alpha \nu_L^j) - \lambda_e^{ij} \, \bar{e}_R^i \left( V_L^\dagger u_{\alpha L}^j + Y_\alpha^\dagger d_{\alpha L}^j \right) + h.c. \,, \quad (2)$$

which after performing the Fierz transformation and then comparing with the SM effective Hamiltonian (1) gives additional Wilson coefficients to the $b \to s l^+ l^-$ processes as

$$C_9^{NP} = C_{10}^{NP} = -\frac{\pi}{2\sqrt{2} G_F \alpha V_{tb} V_{ts}^*} \frac{\lambda_\mu^{32} \lambda_\mu^{22*}}{M_Y^2} \,. \quad (3)$$

Similarly the $X = (3, 2, 1/6)$ LQ also provide new primed Wilson coefficients $C_{9,10}^{\prime NP}$ corresponding to the semileptonic electroweak penguin operators $\mathcal{O}'_{9,10}$. These Wilson coefficients will give additional contributions to the leptonic decay rate $B_s \to \mu\mu$. Now comparing the SM predicted branching ratio of $B_s \to \mu^+\mu^-$ process from [5], with the corresponding experimental result [6], we obtain the constraint on the combination of leptoquark couplings. The detailed formalism of the constraints on leptoquark coupling can be found in [4], therefore, here we will simply quote the results, as

$$0 \le \frac{|\lambda^{32} \lambda^{22*}|}{M_S^2} \le 5 \times 10^{-9} \text{ GeV}^{-2} \quad \text{for} \quad \pi/2 \le \phi^{NP} \le 3\pi/2 \,, \quad (4)$$

where $M_S$ is the mass of scalar leptoquark. Similarly for $B_s \to e^+e^-$ process, the bound on the product of leptoquark coupling is found as $|\lambda^{31}\lambda^{21*}|/M_S^2 < 2.54 \times 10^{-5}$.

## 3. $B \to K^{(*)} \mu^+ \mu^-$ process

In this section, we study the anomalies in the $B \to K^{(*)}\mu^+\mu^-$ process in the scalar leptoquark model. The differential decay distribution with respect to the lepton-pair invariant mass ($q^2$) after integration over all three angles ($\theta_{K^*}$, $\theta_l$ and $\phi$) is given by [7, 8]

$$\frac{d\Gamma}{dq^2} = \frac{3}{4} \Big( J_1 - \frac{J_2}{3} \Big), \quad (5)$$

where the coefficients $J_{1,2}$ are functions of the dilepton invariant mass.

The various combinations of $J_i$, for $i = 1, ...., 9$ coefficients will give additional interesting observables to look for new physics, such as forward-backward asymmetry, isospin asymmetry, polarization fraction of $K^*$ and $P'_{4,5,6}$ observables. The forward backward asymmetry ($A_{FB}$), longitudinal polarization fraction of $K^*$ ($F_L$) and the form-factor-independent (FFI) observable ($P'_5$) are defined as [7, 8, 9]

$$A_{FB}\left(q^2\right) = -\frac{3}{8}\frac{J_6}{d\Gamma/dq^2}, \qquad F_L\left(q^2\right) = \frac{3J_1^c - J_2^c}{4d\Gamma/dq^2}, \qquad P'_5\left(q^2\right) = \frac{J_5}{2\sqrt{-J_2^c J_2^s}}. \tag{6}$$

In Fig. 1, we show the $q^2$ variation of branching ratio (top left panel), $A_{FB}$ (top right panel) and $P'_5$ observable (bottom panel) in the $(3, 2, 7/6)$ LQ model. The integrated values of branching ratio of $B \to K^*\mu^+\mu^-$ process in both SM and LQ model are found to be [7]

$$\mathrm{Br}(B \to K^*\mu^+\mu^-)|_{\mathrm{SM}} = (7.74 \pm 0.46) \times 10^{-7}, \quad \mathrm{Br}(B \to K^*\mu^+\mu^-)|_{\mathrm{LQ}} = (6.88 - 8.73) \times 10^{-7}, \tag{7}$$

and the predicted value of $A_{FB}$, $F_L$ and $P'_5$ in the SM and LQ model are [7]

$$\begin{aligned} &\langle A_{FB}\rangle|_{\mathrm{SM}} = -(0.09 \pm 0.005), \ \langle F_L\rangle_{\mathrm{SM}} = 0.71 \pm 0.043, \ \langle P'_5\rangle_{\mathrm{SM}} = -0.204 \pm 0.012, \\ &\langle A_{FB}\rangle|_{\mathrm{LQ}} = -0.11 \to 0.004, \quad \langle F_L\rangle_{\mathrm{LQ}} = 0.7 \to 0.8, \quad \langle P'_5\rangle_{\mathrm{LQ}} = -0.42 \to 0.13. \end{aligned} \tag{8}$$

Another interesting observable is the lepton non-universality parameter ($R_K$) in the $B \to Kl^+l^-$ process, which is the ratio of branching fractions of $B \to Kl^+l^-$ decays into dimuons over dielectrons. In these processes, the intermediate region is dominated by the charmonium resonance background induced by the decays $B \to K(c\bar{c}) \to Kl^+l^-$, where $c\bar{c} = J/\psi, \psi'$. Therefore, we calculate the value of $R_K$ in both low and high $q^2$ regions except the intermediate region around $q^2 \sim m^2_{J/\psi}$ and $m^2_{\psi'}$. In Fig. 2, we show the variation of lepton non-universality parameter with respect to low $q^2$ (left panel) and high $q^2$ (right panel) in leptoquark model. The predicted values of $R_K$ in $q^2 \in [1, 6]\mathrm{GeV}^2$ bin is $(0.62 - 0.96)$ [4] and in high $q^2$ bin $(q^2 \geq 14.18)$ $\mathrm{GeV}^2$ is $(0.75 \to 1.0)$ [10].

## 4. $B \to K^{(*)}\nu\bar{\nu}$ process

The $B \to K^{(*)}\nu\bar{\nu}$ process is mediated by $b \to s\nu\bar{\nu}$ transition and the effective Hamiltonian in the SM is given by [10, 11]

$$\mathcal{H}_{eff} = \frac{-4G_F}{\sqrt{2}} V_{tb}V_{ts}^* \left(C_L^\nu \mathcal{O}_L^\nu + C_R^\nu \mathcal{O}_R^\nu\right) + h.c.. \tag{9}$$

In the SM, the $C_R^\nu$ Wilson coefficient is zero and can only be generated by the new physics. The new contribution to the SM effective Hamiltonian (9) due to the exchange of $(3, 2, 1/6)$ scalar leptoquark is given by [10]

$$\mathcal{H}_{LQ} = \frac{\lambda_\mu^{32}\lambda_\mu^{22*}}{M_Y^2}(\bar{s}\gamma^\mu P_R b)(\bar{\nu}\gamma_\mu(1-\gamma_5)\nu), \tag{10}$$

which contributes $C_R^\nu$ Wilson coefficient as

$$C_R^\nu|_{LQ} = -\frac{\pi}{2\sqrt{2}G_F\alpha V_{tb}V_{ts}^*}\frac{\lambda_d^{22}\lambda_d^{32*}}{M_V^2}. \tag{11}$$

The decay distribution of $B \to K\nu\bar{\nu}$ process with the di-neutrino invariant mass is given by

$$\frac{d\Gamma}{ds_B} = \frac{G_F^2\alpha^2}{256\pi^5}|V_{ts}^*V_{tb}|^2 m_B^5 \lambda^{3/2}(s_B, \tilde{m}_K^2, 1)|f_+^K(s_B)|^2|C_L^\nu + C_R^\nu|^2, \tag{12}$$

Journal of Physics: Conference Series **770** (2016) 012019 doi:10.1088/1742-6596/770/1/012019

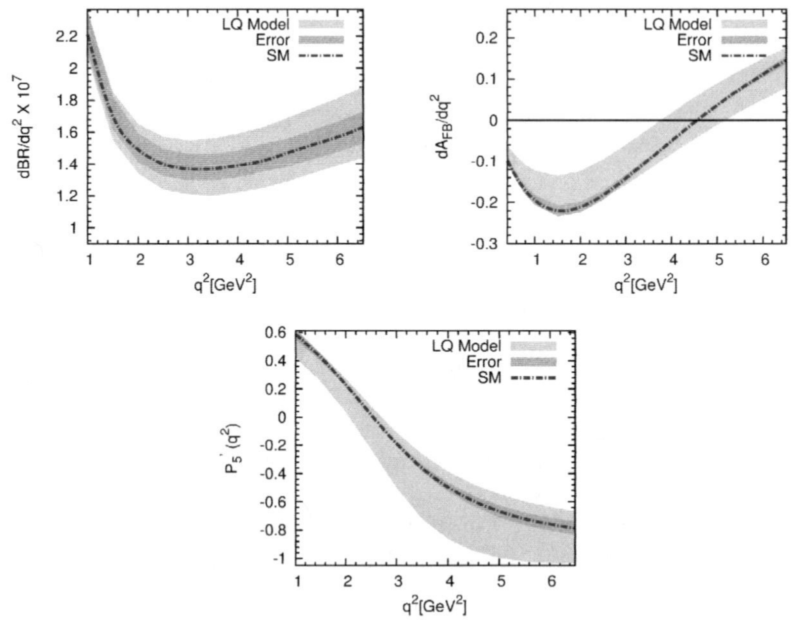


**Figure 1.** The $q^2$ variation of the branching ratio (top left panel), forward-backward asymmetry (top right panel) and $P_5'$ observable (bottom panel) of the $B \to K^* \mu^+ \mu^-$ process in the scalar leptoquark model [7]. The gray bands represent the theoretical uncertainties arising from the input parameters in the SM.

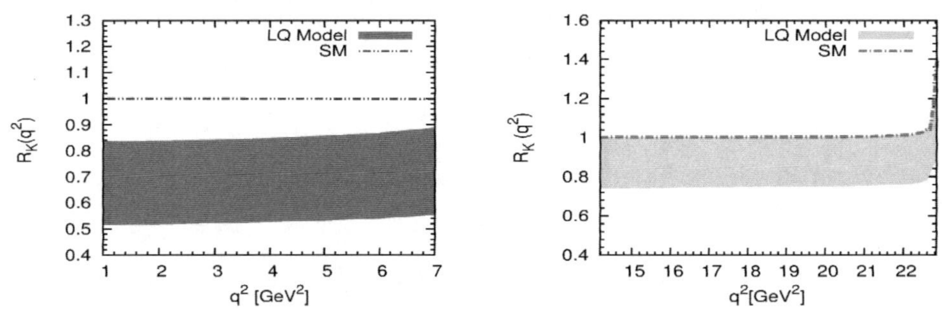


**Figure 2.** The variation of lepton non-universality parameter ($R_K$) with respect to low $q^2$ (left panel) and high $q^2$ (right panel) in the scalar leptoquark model [4].

where $\tilde{m}_i = m_i/m_B$ and $s_B = s/m_B^2$. The differential decay rate for $B \to K^* \nu\bar{\nu}$ process is

$$\frac{d^2\Gamma}{ds_B d\cos\theta} = \frac{9}{4} m_B^2 (|A_\perp|^2 + |A_\parallel|^2) \sin^2\theta + \frac{9}{2} m_B^2 |A_0|^2 \cos^2\theta, \tag{13}$$

where the transversity amplitudes $A_{\perp,\parallel,0}$ are given in [10, 11].

Now using the constrained LQ coupling from Eqn. (4), the predicted branching ratios of $B \to K^{(*)}\nu\bar{\nu}$ processes (in units of $10^{-6}$) both in the SM and LQ model respectively are [10]

$$\mathrm{Br}(B_d \to K\nu\bar{\nu})|_{\mathrm{SM}} = (4.9 \pm 0.29), \qquad \mathrm{Br}(B_d \to K\nu\bar{\nu})|_{\mathrm{LQ}} = (3.6 - 5.2), \tag{14}$$

$$\mathrm{Br}(B_d \to K^*\nu\bar{\nu})|_{\mathrm{SM}} = (9.54 \pm 0.57), \qquad \mathrm{Br}(B_d \to K^*\nu\bar{\nu})|_{\mathrm{LQ}} = (7.0 - 10.13), \tag{15}$$

and the variation of the branching ratios of $B \to K\nu\bar{\nu}$ (left panel) and $B \to K^*\nu\bar{\nu}$ (right panel) processes with $s_B$ in the leptoquark model is shown in Fig. 3.

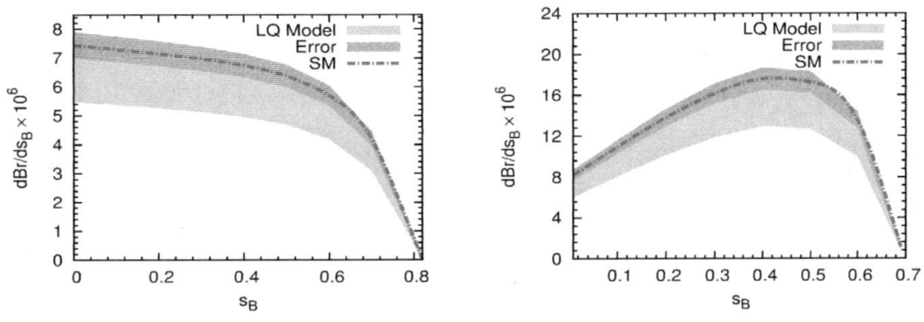


**Figure 3.** The variation of the branching ratios of $B \to K\nu\bar{\nu}$ (left panel) and $B \to K^*\nu\bar{\nu}$ (right panel) processes with respect to $s_B$ in the scalar leptoquark model [10]. The gray bands represent the theoretical uncertainties coming from the input parameters in the SM.

## 5. Lepton flavour violating decays

In this section, we will discuss the lepton flavour violating $B$ meson decays mediated through the exchange of scalar leptoquarks. The LFV processes are extremely rare in the SM as they either proceed through box diagrams or are two-loop suppressed with tiny neutrino mass in one of the loop. However, in the LQ model they can occur at tree level. The effective Hamiltonian for $b \to s l_i^\mp l_j^\pm$ LFV decays in the LQ model is given by [12]

$$\mathcal{H}_{LQ} = \left[ G_{LQ} \left(\bar{q}\gamma^\mu P_L b\right) \left(\bar{l}_i \gamma_\mu (1+\gamma_5) l_j\right) + H_{LQ} \left(\bar{q}\gamma^\mu P_L b\right) \left(\bar{l}_j \gamma_\mu (1+\gamma_5) l_i\right) \right] , \tag{16}$$

where the constant coefficients $G_{LQ}$ and $H_{LQ}$ are

$$G_{LQ} = \frac{\lambda^{3i}\lambda^{kj*}}{8M_Y^2}, \qquad H_{LQ} = \frac{\lambda^{ki}\lambda^{3j*}}{8M_Y^2} . \tag{17}$$

Now, in order to compute the required leptoquark coupling, we used the coupling given in Eqn. (4) as basis value and assumed that the leptoquark couplings between different generation of quark and lepton follow the simple scaling law, i.e. $\lambda^{ij} = (m_i/m_j)^{1/4}\lambda^{ii}$ with $j>i$. Using this ansatz, we compute the branching ratios of LFV decays, such as $B_s \to l_i^+ l_j^-$, $B^+ \to K^{(*)+} l_i^+ l_j^-$ and $B_s \to \phi l_i^+ l_j^-$, and the predicted values are listed in Table-I, which are consistent with present experimental upper limits [6]. We show the plots for the branching ratios of $B^+ \to K^+\mu^+ e^-$ (left panel), $B^+ \to K^+\tau^+ e^-$ (middle panel) and $B^+ \to K^+\tau^+\mu^-$ (right panel) decays in Fig. 4.

## 6. Conclusion

In this paper we have studied the recent anomalies in the rare semileptonic $B$ meson decays in scalar leptoquark model. We constrained the new LQ parameter space using the recent measurements on $B_s \to \mu^+\mu^-$ process. Using such constrained LQ couplings, we computed the branching ratios of the $B \to K^{(*)}\mu^+\mu^-(\nu\bar{\nu})$ processes. We also estimated the forward-backward asymmetry, polarization fractions of $K^*$ and the FFI observable $P_5'$. We found that the observed $R_K$ anomaly can be explained in the LQ model. We then predicted the branching ratios of the

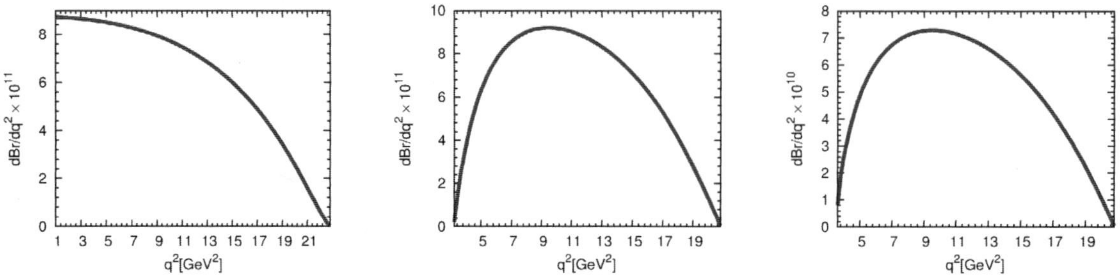


**Figure 4.** The variation of the branching ratios of $B^+ \to K^+\mu^+e^-$ (left panel), $B^+ \to K^+\tau^+e^-$ (middle panel), and $B^+ \to K^+\tau^+\mu^-$ (right panel) processes with respect to $q^2$ in the scalar leptoquark model [12].

**Table 1.** The predicted branching ratios for $B_s \to l_i^+ l_j^-$, $B_{(s)}^+ \to K^{(*)^+}(\phi) l_i^+ l_j^-$ lepton flavour violating decays, where $l = e, \mu, \tau$ in the leptoquark model [12].

| Decay process | Predicted Br | Expt. limit [6] |
|---|---|---|
| $B_s \to \mu^\pm e^\mp$ | $< 1.5 \times 10^{-11}$ | $< 1.1 \times 10^{-8}$ |
| $B_s \to \mu^\pm \tau^\mp$ | $< 1.2 \times 10^{-8}$ | ... |
| $B_s \to e^\pm \tau^\mp$ | $< 8.5 \times 10^{-10}$ | ... |
| $B^+ \to K^+\mu^+e^-$ | $< 1.36 \times 10^{-9}$ | $< 1.3 \times 10^{-7}$ |
| $B^+ \to K^+\tau^+\mu^-$ | $< 8.8 \times 10^{-9}$ | $< 2.8 \times 10^{-5}$ |
| $B^+ \to K^+\tau^+e^-$ | $< 1.12 \times 10^{-9}$ | $< 1.5 \times 10^{-5}$ |
| $B^+ \to K^{*+}\mu^+e^-$ | $<1.4 \times 10^{-9}$ | $< 9.9 \times 10^{-7}$ |
| $B^+ \to K^{*+}\tau^+\mu^-$ | $<1.56 \times 10^{-8}$ | ... |
| $B^+ \to K^{*+}\tau^+e^-$ | $< 2 \times 10^{-9}$ | ... |
| $B_s \to \phi\mu^+e^-$ | $< 8.2 \times 10^{-10}$ | ... |
| $B_s \to \phi\tau^+\mu^-$ | $< 1.1 \times 10^{-8}$ | ... |
| $B_s \to \phi\tau^+e^-$ | $<1.42 \times 10^{-9}$ | ... |

LFV decays such as $B_s \to l_i^+ l_j^-$, $B_{(s)+} \to K^{(*)^+}(\phi) l_i^+ l_j^-$, which are found to be well below the present experimental limits.

## Acknowledgments

We would like to thank Science and Engineering Research Board (SERB), Government of India for financial support through grant No. SB/S2/HEP-017/2013.

## References

[1] Aaij R et al. (LHCb Collaboration) 2014 *Jour High Energy Phys.* **06** 133
[2] Aaij R et al. (LHCb Collaboration) 2013 *Phys. Rev. Lett.* **111** 191801
[3] Aaij R *et al.* (LHCb Collaboration) 2014 *Phys. Rev. Lett.* **113** 151601
[4] Sahoo S and Mohanta R 2015 *Phys. Rev.* D **91** 094019
[5] Bobeth C, Gorbahn M, Hermann T, Misiak M, Stamou E and Steinhauser M 2014 *Phys. Rev. Lett.* **112** 101801
[6] Olive K A et al. (Particle Data Group) 2014 *Chin. Phys.* C **38**, 090001
[7] Sahoo S and Mohanta R 2015 *Phys. Rev.* D **93**, 034018
[8] Bobeth C, Hiller G and Piranishivili G 2008 *Jour High Energy Phys.* **07** 106
[9] Descotes-Genon S, Matias J, Ramon M and Virto J 2013 *Jour. High Energy Phys.* **01** 048
[10] Sahoo S and Mohanta R 2016 *New J. Phys.* **18** 013032
[11] Altmannshofer W A Buras A J, Straub D M and Wick M 2009 *Jour. High Energy Physics* **04** 022
[12] Sahoo S and Mohanta R 2016 *Phys. Rev.* D **93**, 114001

# Search for Rare $B$ Meson Decays at the *BABAR* Experiment

**R Cheaib**
On behalf of the *BABAR* collaboration
Department of Physics, University of Mississippi, 108 Lewis Hall, University MS 38677
E-mail: rachac@phys.olemiss.edu

**Abstract.** $b \to s$ transitions are flavour-changing neutral current (FCNC) processes that play an important role in the search for physics beyond the Standard Model (SM). Contributions from virtual particles in the loop are predicted to deviate observables, such as the branching fraction, from their SM expectations. Using data from the BaBar experiment, we present the first search for the rare decay $B^+ \to K^+ \tau^+\tau^-$. The *BABAR* results on the measurement of the angular asymmetries of $B \to K^* \ell^+\ell^-$, where $\ell = e$ or $\mu$, are also reported. In addition, using a time-dependent analysis of $B \to K^0_S\pi^+\pi^-\gamma$, the mixing induced CP-asymmetry for the radiative FCNC decay, $B \to K^0_S\rho\gamma$, is measured, along with an amplitude analysis of the $m_{K\pi}$ and $m_{K\pi\pi}$ spectrum.

## 1. Introduction

$b \to s$ transitions are highly suppressed in the SM and only occur via loop or box diagrams. Using an effective low-energy theory, the Lagrangian for $b \to s$ transitions can be separated into two distinct parts: the long distance (low-energy) contributions contained in the operator matrix elements and the short-distance (high-energy) physics described by the Wilson coefficients. Measurements of such rare FCNC $B$ meson decays are interesting since they can provide experimental constraints on the associated Wilson coefficients and are thus a stringent test of the SM. Furthermore, contributions from new-physics particles, like a charged Higgs [2], can deviate these Wilson coefficients or require the introduction of new opertaor matrix elements, and thus this is an interesting hunt for physics beyond the SM.

The *BABAR* experiment [3] collected 424 fb$^{-1}$ of data [4] by colliding electrons and positrons at the center-of-mass (CM) energy of the $\Upsilon(4S)$ resonance. The $\Upsilon(4S)$ decays into $B\overline{B}$ pair, resulting in more than 479 million $B\overline{B}$ events to study and analyze. Using the full *BABAR* dataset, a measurement of the $B^+ \to K^+ \tau^+\tau^-$ branching fraction [5], $B \to K^* \ell^+\ell^-$ angular asymmetries [6] and $B \to K^0_S\rho\gamma$ mixing-induced CP-asymmetry [7] is performed.

## 2. Branching fraction measurement of $B^+ \to K^+ \tau^+\tau^-$

$B^+ \to K^+ \tau^+\tau^-$ [8] is a FCNC process with a braching fraction in the range 1-2$\times 10^{-7}$ [9]. It is the third-lepton generation equivalent of $B \to K \ell^+\ell^-$, where $\ell = e$ or $\mu$. The large mass of the $\tau$ lepton may provide improved sensitivity to new-physics contributions as compared to its light lepton counterparts [10, 11]. For instance, in two-Higgs-doublet-models [2], the Higgs-lepton-lepton vertex is proportional to $m_\tau^2$ and thus contributions to the total decay rate, as well as

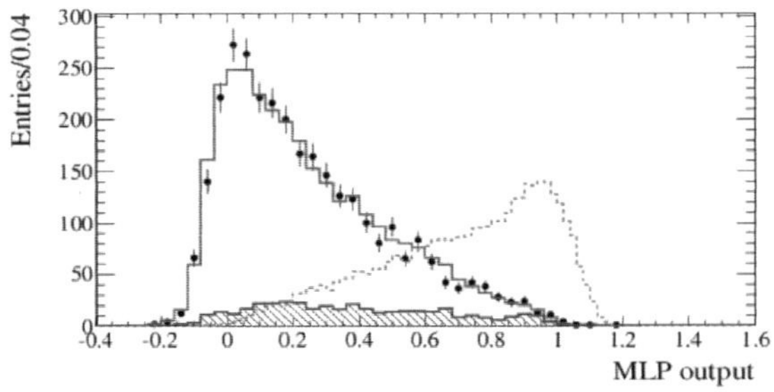


**Figure 1.** MLP output distribution for the $B^+ \to K^+\tau^+\tau^-$ analysis. The signal MC distribution is shown (dashed) with arbitrary normalization, along with the data (points) and the expected combinatorial (shaded) plus $m_{\rm ES}$-peaking (solid) background contributions.

other observables, can be significant. The branching fraction of $B^+ \to K^+ \tau^+\tau^-$ is measured by exclusively reconstructing one $B$ meson, referred to as the $B_{\rm tag}$, in the $\Upsilon(4S) \to B\overline{B}$ decay using hadronic modes, and then examining the rest of the event for the $B^+ \to K^+ \tau^+\tau^-$ signal. This technique is referred to as the hadronic $B_{\rm tag}$ reconstruction, and is ideal for decays with missing energy. With exclusive reconstruction of the $B_{\rm tag}$, the four-vector of the other $B$, the $B_{\rm sig}$, can also be fully determined and thus the kinematics of the event are better constrained. The $\tau$ is required to decay only via leptonic modes: $\tau^- \to e^-\overline{\nu}_e\nu_\tau$ or $\tau^- \to \mu^-\overline{\nu}_\mu\nu_\tau$, and thus there are three possible final states: $e^+e^-$, $\mu^+\mu^-$ or $e^+ \mu^-$ with their associated neutrinos.

To select for $B^+ \to K^+ \tau^+\tau^-$, every event is required to have exactly one properly-reconstructed charged $B_{\rm tag}$ with an energy substituted mass, $m_{\rm ES}$, that lies within the range of the mass of a $B$ meson. The $m_{\rm ES}$ of a $B_{\rm tag}$ candidate is given by $m_{\rm ES} = \sqrt{E^2_{\rm CM} - \vec{p}^{\,2}_{B_{\rm tag}}}$, where $E_{\rm CM}$ is half the total colliding energy and $\vec{p}_{B_{\rm tag}}$ is the 3-momentum of the reconstructed $B_{\rm tag}$, in the CM frame. Furthermore, $B^+ \to K^+ \tau^+\tau^-$ events are required to have nonzero missing energy, which is calculated by subtracting all signal side tracks and clusters from the $B_{\rm sig}$ four-vector. A signal event is also required to have 3 tracks, one satisfying the particle identification (PID) criteria of a $K$ and two of an electron or muon. In addition, the presence of massive $\tau$ leptons imposes an upper limit on the $K$ momentum, which restricts the $s_B$ distribution, given by $s_B = (p_{B_{\rm sig}} - p_K)^2/m_B^2$, of signal events to higher values as compared with background events. Therefore, a requirement of $s_B > 0.45$ is applied. At this point, the main source of background is from combinatorial events with semi-leptonic charmed $B$ decays, such as $B \to D^{(*)}\ell\nu_\ell, D^{(*)} \to K\ell\nu_\ell$. To suppress this background, a multi-layer perceptron (MLP) neural network, consisting of eight discriminating variables such as the angle between the lepton and the oppositely charged kaon in the $\tau^+\tau^-$ rest frame , is used. The neural network is then trained and tested for each of the three signal channels, and the combined MLP output is shown in Fig. 1. The final step in the signal selection is a requirement on the MLP output for each signal channel.

The data yield, after the MLP output cut, is determined seperately for each of the three signal channels and then combined into one $B^+ \to K^+ \tau^+\tau^-$ result. No significant excess is observed in the $e^+e^-$ or $\mu^+\mu^-$ channels, while a 3.7 $\sigma$ excess is obesrved in the $e^+ \mu^-$ channel. Examination of the input and output distributions of the $e^+ \mu^-$ channels does not show any clear evidence of signal-like behaviour or mis-modelling of the background. Furthermore, the combined excess for all three modes is less than 2 $\sigma$. In the absence of signal, the combined upper limit at the 90% confidence level is measured to be $\mathcal{B}(B^+ \to K^+\tau^+\tau^-) < 2.6 \times 10^{-3}$. This is the first search for $B^+ \to K^+ \tau^+\tau^-$.

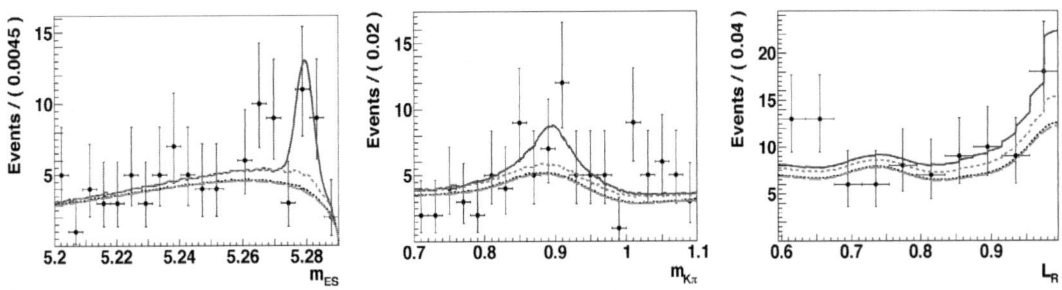


(a) $m_{\rm ES}$: $B^0 \to K^+\pi^-e^+e^-$. (b) $m_{K\pi}$: $B^0 \to K^+\pi^-e^+e^-$. (c) $L_R$: $B^0 \to K^+\pi^-e^+e^-$.

**Figure 2.** 3-D fit projections for $B^0 \to K^+\ \pi^-\ e^+e^-$ in $q_5^2$. Each event class is shown: combinatorial (magenta long dash) and charmonium background events (black dots), crossfeed signal events (red short dash) and total pdf (solid blue).

## 3. Angular asymmetries in $B \to K^*\ \ell^+\ell^-$

$B \to K^*\ \ell^+\ell^-$ is also a FCNC process, with an amplitude expressed in terms of hadronic form factors and the $C_7, C_9$, and $C_{10}$ Wilson Coefficients[6]. The angular distributions of $B \to K^*\ \ell^+\ell^-$, specifically the $K^*$ longitudinal polarization, $F_L$, and the forward-backward asymmetry, $A_{FB}$, are notably sensitive to physics beyond the SM [12]-[13] and have been previously measured by various experiments [14]-[18].

For any given value of the $q^2$, the kinematic distribution of the $B \to K^*\ell^+\ell^-$ decay products can be expressed in terms of three distinct angles: $\theta_K$, the angle between the $K$ and the $B$ in the $K^*$ rest frame, $\theta_l$, the angle between the lepton and the $B$ in the $\ell^+\ell^-$ rest frame, and $\phi$, the angle between the $\ell^+\ell^-$ and $K\pi$ decay planes in the $B$ rest frame [6]. After integrating out $\phi$ and $\theta_l$, $F_L$ can be determined using a fit to $\cos\theta_K$ of the form $\frac{1}{\Gamma(q^2)}\frac{d\Gamma}{d(\cos\theta_K} = \frac{3}{2}F_L(q^2)\cos^2\theta_K + \frac{3}{4}(1 - F_L(q^2))(1 - cos^2\theta_K)$. Similarly, $A_{FB}$ can be extracted after integrating over $\phi$ and $\theta_K$ using a fit to $\theta_l$ of the form $\frac{1}{\Gamma(q^2)}\frac{d\Gamma}{d(\cos\theta_l} = \frac{3}{4}F_L(q^2)(1 - cos^2\theta_l) + \frac{3}{8}(1 - F_L(q^2))(1 + cos^2\theta_l) + A_{FB}(q^2)\cos\theta_l$ [19].

To measure $F_L$ and $A_{FB}$, $B \to K^*\ \ell^+\ell^-$ signal events are reconstructed in one of the following final states: $K^{*+}\ (\to K_S^0\ \pi^+)\mu^+\mu^-$, $K^{*0}\ (\to K^+\ \pi^-)\mu^+\mu^-$, $K^{*+}\ (\to K^+\ \pi^0)e^+e^-$, $K^{*+}\ (\to K_S^0\ \pi^+)e^+e^-$, $K^{*0}\ (\to K^+\ \pi^-)e^+e^-$. Each $K^*$ candidate is required to have an invariant mass ranging between 0.72 and $1.10\,{\rm GeV}/c^2$, and is combined with a pair of leptons, each with an individual momenta greater than 0.3 GeV/$c$. The $m_{\rm ES}$ and $\Delta E$ of the resulting $B$ candidate is then determined and used to separate between signal and background events. Here, $\Delta E = E_B^* - E_{CM}/2$, where $E_B^*$ is the CM energy of the $B$ and $E_{CM}$ is total CM energy.

The main source of background is from semileptonic $B$ and $D$ decays, as well as continuum background events with random combinations of leptons. Eight bagged decision trees (BDT) are trained to suppress these $B\overline{B}$ and $q\overline{q}$ backgrounds and a final requirement on $\Delta E$ and $L_R$ is applied at the end of the signal selection. Here, $L_R$ is a likelihood ratio which uses the output of the $B\overline{B}$ BDT to determine how likely a given event is signal vs background.

To extract the angular observables, the $q^2$ spectrum is divided into five disjoint bins ($q_1 - q_5$) of varying size, and an additional bin $q_0$, ranging between 1.0 and 6.0 GeV$^2/c^4$. An initial unbinned maximum likelihood fit to the $m_{\rm ES}$, $m(K\pi)$, and $L_R$ spectrums is performed to fix the normalizations and shapes of all probability density functions (pdfs) dependent on these three variables. Then, for each mode and each $q^2$ bin, the 3-D likelihood fit is used to fix the normalizations of events with $m_{\rm ES} > 5.27\,{\rm GeV}/c^2$. Third, $\cos\theta_K$ is added as a fourth dimension

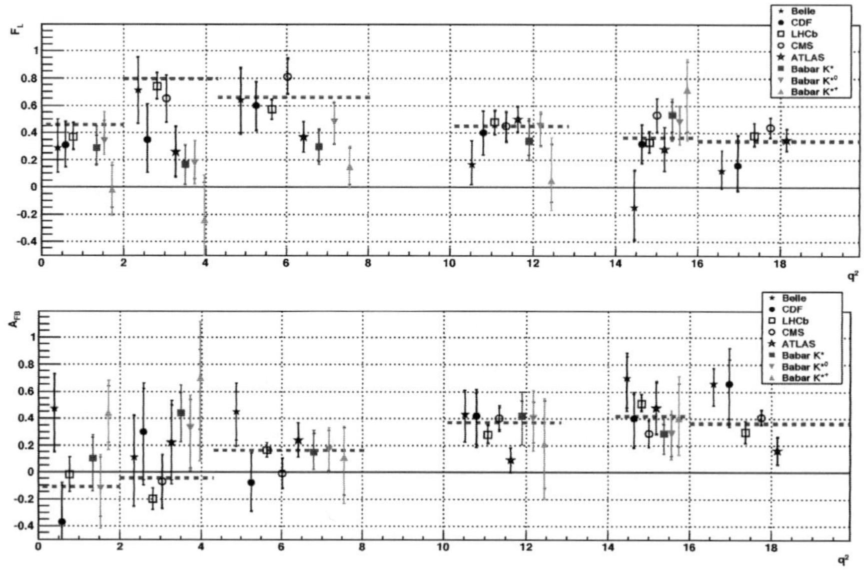


**Figure 3.** $F_L$ and $A_{FB}$ results for charged (magenta filled pointing-up triangle), neutral (red filled down-pointing triangle) and total $B \to K^*\ \ell^+\ell^-$ (blue filled square) in disjoint $q^2$ bins. The SM expectations are shown as blue dashed lines along with results from other experiments: Belle [14] (black-filled star), CDF [15] (black-filled circle), LHCB [16] (black open square), CMS [17] (black open circle), and ATLAS [18] (black open star).

to the likelihood function, and four-dimensional likelihoods are defined for each signal mode and each $q^2$ bin with $F_L$ as the only free parameter. Finally, the fitted value of $F_L$ is then used as input to a similar 4-D fit, where $\cos\theta_l$ has been added as a fourth dimension instead of $\cos\theta_K$. The pdfs in the likelihood fit are defined for five different event classes: true signal events, crossfeed signal events, combinatorial backgrounds, backgrounds from charmonium decays, and finally backgrounds from hadronic decays which are only prominent for $\mu^+\mu^-$ modes. Fig. 2 shows the initial 3-D fit projections for $B^0 \to K^+\ \pi^-\ e^+e^-$ in $q_5^2$ with the different event classes. $F_L$ and $A_{FB}$ are extracted in each $q^2$ bin for the charged, $B^+ \to K^{*+}\ \ell^+\ell^-$, neutral, $B^0 \to K^{*0}\ \ell^+\ell^-$, and total $B \to K^*\ \ell^+\ell^-$ modes. The results are shown in Fig. 3 along with previous ones from Belle [14], CDF [15], LHCb [16], and CMS [17].

As can be readily seen, the $B^0 \to K^{*0}\ \ell^+\ell^-$ results show good agreement with the SM expectations and other experiments. For the charged mode, the value of $F_L$ is relatively small in the low $q^2$ region and thus exhibits tension with the SM expecation. An additional angular observable $P_2$ is defined such that $P_2 = (-2/3) \times A_{FB}/(1 - F_L)$. $P_2$ has diminished theoretical uncertainty and thus higher sensitvity to non-SM contributions. The tension at low $q^2$ is still found in the $P_2$ distribution and can be a hint of new physics, specifically the results are consistent with the existence of right-handed currents. This is the first measurement of angular asymmetries in $B^+ \to K^{*+}\ \ell^+\ell^-$.

## 4. Mixing induced CP-asymmetry, $S_{fCP}$, in $B \to K_S^0\rho\gamma$

In the SM, the photon emitted in $b \to s\gamma$ transitions is predominantly left-handed, with contamination from right-handed photons suppressed by a factor of $m_s/m_b$ [20]. This implies that $B^0$ ($\overline{B}^0$) mesons decay predominantly to right-handed (left-handed) photons and the mixing-induced CP-asymmetry in $B \to f_{CP}\gamma$ decays is expected to be small. However, various new physics models [21, 22] introduce enhanced contributions from right-handed photons and thus alter the prediction of a small CP-asymmetry.

In this analysis, the mixing-induced CP-asymmetry of $B^0 \to K^0_S\rho\gamma$, $S_{K^0_S\rho\gamma}$, is measured using a time-dependent analysis of $B^0 \to K^0_S\pi^+\pi^-\gamma$. Due to the large natural width of the $\rho$ meson, there is an irreducible contribution from the non-CP eigenstate $B^0 \to K^{*\pm}(\to K^0_S\pi^\pm)\pi^\mp\gamma$ which affects $S_{K^0_S\rho\gamma}$, and thus a dilution factor, $D_{K^0_S\rho\gamma} \equiv S_{K^0_S\pi^+\pi^-\gamma}/S_{K^0_S\rho\gamma}$, must be determined. Here, $S_{K^0_S\pi^+\pi^-\gamma}$ is the effective value of the mixing-induced CP asymmetry for the full $B^0 \to K^0_S\pi^+\pi^-\gamma$ dataset. To determine $D_{K^0_S\rho\gamma}$, an amplitude analysis of the $m_{K\pi}$ spectra must be performed. Given the small number of events expected in the $B^0 \to K^0_S\pi^+\pi^-\gamma$ sample, the amplitudes of the resonant modes are determined from the charged $B^+ \to K^+\pi^+\pi^-\gamma$ mode instead and then extracted, under the assumption of isospin asymmetry, to the neutral mode. Furthermore, because the decay to the $K^+\pi^+\pi^-\gamma$ final state proceeds in general through three-body resonances first which then further decay into their $K^*\pi$ or $K\rho$ components, it is necessary to determine the three-body resonance content of the $m_{K\pi\pi}$ spectrum as well.

$B^+ \to K^+\pi^+\pi^-\gamma$ events are reconstructed from one high energy photon with $1.5 < E_\gamma < 3.5$ GeV, two oppositely-charged pions, and one charged kaon. These are combined to form a $B$ candidate, whose $m_{\rm ES}$ should lie with 5.20 and 5.92 GeV/$c^2$ and $|\Delta E| < 0.200\,$GeV. A Fisher discriminant, consisting of six discriminating variables, is trained to suppress continuum background events. Furthermore, to reduce backgrounds from photons that originate from $\pi^0$ or $\eta$ decay, a likelihood ratio, $L_R$, is constructed. To extract the $B^+ \to K^+\pi^+\pi^-\gamma$ yield, an unbinned extended maximum likelihood fit to the $m_{\rm ES}$, $\Delta E$, and Fisher discriminat output $F$ is performed. The resulting branching fraction is measured to be $\mathcal{B}(B^+ \to K^+\pi^+\pi^-\gamma) = (24.5 \pm 0.9 \pm 1.2) \times 10^{-6}$.

The $m_{K\pi\pi}$ spectrum is then extracted from the maximum likelihood fit, and modeled as the coherent sum of five kaonic Breit-Wigner resonances: $K_1(1270), K_1(1400), K^*(1410), K^*(1680)$ and $K^*_2(1430)$. The fit fraction of each resonance is determined and its corresponding branching fraction, given by $\mathcal{B}(B^+ \to K_{res}(\to K^+\pi^+\pi^-)\gamma)$, is computed. Furthermore, a binned maximum likelihood fit is further performed to the efficiency-corrected $m_{K\pi}$ spectrum. Using the phase space decay of the three-body resonances $m_{K\pi\pi}$, an efficiency map is determined and applied to the $m_{K\pi}$ spectrum. The latter is modeled as the projection of two $1^-$ P-wave components, $K^*(892)$ and $\rho(770)$, and one $0^+$ S-wave component, $(K\pi)_0^{(*0)}$. The branching fractions $\mathcal{B}(B^+ \to K_{res}\pi^+\gamma)$ are also determined. Many of the measured branching fractions in this analysis are the first to be done or more accurate that previous world averages. Using the results of the $m_{K\pi}$ spectrum, the dilution factor is computed and yields $D_{K^0_S\rho\gamma} = -0.78^{+0.19}_{-0.17}$.

To measure the time-dependent CP asymmetry, the proper-time difference, given by $\Delta t = t_{rec} - t_{tag}$, is determined, between a fully reconstructed $B^0 \to K^0_S\rho\gamma$ decay ($B^0_{rec}$) and the other $B$ in the event $B_{tag}$, which is partially reconstructed. The distance between the decay-vertex positions of $B_{\rm tag}$ and $B_{rec}$ is measured and transformed to $\Delta E$ using the boost, $\beta\gamma = 0.56$l, of the $e^+e^-$ beams. To reconstruct the $B_{\rm tag}$, a $B$-flavor tagging algorithm [23] is used, which combines various event variables to achieve optimal separation between the two $B$ candidates in a signal event. $B^0 \to K^0_S\pi^+\pi^-\gamma$ events are reconstructed using the same signal selection as $B^+ \to K^+\pi^+\pi^-\gamma$, but with $K^0_S \to \pi^+\ \pi^-$. An unbinned maximum likelihood fit is then performed to the $m_{\rm ES}$, $\Delta E$, Fisher discriminant output, $\Delta t$ and $\sigma_{\Delta t}$ distributions to extract the signal yield. The fit is shown in Fig 4 and yields a branching fraction $\mathcal{B}(B^0 \to K^0\pi^+\pi^-\gamma) = (20.5 \pm 2.0^{+2.6}_{-2.2}) \times 10^{-6}$. The CP-violation parameter is then determined to be $S_{K^0_S\pi^+\pi^-\gamma} = 0.14 \pm 0.25 \pm 0.03$. Using the calculated dilution factor and assuming isospin asymmetry, the resulting time-dependent CP asymmetry for $B^0 \to K^0_S\rho\gamma$ is calculated to be: $S_{K^0_S\rho\gamma} = -0.18 \pm 0.32^{+0.06}_{-0.05}$. This measurement is in agreement with previously published results [24]-[26] and shows no deviation from the SM prediction.

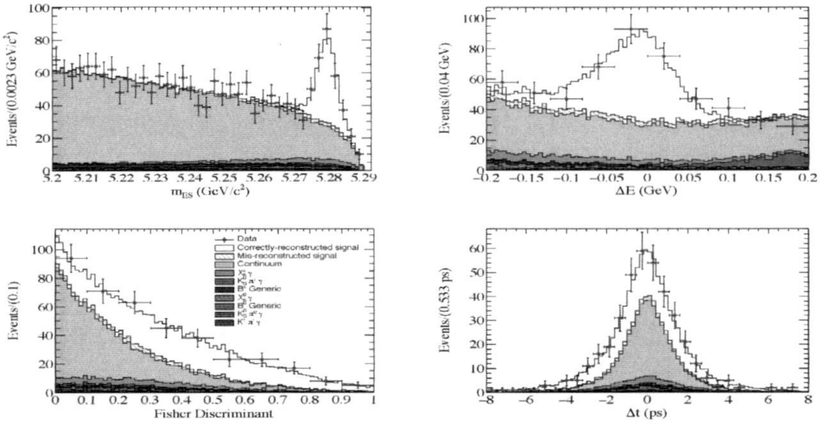


**Figure 4.** Distributions of $m_{\rm ES}$ (top left), $\Delta E$, Fisher discriminant output $F$, and $\Delta t$ with the fit results for the $B^0 \to K^0_S\pi^+\pi^-\gamma$ data sample. The data is shown as points with error bars and the stacked histograms represent the different background contributions.

## 5. Conclusion

Various interesting and leading results are still being produced using the *BABAR* dataset. An upper limit on th $B^+ \to K^+ \, \tau^+\tau^-$ has been done for the first time. Furthermore, the angular asymmetries of $B \to K^* \, \ell^+\ell^-$ are measured and display tension with the SM expectations in the low $q^2$ region. In addition, the time-dependent CP-asymmetry in $B \to K^0_S\rho\gamma$ has been measured and shows consistency with the SM. $b \to s$ transitions continue to be a promising probe of physics beyond the SM and a point of interest for current and future $B$-factories.

[1] A. J. Bevan *et al.* [*BABAR* and Belle Collaborations], Eur. Phys. J. C**74**, 3026 (2014).
[2] T. M. Aliev , M. Savci and A. Ozpineci, J. Phys. G **24**, 49 (1998).
[3] B. Aubert *et al.* [*BABAR* Collaboration], Nucl. Instrum. Meth. A **729**, 615 (2013).
[4] J. P. Lees *et al.* [*BABAR* Collaboration], Nucl. Instrum. Meth. A **726**, 203 (2013).
[5] J. P. Lees *et al.*[*BABAR* Collaboration], arXiv:1605.09637
[6] J. P. Lees *et al.*[*BABAR* Collaboration], Phys. Rev. D **93**, 052015 (2016).
[7] J. P. Lees *et al.* [*BABAR* Collaboration], Phys. Rev. D **93**, 052013 (2016).
[8] The charge conjugate mode $B^- \to K^- \, \tau^+\tau^-$ is also implied.
[9] C. Bouchard, G. P. Lepage, C. Monahan, H. Na and J. Shigemitsu, Phys. Rev. Lett. **111**, 162002 (2013).
[10] Q. Yan *et al.*, Phys. Rev. D **62**, 094023 (2000).
[11] D. Guetta and E. Nardi, Phys. Rev. D **58**, 012001 (1998).
[12] T. Feldmann and J. Matias, JHEP **0301**, 074 (2003).
[13] Y. G. Xu, R. M. Wang and Y.D. Yang, Phys. Rev. D **74**, 114019 (2006).
[14] J. T. Wei *et al.* [Belle Collaboration], Phys. Rev. Lett. **103**, 171801 (2009).
[15] T. Aaltonen *et al.* [CDF Collaboration], Phys. Rev. Lett. **108**, 081807 (2012).
[16] R. Aaij *et al.* [LHCb Collaboration], JHEP **1308**, 131 (2013).
[17] S. Chatrchyan *et al.* [CMS collaboration], Phys. Lett. B **727**, 77 (2013).
[18] G. Aad *et al.* [ATLAS Collaboration], ATLAS-CONF-2013-038.
[19] F. Kruger and J. Matias, Phys. Rev. D **71**, 094009 (2005).
[20] D. Atwood, M. Gronau and A. Soni, Phys. Rev. Lett. **79**, 185 (1997).
[21] K. Fujikawa and A. Yamada, Phys. Rev. D **49**, 5890 (1994).
[22] P. L. Cho and M. Misiak, Phys. Rev. D **49**, 5894 (1994).
[23] B. Aubert *et al.* [*BABAR* Collaboration], Phys. Rev. Lett. **99**, 171803 (2007).
[24] J. Li *et al.* [Belle Collaboration], Phys. Rev. Lett. **101**, 251601 (2008).
[25] B.Aubert *et al.* [*BABAR* Collaboration], Phys. Rev. D **78**, 071101 (2008).
[26] Y. Ushiroda *et al.* [Belle Collaboration], Phys. Rev. D **74**, 111104 (2006).

# Antihydrogen studies in ALPHA

**N Madsen**

Department of Physics, College of Science, Swansea University, Swansea SA2 8PP, UK

E-mail: n.madsen@swansea.ac.uk

**for the ALPHA collaboration**[1]

**Abstract.** The ALPHA experiment studies antihydrogen as a means to investigate the symmetry of matter and antimatter. Spectroscopic studies of the anti-atom hold the promise of the most precise direct comparisons of matter and antimatter possible. ALPHA was the first to trap antihydrogen in a magnetic trap, allowing the first ever detection of atomic transitions in an anti-atom. More recently, through stochastic heating, we have also been able to put a new limit on the charge neutrality of antihydrogen. ALPHA is currently preparing to perform the first laser-spectroscopy of antihydrogen, hoping to excite the 2s state using a two-photon transition from the 1s state. We discuss the recent results as well as the key developments that led to these successes and discuss how we are preparing to perform the first laser-spectroscopy. We will also discuss plans to use our novel technique for gravitational tests on antihydrogen for a direct measurement of the sign of the gravitational force on antihydrogen.

## 1. Introduction

Antihydrogen, the bound state of an antiproton and a positron, is a promising test-bed for high precision comparisons of matter and antimatter and thus for testing fundamental symmetries. However, a fundamental experimental challenge is that antihydrogen has to be made in the laboratory starting from high energy antiprotons, but high precision measurements require low energies and ultimately trapping due to the low quantities available. The neutral nature of antihydrogen makes it difficult to trap and manipulate, but is also the reason why it holds the potential for gravitational measurements, and for the use of high precision atomic spectroscopy techniques for comparisons of matter and antimatter. In recent years the ALPHA experiment has noted a number of breakthroughs that are helping pave the way for such comparisons. These include the trapping and holding of antihydrogen for extended periods [1, 2], the first observation of atomic transitions in the from of a positron spin-flip (hyperfine transition) in the ground state [3], demonstration of a novel technique for studying the effects of gravity [4] and measurements of the charge neutrality of antihydrogen [5]. These feats have put ALPHA at the lead of the

[1] ALPHA collaboration: M. Ahmadi, M. Baquero-Ruiz, W. Bertsche, E. Butler, A. Capra, C. Carruth, C. L. Cesar, M. Charlton, S. Eriksson, L. T. Evans, N. Evetts, J. Fajans, T. Friesen, M. C. Fujiwara, D. R. Gill, A. Gutierrez, J. S. Hangst, W. N. Hardy, M. E. Hayden, C. A. Isaac, A. Ishida, S. A. Jones, S. Jonsell, L. Kurchaninov, N. Madsen, D. Maxwell, J. T. K. McKenna, S. Menary, J. M. Michan, T. Momose, J. J. Munich, P. Nolan, K. Olchanski, A. Olin, A. Povilus, P. Pusa, C. Ø. Rasmussen, F. Robicheaux, R. L. Sacramento, M. Sameed, E. Sarid, D. M. Silveira, C. So, T. D. Tharp, R. I. Thompson, D. P. van der Werf, J. S. Wurtele and A. I. Zhmoginov

field of antihydrogen physics and we present here some of the highlights as well as upcoming challenges for continuing from these successes.

## 2. Synthesising and trapping antihydrogen

Antihydrogen is now synthesised and trapped in an upgraded version of the original ALPHA apparatus that has been operational since 2013. Figure 1 shows an overview of the setup. The spin-flip and gravitational results were performed using the previous version of ALPHA, which for the purposes of this discussion, was mostly identical. The upgrade has served to separate out the antiproton manipulations at keV energies in preparation for the ELENA ring [6] and as a means to allow laser-access to the antihydrogen trapping volume as discussed in Ref. [7].

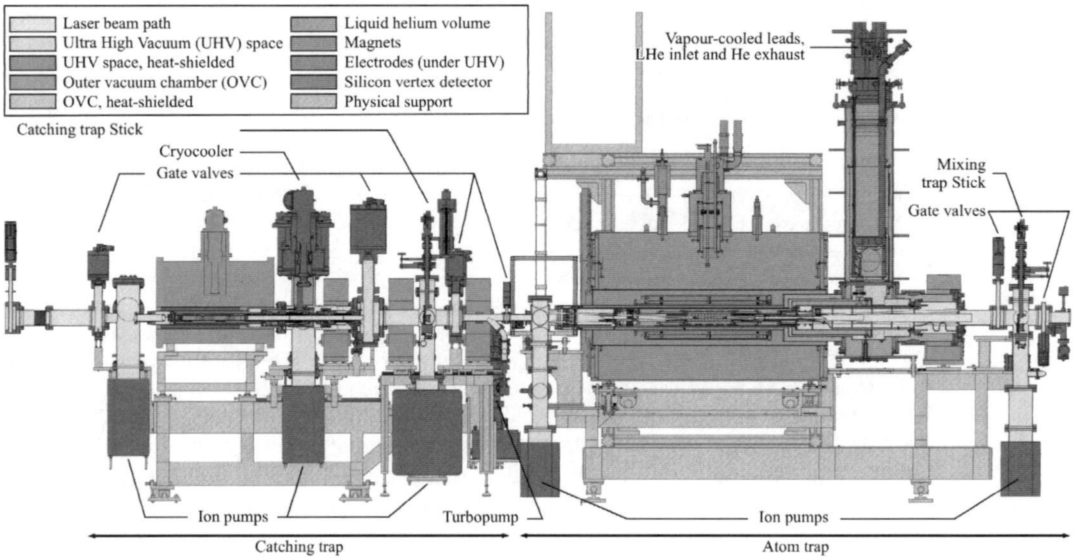


**Figure 1.** ALPHA setup as installed since 2013. The catching trap (left section) captures and prepares antiprotons for transfer to the atom trap (right). Positrons originate from the positron accumulator (not shown) to the right of the atom trap. There is access on the left and the right side of the atom trap for up to four lasers to enter at a grazing angle (2.34°) to the axis, paths crossing at the centre of the apparatus.

Antiprotons are sourced from the CERN antiproton decelerator (AD) at 5.3 MeV. The AD delivers a $\sim$130 ns bunch of $\sim 3\times 10^7$ antiprotons every $\sim$2 min. The antiprotons are degraded to keV energies and are dynamically trapped in a Penning-Malmberg trap. All charged-particle traps in the ALPHA apparatus are Penning-Malmberg traps where axial magnetic fields confine the particles transversely, and a large number of independently excitable co-axial cylinders provide axial electric fields that ensure axial confinement. The antiprotons are cooled to $\sim$100 K by being merged with a cloud of pre-loaded electrons that cool through the emission of cyclotron radiation towards the ambient temperature. The traps are all cooled to cryogenic temperatures (5-7 K). The antiproton plasma undergoes a number of manipulations, such as being transferred from the catching trap to the atom trap (Figure 1), before eventually being injected into a plasma of cold positrons for antihydrogen formation. Positrons are sourced from a solid-neon moderated $^{22}$Na source and accummulated in a Surko-type buffer gas accummulator [6]. For antihydrogen formation typically $2\times 10^4$ antiprotons and $2\times 10^6$ positrons are merged yielding of order $10^4$ antihydrogen atoms of which on average one is trapped [3]. The depth of the magnetic-minimum neutral trap is about 50 $\mu$eV.

## 3. First atomic transitions

While only about one anti-atom is currently available per experiment, it can be held for long times, up to ~15 min. [2]. This allows for a long interaction time which somewhat compensates for the low number. Profiting from this we have carried out a first measurement of an atomic transition in antihydrogen using microwaves to flip the spin of the positron in the trapped ground state to an untrapped state [3]. Figure 2a shows the Breit-Rabi diagram for ground state antihydrogen, and shows how in a magnetic field, the two hyperfine ground states split up in four, whereof two are trappable in that they are so-called low-field seekers (diamagnetic).

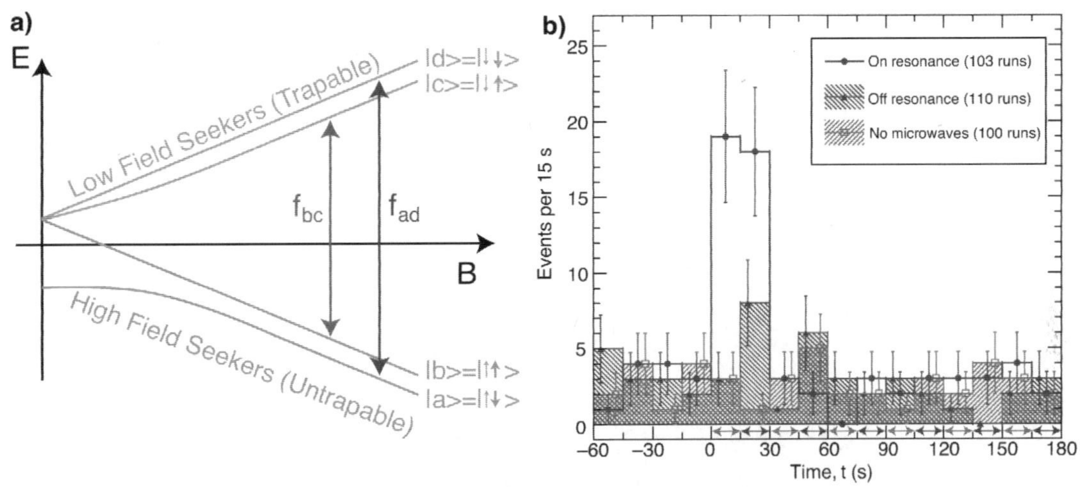


**Figure 2.** (a) Breit-Rabi diagram of (anti)hydrogen indicating the two transitions addressed by our experiment. The fat (thin) arrows indicate the direction of the positron (antiproton) spin relative to the external field. (b) Experimental observation of annihilations during microwave illumination. The microwave illumination starts at 0 s, and every 15 s the frequency is shifted to address either $f_{bc}$ or $f_{ad}$ as illustrated. During each 15s interval the frequency is slowly swept 15 MHz, to ensure overlap with the sharp resonance peak [3].

By injecting microwaves of the frequency expected for hydrogen in our magnetic fields, we succeeded in detecting the resonant ejection of antihydrogen atoms. Figure 2b shows how we detected the appearance of annihilations during microwave illumination when the frequency was tuned to be on-resonance. On-resonance, in this case, includes a 15 MHz sweep of the microwave frequency that was around 29 GHz in order to ensure that we did not miss the transition completely, and illumination of both transitions (blue and red arrows on Figure 2) to increase statistics. In the measurement we compare on-resonance and off-resonance experiments. For the off-resonance experiments the atomic transition frequency was shifted 100 MHz up relative to nominal by increasing the magnetic field by 3.5 mT while keeping the microwave frequency constant. This means that roughly speaking the observed resonance is within 100 MHz of that expected from hydrogen, and as the frequencies observed were around 29 GHz, this corresponds to a relative precision of about $4\times10^{-3}$, thus marking the advent of antimatter spectroscopy.

## 4. Gravitational measurements

The ALPHA experiment uses a silicon strip detector to register antiproton annihilation by tracking the charged pions stemming from the annihilations. This allows reconstruction of both the position and time of the annihilation and is used to separate antiproton annihilations from cosmic rays passing through the detector. We have demonstrated how this position and time knowledge can also be used to probe the force of gravity on antihydrogen. In current experiments antihydrogen is released from the trap by essentially initiating a quench of the super-conducting trap magnets which allows the trap fields to decay with a time constant of about 9 ms. The trap is thus at about 1% of its full depth only 30 ms after the quench is initiated. This is the window during which we look for annihilations, and the cosmic ray induced background during this time is about one fake annihilation event every 700 experiments. The last antihydrogen atoms to escape while de-energising the trap are also the coldest, and those whose annihilation position is most sensitive to the influence of gravity [4]. Figure 3 shows a simulation of the release of antihydrogen from our trap with a hypothetical gravitational to inertial mass ratio of 100 and demonstrates the principle. We have done the exercise on actual data, but as the vertical extent of our trap is only 4 cm, and the experiment was not done with this test in mind, there is limited sensitivity and the result was that we could reject ratios of the gravitational to inertial mass of antihydrogen >75 with a statistical significance at the 95% confidence level, with worst-case systematic errors increasing the minimum rejection ratio to 110, and with similar bounds for negative gravitational mass.

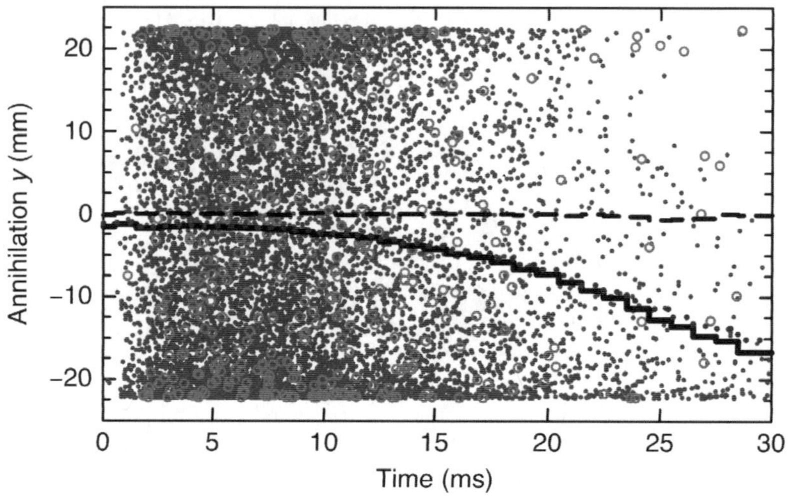


**Figure 3.** Annihilation locations. The times and vertical (y) annihilation locations (green dots) of 10,000 simulated antihydrogen atoms during de-energisation of the magnetic trap assuming that gravitational mass is 100 times the inertial mass. We observe a clear tendency for annihilation in the bottom half (y<0) as illustrated by the black line, that plots the average annihilation locations in 1 ms bins. The average was taken by simulating 900,000 anti-atoms of which the green dots are a sub-sample. The black dashed line shows the average for identical gravitational and inertial mass. The red circles are the annihilation times and locations for 434 real anti-atoms, as measured by our particle detector [4].

In spite of the rather wide bounds, this exercise demonstrated the potential for conducting gravity measurements starting from trapped antihydrogen, and as such has motivated us to work towards a vertical version of our setup that, using this technique, should be able to measure the gravitational acceleration of antimatter. This future upgrade is currently in the early design stages.

## 5. Neutrality of antihydrogen

It is supposed that antihydrogen is charge neutral. However, assuming charges are simply additive, our current knowledge of the charge of antihydrogen made from an antiproton and a positron is limited by our knowledge of the charge of the positron such that combining the two leads us to infer a bound of $|Q| \leq 25$ p.p.b. on the charge $Qe$ of antihydrogen where $e$ is the elementary charge. We have implemented a new technique to measure the charge of antihydrogen [5].

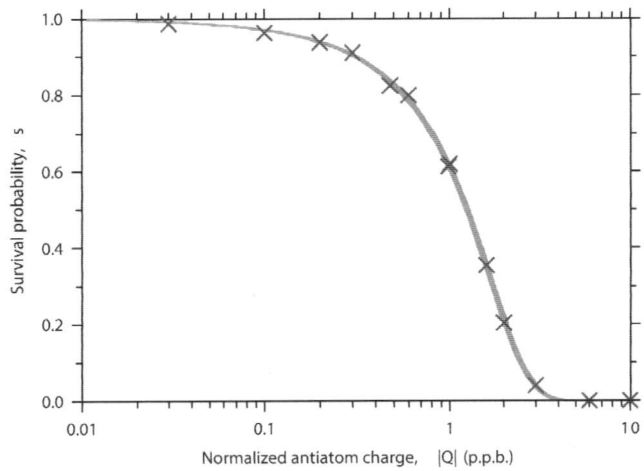


**Figure 4.** Simulated survival probability $s$ as a function of $|Q|$ for the stochastic trials [5].

We apply a series of randomly timed electrical fields across our trap while holding the trapped antihydrogen inside. If antihydrogen had a charge $Qe$, and the average kick of each pulse is $\Delta\Phi$, such pulses will increase the kinetic energy of the antihydrogen atoms in our trap by about $|Q|e\Delta\Phi\sqrt{N}$, where $N$ is the number of pulses. The antihydrogen would escape from the trap if it gained about the energy of the trap depth $E_{well}$, i.e. if

$$|Q| \gtrsim \frac{E_{well}}{e\Delta\Phi\sqrt{N}}. \tag{1}$$

Thus by measuring if antihydrogen survives this treatment for a fixed number of kicks we can set an upper limit on its charge. The simple estimate based on the equation above does not include the averaging of the motion of the antihydrogen atoms in the neutral trap, nor the time to raise a given field and the randomness of the antihydrogen atom's position at a given time. These can all be included in a more elaborate simulation of our full apparatus and the exact responses of our electrodes etc. This simulation showed that, with our particular set of experiments, the survival probability of antihydrogen would be correlated with its charge as depicted in Figure 4. In our measurements we did not succeed in kicking any antihydrogen atoms out of the trap in 10 trials with 12 atoms observed in total, and we concluded that the upper limit on the charge of antihydrogen was $|Q| < 0.71$ parts per billion (one standard deviation), or about 20 times improved compared to the bound based on adding the positron and antiproton charges.

## 6. Laser-spectroscopy

The ALPHA apparatus (Figure 1) was upgraded in particular to allow laser spectroscopy of antihydrogen. It allows for up to four independent laser-paths to cross through the antihydrogen trapping region. The 1S-2S transition of interest is dipole forbidden and is therefore a two-photon transition. By illuminating with retro-reflected light the first order Doppler shift is cancelled and Doppler broadening eliminated, resulting in both a narrower line and a higher transition rate for a given power. However, as pointed out in Ref. [2] the trapped antihydrogen atoms have an energy distribution consistent with a truncated high temperature distribution. This means that the antihydrogen atoms essentially explore the full 0.4 L volume of the magnetic minimum trap. In order to ensure a transition rate high enough to be detectable we have therefore implemented an internal enhancement cavity for the 243 nm laser light needed for this transition. State-of-the art continuous wave (cw) lasers can deliver up to around 200 mW at this wavelength, but due to the unfortunate details of our experiment, in particular the only-one-atom-at-a-time, we estimate that we need around 2 W of laser light to ensure a detectable signal. We expect to see a first measurement in 2016.

As pointed out above a limiting factor (but not detrimental) in these experiments is the low number of trapped anti-atoms. Fundamentally the challenge is that our trap depth is around 50 $\mu$eV, or equivalent to about 0.5 K. The antihydrogen must be made at lower energy than this depth and inside the trap to be held. We have previously demonstrated cooling of the antiprotons, carrying almost all the momentum of the antihydrogen, to around 9 K [8]. However, we observe that, upon injection into the positron plasma, they thermalise with the positrons faster than they recombine with them. It is thus, currently, the positron temperature that sets the antihydrogen temperature. In order to cool the positrons further we have made an effort to eliminate external noise and improve the cryostat, but their temperature will also be limited by magnetic field-inhomogeneities, which are intrinsic to a magnetic minimum trap, so we have started working towards active cooling using laser-cooled Beryllium ions [9]. Simulations show that this strategy holds the promise of reducing our positron temperatures to about 5 K, or about an order of magnitude lower than the temperatures at which we are currently operating. Such an improvement should result in an up to two orders of magnitude increase in the number of trapped antihydrogen, a potential game-changer for probing the anti-atoms.

## 7. Conclusions

We here presented the latest results of the ALPHA antihydrogen experiment and discussed some exciting upcoming directions. The ALPHA programme is concerned with comparing antihydrogen and hydrogen in all ways possible in order to help understand the fundamental asymmetry of our Universe. We discussed how we now routinely trap antihydrogen in our trap and demonstrated how this feat is a powerful platform for further experimentation, already having measured the first atomic transition, tested the neutrality of antihydrogen and demonstrated a technique for measuring the gravitational force on antihydrogen. We also discussed how we are proceeding towards laser-spectroscopy and how we plan to increase the antihydrogen trapping rates significantly.

## 8. References

[1] G. B. Andresen, et al. Trapped antihydrogen. *Nature*, 468(7324):673–676, 12 2010.
[2] G. B. Andresen, et al. Confinement of antihydrogen for 1,000 seconds. *Nat Phys*, 7(7):558–564, 07 2011.
[3] C. Amole, et al. Resonant quantum transitions in trapped antihydrogen atoms. *Nature*, 483:439–443, 03 2012.
[4] ALPHA Collaboration (Madsen) and A. E. Charman. Description and first application of a new technique to measure the gravitational mass of antihydrogen. *Nat Commun*, 4:1785, 04 2013.
[5] M. Ahmadi, et al. An improved limit on the charge of antihydrogen from stochastic acceleration. *Nature*, 529(7586):373–376, 01 2016.
[6] W A Bertsche, E Butler, M Charlton, and N Madsen. Physics with antihydrogen. *Journal of Physics B: Atomic, Molecular and Optical Physics*, 48(23):232001, 2015.
[7] Niels Madsen. Alpha: antihydrogen and fundamental physics. *Hyperfine Interactions*, pages 1–9, February 2014.
[8] G. B. Andresen, et al. Evaporative cooling of antiprotons to cryogenic temperatures. *Phys. Rev. Lett.*, 105:013003, Jul 2010.
[9] N Madsen, F Robicheaux, and S Jonsell. Antihydrogen trapping assisted by sympathetically cooled positrons. *New Journal of Physics*, 16(6):063046, June 2014.

Journal of Physics: Conference Series **770** (2016) 012022 doi:10.1088/1742-6596/770/1/012022

# Latest (Anti) Neutrino Oscillation Results from T2K

**J Zalipska for the T2K Collaboration**

National Centre for Nuclear Research, Hoża 69, 00-681 Warsaw, Poland

joanna.zalipska@ncbj.gov.pl

**Abstract**. The T2K long-baseline neutrino oscillation experiment has been running since 2010; it collects data in both neutrino and anti-neutrino modes. The results presented here refer to a joint analysis of $\nu_\mu$ disappearance and $\nu_e$ appearance for neutrino data giving the world-leading measurements of $\delta_{CP}$, $\theta_{23}$ and $\Delta m^2_{32}$. The first anti-neutrino beam results are also discussed here which allows for testing of CPT symmetry and an electron anti-neutrino appearance search.

## 1. Introduction

Neutrinos are the only elementary particles which can spontaneously change flavour. This phenomenon is called "neutrino oscillation". Since we have three neutrino types of different flavour ($\nu_e$, $\nu_\mu$, and $\nu_\tau$), we can observe mixing of all three types of neutrinos. Neutrino oscillations were proven to exist by two experiments. The Super-Kamiokande experiment first announced discovery of neutrino oscillations in 1998, using measurements of atmospheric neutrinos [1]. A few years later their results were followed by proof of neutrino oscillations in the solar neutrino sector by the SNO experiment [2]. Both of those discoveries were awarded the Nobel Prize in Physics in 2015. Those studies have been followed by measurements of neutrino oscillations under controlled conditions using artificially produced neutrino beams. The first experiment of that type was K2K [3], later followed by the MINOS [4] and OPERA [5] experiments. The T2K experiment (which results are discussed here) [6] and NOvA are the currently running long-baseline neutrino oscillation experiments. They are dedicated to answer still unknown questions in neutrino physics. On the one side they provide precise measurement of oscillation parameters such as $\theta_{23,}$ $\theta_{13}$ and $\Delta m^2_{32}$ [7,8]. They also try to answer the questions of wether CP symmetry is violated or conserved in neutrino sector, and what is the neutrino mass hierarchy.

Neutrino oscillations are possible because neutrino flavour states are not identical to neutrino mass eigenstates. They are mixed as described by the Pontecorvo-Maki-Nakagawa-Sakata mixing matrix:

$$\begin{pmatrix} \nu_e \\ \nu_\mu \\ \nu_\tau \end{pmatrix} = \begin{pmatrix} U_{e1} & U_{e2} & U_{e3} \\ U_{\mu 1} & U_{\mu 2} & U_{\mu 3} \\ U_{\tau 1} & U_{\tau 2} & U_{\tau 3} \end{pmatrix} \begin{pmatrix} \nu_1 \\ \nu_2 \\ \nu_3 \end{pmatrix}$$

That matrix can be parameterized by the three mixing angles $\theta_{12}$, $\theta_{13}$, and $\theta_{23}$ and one CP violating phase $\delta_{CP}$ as shown here:

Journal of Physics: Conference Series **770** (2016) 012022 doi:10.1088/1742-6596/770/1/012022

$$U = \begin{pmatrix} 1 & 0 & 0 \\ 0 & cos\theta_{23} & sin\theta_{23} \\ 0 & -sin\theta_{23} & cos\theta_{23} \end{pmatrix} \begin{pmatrix} cos\theta_{13} & 0 & sin\theta_{13}e^{-i\delta_{CP}} \\ 0 & 1 & 0 \\ -sin\theta_{13}e^{i\delta_{CP}} & 0 & cos\theta_{13} \end{pmatrix} \begin{pmatrix} cos_{12} & sin\theta_{12} & 0 \\ -sin\theta_{12} & cos\theta_{12} & 0 \\ 0 & 0 & 1 \end{pmatrix}$$

The probability for neutrinos to change flavour can be expressed as:

$$P(\alpha \to \beta) = |\sum_{ij} U^*_{\alpha i} U_{\beta i} U_{\alpha j} U^*_{\beta j} exp(-i\frac{L\Delta m^2_{ij}}{2E})|$$

In general, neutrino oscillations can occur only when two conditions are fulfilled: the mixing angle is not equal zero, and $\Delta m^2=m_i^2-m_j^2$ is not zero. Therefore, observation of neutrino oscillations proves that neutrinos are not massless. Measurements of neutrino mixing parameters allow us to experimentally probe neutrino masses by measuring the mass squared differences ($\Delta m^2$). The T2K experiment studies neutrino oscillation using an accelerator-produced $\nu_\mu$ beam by measuring probability of $\nu_\mu$ to disappear, so called $\nu_\mu \to \nu_\mu$ oscillations. It was also the first experiment in the world to measure the appearance of $\nu_e$ from the $\nu_\mu$ beam ($\nu_\mu \to \nu_e$ oscillation). T2K is currently measuring neutrino oscillations in an anti-neutrino beam, which could lead to discovery of CP violation in the neutrino sector. Consequently it is able to study the disappearance on anti-neutrinos via anti-$\nu_\mu \to$ anti-$\nu_\mu$ oscillation and tries to measure the appearance of anti-$\nu_e$ in anti-$\nu_\mu$ beam (anti-$\nu_\mu \to$ anti-$\nu_e$ oscillations). The first of these transformations can provide a test of the CPT symmetry. This is due to the fact that the probability of muon neutrinos to survive, $P(\nu_\mu \to \nu_\mu)$, depends on $cos\delta_{CP}$ only; therefore all terms are CP conserving. The situation is different when we try to probe the appearance of electron neutrinos, $\nu_\mu \to \nu_e$. The probability which describes that transformation has CP violating terms which depend on $sin\delta_{CP}$. If CP is violated, we expect a different probability for neutrinos to oscillate than for the situation when $\delta_{CP}$ equals 0 or $\pi$. On the other hand, the probability of anti-neutrinos to oscillate as anti-$\nu_\mu \to$ anti-$\nu_e$ will be different that for neutrino oscillations of $\nu_\mu \to \nu_e$ if $\delta_{CP}$ is not zero or $\pi$. The maximum effect of the CP violation corresponds to the value of $\delta_{CP}$ equals $+/-\pi/2$. That paper will summarize results of neutrino and anti-neutrino oscillations measurements performed by the T2K experiment.

## 2. Overview of the T2K experiment

The T2K experiment is an accelerator neutrino experiment located in Tokai, Japan which uses a 30 GeV proton beam produced by the JPARC accelerator to create an almost pure $\nu_\mu$ beam. Protons are collided with a graphite target where they produce positive and negative pions and kaons as a result of the interaction. Those charged particles are focused by three horn magnets which select either positive or negative pions. When $\pi^+$ are selected then they produce a $\nu_\mu$ beam. Alternatively, when the horns are operated in the so-called "reverse horn current mode" the $\pi^-$ are focused and sent to the decay pipe, where they decay to produce the anti-$\nu_\mu$ beam. The T2K experiment started operation in 2010 and was analyzed $6.2x10^{20}$POT (protons on target) of neutrino beam data. Since 2014, we have been collecting data for the anti-neutrino mode. The results presented here are based on the analysis of the $4.0x10^{20}$POT collected statistics from the anti-neutrino run.

The experiment is composed of near and far detector stations. One of the near detectors, INGRID, is placed on the beam axis and used to monitor the beam intensity and stability during operation. The other near detector station, called ND280, is placed 280m from the neutrino production point to study the unoscillated neutrino beam. The phenomena of oscillations are measured by the Super-Kamiokande (SK) far detector which is located 295km from the beginning of the neutrino beam. SK is a water Cherenkov detector which records muons and electrons coming out of charged current interactions of (anti-)$\nu_\mu$ and (anti-)$\nu_e$ [9]. It is a cylindrical water tank 41m high and 39m diameter placed 1km underground. SK has the ability to distinguish between $\nu_\mu$ and $\nu_e$ events by studying the pattern of the Cherenkov ring seen by the photosensors. The misidentification probability is less than 1%. Consequently we can study $\nu_\mu \rightarrow \nu_\mu$ and $\nu_\mu \rightarrow \nu_e$ oscillations separately.

## 3. Overview of oscillation analysis

In order to perform oscillation analysis, the initial inputs of neutrino flux prediction and cross-sections of $\nu$ interactions are necessary. The NA61/SHINE experiment at CERN delivers spectra of neutrino parents, mostly $\pi$ and $K$, coming from interactions of a proton beam on a carbon target. Such measurements are used to tune the multiplicity of corresponding particles in the T2K beam simulation. This allows prediction of neutrino flux with precision as good as 10% [10]. External data from MiniBooNE and Minerva are used to model neutrino interaction cross-sections. As a next step, the near detector data of the T2K experiment are used to perform a fit which allows us to constrain flux and cross-section parameters (as described in Section 4). Using that information, predictions of event rates in the far detector Super-Kamiokande, are made. Selection of neutrino data in SK is described in Ref.[11]. The predicted number of events and their spectra are compared with the detected $\nu_\mu$ and $\nu_e$ induced events in the far detector to derive the oscillation parameters.

## 4. Near Detector measurement

The near detector, ND280, has an important role to provide the reference measurement of unoscillated neutrino flux and measure cross-sections of neutrino interactions. The ND280 is a tracker detector which is placed inside the former UA1/NOMAD magnet creating a 0.2T magnetic field. For the analysis presented in this paper, neutrino interactions in the Fine-Grained Detector (FGD) – a scintillator detector, were analyzed. The FGD is interlayed with Time Projection Chamber detectors (TPC). These ensure tracking of the particles coming out of $\nu$ interactions, their identification and momentum reconstruction. The Electromagnetic Calorimeter and Side Muon Range Detector are used to measure particles escaping the tracker volume.

Thanks to the fact that detector is placed in a magnetic field, we are able to distinguish interactions of $\nu$ from anti-$\nu$ by identifying the sign of the outgoing lepton. The analysis of near detector data relies on information about the presence of the $\pi^+$ in the interaction. The selected sample of $\nu_\mu$ charged current events are split into sub-categories based on whether $\pi^+$ were present in the interaction, or whether there were more charged or neutral pions present. It allows us to be sensitive to various types of neutrino interaction processes such as charged current quasi elastic interactions (with no pions coming out of the interaction), processes where pions are produced through the resonanses, and deep inelastic interactions. Consequently, the near detector data measurement provides a detailed $\nu_\mu$ and $\nu_e$ flux measurement and allows us to reduce the predicted flux uncertainty in the far detector [11]. It also allows us to constrain parameters describing different models of neutrino interaction cross-sections. To summarize: the flux and cross-section related systematic uncertainty on predicted number of $\nu_\mu$ events in SK is reduced from 9.2% to 3.4% when the near detector measurement is taken into account.

## 5. Oscillation analysis results

### 5.1 Neutrino data results

The neutrino data were collected from 2010 until 2013 and correspond to $6.2\text{x}10^{20}$POT of analyzed statistics [11]. For that period, 120 $\nu_\mu$-induced events were observed in the Super-Kamiokande far detector when 446.0+-22.5 are expected in case of no oscillation (for $\sin^2 2\theta_{32}$=0.0). Therefore, a clear disappearance signal for $\nu_\mu$ is observed. In case of the maximal mixing ($\sin^2 2\theta_{23}$=1.0) the expected number of $\nu_\mu$ events is 125.85 ($\Delta m^2_{32}=2.4\text{x}10^{-3}\text{eV}^2/\text{c}^4$ is assumed).

When the $\nu_e$ selection is applied in the Super-Kamiokande data, then 28 data events are found. In the case of no $\nu_\mu \rightarrow \nu_e$ oscillation ($\sin^2 2\theta_{13}$=0.0), the predicted background is 4.92+-0.55 which is estimated based on assumption of other oscillation parameters ($\sin^2 2\theta_{23}$=1.0, $\Delta m^2_{32}=2.4\text{x}10^{-3}\text{eV}^2/\text{c}^4$ and

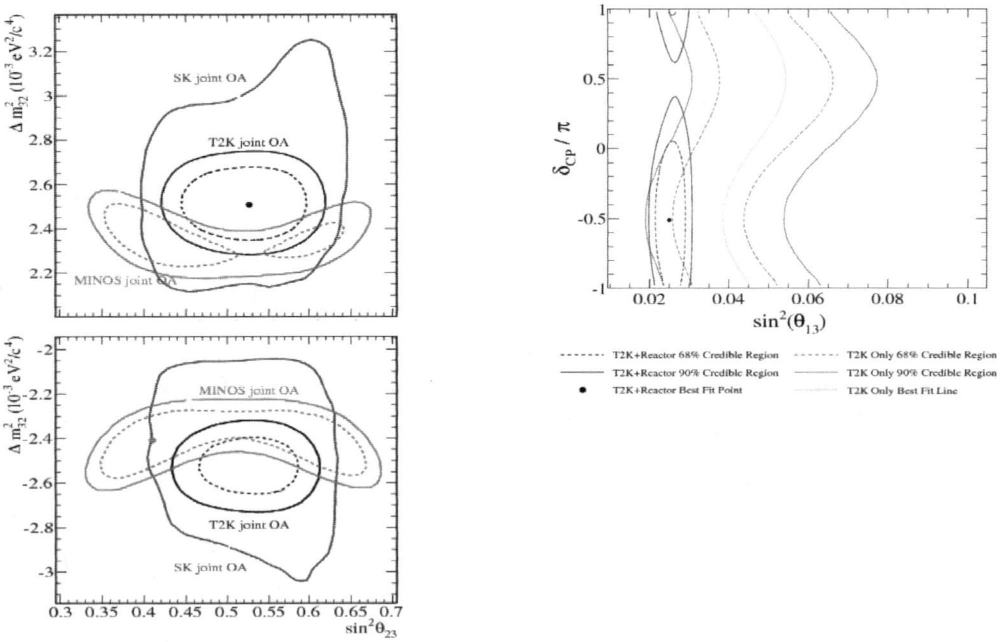


Fig.1 Left – allowed region of oscillation parameters $\sin^2\theta_{23}$, $\Delta m^2_{32}$ for normal (upper plot) and inverted (bottom plot) mass hierarchy; Right – allowed region of $\sin^2\theta_{13}$, $\delta_{CP}$ assuming normal mass hierarchy.

$\delta_{CP}$=0.0). For $\sin^2 2\theta_{13}$=0.1 we expect to find 21.59 $\nu_e$ events. Consequently the evidence of the $\nu_\mu \rightarrow \nu_e$ transition is observed.

We perform a combined 3-flavour simultaneous fit to the energy spectra of $\nu_\mu$-induced and $\nu_e$-induced events. It allows us to be sensitive to correlations between oscillation parameters. During the fit, the four parameters $\Delta m^2_{23}$, $\sin^2\theta_{23}$, $\sin^2\theta_{13}$ and $\delta_{CP}$ are simultaneously determined. As a result, we obtain the world best and most precise measurement of $\sin^2\theta_{23}$=0.524+0.057-0.059 (normal hierarchy) and 0.523+0.055-0.065 (inverted hierarchy). The allowed region of mixing parameters are shown on the left plot in Fig.1 alongside the MINOS and Super-Kamiokande atmospheric neutrino results. The results for the other mixing angle gives $\sin^2\theta_{13}$=0.042+0.013-0.021 (normal hierarchy) and 0.049+0.015-0.021 (inverted hierarchy). The allowed region of parameters in $\delta_{CP}$ and $\sin^2\theta_{13}$ phace space in shown in the right plot of Fig.1. When we combine the T2K results with the most precise measurement of the $\theta_{13}$ mixing parameter (obtained by reactor experiment [12]), then the allowed region of these parameters shrinks. Based on analyzed statistics of data reported here T2K-only could not claim sensitivity to $\delta_{CP}$. However, when the reactor results are used, then both experiments together start to be sensitive to $\delta_{CP}$. As a result we can exclude $\delta_{CP}$=[0.15,0.83]$\pi$ for the normal hierarchy and $\delta_{CP}$=[-0.08,1.09]$\pi$ for the inverted hierarchy at 90% confidence level. We can also conclude that the favored value of $\delta_{CP}$ is $-\pi/2$.

### 5.2 Anti-neutrino data results

The anti-neutrino data were collected from 2014 and the results presented in this paper correspond to 4.0x$10^{20}$POT. For the anti-neutrino run, the anti-$\nu_\mu$ disappearance and anti-$\nu_e$ appearance analyses were performed separately [13, 14].

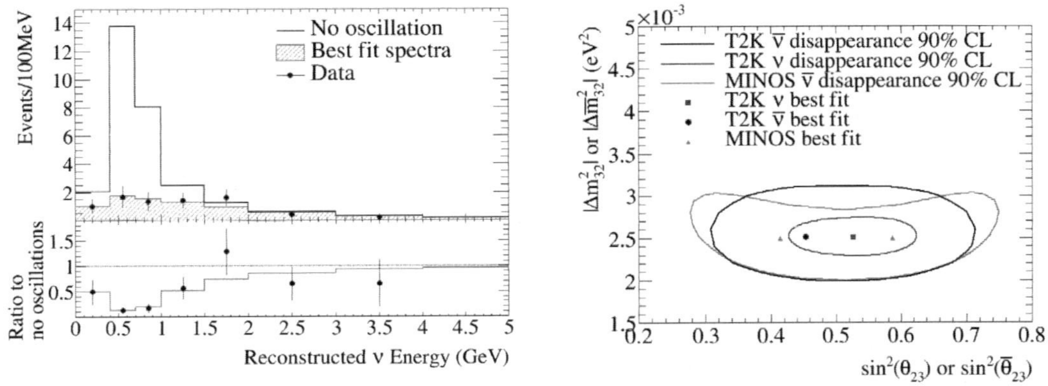


Fig.2 Left – Distribution of reconstructed neutrino energy for anti-$\nu_\mu$ data and MC; Right – Allowed region of parameters for anti-$\nu_\mu \rightarrow$ anti-$\nu_\mu$ and $\nu_\mu \rightarrow \nu_\mu$ disappearance oscillations.

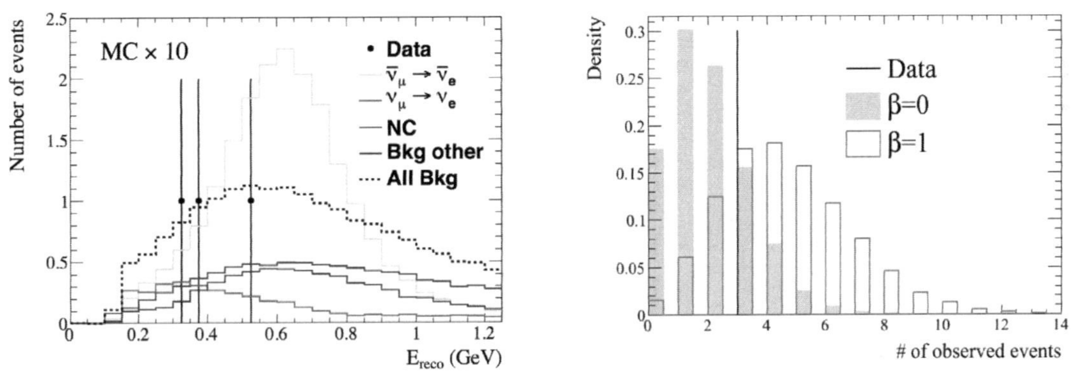


Fig.3 Left – Distribution of reconstructed neutrino energy for anti-$\nu_e$ data and MC; Right – Distribution of expected number of events for set of toy MC experiments.

There are 34 muon-like events found in the anti-neutrino run at the far detector. Super-Kamiokande detector cannot distinguish $\mu^-$ from $\mu^+$ and, therefore, can not specify if it was $\nu_\mu$ or anti-$\nu_\mu$ interacting. Here we refer to the accelerator produced neutrino or anti-neutrino beam for the specific run which was analyzed. In case of no oscillation, the expected number of events is 103.6 (for $\sin^2 2\bar{\theta}_{32}=0.0$). The clear signature of observed anti-$\nu_\mu$ disappearance is shown on the energy spectrum distribution of muon events in the left plot of Fig.2. During the fitting of anti-$\nu_\mu$ data, all oscillation parameters were fixed except $\sin^2\theta_{23}$ and $\Delta m^2_{32}$. The results, presented on the right plot of Fig.2 shows that the obtained mixing parameters for anti-neutrinos are consistent with those measured by T2K for the neutrino beam (indicating that CPT is conserved). The results are also consistent with other measurements of anti-neutrinos performed by MINOS and Super-Kamiokande. The measured values for the oscillation parameters were found to be $\sin^2\bar{\theta}_{23}$=0.45+0.38-0.64 and $|\Delta\bar{m}^2_{32}|$=2.51+0.29-0.26 x $10^{-3}$ $eV^2/c^4$.

For the T2K anti-$\nu_e$ appearance search we detected 3 events when 1.3 events are expected in case of no oscillation. 3.7 events are expected if anti-neutrinos oscillate with the same parameters as neutrinos assuming normal mass hierarchy and $\delta_{CP}=-\pi/2$ (see the left plot of Fig.3 for the energy distribution of those events). The expected number of events depends on $\delta_{CP}$ phase and the mass hierarchy. For the normal mas hierarchy, we expect 3.7, 4.3 and 4.9 events for $\delta_{CP}$ equals $-\pi/2$, 0 and

$\pi/2$ respectively. For the inverted mass hierarchy, those numbers change to 4.2, 4.9 and 5.5 expected events. Since the statistics of anti-$\nu$ accumulated data is low, we introduce the $\beta$ parameter which rescales the probability of anti-$\nu_e$ appearance:

$$P(\bar{\nu}_\mu \rightarrow \bar{\nu}_e) = \beta \times P_{PMNS}(\bar{\nu}_\mu \rightarrow \bar{\nu}_e)$$

$\beta$=0 corresponds to the situation when no anti-$\nu_e$ oscillation occurs while $\beta$=1 represents the case where anti-$\nu_e$ appearance occurs with the same oscillation parameters as $\nu_e$ appearance. Toy experiments were performed which throw parameters $\theta_{23}$, $\theta_{13}$, $\Delta m^2_{32}$ and $\delta_{CP}$ to obtain the expected number of events for each toy MC for two values of parameter $\beta$ (0 or 1). The obtained distribution of the expected number of events for those two scenarios is shown in right plot of Fig.3. This shows that, with only 3 detected events, those two scenarios can not be distinguished. More statistics are required to judge whether anti-$\nu_\mu$->anti-$\nu_e$ oscillation occurs.

## 6. Summary remarks

T2K is a world-leading experiment of neutrino physics, which has provided the most precise measurement of the $\theta_{23}$ mixing angle and discovered the appearance of $\nu_e$ in $\nu_\mu$ beam. We have also started to be sensitive to the unknown $\delta_{CP}$ violating phase, and to exclude some of its values when we combine our results with these from reactor neutrino experiments. T2K has begun to analyze its anti-neutrino data, and has proven that anti-$\nu_\mu$ disappearance is consistent with neutrino results. Results based on statistics discussed here are insufficient to allow us to draw conclusions regarding anti-$\nu_e$ appearance. For the most up-to-date situation, the reader may refer to the results announced during Neutrino 2016 and ICHEP 2016 conferences.

## Acknowledgements

The author acknowledge support of the National Science Center, Poland grant n.2014/14/M/ST2/00850 and Horizon 2020 MSCA-RISE project JENNIFER, under grant agreement n.644294.

## References

[1] The Super-Kamiokande Collaboration, *Phys. Rev. Lett. 81 (1998) 1562-1567*
[2] Q. R. Ahmad *et al., Phys. Rev. Lett. 89 (2002) 011301*
[3] M. H. Ahn *et al., Phys. Rev. D74 (2006) 072003*
[4] P. Adamson *et al. Phys. Rev. Lett.112.191801(2014)*, and P. Adamson *et al., Phys. Rev. Lett. 110.2518011(2013)*
[5] N. Agafonova *et al., Phys. Rev. D89 (2014) 5, 051102*
[6] K. Abe *et al., Nucl. Instrum. Meth. A659 (2011), 106-135*
[7] K. Abe *et al., Phys. Rev. Lett. 112 (2014) 181801*
[8] K. Abe *et al., Phys. Rev. Lett. 112 (2014) 061802*
[9] S. Fukuda *et al., Nucl. Instrum. Meth. A501 (2003) 418-462*
[10] D. Sgalaberna *et al., Nucl. Part. Phys. Proc. 273-275 (2016) 2672-2674*
[11] K. Abe *et al., Phys. Rev. D91 (2015) 7, 072010*
[12] P. F. An *et al., Phys. Rev. Lett. 115 (2015) 111802*
[13] K. Abe *et al., Phys. Rev. Lett. 116 (2016) 18, 181801*
[14] M. Ravonel *et al., PoS EPS-HEP2015 (2015) 047*

Journal of Physics: Conference Series **770** (2016) 012023 doi:10.1088/1742-6596/770/1/012023

# Recent Results from the Daya Bay Reactor Neutrino Experiment

**En-Chuan Huang**

Physics Department, University of Illinois at Urbana-Champaign
1110 W Green St, Urbana, IL 61801, U.S.A.

E-mail: ehuang7@illinois.edu

**Abstract.** The Daya Bay Reactor Neutrino Experiment is designed to precisely measure the mixing parameter $\sin^2 2\theta_{13}$ via relative measurements with eight functionally identical antineutrino detectors (ADs). In 2012, Daya Bay has first measured a non-zero $\sin^2 2\theta_{13}$ value with a significance larger than $5\sigma$ with the first six ADs. With the installation of two new ADs to complete the full configuration, Daya Bay has continued to increase statistics and lower systematic uncertainties for better precision of $\sin^2 2\theta_{13}$ and for the exploration of other physics topics. In this proceeding, the latest analysis results of $\sin^2 2\theta_{13}$ and $|\Delta m^2_{ee}|$, including a measurement made with neutron capture on Gadolinium and an independent measurement made with neutron capture on hydrogen are presented. The latest results of the search for sterile neutrino in the mass splitting range of $10^{-3}$ eV$^2$ < $|\Delta m^2_{41}|$ < 0.3 eV$^2$ and the absolute measurement of the rate and energy spectrum of reactor antineutrinos will also be presented.

## 1. Introduction

Neutrino oscillation between three active neutrinos has been well established by experiments. Among them, Daya Bay first observed the non-zero value of the mixing angle $\theta_{13}$ [1]. Encouraged by several experimental anomalies, the possible existence of sterile neutrino, neutrinos that do not interact via weak interaction, is also actively considered. The "Reactor Neutrino Anomaly" [2] found that the predicted reactor neutrino flux, derived from the ILL beta spectra measurements [3],[4],[5], is higher than the measured rate. Precision measurement at Daya Bay can precisely determine the oscillation parameters $\theta_{13}$ and $\Delta m^2_{ee}$, search for sterile neutrino, and provide reactor neutrino flux and energy spectrum.

In the standard framework of neutrinos mixing, the three flavor eigenstates $(\nu_e, \nu_\mu, \nu_\tau)$ mix with the three mass eigenstates $(\nu_1, \nu_2, \nu_3)$ via the PMNS matrix [6],[7], which contains three mixing angles ($\theta_{12}$, $\theta_{13}$, $\theta_{23}$) and one CP phase. The survival probability $P_{\bar{\nu}_e \to \bar{\nu}_e}$ for electron antineutrino can be written as

$$P_{\bar{\nu}_e \to \bar{\nu}_e} = 1 - \cos^4\theta_{13} \sin^2 2\theta_{12} \sin^2\Delta_{21} - \sin^2 2\theta_{13} \left(\cos^2\theta_{12} \sin^2\Delta_{31} + \sin^2\theta_{12} \sin^2\Delta_{32}\right), \quad (1)$$

where $\Delta_{ij} = \Delta m^2_{ij} L/E_\nu$. The $\Delta m^2_{ij}$ is the mass-squared difference between mass eigenstates $i$ and $j$, while $L$ and $E_\nu$ are the baseline and $\bar{\nu}_e$ energy, respectively. For Daya Bay, the oscillation due to $|\Delta m^2_{32}|$ is indistinguishable from the one due to $|\Delta m^2_{31}|$; therefore, the term in the parenthesis in Eq. (1) can be approximated as $\sin^2\Delta_{ee}$. For reactor neutrinos with $E_{\bar{\nu}_e} \approx 3$ MeV, the first oscillation minimum happens at $L \approx 1.6$ km.

Journal of Physics: Conference Series **770** (2016) 012023 doi:10.1088/1742-6596/770/1/012023

## 2. Daya Bay Experiment

The Daya Bay reactor neutrino experiment [8] is designed to precisely determine $\sin^2 2\theta_{13}$ via relative measurement. Daya Bay nuclear power plant has six 2.9 $GW_{th}$ reactor cores, and the reactor cores are grouped by two into the Daya Bay cores, Ling Ao I cores, and Ling Ao II cores. Eight identically-designed antineutrino detectors (ADs) are placed at three experimental halls (EHs). EH1 (EH2), the near site to Daya Bay (Ling Ao) cores, has two ADs, while EH3, the far site located around the first oscillation minimum, has four ADs. Figure 1 shows the position of Daya Bay nuclear power plant near Hong Kong and the configuration of the Daya Bay experiment.

Each AD consists of three concentric cylindrical vessels. The innermost vessel, having a diameter of 3.1 meters, is filled with 20 tons of Gadolinium-doped liquid scintillator (LS), serving as the $\bar{\nu}_e$ target. The intermediate vessel with a diameter of 4 meters is filled with LS to catch gammas escaping from the target zone. The outermost vessel with a diameter of 5 meters contains mineral oil to shield radiation from surrounding materials. Eight rows of PMTs, each row with 24 equally spaced PMTs, are installed on the vertical wall within each AD. Optical reflectors are installed on top and bottom of the AD to maximize the photoelectrons collected. On top of each AD, three automatic calibration units can deploy calibration sources into different regions of the AD. All the ADs in one EH are immersed in a water pool, acting as a water Cherenkov detector. Figure 2 shows the schematic of an AD.

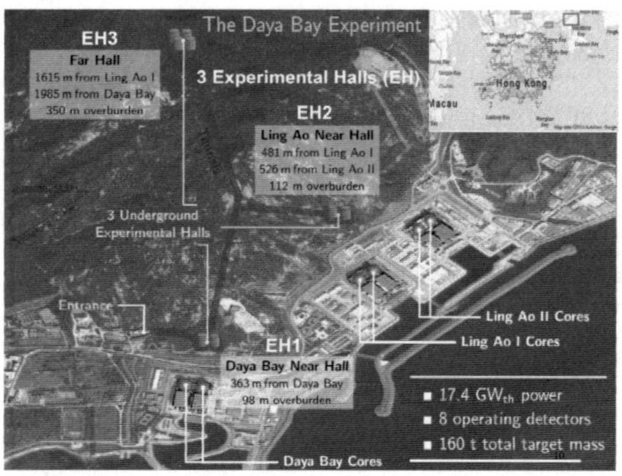


**Figure 1.** The configuration of the Daya Bay reactor neutrino experiment and the relative position to Hong Kong. The Daya Bay nuclear power plant, consisting six 2.9-$GW_{th}$ reactor cores, sits northwest to Hong Kong. Eight identically-designed antineutrino detectors (ADs) are placed in three experimental halls (EHs).

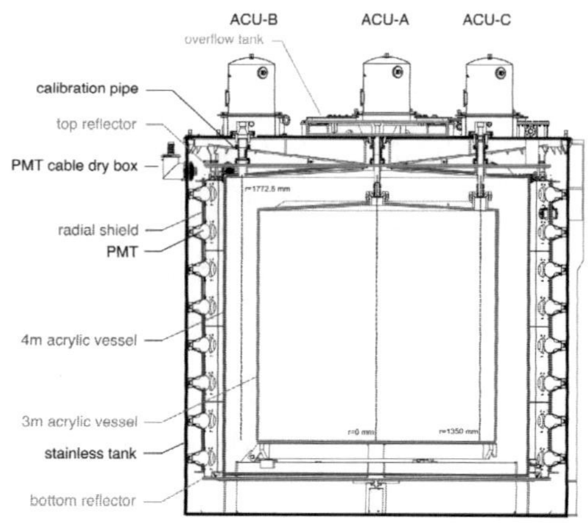


**Figure 2.** The design of Daya Bay antineutrino detector (AD). Each AD is separated into three concentric zones. The 3m innermost zone is filled with Gd-doped liquid scintillator, and the intermediate zone is filled with LS. The outermost zone is filled with mineral oil.

## 3. Inverse Beta Decay Events

The reactor $\bar{\nu}_e$ is observed via inverse beta decay (IBD) reaction, $\bar{\nu}_e + p \to e^+ + n$. The $e^+$ annihilates with an electron within a few nanoseconds, generating a prompt signal. The prompt energy $E_p$ can be related to the $\bar{\nu}_e$ energy $E_\nu$ as $E_p \approx E_\nu - 0.8$ MeV. The $n$ can be captured by either gadolinium or hydrogen. For neutron captured by gadolinium, which is 0.1% by weight in the target zone, the capture time is about $30\mu s$ with a delayed energy, $E_d$, around 8 MeV. For

Journal of Physics: Conference Series **770** (2016) 012023 doi:10.1088/1742-6596/770/1/012023

**Table 1.** Number of IBD and background events for each data set.

| | IBD Candidates | | | Backgrounds | | |
|---|---|---|---|---|---|---|
| Data set | EH1 | EH2 | EH3 | EH1 | EH2 | EH3 |
| 217 days(6AD) | 203809 | 92912 | 41589 | $4076.6 \pm 462.4$ | $1580.3 \pm 147.8$ | $1878.9 \pm 94.6$ |
| 621 days | 613813 | 477144 | 150255 | $11624.5 \pm 968.8$ | $7371.5 \pm 530.0$ | $3984.1 \pm 152.3$ |
| 1230 days | 1203969 | 1033209 | 308150 | $21451.4 \pm 1359.9$ | $14911 \pm 944.2$ | $6011.9 \pm 154.1$ |

neutron captured by hydrogen in the LS zone, capture time is ~200 $\mu$s with a delayed energy at ~2.2 MeV.

### *3.1. Selection Criteria*
The selection criteria for IBD events with neutron captured on Gd requires that $E_p$ is within $(0.7, 12)$ MeV, $E_d$ is within $(6, 12)$ MeV, and the capture time is in the $(1, 200)$ $\mu$s window. For neutron captured on hydrogen, the IBD selection criteria requires that $E_p$ is within $(1.5, 12)$ MeV, $E_d$ is within $3\sigma$ of the hydrogen 2.2 MeV peak, the capture time is in the $(1, 400)$ $\mu$s window, and distance between the prompt and the delayed signal is within 50 cm. The higher $E_p$ threshold and the vertex cut for n-H events are to avoid accidental events, events that are physically-unrelated but accidentally appear in the selection window. Furthermore, an event is rejected if it is within the cosmic muon window determined by the water pool or the AD, or if it is determined to be a flasher event, that is, spontaneous PMT light emission, or if there is more than one coincidence pair.

### *3.2. Data Sets*
Daya Bay started data-taking with 6 ADs installed, 2 in EH1, 1 in EH2, and 3 in EH3. With 217 days of data taken in the 6AD period, Daya Bay conducted the measurement of $\sin^2 2\theta_{13}$ and $\Delta m^2_{ee}$ using spectral information [9], the search for sterile neutrino [10], and the measurement of reactor flux [11]. The last two ADs were installed in the summer of 2012. The 621 days of data, combining both 6-AD period and 8-AD period, was analyzed for the 3-$\nu$ oscillation. In this proceeding, the results for the 3-$\nu$ oscillation analysis is updated with the 1230 days of data from the previous 621-day analysis. For sterile neutrino search and the measurement reactor antineutrino flux and spectrum, the results are updated with the 621 days of data from the 6-AD analysis. Table 1 lists the number of IBD events and backgrounds for each data period.

### *3.3. Energy Calibration*
The relative energy scale, the energy difference between ADs, is determined by measuring peaks from calibration sources, spallation neutrons, and natural radioactivity. The uncertainty of relative energy scale was found to be less than 0.2%. The absolute energy scale, connecting the reconstructed energy to true energy, includes the nonlinearity from both LS and electronic readouts. The absolute energy scale is determined by the measured $\gamma$ peaks and the continuous $\beta$ spectrum from $^{12}$B, validated with Michel electrons and the continuous $\beta + \gamma$ spectra from $^{212/214}$Bi and $^{208}$Tl.

## 4. Recent Results
### *4.1. Measurement of* $\sin^2 2\theta_{13}$ *and* $\Delta m^2_{ee}$
With the 1230 days of data, the measurement of $\sin^2 2\theta_{13}$ and $|\Delta m^2_{ee}|$ yielded

$$\begin{aligned} \sin^2 2\theta_{13} &= [8.41 \pm 0.27(\text{stat.}) \pm 0.19(\text{syst.})] \times 10^{-2} \\ |\Delta m^2_{ee}| &= [2.50 \pm 0.06(\text{stat.}) \pm 0.06(\text{syst.})] \times 10^{-3}\ \text{eV}^2 \end{aligned} \quad (2)$$

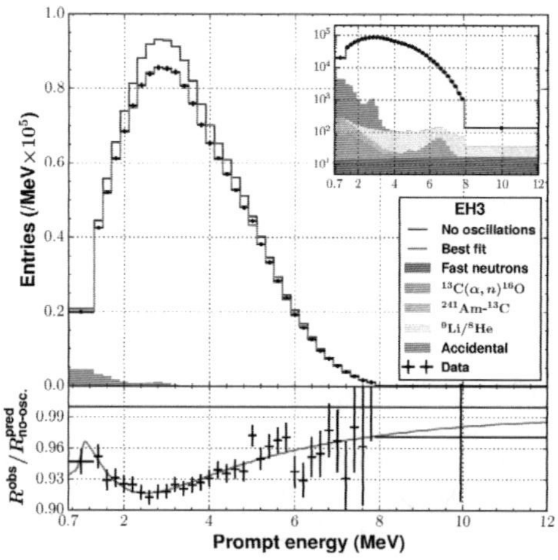


**Figure 3.** (1230 days) Observed energy spectra at EH3 and the best-fit curve.

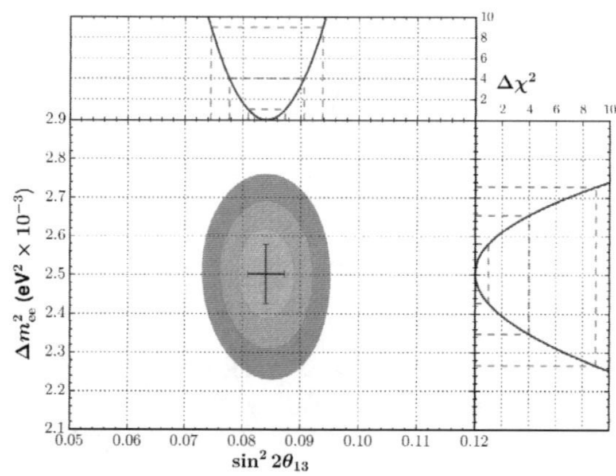


**Figure 4.** (1230 days) Allowed regions at the 68.3%, 95.5%, and 99.7% confidence levels in the $|\Delta m^2_{ee}| - \sin^2 2\theta_{13}$ plane.

Figure 3 shows the observed energy spectra at the far site, EH3. A clear deficit compared with the no-oscillation prediction and the energy dependence of the oscillation probability can be observed. Figure 4 shows the allowed region at 1, 2, and 3 $\sigma$ in the plane of $|\Delta m^2_{ee}| - \sin^2 2\theta_{13}$. Daya Bay has the most precise measurement on $\sin^2 2\theta_{13}$. The experiment is still statistically dominated, while the largest systematic contribution is the uncertainty in relative energy scale. For $|\Delta m^2_{32}|$, Daya Bay has achieved similar precision as those measured in the muon neutrino channel, as shown in Figure 5.

For neutron captured on hydrogen, the rate analysis, updated with the 621 days of data from the 6-AD only data, yielded $\sin^2 2\theta_{13} = 0.071 \pm 0.011$, consistent with the analysis with neutron captured on gadolinium.

### *4.2. Search for Sterile Neutrino*

With a minimum extension to the standard three active neutrino framework, Daya Bay searched for sterile neutrino in a 3(active)+1(sterile) framework. The $\bar{\nu}_e$ survival probability can be approximated with the convention in Ref. [12] as:

$$P_{\bar{\nu}_e \to \bar{\nu}_e} \approx 1 - \sin^2 2\theta_{13} \sin^2 \Delta_{31} - \sin^2 2\theta_{14} \sin^2 \Delta_{41}. \tag{3}$$

Therefore, if light sterile neutrino exists, it would distort the observed energy spectra. With ADs at multiple baselines, Daya Bay compares the difference in the measured energy spectra to exclude a large range of $|\Delta m^2_{41}|$ in the sub-eV region. Figure 6 shows that two independently-developed analyses at Daya Bay show consistent exclusion region [13]. The constraints set on $\sin^2 2\theta_{14}$ for $2 \times 10^{-4}$ eV$^2 \lesssim \Delta m^2_{41} \lesssim 0.3$ eV$^2$ is about 50% smaller than the constraints set by the analysis using the 6-AD data set [10].

### *4.3. Absolute Neutrino Flux*

With the 621 days of data, Daya Bay has also released a new measured reactor antineutrino flux and energy spectrum. The measured IBD yield $Y$ was found to be $Y(\mathrm{cm}^2/\mathrm{GW/day}) = (1.55 \pm 0.03) \times 10^{-18}$ [14]. Compared with the Huber+Mueller [15],[16] (ILL+Vogel [3],[4],[5],[17]) model,

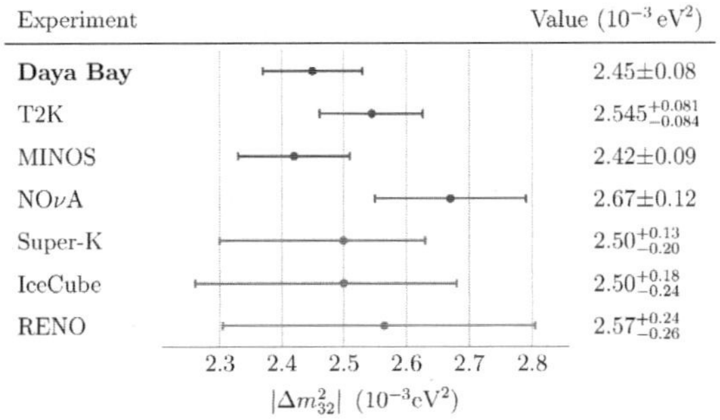


**Figure 5.** (1230 days) The global measurements on $|\Delta m^2_{32}|$. Daya Bay has achieved similar precision on $|\Delta m^2_{32}|$ with experiments in the muon neutrino channel.

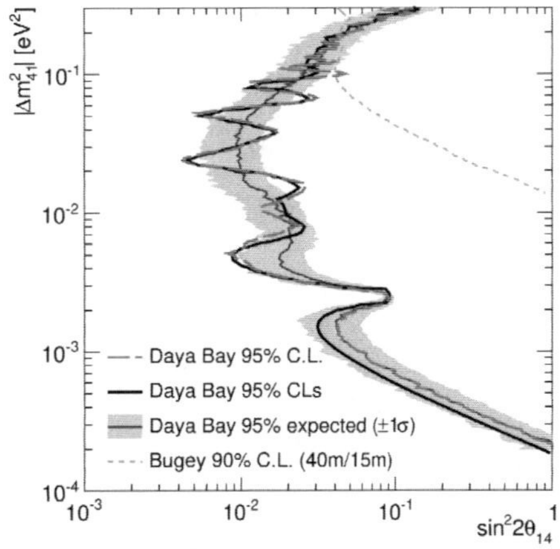


**Figure 6.** (621 days) The exclusion contour [13] for the sterile neutrino search at Daya Bay in the plane of $|\Delta m^2_{41}| - \sin^2 2\theta_{14}$. Figure is extracted from Ref. [13].

the ratio of the measured flux to predicted flux was found to be $0.946\pm0.020$ ($0.992\pm0.021$), a 2.9 $\sigma$ deviation from Huber+Mueller prediction, consistent with the global average of measured reactor neutrino flux [2]. The measured yields for each AD at 6-AD only and 6 plus 8 AD periods are consistent, as shown in Figure 7. Compared with the Huber+Mueller model, the measured energy spectra was found to have a 4.4 $\sigma$ excess in the region of 4-6 MeV, as shown in Figure 8.

## 5. Conclusion

With the new 1230 days of data, Daya Bay has the most precise measurement of $\sin^2 2\theta_{13}$ to date. The energy spectrum and flux of reactor antineutrinos and a search of sterile neutrino are reported with the 621 days of data.

## 6. Acknowledgment

The Daya Bay experiment is supported in part by the Ministry of Science and Technology of China; the U.S. Department of Energy; the Chinese Academy of Sciences; the CAS Center for Excellence in Particle Physics; the National Natural Science Foundation of China; the Guangdong provincial government; the Shenzhen municipal government; the China General Nuclear Power Group; the Research Grants Council of the Hong Kong Special Administrative Region of China; the Ministry of Education in Taiwan; the U.S. National Science Foundation; the Ministry of Education, Youth and Sports of the Czech Republic; the Joint Institute of Nuclear Research in Dubna, Russia; the NSFC-RFBR joint research program; and the National Commission for Scientific and Technological Research of Chile. We acknowledge Yellow River Engineering Consulting Co., Ltd. and China Railway 15th Bureau Group Co., Ltd. for building the underground laboratory. We are grateful for the ongoing cooperation from the China Guangdong Nuclear Power Group and China Light & Power Company.

## References

[1] An F *et al.* (Daya Bay Collaboration) 2012 *Phys. Rev. Lett.* **108** 171803

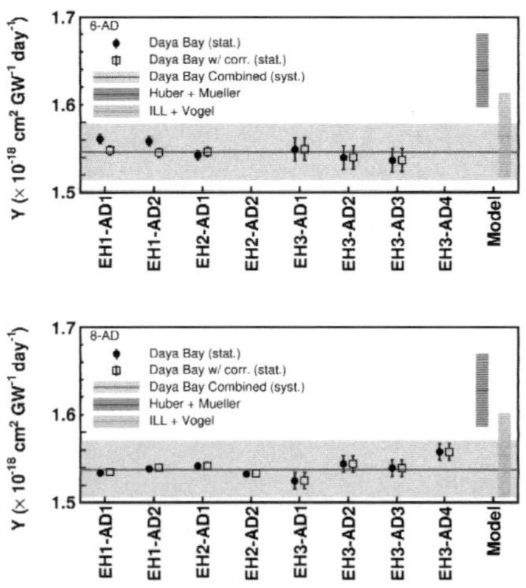

**Figure 7.** (621 days) The measured reactor flux for each AD at both 6-AD (top) and 6+8 AD (bottom) period. The closed circles show the measured rates corrected by the 3-$\nu$ oscillation, while the open squares show the results corrected by flux-weighted fission fraction. The horizontal lines show the average for the period, and the gray band shows the 1-$\sigma$ uncertainty. Figure is extracted from Ref. [14].

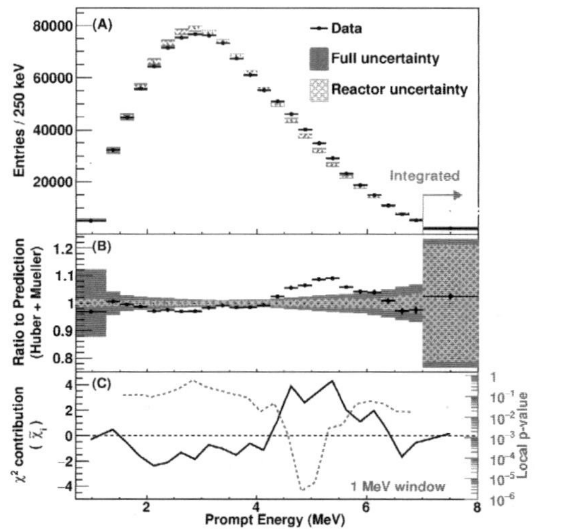

**Figure 8.** (621 days) (A) The measured reactor neutrino flux compared with the Huber+Mueller model. (B) The ratio of the measured energy spectrum to the Huber+Mueller prediction. (C) The $\chi^2$ distribution and local p-value. A 4.4 $\sigma$ excess is observed in the 4 to 6 MeV window. Figure is extracted from Ref. [14].

[2] Mention G *et al.* 2011 *Phys. Rev. D* **83** 073006
[3] Schreckenbach K *et al.* 1985 *Phys. Lett. B* **160** 325
[4] von Feilitzsch F, Hahn A and Schreckenbach K 1982 *Phys. Lett. B* **118** 162–166
[5] Hahn A *et al.* 1989 *Phys. Lett. B* **218** 365–368
[6] Pontecorvo B 1968 *Sov. Phys. JETP* **26** 984–988
[7] Maki Z, Nakagawa M and Sakata S 1962 *Prog. Theor. Phys.* **28** 870–880
[8] An F P *et al.* (Daya Bay Collaboration) 2016 *Nucl. Instrum. Meth.* **A811** 133–161
[9] An F *et al.* (Daya Bay Collaboration) 2014 *Phys. Rev. Lett.* **112** 061801
[10] An F *et al.* 2014 *Phys. Rev. Lett.* **113** 141802
[11] An F P *et al.* (Daya Bay Collaboration) 2016 *Phys. Rev. Lett.* **116** 061801
[12] Palazzo A 2013 *JHEP* **1310** 172
[13] An F P *et al.* (Daya Bay Collaboration) 2016 *Phys. Rev. Lett.* **117** 151802
[14] An F P *et al.* (Daya Bay Collaboration) 2016 (arXiv:1607.05378 [hep-ex])
[15] Huber P 2011 *Phys. Rev. C* **84** 024617
[16] Mueller T *et al.* 2011 *Phys. Rev. C* **83** 054615
[17] Vogel P, Schenter G K, Mann F M and Schenter R E 1981 *Phys. Rev. C* **24**

Journal of Physics: Conference Series **770** (2016) 012024 doi:10.1088/1742-6596/770/1/012024

# Hadron scattering, resonances, and QCD

**R A Briceño**

Thomas Jefferson National Accelerator Facility,
12000 Jefferson Avenue, Newport News, VA 23606, USA

E-mail: rbriceno@jlab.org

**Abstract.** The non-perturbative nature of quantum chromodynamics (QCD) has historically left a gap in our understanding of the connection between the fundamental theory of the strong interactions and the rich structure of experimentally observed phenomena. For the simplest properties of stable hadrons, this is now circumvented with the use of lattice QCD (LQCD). In this talk I discuss a path towards a rigorous determination of few-hadron observables from LQCD. I illustrate the power of the methodology by presenting recently determined scattering amplitudes in the light-meson sector and their resonance content.

## 1. Introduction

The vast majority of hadronic states observed in nature are not stable under the strong interaction, but rather decay in the timescale of the interactions. As a result, these resonances do not live long enough to propagate and interact with particle detectors. Instead, their existence is manifested as dynamical enhancement of scattering amplitudes of their by-products - states composed of two or more hadrons. Therefore, to study the spectrum of the fundamental theory of the strong interactions, quantum chromodynamics (QCD), necessarily requires one to study few-body, strongly-interacting systems. In this talk I review some of recent progress towards achieving this goal directly from QCD and I focus my attention in the light sector of QCD [9, 8, 7, 14, 29, 28, 30, 13, 12], where calculations have reached a high level of maturity. In particular, I present the very first scattering amplitude in the isoscalar sector of QCD [9] and the first form factor of a resonance [8, 7]. These studies partly stem from formal theoretical developments which I briefly discuss [4, 17, 2, 5, 6].

Let me begin by focusing even further to two key examples. The first is perhaps the best understood hadronic resonance, namely the $\rho$. This resonance manifests itself as a "*bump*" in the isovector $\pi\pi$ scattering amplitude (or cross section), making it a perfect example of the common understanding of a resonance. This is nicely illustrated in Fig. 1, where one sees a clear enhancement in the $\pi\pi$ amplitude when the center of mass (cm) energy is in the proximity of 770 MeV. Unitarity dictates that the amplitude for the $\ell th$ partial wave can be written in terms of the scattering phase shift, $\delta_\ell$, $\mathcal{M}_\ell = \frac{8\pi E_{cm}}{p\cot\delta_\ell - ip}$, where $E_{cm}$ and $p$ are the c.m. energy and relative momentum. Using this, we find that the peak of this amplitude corresponds to approximately when the phase shift goes through 90° as seen in Fig. 1.

Although it is easy to understand resonances as bumps in amplitudes, this definition fails to describe some of the most interesting resonances in nature. To find a counterexample one can look in the isoscalar sector of QCD. This is perhaps one of the most interesting corners of QCD, as it includes tentative glueballs, molecules, tetraquarks, etc. Also, this sector hosts the lightest of all hadronic resonances, namely $\sigma/f_0(500)$. Although this plays an important role

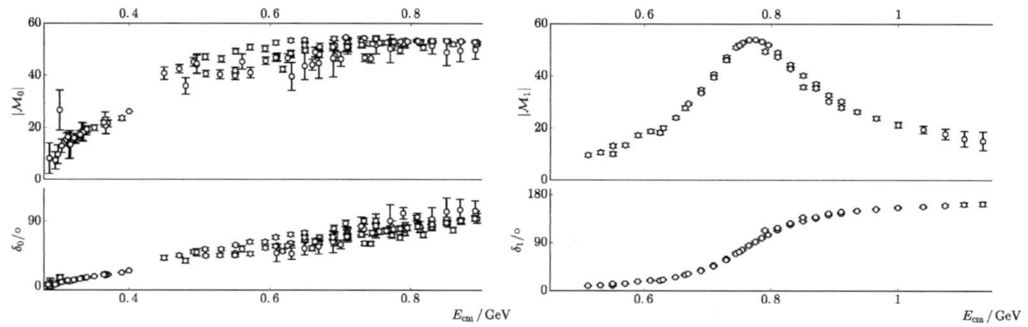


**Figure 1.** The top left and right panels respectively show the $\pi\pi$ isoscalar and isovector scattering amplitudes obtained from experiment [26, 20, 16, 15]. The bottom panels show the corresponding phase shifts.

in a wide range of phenomenology (see Ref. [25] for a recent review), it was not until recently that its existence was demonstrated with reliable certainty. As shown in Fig. 1, the reason the evidence for this state has been controversial is due to the absence of a bump-like signature in the low-energy region of the isoscalar $\pi\pi$ scattering amplitude. This is perhaps most striking when one realizes that its mass is $m_\sigma = 449(^{22}_{16})$ MeV, which is smaller than its decay width $\Gamma_\sigma = 550(24)$ MeV [25].

Resonances can be rigorously defined as complex poles in scattering amplitudes. The imaginary and real components of these poles are directly related to their mass and width, $E_{pole} = \sqrt{s_{pole}} = m_R \pm i\Gamma_R/2$. It is easy to convince oneself that this definition is consistent with the picture of a resonance as a bump in a scattering amplitude when the width of the resonance is small compared to its mass. In the remainder of this talk, I review how resonant and non-resonant scattering amplitudes can be determined from lattice QCD. In Sec. 2, we will implement this methodology to determine the isovector and isoscalar $\pi\pi$ scattering amplitude. Having determined these amplitudes we will examine their pole structure and consequently their resonance content. In Sec. 3, I explain how these ideas are being extended to processes involving electroweak probes and present the first determination of a resonance form factor from lattice QCD [8, 7].

## 2. Scattering amplitudes from lattice QCD

In order to define lattice QCD, it is necessary to make four important compromises: nonzero lattice spacing, Wick rotation to a Euclidean spacetime, finite spacetime, and unphysical values for the quark masses. The first of these introduces a separation between all points in spacetime, and as a result spacetime resembles a crystal lattice. The second is done in order to use Monte Carlo sampling. The third is necessary to be able to store gauge fields, propagators, etc, which scale with the volume $(L^3 \times T)^2$ assuming a cubic volume, where $L$ and $T$ are the spatial and temporal extents. The last one has been a historical limitation that has been circumvented for the simplest of observables. In this talk I focus on few-body systems, where calculations are still in the exploratory stages and still being performed using unphysically heavy values for the quark masses.

For the physics we are after discretization effects are negligible and will be ignored. Naïvely one might consider the second and third compromises above as limitations, but these play a crucial role in the determination of few-body observables. First, the fact that the spacetime is finite leads to a discrete spectrum. Second, the fact that calculations are performed in a Euclidean spacetime allows us to identify correlation functions as sums of exponential whose arguments depend on the discrete spectrum. For example, the two-point correlation

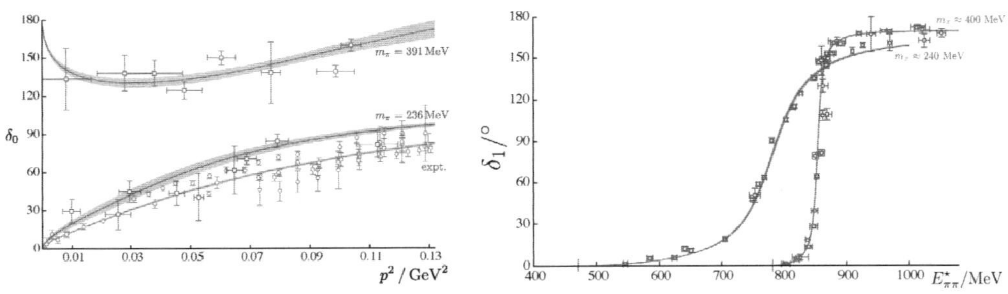


(a) (b)

**Figure 2.** Shown are the lattice results for the elastic $\pi\pi$ scattering phase shifts in the (a) isoscalar [9] and (b) isovector channels [29, 12]. The isoscalar and isovector channels are plotted as a function of the c.m. relative momentum and energy respectively. These have been determined using two values of the quark masses, and the isoscalar is compared with the experimental values [26, 20, 16, 15].

function for a source $\mathcal{O}_a^\dagger(0)$ at time $t = 0$ and a sink operator $\mathcal{O}_b(t)$ can be written as, $C_{ab}^{2pt.}(t) \equiv \langle 0|\mathcal{O}_b(t)\mathcal{O}_a^\dagger(0)|0\rangle = \sum_n Z_{b,n}Z_{a,n}^* e^{-E_n t}$, where $E_n$ is the $nth$ eigenvalue of the finite-volume Hamiltonian, $Z_{b,n}$ and $Z_{a,n}^*$ are the overlap factors with the sink and source operators respectively.

For a state whose energy lies well below multiparticle thresholds, the finite-volume energy can be identified as its infinite-volume counterpart up to corrections of the order of $\mathcal{O}(e^{-m_\pi L})$ [21]. For volumes satisfying $m_\pi L \gg 1$ one can safely neglect such corrections. For states above multiparticle thresholds there is no direct relation between finite-volume and infinite-volume states. This is easy to understand when one remembers that in the infinite-volume limit, states above thresholds are necessarily resonances and they correspond to complex poles on the second Riemann sheet. In a finite-volume, since there is not a continuum of states there is no branch cut. This means that there is a single Riemann sheet, and due to causality all states must lie on the real axis.

Instead one can non-perturbatively relate the finite-volume spectrum to the infinite-volume scattering amplitude. For two-particle systems such a relation was first found by Martin Lüscher [22, 23] and has been generalized to increasingly complex two-particle systems [22, 23, 4, 17, 2]. In general, one can find that an energy level in between the two- and three-particle thresholds satisfies, $\det[F^{-1}(P,L) + \mathcal{M}(E_L)] = 0$, where $F(E_L, L)$ is a known function of the spectrum and the volume [2] and $\mathcal{M}$ is the scattering amplitude we are after. The determinant acts in the space of open channels and partial waves. For sufficiently low energies where a single partial wave dominates, it is straightforward to show that this equation simplifies down to where $\cot\phi(P,L)$ is related to $F(E_L, L)$.

Using this equation one can then directly map a finite-volume energy obtained from lattice QCD onto the infinite-volume scattering phase shift of the given channel evaluated at that same energy. This is what has been done in Refs. [9, 29, 12] for the isoscalar and isovector $\pi\pi$ channels using two different values of the light quark-masses corresponding to $m_\pi \approx 236, 391$ MeV. The resulting phase shifts and and fits are shown in Fig. 2. From the isovector phase shifts, it is evident that for these quark-masses the $\rho$ resonance is fairly narrow and broadens as the quark-masses approach their physical value. This is as one would expect, given that as the quarks becomes increasingly light the phase space for the $\rho$ to decay opens up. For the isoscalar phase shifts, the $\sigma$ is bound for the heavy quark-masses, which is manifested by the phase shift starting at 180°. For the lighter ensemble, just like in experiment, it is less clear what the manifestation

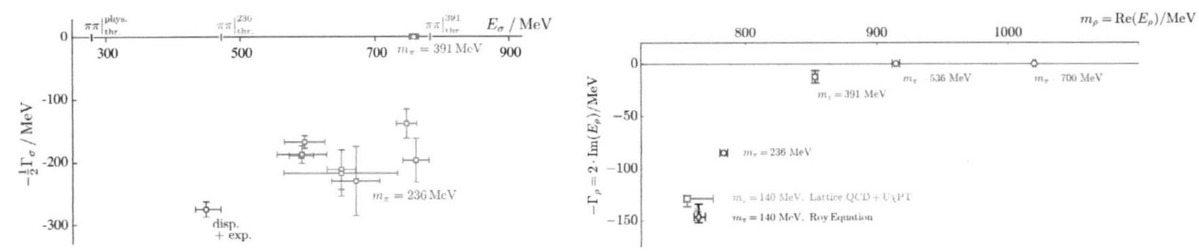


(a) (b)

**Figure 3.** (a) Shown is the mass and width of the $\sigma$ resonance for the two values of the quark masses studied via lattice QCD [9] and it is compared to the results from dispersive analysis of experimental data [25]. (b) Shown are the same quantities for the $\rho$ [29, 12, 11], including a chiral extrapolation in red [1] and the result from dispersive analysis in black.

of the $\sigma$ is, except one can conclude it is not a narrow resonance.

In Ref. [1] a chiral extrapolation of the isovector scattering amplitude was performed. The resulting amplitude was in good agreement with experimental values up to c.m. energies of 1.2 GeV, well above several inelastic thresholds. This confirms that inelastic effects can be safely ignored for this channel for a large kinematic region. Although a chiral extrapolation of the isoscalar amplitude is presently missing, in Fig. 2(a) one sees a natural trend for the phase shift from heavy quark masses down to the experimental point [26, 20, 16, 15].

To rigorously study the resonant content of these amplitudes, one must investigate their pole structure in the complex plane. Since lattice QCD energies are real, the scattering amplitude is only constrained on the real axis. To analytically continue onto the complex plane, one must rely on parametrizations of the scattering amplitude. Using a wide range of these, described in Ref. [30], one sees in Fig. 3 the resulting pole locations described in terms of the masses and widths of the $\sigma$ and $\rho$ resonances. First, let's focus on the $\rho$ poles, which also include values of quark-masses where it is bound [11]. The red point depicts the pole position obtained after performing a chiral extrapolation of the $m_\pi = 236$ MeV ensembles [1], and it is in good agreement with solutions to the Roy equation constrained from experimental data. This figure reemphasizes the transition of the $\rho$ from a stable particle to a an increasingly broad resonance.

By performing calculations at sufficiently heavy values of the quark-masses, the $\sigma$ is found to be bound [9]. This is perhaps the first clear evidence of the existence of the $\sigma$ from QCD. As $m_\pi$ approaches its physical value, the phase space for the $\sigma$ to decay quickly opens and it becomes a broad resonance resembling the experimental situation. Given that this corresponds to a pole far from the real axis where the scattering amplitude has been constrained, it should not be a surprise that there might be a large systematic error associated with the parametrization chosen to perform such an extrapolation. This is represented by the scattered red points in Fig. 3(a). This systematic error can be dramatically reduced by using dispersive techniques, as has been done to analyze experimental data [25].

I end this section by making a couple of remarks. First, it is important to emphasize that determining the isoscalar finite-volume spectra, and even more the corresponding scattering amplitudes, directly from QCD is a technological achievement. Having obtained these amplitudes, the determination of resonances closely mirrors experimental analysis. As we approach the physical point, the analysis becomes increasingly complex, just as in experiment. Lattice QCD does possess one advantage over experiment. This is the fact that one can '*dial*' the parameters of the standard model, in particular the quark masses. This allows one to see direct evidence of states whose experimental signature might be small, the $\sigma$ being a perfect example. Second, although I have only discussed kinematic regions where a single channel is open, these

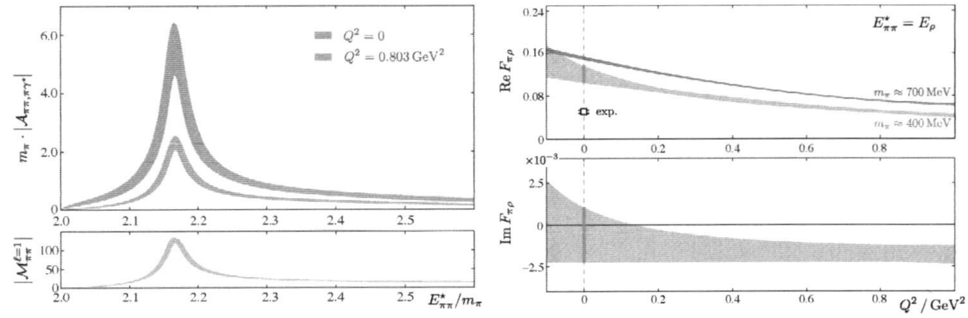


**Figure 4.** (a) Shown is the absolute values of the $\pi^+\gamma^\star \to \pi^+\pi^0$ amplitude, $\mathcal{A}_{\pi\pi,\pi\gamma^\star}$ as a function of the $\pi\pi$ c.m. energy for two different $Q^2$ [8, 7], and it is compared to the elastic $\pi\pi$ amplitude below [12]. (b) Shown are the real and imaginary components of the $\pi \to \rho$ form factor for $m_\pi \approx 400$ MeV [8, 7], and it is compared to its value at $m_\pi \approx 700$ MeV [28].

ideas are also applicable for energies more channels are open. In fact the very first calculations involving more than open channel have been carried recently [14, 29, 30, 13]. Lastly, these ideas are also applicable in the heavy sector, and the very first coupled-channels calculations involving a heavy quark has been performed [24].

## 3. Electroweak processes from lattice QCD

In the previous section, I discussed how scattering amplitudes can be determined from lattice QCD. From these one can determine masses, decay widths, and strong couplings of hadronic resonances. As discussed above, these amplitudes are only accessible from lattice QCD, in part, due to the finite-volume formalism discussed above. It has recently been demonstrated how these ideas can be extended to the study of electroweak processes [5, 6, 27]. In particular, Refs. [5, 6] demonstrated how resonant $0 \to 2$ and $1 \to 2$ amplitudes in the presence of an external electroweak current can be extracted from lattice QCD, and Ref. [27] demonstrated how $2 \to 2$ amplitudes in the presence of an external current can be studied. From the residue of these amplitudes evaluated at the resonance poles, one can then rigorously define and determine the form factors of hadronic resonances.

In order to test these ideas, in Ref. [8, 7] an exploratory calculation of the $\pi^+\gamma^\star \to \pi^+\pi^0$ amplitude was performed using a value of $m_\pi \approx 396$ MeV. This amplitude was determined for a range of values of the c.m. energy and virtuality of the photon. Figure 4 shows a result of the global fit to the amplitude as a function of energy and two specific values of the virtuality. One sees a clear enhancement due to the presence of the $\rho$ resonance. By analytically continuing the amplitude onto the resonance pole, the real and imaginary components of the $\pi \to \rho$ form factor were obtained and are shown in the same figure. These are compared with a previous calculation performed using $m_\pi \approx 700$ MeV, where the $\rho$ is stable [28]. Beyond testing these formal ideas, this calculation constitutes the first determination of a resonance form factor from lattice QCD.

## 4. Final Remarks and Outlook

In this talk I have discussed some recent developments in the study of scattering and resonances from lattice QCD. Two-body elastic and inelastic scattering amplitudes are now being determined, albeit at unphysically heavy quark masses. From these one can unambiguously determine masses and widths of hadronic resonances, and I have presented the very first rigorous study of the lightest of all hadronic resonances, the $\sigma$. Also, I discussed how these ideas are being extended to the study of few-body amplitudes involving electroweak currents, and I focused its application on the resonant $\pi^+\gamma^\star \to \pi^+\pi^0$ amplitude.

Beyond the application of these ideas to other sectors of QCD, for example the charm sector [24], these ideas will allow for the determination of phenomenologically interesting observables that are not directed accessible from experiment. One such example is in the study of elastic form factors of resonances [27]. The implementation of this formalism will allows us to peer inside resonances and be able to describe their inner structure, very much as is already being done for the study of stable states. Presently these ideas are being extended to allow the study of systems where three particles can go on-shell [19, 18, 10, 3]. This will allow for a QCD determination of, for example, the three-neutron force as well as resonances that couple strongly to three-body final states, such as the Roper.

### Acknowledgments

I would like to thank my colleagues within the Hadron Spectrum Collaboration, in particular Jozef Dudek, Robert Edwards, Christian Shultz, Christopher Thomas and David Wilson, as well as my other collaborators, Daniel Bolton and Maxwell Hansen, whose hard work made the results being presented possible. The work presented was partly supported by the U.S. Department of Energy contract DE-AC05-06OR23177, under which Jefferson Science Associates, LLC, manages and operates Jefferson Lab.

### References

[1] Daniel R. Bolton, Raul A. Briceno, and David J. Wilson. *Phys. Lett.*, B757:50–56, 2016.
[2] Raul A. Briceño. *Phys.Rev.*, D89(7):074507, 2014.
[3] Raul A. Briceño and Zohreh Davoudi. *Phys.Rev.*, D87:094507, 2012.
[4] Raul A. Briceño and Zohreh Davoudi. *Phys. Rev. D. 88,*, 094507:094507, 2013.
[5] Raúl A. Briceño and Maxwell T. Hansen. *Phys. Rev.*, D92(7):074509, 2015.
[6] Raul A. Briceño, Maxwell T. Hansen, and Andr Walker-Loud. *Phys.Rev.*, D91(3):034501, 2015.
[7] Ral A. Briceño, Jozef J. Dudek, Robert G. Edwards, Christian J. Shultz, Christopher E. Thomas, and David J. Wilson. *Phys. Rev.*, D93(11):114508, 2016.
[8] Raul A. Briceno, Jozef J. Dudek, Robert G. Edwards, Christian J. Shultz, Christopher E. Thomas, and David J. Wilson. *Phys. Rev. Lett.*, 115:242001, 2015.
[9] Raul A. Briceno, Jozef J. Dudek, Robert G. Edwards, and David J. Wilson. 2016.
[10] Raul A. Briceno, Maxwell T. Hansen, and Stephen R. Sharpe. 2016.
[11] Jozef J. Dudek, Robert G. Edwards, Peng Guo, and Christopher E. Thomas. *Phys. Rev.*, D88(9):094505, 2013.
[12] Jozef J. Dudek, Robert G. Edwards, and Christopher E. Thomas. *Phys.Rev.*, D87:034505, 2013.
[13] Jozef J. Dudek, Robert G. Edwards, Christopher E. Thomas, and David J. Wilson. *Phys. Rev. Lett.*, 113(18):182001, 2014.
[14] Jozef J. Dudek, Robert G. Edwards, and David J. Wilson. *Phys. Rev.*, D93(9):094506, 2016.
[15] P. Estabrooks and Alan D. Martin. *Nucl. Phys.*, B79:301, 1974.
[16] G. Grayer et al. *Nucl. Phys.*, B75:189–245, 1974.
[17] Maxwell T. Hansen and Stephen R. Sharpe. *Phys.Rev.*, D86:016007, 2012.
[18] Maxwell T. Hansen and Stephen R. Sharpe. *Phys.Rev.*, D90(11):116003, 2014.
[19] Maxwell T. Hansen and Stephen R. Sharpe. *Phys. Rev.*, D92(11):114509, 2015.
[20] B. Hyams et al. *Nucl. Phys.*, B64:134–162, 1973.
[21] M. Luscher. *Commun.Math.Phys.*, 104:177, 1986.
[22] M. Luscher. *Commun.Math.Phys.*, 105:153–188, 1986.
[23] Martin Luscher. *Nucl.Phys.*, B354:531–578, 1991.
[24] Graham Moir, Michael Peardon, Sinad M. Ryan, Christopher E. Thomas, and David J. Wilson. 2016.
[25] J. R. Pelaez. 2015.
[26] S.D. Protopopescu, M. Alston-Garnjost, A. Barbaro-Galtieri, Stanley M. Flatte, J.H. Friedman, et al. *Phys.Rev.*, D7:1279, 1973.
[27] Raúl A. Raúl and Maxwell T. Hansen. *Phys. Rev.*, D94(1):013008, 2016.
[28] Christian J. Shultz, Jozef J. Dudek, and Robert G. Edwards. *Phys. Rev.*, D91(11):114501, 2015.
[29] David J. Wilson, Raul A. Briceño, Jozef J. Dudek, Robert G. Edwards, and Christopher E. Thomas. *Phys. Rev.*, D92(9):094502, 2015.
[30] David J. Wilson, Jozef J. Dudek, Robert G. Edwards, and Christopher E. Thomas. *Phys. Rev.*, D91(5):054008, 2015.

Journal of Physics: Conference Series **770** (2016) 012025 doi:10.1088/1742-6596/770/1/012025

# Neutral *B*-meson Mixing and *CP* Violation at LHCb

**A Oblakowska-Mucha[1] (on behalf of the LHCb Collaboration)**

[1]AGH University of Science and Technology,
Faculty of Physics and Applied Computer Science,
Al. Mickiewicza 30, 30-059 Krakow, Poland

Agnieszka.Mucha@agh.edu.pl

**Abstract**. The LHCb detector is a single-arm forward spectrometer that collects data at the LHC, designed for studies of flavour physics with high precision. We present a selection of recent measurements of mixing and *CP*-violating parameters, including $\sin 2\beta$ and weak phase $\phi_s$, using several decays. A good understanding of the pollution from sub-leading penguin topologies in these measurements can be achieved by measuring *CP* violation and polarization in the decay $B_s^0 \to J/\psi\, \bar{K}^{0*}$ and $B^0 \to J/\psi\, \rho^0$. All results here presented are obtained using the full Run I dataset.

## 1. Introduction

Studies of *CP* violation with heavy-flavoured hadrons are amongst the main interests of the LHCb collaboration. They can unveil indirect evidence of New Physics, revealing inconsistencies with respect to theoretical predictions based on the validity of the Standard Model (SM). Subtle *CP* violating effects can be searched for by overconstraining the angles and the sides of the CKM unitary triangle.

In this summary, selected results regarding *B* mesons mixing and *CP* violation are presented, including the measurements of the weak phases $\varphi_d$ and $\varphi_s$ and measurements to constrain penguin pollution.

## 2. LHCb spectrometer

The LHCb experiment (Figure 1) is a dedicated apparatus for studying flavour physics at the LHC at CERN. In particular, the experiment is designed to study *CP* violation and rare decays of beauty and charm particles. It is a single-arm forward spectrometer covering the pseudorapidity range $2<\eta<5$. Thus the LHCb programme is highly complementary to the direct searches performed by the ATLAS and CMS experiments.

The $b\bar{b}$ pair production in proton-proton collisions at the LHC is spatially correlated and occurs predominantly at small angles with respect to the beam axis.

*Figure 1. The LHCb forward spectrometer at the LHC with a sample event superimposed.*

The interaction point is surrounded by the Vertex Locator sub-detector, the role which is to

Journal of Physics: Conference Series **770** (2016) 012025 doi:10.1088/1742-6596/770/1/012025

precisely determine both the position of primary and secondary vertices and allow study of the decays of beauty and charm particles.

In the Run I data taking, during the years 2010-12, LHCb collected data corresponding to an integrated luminosity of 3 fb$^{-1}$ at $\sqrt{s}$ =7 TeV and 8 TeV. The spectrometer achieved an excellent vertex resolution, momentum determination with a precision of $\delta p/p \sim$0.4-0.6% and very good particle identification of hadrons in the range of 2-100 GeV. A complete description of the LHCb experiment may be found in [1].

## 3. Weak phase in *B* meson mixing

The phase differences between the amplitude for a direct decay $B^0_{(s)} \to f$ and the amplitude for decay after oscillation $B^0_{(s)} \to \bar{B}^0_{(s)} \to f$ can be expressed by the CKM angles $\phi_d \approx 2\beta$, $\phi_s \approx -2\beta_s$, where: $\beta = \arg\left(-\frac{V_{cd}V^*_{cb}}{V_{td}V^*_{tb}}\right)$ and $\beta_s = \arg\left(-\frac{V_{cb}V^*_{cs}}{V_{tb}V^*_{ts}}\right)$ for the $B^0$ and $B^0_s$ mesons respectively.

From the experimental point of view, main aspects in the analysis are the tagging of the initial $B^0$ flavour at production and a very good decay-time resolution.

A major improvement in most of the LHCb analyses has been achieved due to the inclusion of a new tagging procedure (so-called "same-side pion" tagger) which deduces the production flavor by exploiting pions produced in the fragmentation of *b* quark in association with the signal *B* meson. All time-dependent studies require a very good decay-time resolution that is related not only to momentum resolution but also to an excellent precision on both primary and secondary vertex reconstruction. In the analyses presented below, the decay time distribution was extensively modeled taking into account different detector conditions (e.g. trigger, reconstruction, beam energy) using real and Monte Carlo data samples. The production and detector asymmetries were determined in separate analyses [2] and taken into account as well. All shown results are obtained using the whole Run I data sample.

### *3.1. Measurement of* $\sin 2\beta$

The $B^0 \to J/\psi\, K^0_S$ decay is considered as a "golden mode" for the measurement of $\sin 2\beta$. As the $J/\psi\, K^0_S$ final state is common to both $B^0$ and $\bar{B}^0$ meson decays, the interference between the amplitudes for the direct decay and for the decay after $B^0$ oscillation results in a decay-time dependent asymmetry between the decay rates: $A_{CP}(t) = \frac{\Gamma\{B^0 \to J/\psi\, K^0_S\} - \Gamma\{\bar{B}^0 \to J/\psi\, K^0_S\}}{\Gamma\{B^0 \to J/\psi\, K^0_S\} + \Gamma\{\bar{B}^0 \to J/\psi\, K^0_S\}}$, which, neglecting penguin pollution, in the SM can be written as: $A_{CP}(t) \approx -\sin 2\beta \sin \Delta m t \equiv S \sin \Delta m t$, where $\Delta m$ is the mass differences between the heavy and light mass eigenstates.

The reconstructed mass of the signal events is presented in Figure 2 together with the decay-time asymmetry of tagged *B* mesons [3].

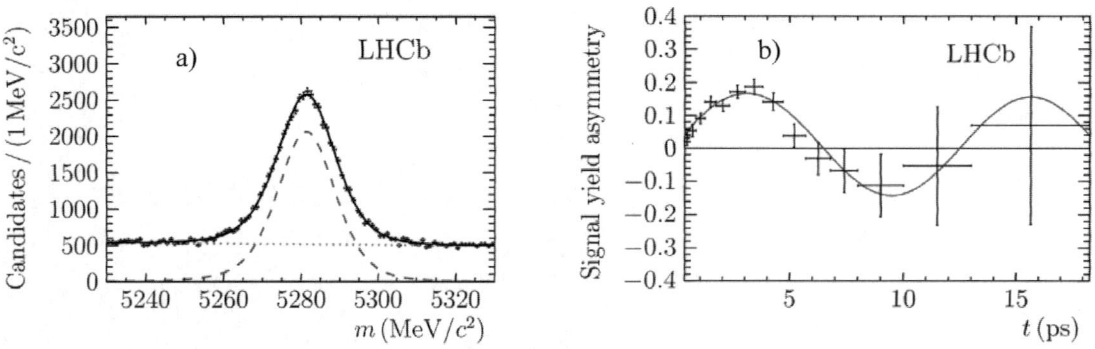


*Figure 2. a) Mass distribution of $B^0 \to J/\psi\, K^0_S$ candidates. b) Time dependent asymmetry of signal events [3].*

The likelihood fit to the time-dependent *CP* asymmetry, whose model takes into account different experimental conditions, signal with background and tagging decisions, yields the result: $S \equiv \sin 2\beta = 0.731 \pm 0.035(stat) \pm 0.020(syst)$. The value is consistent with the current world average and with the SM and is the most precise time-dependent *CP* violation measurement at hadron colliders obtained to date, with a precision competitive with *B*-factories [4].

### 3.2. Results on weak phase $\phi_s$

The weak phase $\phi_s$ within the SM is predicted to be very small. The *CP*-violating phase $\phi_s$ arises from the interference between the amplitude of the mesons $B_s^0$ decaying directly via $b \to c\bar{c}s$ to *CP* eigenstates and after oscillation $B_s^0 - \bar{B}_s^0$. The LHCb experiment pioneered this measurement with a number of final states and currently dominates the world average [4].

*3.2.1* $B_s^0 \to J/\psi\,\phi$. The "golden mode" for the $\phi_s$ measurement is the decay $B_s^0 \to J/\psi\,\phi$, counterpart of $B^0 \to J/\psi\,K_S^0$. The decays of $B_s^0$ and $\bar{B}_s^0$ mesons proceed through tree (which is dominant) and penguin (that is suppressed) diagrams.
The final state includes the vector meson $\phi(1020)$ that is in *P*-wave configuration and, depending on the relative momentum of the $\phi$ and $J/\psi$, *CP*-even and *CP*-odd components are expected. Their disentanglement is done by means of an angular analysis in the helicity basis. The mass distribution for $J/\psi\,(\to \mu^+\mu^-)K^-K^+$ candidates is shown in Figure 3a, when a huge number of about $96 \cdot 10^3$ signal events enables a precise angular analysis ($\cos\theta_K$ distribution is shown in Figure 3b) [5].

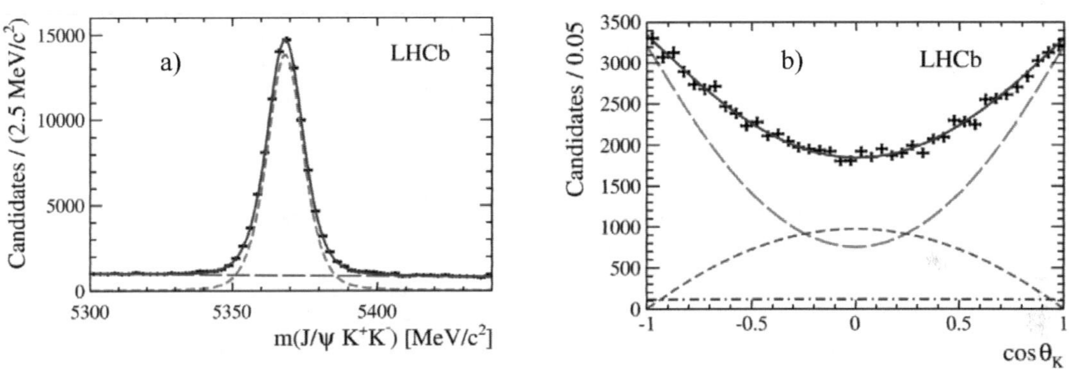


*Figure 3. a) Mass distribution of the J/ψKK events. The $B^0$ signal component is shown by the red dashed line and the combinatorial background by the green long-dashed line. b) Distribution of $\cos\theta_K$ with the fit projection. The solid blue line shows the total signal contribution, which is composed of CP-even (long-dashed red), CP-odd (short-dashed green) and S-wave (dotted-dashed purple) components [5].*

The main physics parameters ($\Delta m_s$, $\Gamma_S$, $\Delta\Gamma_S$, $\phi_S$, $|\lambda|$ and polarization amplitudes) were obtained from a maximum likelihood fit to the decay-time and angle distributions, yielding: $\phi_s = -0.058 \pm 0.049 \pm 0.006$ rad, $\Delta\Gamma_S = 0.805 \pm 0.0027 \pm 0.0015\, ps^{-1}$, $\Delta m_s = 17.711^{+0.055}_{-0.057} + 0.011\, ps^{-1}$. The parameter $|\lambda|$ is consistent with unity, implying no evidence for direct *CP* violation.
The combination of the results obtained in the $B_s^0 \to J/\psi K^-K^+$ and $B_s^0 \to J/\psi\,\pi^-\pi^+$ analyses [6] makes the measurement of $\phi_s$ the most precise result to date, in agreement with the SM prediction [7].

*3.2.1* $B_s^0 \to D_s^+ D_s^-$. This channel also proceeds via $b \to c\bar{c}s$ transitions. As the final state does not include vector mesons, an angular analysis is unnecessary. From the Run I data sample about 3350 flavour tagged signal events were reconstructed. The mass distribution and decay time of the candidates are presented in Figure 4a and 4b respectively [8].

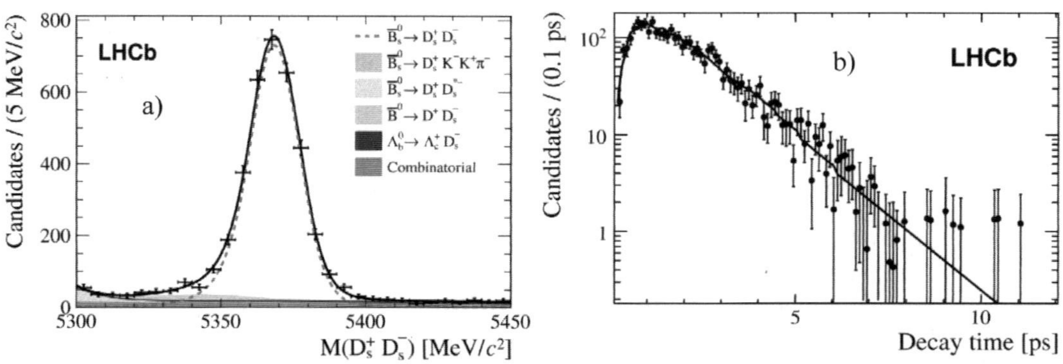


*Figure 4. a) Invariant mass distribution of $D_s^+ D_s^-$ events. b) Distribution of the decay time for signal events along with the fit. Discontinuities in the fit line shape are a result of the binned acceptance [8].*

The first measurement with this decay mode, obtained by means of a time-dependent study, yields the result: $\phi_s(D_s^+ D_s^-) = 0.02 \pm 0.17(stat) \pm 0.02(syst)$ and $|\lambda| = 0.91^{+0.18}_{-0.15}(stat) \pm 0.02(syst)$, that is consistent with the SM.

*3.2.1* $B_s^0 \to \phi\phi$. This is a penguin-dominated process that proceeds via $b \to s s \bar{s}$ transition. As such, it is an excellent probe for New Physics, since new, heavy particles may enter the quantum loops.

More than 4000 signal events were observed in the LHCb Run I data sample [9]. The presence of vector mesons in the final state requires a time-dependent angular analysis. The result yields: $\phi_s(\phi\phi) = -0.17 \pm 0.15(stat) \pm 0.03(syst)$. It's worth noting, that this channel will greatly benefit from the new LHCb trigger in Run II and the obtained precision is expected to improve considerably.

## 4. Constraining the penguin pollution

The measured value of weak phases, arising from the interference between neutral $B$ meson mixing and decay, is effectively the sum of a term that contains CKM angle $\beta_s$, a hadronic term related to gluonic penguin diagrams $\delta_P$ and a possible New Physics (NP) contribution $\delta_{NP}$: $\phi_s(measured) = -2\beta_s + \delta_P + \delta_{NP}$. For this reason, before claiming NP discovery, we need to constrain the SM gluonic part, which is colloquially called "penguin pollution". A novel method of constraining the gluonic part $\delta_P$ directly from the measurement was proposed [10] and firstly performed by LHCb [11].

In order to constrain penguin contribution in $B_s^0 \to J/\psi\, \phi$ and $B^0 \to J/\psi\, K_S^0$ we need to find control channels, where processes with gluonic loop are not suppressed with respect to dominant tree diagram (see the Figure 5).

This method of obtaining the estimator for $\delta_P$, referred to as "effective approach" can be described by following steps:

- identifying penguin parameters related to the $B_s^0 \to J/\psi\, \phi$ and $B^0 \to J/\psi\, K_S^0$ decays,
- measurement of certain observables in control channels, such as $B^0 \to J/\psi\, \rho^0$ and $B_s^0 \to J/\psi\, \overline{K}^{0*}$,
- relating them with the penguin parameters by assuming SU(3) flavour symmetry.

This approach requires measurement of the branching fractions, direct asymmetries and polarisation fractions, which all depend on the penguin parameters. The analysis is based on the helicity basis, whose amplitudes are rotated into transversity ones that correspond to the different parity eigenstates.

The angular distributions for the $B_s^0 \to J/\psi\, \overline{K}^{0*}$ signal events are presented in Figure 6. The transversity amplitudes of the angular method depend on the $K^-\pi^+$ mass, so the analysis is performed in different bins of $m_{K^-\pi^+}$ and finally summed over all contributions.

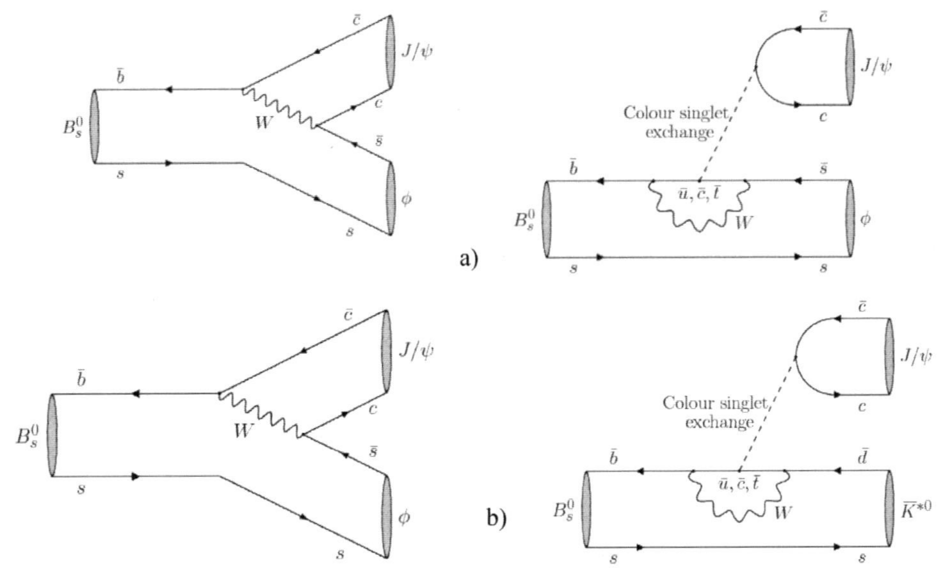


*Figure 5. a) The dominant tree and suppressed loop diagrams contributing to the weak phase $\phi_s$ measurement via $B_s^0 \to J/\psi\,\phi$. b) The diagrams of the control channel $B_s^0 \to J/\psi\,\bar{K}^{0*}$ which contains amplitudes of comparable order.*

The penguin parameters are measured and translated into phase shift $\delta_P$. The value obtained is consistent with zero but with large systematic uncertainties coming from theoretical assumptions. They can be further reduced by using the $B^0 \to J/\psi\,\rho^0$ decay mode. The mass distributions for the two mentioned channels are shown in Figure 7.

Direct and mixing-induced time-dependent *CP* violation is observed in the case of $B^0 \to J/\psi\,\rho^0$ decay which parameters depend on the phase $\phi_d$. So quantifying the gluonic contribution requires an interplay between the high precision determination of $\phi_d$ and $\phi_s$.

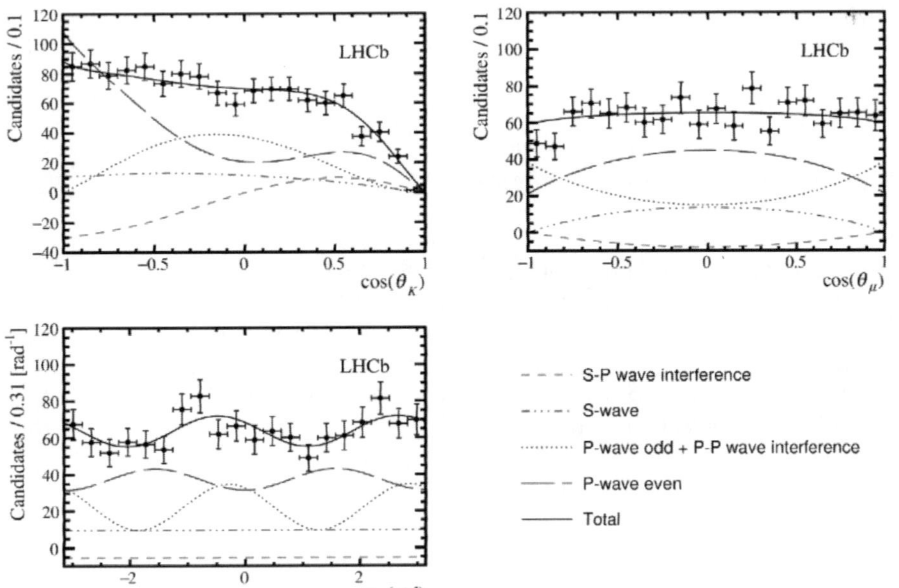


*Figure 6. The angular distributions of the $B_s^0 \to J/\psi\,\bar{K}^{0*}$ signal events. The fit result is represented by the solid black line and the contributions from the different amplitude components are described in the legend [11].*

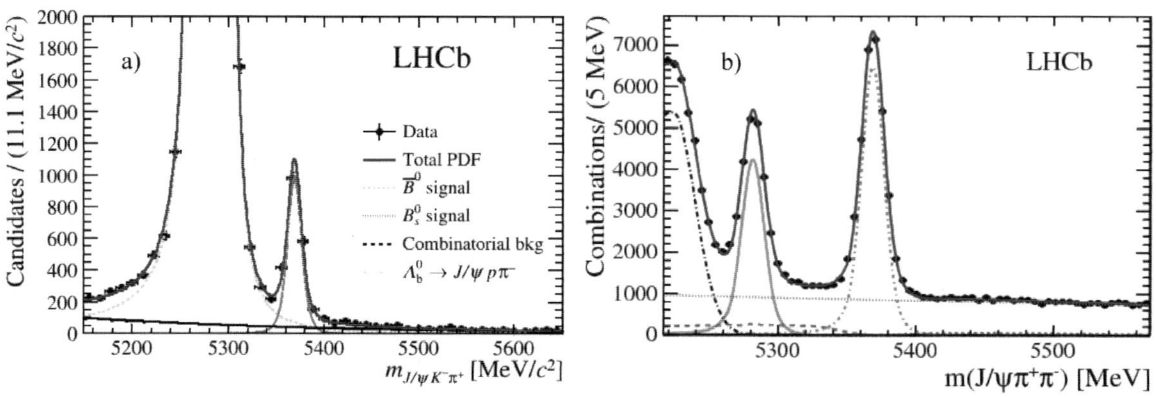


*Figure 7. a) The $J/\psi\, K^-\pi^+$invariant mass distributions with the sum of fit projections (blue line) in $m_{K^-\pi^+}$ and $\cos\theta_\mu$ bins. The contributions of different components are detailed in the legend. b) The $J/\psi\,\pi^+\pi^-$invariant mass distributions. The purple solid line represents the $B^0$signal, the red dot-dash is $B_s^0 \to J/\psi\,\pi^+\pi^-$, the dotted brown – combinatorial background.*

Combining the control channels, we measured the weak phase $\phi_s$ in the different polarisation states and the shift that comes from penguin pollution. For each polarization states $(0, \parallel, \perp)$ the phase shifts in $B_s^0 \to J/\psi\,\phi$ due to penguin pollution are: $\delta_p^0 = 0.000^{+0.009}_{-0.011}\,(stat)\,^{+0.004}_{-0.009}\,(syst)$ rad, $\delta_p^{\parallel} = 0.001^{+0.010}_{-0.014}\,(stat) \pm 0.008\,(syst)$ rad and $\delta_p^{\perp} = 0.003^{+0.010}_{-0.014}\,(stat) \pm 0.008\,(syst)$ rad. The results are in agreement with the SM and gluonic term appeared to be small.

## 5. Summary

The LHCb experiment, designed to study *CP* violation and rare decays of beauty and charmed hadrons produced in proton-proton collisions at the LHC, has performed a plethora of high-precision measurements, largely surpassing in many cases the knowledge from previous experiments.
Especially the principal *CP*-violating parameters, such as the weak mixing phases in both $B^0$ and $B_s^0$ sectors, yields the results: $\sin 2\beta = 0.731 \pm 0.035(stat) \pm 0.020(syst)$ and $\phi_s = -0.058 \pm 0.049(\text{stat}) \pm 0.006(\text{syst})$ rad. Sensitivity for $\phi_s$ will further increase after Run II and the LHCb Upgrade.

Constraints on penguin pollution were put using measurements of the hadronic parameters in the control channels. Such pioneering results showed that they are small within the present uncertainties, but further data will be needed to constrain such quantities to the required level of precision.

### References

[1] Alves A A Jr. *et al* (LHCb Collaboration) 2008 *JINST* **3** S08005
[2] Aaij R *et al* (LHCb Collaboration) 2014 *Phys. Lett. B* **739** 218
[3] Aaij R *et al* (LHCb Collaboration) 2015 *Phys. Rev. Lett.* **115** 031601
[4] Amhis Y *et al* Heavy Flavor Averaging Group (HFAG) collaboration arXiv:1412.7515
[5] Aaij R *et al* (LHCb Collaboration) 2015 *Phys. Rev. Lett.* **114** 041801
[6] Aaij R *et al* (LHCb Collaboration) 2014 *Phys. Lett. B* **736** 186
[7] Charles J Deschamps O *et al* 2011 *Phys. Lett. D.* **84** 033005
[8] Aaij R *et al* (LHCb Collaboration) 2014 *Phys. Lett. B* **113** 211801
[9] Aaij R *et al* (LHCb Collaboration) 2014 *Phys. Rev. D* **90** 052011
[10] De Bruyn K and Fleischer R 2015 *JHEP* ***03*** 145
[11] Aaij R *et al* (LHCb Collaboration) 2015 *JHEP* **11** 082
Aaij R *et al* (LHCb Collaboration) 2015 *Phys. Lett. B* **742** 38

Journal of Physics: Conference Series **770** (2016) 012026 doi:10.1088/1742-6596/770/1/012026

# Mixing and CP Violation in the $B_d$ and $B_s$ Systems at ATLAS

**A E Barton**

Physics Department, Lancaster University, LANCASTER, LA1 4YB

E-mail: abarton@cern.ch

**Abstract.** This note includes a measurement of the $B_s^0$ decay parameters in the $B_s \to J/\psi\phi$ channel and a measurement of the $B^0$ meson proper decay time and decay width using the statistics collected by the ATLAS experiment in Run I of the LHC. The first result presents the measurement of the $CP$-violating phase $\phi_s$, the decay width $\Gamma_s$ and the width difference between the mass eigenstates $\Delta\Gamma_s$. The second result presents the measurement of the width difference $\Delta\Gamma_d$, which is extracted from the measurement of the lifetime dependence of $B^0 \to J/\psi K_S$ and $B^0 \to J/\psi K^{*0}$ decays. The obtained results are $\phi_s = 0.098 \pm 0.084(\text{stat.}) \pm 0.040(\text{syst.})$ rad. and $\Delta\Gamma_d/\Gamma_d = (-0.1 \pm 1.1(\text{stat.}) \pm 0.9(\text{syst.})) \times 10^{-2}$.

## 1. Introduction

ATLAS [1] is a general purpose detector that measures heavy flavour properties using its inner dectectors, muon spectrometers and electromagnetic calorimeters (for tagging). Measuring the properties of heavy-flavour particles has been part of the B physics program of the ATLAS experiment since the start of the proton-proton collisions at LHC in 2010. This note presents an overview of recent results obtained using data collected at $\sqrt{s} = 7$ TeV during 2011 and $\sqrt{s} = 8$ TeV during 2012. The analysis of CP violation and mixing can still access physics beyond the standard model; precise measurements may constrain new physics scenarios such as supersymmetry or advance $b$ and $c$ hadron spectroscopy and test QCD.

## 2. The $B_q^0$ system

The time evolution of the neutral $B_q^0 - \bar{B}_q^0$ system is described by:

$$\begin{aligned} \mathrm{i}\frac{\mathrm{d}}{\mathrm{d}t}\begin{pmatrix} B_q^0(t) \\ \bar{B}_q^0(t) \end{pmatrix} &= \boldsymbol{M_q}\begin{pmatrix} B_q^0(t) \\ \bar{B}_q^0(t) \end{pmatrix}, \\ \boldsymbol{M_q} &= \begin{pmatrix} m_q & m_q^{12} \\ (m_q^{12})^* & m_q \end{pmatrix} - \frac{i}{2}\begin{pmatrix} \Gamma_q & \Gamma_q^{12} \\ (\Gamma_q^{12})^* & \Gamma_q \end{pmatrix}. \end{aligned} \quad (1)$$

The time dependence of the decay rate $B_q^0 \to f$ is sensitive to $f$. The time-dependent decay rate is given by:

$$\Gamma(B_q^0(t) \to f) \propto \mathrm{e}^{-\Gamma_q t}\left[\cosh\frac{\Delta\Gamma_q t}{2} + A_{\text{CP}}^{\text{dir}}\cos(\Delta m_q t) + A_{\Delta\Gamma}\sinh\frac{\Delta\Gamma_q t}{2} + A_{\text{CP}}^{\text{mix}}\sin(\Delta m_q t)\right]. \quad (2)$$

Journal of Physics: Conference Series **770** (2016) 012026 doi:10.1088/1742-6596/770/1/012026

Here $t$ is the proper decay time of the $B_q^0$ meson.

The parameters $A_{\text{CP}}^{\text{dir}}$, $A_{\Delta\Gamma}$ and $A_{\text{CP}}^{\text{mix}}$ depend on the final state $f$. The abbreviations "dir" and "mix" stand for "direct" and "mixing". By definition:

$$|A_{\text{CP}}^{\text{dir}}|^2 + |A_{\Delta\Gamma}|^2 + |A_{\text{CP}}^{\text{mix}}|^2 \equiv 1. \tag{3}$$

Assuming that the CP-violating phase $\phi_q^{12}$ is small, which is experimentally confirmed for both the $B^0$ and $B_s$ mesons [2]:

$$\Gamma(\bar{B}_q^0(t) \to f) \propto \mathrm{e}^{-\Gamma_q t}\left[\cosh\frac{\Delta\Gamma_q t}{2} - A_{\text{CP}}^{\text{dir}}\cos(\Delta m_q t) + A_{\Delta\Gamma}\sinh\frac{\Delta\Gamma_q t}{2} - A_{\text{CP}}^{\text{mix}}\sin(\Delta m_q t)\right]. \tag{4}$$

The parameters $A_{\text{CP}}^{\text{dir}}$, $A_{\Delta\Gamma}$ and $A_{\text{CP}}^{\text{mix}}$ are theoretically well defined for flavour-specific final states and CP eigenstates [3]. For a flavour-specific final state $f_{\text{fs}}$, such that only the decay $B_q^0 \to f_{\text{fs}}$ is allowed while $\bar{A}_f = \langle f_{\text{fs}}|\bar{B}_q^0\rangle = 0$, the parameters are:

$$A_{\text{CP}}^{\text{dir}} = 1, \quad A_{\Delta\Gamma} = 0, \quad A_{\text{CP}}^{\text{mix}} = 0. \tag{5}$$

For a flavour-specific final state $\bar{f}_{\text{fs}}$, such that $A_f = \langle \bar{f}_{\text{fs}}|B_q^0\rangle = 0$, i.e. only the decay $\bar{B}_q^0 \to \bar{f}_{\text{fs}}$ is allowed, the parameters are:

$$A_{\text{CP}}^{\text{dir}} = -1, \quad A_{\Delta\Gamma} = 0, \quad A_{\text{CP}}^{\text{mix}} = 0. \tag{6}$$

For the $B^0$ decay to the CP eigenstate $J/\psi K_s$ the parameters are:

$$A_{\text{CP}}^{\text{dir}} = 0, \quad A_{\Delta\Gamma} = \cos(2\beta), \quad A_{\text{CP}}^{\text{mix}} = -\sin(2\beta), \tag{7}$$

$$\beta = \arg\left(-\frac{V_{cd}V_{cb}^*}{V_{td}V_{tb}^*}\right). \tag{8}$$

If the initial flavour of the $B_q^0$ meson is not tagged and the mixture of the states is equal and unbiased, the decay rates given by equations (2) and (4) are added together. In this case, the production asymmetry $A_{\text{P}}$ of the $B_q^0$ meson in $pp$ collisions should be taken into account. This asymmetry is defined as:

$$A_{\text{P}} = \frac{\sigma(B_q^0) - \sigma(\bar{B}_q^0)}{\sigma(B_q^0) + \sigma(\bar{B}_q^0)}. \tag{9}$$

The oscillation rates $\Delta m_d$ and $\Delta m_s$ have not been measured at ATLAS so the world averages [4] are used in our analyses.

$$\Delta m_d = 0.510 \pm 0.003 \times 10^{12}\text{s}^{-1} \tag{10}$$

$$\Delta m_s = 17.757 \pm 0.021 \times 10^{12}\text{s}^{-1} \tag{11}$$

## 3. CP Violation

ATLAS has performed a measurement of the CP-violating phase $\phi_s$ using the 2012 proton-proton dataset [5]. This is done using the exclusive decay $B_s \to J/\psi(\mu^+\mu^-)\phi(K^+K^-)$, which is a mixture of CP-even and CP-odd eigenstates. CP violation arises from interference between the direct decay and the mixing. From existing measurements, the standard model constrains this to be $\phi_s \approx -0.04$ rad.

Table 1: Summary of tagging performance for the different flavour tagging methods described in the text. Uncertainties shown are statistical only. The efficiency and tagging power are each determined by summing over the individual bins of the charge distribution. The effective dilution is obtained from the measured efficiency and tagging power. For the efficiency, dilution and tagging power, the corresponding uncertainty is determined by combining the appropriate uncertainties in the individual bins of each charge distribution. [5]

| Tagger | Efficiency [%] | Dilution [%] | Tagging Power [%] |
|---|---|---|---|
| Combined $\mu$ | $4.12 \pm 0.02$ | $47.4 \pm 0.2$ | $0.92 \pm 0.02$ |
| Electron | $1.19 \pm 0.01$ | $49.2 \pm 0.3$ | $0.29 \pm 0.01$ |
| Segment-tagged $\mu$ | $1.20 \pm 0.01$ | $28.6 \pm 0.2$ | $0.10 \pm 0.01$ |
| Jet-charge | $13.15 \pm 0.03$ | $11.85 \pm 0.03$ | $0.19 \pm 0.01$ |
| Total | $19.66 \pm 0.04$ | $27.56 \pm 0.06$ | $1.49 \pm 0.02$ |

The analysis measures the mass, lifetime, three transversity angles and tagging information. These are placed in an unbinned maximum likelihood fit function:

$$\ln\ \mathcal{L} = \sum_{i=1}^{N} \{w_i \cdot \ln(f_S(\mathcal{F}_S + f_{B_d^0} \cdot \mathcal{F}_{B_d^0}) + (1 - f_s \cdot (1 + f_{B_d^0}))\mathcal{F}_{\text{bkg}})\} \tag{12}$$

$w_i$ is a per-candidate weight for trigger lifetime efficiency.

The fitted PDF has four components:
$\mathcal{F}_s$ = Mass · Lifetime+Angles+Tagging · Acceptance
$\mathcal{F}_{B_d^0}$ = Mass · Lifetime · Angles
$\mathcal{F}_{\text{bkg}}$ = Mass · Lifetime · Angles

The probability density function contains a number of symmetries. As the sign of the $\Delta\Gamma_S$ is not resolved by the data, it is assumed to be $> 0$ and a measurement from LHCb confirms this [6]. A second ambiguity exists in $\phi_s$ and the strong phases, this can be resolved by the flavour tagging information.

### 3.1. *Tagging*

The analysis uses opposite-side tagging which was calibrated from the $B^{\pm}$ self-tagging channels. A combination of four tagging methods are used and they can be seen along with their efficiency $\epsilon$, dilution $D = (1 - \omega)$ and power $P = \epsilon D^2$ in the table 1.

### 3.2. *Results*

The results from the 2012 data set demonstrate compatability with the standard model and the 2011 dataset previously published [7]. The combined parameters are calculated using the Best Linear-Unbiased Estimate (BLUE) taking into account parameter correlations. The results can be seen in table 2, with figures 1 and 2 showing the likelihood contours for the result. Comparisons with other experiments can be seen in figure 3, it shows the ATLAS result is in agreement with the other measurements.

### 3.3. *Future performance of detector*

ATLAS has released a note which gives estimates of the detector performance after the planned upgrades in the coming years [8]. This includes the IBL (insertable-B-layer) that has been

Table 2: Current measurements using data from 8 TeV $pp$ collisions, the previous measurement using data taken at centre of mass energy of 7 TeV and the values for the parameters of the two measurements, statistically combined. [5]

| | 8 TeV data | | | 7 TeV data | | | Run1 combined | | |
|---|---|---|---|---|---|---|---|---|---|
| Par | Value | Stat | Syst | Value | Stat | Syst | Value | Stat | Syst |
| $\phi_s$[rad] | −0.109 | 0.082 | 0.042 | 0.12 | 0.25 | 0.05 | −0.098 | 0.084 | 0.040 |
| $\Delta\Gamma_s$[ps$^{-1}$] | 0.101 | 0.013 | 0.007 | 0.053 | 0.021 | 0.010 | 0.083 | 0.011 | 0.007 |
| $\Gamma_s$[ps$^{-1}$] | 0.676 | 0.004 | 0.004 | 0.677 | 0.007 | 0.004 | 0.677 | 0.003 | 0.003 |
| $\|A_{\|\|}(0)\|^2$ | 0.229 | 0.005 | 0.006 | 0.220 | 0.008 | 0.009 | 0.227 | 0.004 | 0.006 |
| $\|A_0(0)\|^2$ | 0.521 | 0.004 | 0.007 | 0.529 | 0.006 | 0.012 | 0.514 | 0.004 | 0.003 |
| $\|A_S\|^2$ | 0.098 | 0.008 | 0.022 | 0.024 | 0.014 | 0.028 | 0.071 | 0.007 | 0.017 |
| $\delta_\perp$ [rad] | 4.50 | 0.45 | 0.30 | 3.89 | 0.47 | 0.11 | 4.13 | 0.33 | 0.16 |
| $\delta_{\|\|}$ [rad] | 3.15 | 0.13 | 0.05 | [3.04, 3.23] | | 0.09 | 3.15 | 0.13 | 0.05 |
| $\delta_\perp - \delta_S$ [rad] | −0.08 | 0.04 | 0.01 | [3.02, 3.25] | | 0.04 | −0.08 | 0.04 | 0.01 |

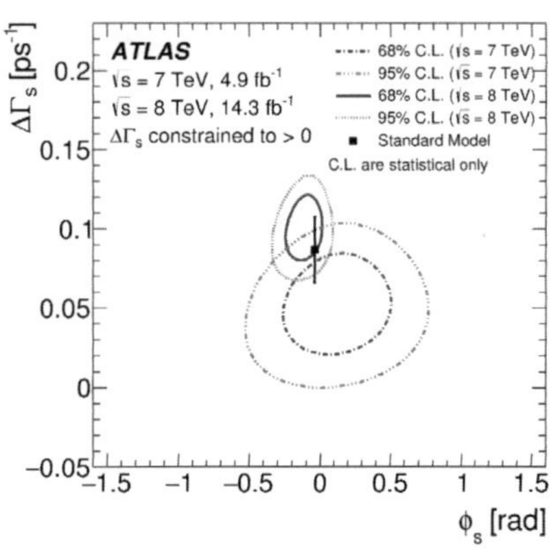


**Figure 1.** Likelihood contours for the 2011 and 2012 datasets. [5]

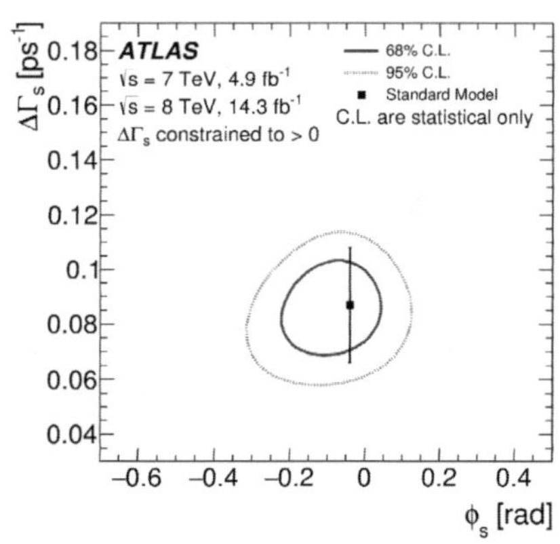


**Figure 2.** Likelihood contours for the combined result [5]

installed for 2015 and the planned ITK detector that will be installed alongside the High-Luminosity LHC upgrade. Figure 4 shows the proper decay time uncertainty vs the $p_T$ of the decaying $B_s$ meson for the tested detector layouts. Table 3 includes information on the estimated signal yields and the estimated precision of the $\phi_s$ measurements.

## 4. Measurement of the relative width difference of the $B_0$–$\bar{B}_0$ system with the ATLAS detector

$\Delta\Gamma_d$ is one of the least measured parameters in the B mass system. The standard model makes a precise prediction of a small value for $\Delta\Gamma_{\text{d SM}} = 0.1 \pm 1.0 \times 10^{-2}$. The eigenstate information outlined in section 2 indicates we can measure this using a ratio of events from the decays $B_d^0 \to J/\psi K^*$ and $B_d^0 \to J/\psi K_s^0$. This is measured with the 2011 and 2012 datasets recorded by the ATLAS experiment [10].

$A_p$ is the particle/anti-particle production asymmetry. The observed asymmetry $A_{obs}$ is calculated in bins of proper decay length, the asymmetry $A_P$ is then obtained from a $\chi^2$

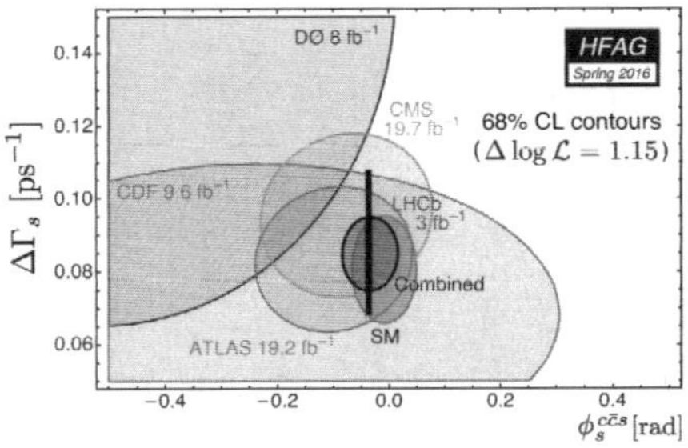


**Figure 3.** Likelihood contours comparing results from different experiments. [2]

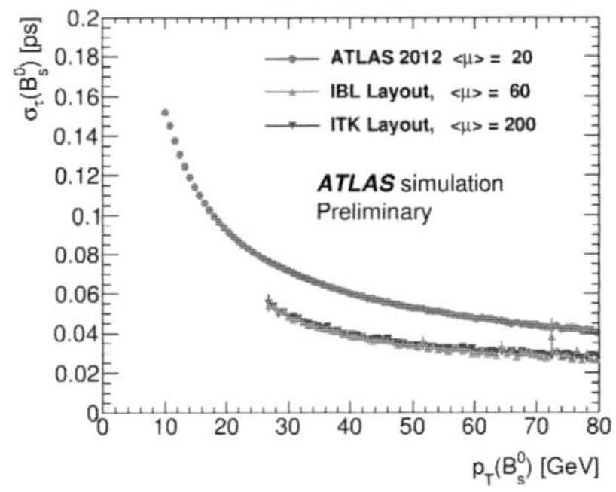


**Figure 4.** A plot showing the proper decay time uncertainty vs the $p_T$ of the decaying $B_s$ meson of the current and future ATLAS detectors. [8]

Table 3: Estimated ATLAS statistical precisions $\phi_s$ for considered LHC periods. Values for 2011 and 2012 in this table are derived using the same method as for future periods. [8]

| | 2011 | 2012 | 2015-17 | | 2019-21 | 2023-30+ |
|---|---|---|---|---|---|---|
| Detector | current | current | IBL | | IBL | ITK |
| Average interaction per BX $< \mu >$ | 6-12 | 21 | 60 | | 60 | 200 |
| Luminosity, fb$^-$1 | 4.9 | 20 | 100 | | 250 | 3 000 |
| Di-$\mu$ trigger $p_T$ thresholds, GeV | 4-4(6) | 4-6 | 6-6 | 11-11 | 11-11 | 11-11 |
| Signal events per fb$^{-1}$ | 4 400 | 4 320 | 3 280 | 460 | 460 | 330 |
| Signal events | 22 000 | 86 400 | 327 900 | 45 500 | 114 000 | 810 000 |
| Total events in analysis | 130 000 | 550 000 | 1 874 000 | 284 000 | 758 000 | 6 461 000 |
| MC $\sigma(\phi_s)$((stat.)), rad | 0.25 | 0.10 | 0.045 | 0.083 | 0.053 | 0.018 |

minimisation where $A_{det}$ is the detector reconstruction asymmetry:

$$A_{i,obs} \equiv \frac{N_i(J/\psi K^{*0}) - N_i(J/\psi \bar{K}^{*0})}{N_i(J/\psi K^{*0}) + N_i(J/\psi \bar{K}^{*0})}, \tag{13}$$

$$\chi^2[A_{\text{det}}, A_p] = \sum_{i=2}^{10} \frac{(A_{i,obs} - A_{i,exp})^2}{\sigma_i^2}. \tag{14}$$

The detector asymmetry $A_{\text{det}}$ and $A_p$ are fit to obtain values of $A_{\text{det}} = (+1.33 \pm 0.24 \pm 0.22) \times 10^{-2}$ and $A_p = (+0.25 \pm 0.48 \pm 0.05) \times 10^{-2}$ The $A_P$ is found to be consistent with LHCb, although since ATLAS is observing a different region it does not need to be equivalent.

The mass spectra are fitted in each proper decay length bin to extract the yields. Examples of the fits can be seen in figures 5 and 6. Each bin has an efficiency correction ratio applied; this ratio is determined using Monte Carlo data.

The final results are $\Delta\Gamma_d/\Gamma_d = (-2.8 \pm 2.2(\text{stat.}) \pm 1.5(\text{MC stat.})) \times 10^{-2}$ for the 2011 dataset and $\Delta\Gamma_d/\Gamma_d = (+0.8 \pm 1.3(\text{stat.}) \pm 0.5(\text{MC stat.})) \times 10^{-2}$ for the 2012 dataset. These results are combined using the $\chi^2$ method. Correlations between sources of systematics common to the two datasets are taken into account. The systematic uncertainty due to the background description and the MC are considered to be uncorrelated. This combined result is

$\Delta\Gamma_d/\Gamma_d = (-0.1 \pm 1.1(\text{stat.}) \pm 0.9(\text{syst.})) \times 10^{-2}$. This combined result is currently the most precise single measurement of this quantity; it agrees well with the standard model prediction and the indirect measurement by D0 [9].

The fact the asymmetry is a ratio cancels out most biases from the trigger, time resolution or B production properties. However, differences between the channels and simulation inaccuracies could remain. The systematics uncertainties are thus estimated and shown in table 4.

Table 4: Sources of systematic uncertainty in the $\Delta\Gamma_d/\Gamma_d$ measurement and their values for the 2011 and 2012 data sets. [10]

| Source | $\delta(\Delta\Gamma_d/\Gamma_d)$, 2011 | $\delta(\Delta\Gamma_d/\Gamma_d)$, 2012 |
|---|---|---|
| $K_S$ decay length | $0.21 \times 10^{-2}$ | $0.16 \times 10^{-2}$ |
| $K_S$ pseudorapidity | $0.14 \times 10^{-2}$ | $0.01 \times 10^{-2}$ |
| $B^0 \to J/\psi K_s$ mass range | $0.47 \times 10^{-2}$ | $0.59 \times 10^{-2}$ |
| $B^0 \to J/\psi K^{*0}$ mass range | $0.30 \times 10^{-2}$ | $0.15 \times 10^{-2}$ |
| Background description | $0.16 \times 10^{-2}$ | $0.09 \times 10^{-2}$ |
| $B_s \to J/\psi K_S$ contribution | $0.11 \times 10^{-2}$ | $0.08 \times 10^{-2}$ |
| $L^B_{\text{prop}}$ resolution | $0.29 \times 10^{-2}$ | $0.29 \times 10^{-2}$ |
| Fit bias (Toy MC) | $0.07 \times 10^{-2}$ | $0.07 \times 10^{-2}$ |
| $B^0$ production asymmetry | $0.01 \times 10^{-2}$ | $0.01 \times 10^{-2}$ |
| MC sample | $1.54 \times 10^{-2}$ | $0.45 \times 10^{-2}$ |
| Total uncertainty | $1.69 \times 10^{-2}$ | $0.84 \times 10^{-2}$ |

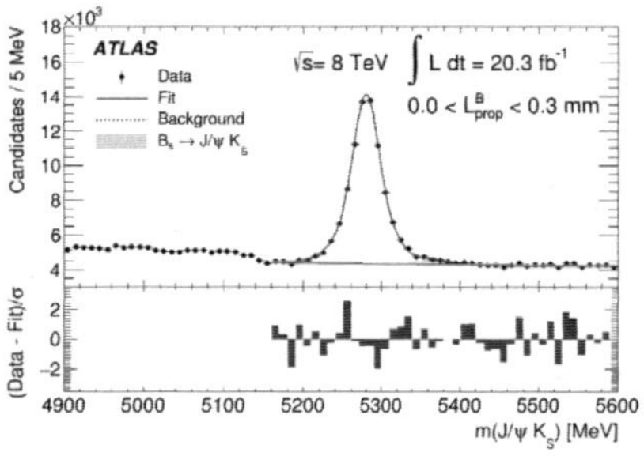


**Figure 5.** The $J/\psi K_S$ mass fit [10]

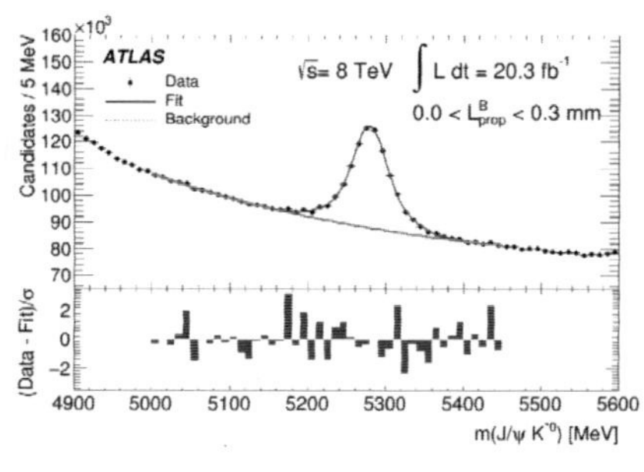


**Figure 6.** The $J/\psi K^{*0}$ mass fit [10]

## References

[1] ATLAS Collaboration, *J. Instrum. 3* (2008) S08003.
[2] Amhis Y *et al.* (Heavy Flavour Averaging Group), Preprint arXiv:1412.7515 [hep-ex]
[3] Dighe A S *et al. Eur. Phys. J. C* 6 647 1999.
[4] Olive K A *et al.* (Particle Data Group), Chin. Phys. C, 38, 090001 (2014).
[5] ATLAS Collaboration, Preprint arXiv:1601.03297 [hep-ex]
[6] LHCb Collaboration, *Phys. Rev. Lett. 108* (2012) 241801
[7] ATLAS Collaboration, *Phys.Rev.* D90 (2014) 052007, (Preprint arXiv:1407.1796 [hep-ex])
[8] ATLAS Collaboration, ATL-PHYS-PUB-2013-010.
[9] D0 Collaboration, *Phys. Rev. D 89 (2014)* 012002, (Preprint arXiv:1310.0447 [hep-ex])
[10] ATLAS Collaboration, *JHEP06 (2016)* 81 (Preprint arXiv:1605.07485 [hep-ex])

Journal of Physics: Conference Series **770** (2016) 012027 doi:10.1088/1742-6596/770/1/012027

# Higher order effects in $\varepsilon'/\varepsilon$

**M Cerdà-Sevilla**

*Department of Mathematical Sciences, University of Liverpool, L69 7ZL Liverpool, United Kingdom*

E-mail: Maria.Cerda-Sevilla@liv.ac.uk

**Abstract.** The quantity $\varepsilon'/\varepsilon$ measures direct CP violation in Kaon decays. Recent analysis of this ratio resulted in a 2.9 sigma discrepancy between the Standard Model predictions and the experimental data. Our ability to observe or constrain New Physics depends on the accuracy of determining the SM "background", hence a precise evaluation of $\varepsilon'/\varepsilon$ is particularly important. We discuss the Standard Model prediction and the relevant matching calculations at NNLO for this observable.

## 1. Introduction

The ratio $\varepsilon'/\varepsilon$ parametrises the size of direct CP violation with respect to indirect CP violation in Kaon decays. Over the last decades, this observable has been subject to very intensive experimental and theoretical studies. After enormous efforts, on the experimental side the world average based on the recent results from NA48 [1] and KTeV [2,3] collaborations reads

$$\left(\frac{\varepsilon'}{\varepsilon}\right)_{\text{exp}} = (16.6 \pm 2.3) \times 10^{-4}. \tag{1}$$

Theoretical predictions of this quantity are obtained through a weak effective Hamiltonian characterised by local operators and the corresponding Wilson coefficients. The latter can be computed in perturbation theory, whereas the matrix elements of the operators have to be evaluated within some non-perturbative approach like lattice QCD or the large-N approach [4,5]. For Kaon decays, the Wilson coefficients are known at the Next-to-leading order (NLO) accuracy [7–12] and some pieces are available at the Next-to-Next-to-leading Order (NNLO) [13–15]. In the meantime an important progress has been achieved in the non-perturbative sector by the RBC-UKQCD lattice collaboration [16, 17]. Even though considerable improvement in calculating $\varepsilon'/\varepsilon$ has been made, theoretical estimates of this ratio are still subject to very large hadronic uncertainties which interfere in the prediction of $\varepsilon'/\varepsilon$. A recent theoretical determination [6] of this observable at NLO within the Standard Model has found a 2.9 sigma tension between theory and experiment

$$\left(\frac{\varepsilon'}{\varepsilon}\right)_{\text{SM}} = (1.9 \pm 4.5) \times 10^{-4}. \tag{2}$$

The main point here is the identification of a possible new anomaly in flavor physics, this time in the K sector. It is very important to disentangle the origin of this discrepancy. Kaon decays, within the SM, are loop induced and highly suppressed, consequently physics Beyond the Standard Model can easily contribute. To be sure that this effect is due to the presence of New Physics (NP) we need a reliable SM prediction.

Motivated by this inconsistency and the fact there are realistic prospects for improvements in the non-perturbative sector via Lattice QCD that would render NNLO accuracy essential, we are working on the computation of these higher order corrections to the Wilson coefficients. These new contributions will have an impact on the theory prediction for $\varepsilon'/\varepsilon$. Although it seems greatly improbable that they can bring the SM prediction into reconciliation with experiment, it is still essential to study them in more detail. These perturbative corrections could be potentially important due to the large value of $\alpha_s$ at the low energy scale $\mu_c$. In this talk, we cover the several steps involved in the calculation of higher order QCD corrections to this observable.

## 2. CP-Violation in Kaon Decays

CP violation emerges naturally in the three generation Standard Model. Its origin lies solely in Yukawa-type interactions of the quark fields with the complex Higgs fields. Within this model, CPV is only an effect of a complex phase. To explain matter-antimatter asymmetry in the Universe, new sources are required.

In neutral K-meson decays CP violation has been accommodated in the Standard Model in a simple way. However, this phenomenon is one of the least tested aspects of the Standard Model. CP violating effects are of the order $\mathcal{O}(10^{-4})$. The small prediction for CPV in this framework is due to flavor suppression (CKM factors).

The study of Kaon decays can bring some light on this puzzle. However, in the framework of K mesons it is not easy to estimate CP violating observables with great precision, since strong interactions are in a non-perturbative regime. Nevertheless, the use of Effective Field Theories (EFT) simplifies the calculations. This framework introduces a natural way to separate the different energy scales and allows for a convenient method to sum large logarithms to all orders in perturbation theory. Below the charm mass scale, $\mu < m_c$, the effective Hamiltonian looks like [7–12],

$$\mathcal{H}_{\text{eff}} = \frac{G_F}{\sqrt{2}} V_{ud} V_{us}^* \sum_{i=1}^{10} (z_i(\mu) + \tau y_i(\mu)) \mathcal{O}_i, \quad \tau \equiv -\frac{V_{td} V_{ts}^*}{V_{ud} V_{us}}, \tag{3}$$

where the functions $z_i(\mu)$ and $y_i(\mu)$ are the Wilson coefficients which can be calculated perturbatively, $V_{ij}$ are the Cabibbo-Kobayashi-Maskawa (CKM) matrix elements, $\mathcal{O}_{1,2}$ are current-current operators, $\mathcal{O}_{3-6}$ the QCD penguin operators, and $\mathcal{O}_{7-10}$ the electroweak penguin operators.

In order to study this CP-violating observable, it is important to improve both the theoretical calculations of short-distance contributions [7–15] (the Wilson coefficients functions $z_i$ and $y_i$) and the estimation of long-distance contributions (the hadronic matrix elements $\langle \mathcal{O}_i(\mu) \rangle$) [16,17]. For this reason we aim to complete the elaborate perturbative calculation. Our future results will address whether the complete NNLO corrections enhance or suppress the Wilson coefficients and whether they will have an impact on $\varepsilon'/\varepsilon$.

Theoretical predictions for CPV are quite involved. In terms of the isospin amplitudes this observable is parametrised as

$$\frac{\varepsilon'}{\varepsilon} = -i \frac{\omega_+}{\sqrt{2}|\varepsilon_K|} e^{i(\delta_2 - \delta_0 - \phi_{\varepsilon_K})} \left[ \frac{\text{Im}(\mathcal{A}_0)}{\text{Re}(\mathcal{A}_0)} (1 - \Omega_{\text{eff}}) - \frac{1}{a} \frac{\text{Im}(\mathcal{A}_2)}{\text{Re}(\mathcal{A}_2)} \right], \tag{4}$$

where $a$ and $\Omega_{\text{eff}}$ comprise isospin-breaking correction, $\delta_{0,2}$ denote the isospin strong phase-shifts, $\phi_{\varepsilon_K}$ stands for the phase of $\varepsilon_K$ and $\omega_+$ is determined from the charged decay mode.

Journal of Physics: Conference Series **770** (2016) 012027 doi:10.1088/1742-6596/770/1/012027

For a precision SM prediction it is required to obtain the real and imaginary parts of the isospin amplitudes $\mathcal{A}_I \equiv \langle(\pi\pi)_I|\mathcal{H}_{\text{eff}}|K\rangle$ entering Eq.(4) in terms of the Wilson coefficients and hadronic matrix elements of the operators in the weak Hamiltonian, Eq.(3).

## 3. Computation of Im$(A_i)$/Re$(A_i)$

The formalism used to determine these important pieces is based on reference [6]. They work under the hypothesis that the amplitudes Re$(\mathcal{A}_0)$ and Re$(\mathcal{A}_2)$, in the SM, originate already at tree-level. Therefore these two quantities are expected to be only marginally affected by NP contributions,

$$\text{Re}(\mathcal{A}_0) = \frac{G_F}{\sqrt{2}} V_{ud}V_{us}^*(z_+\langle\mathcal{O}_+\rangle_0 + z_-\langle\mathcal{O}_-\rangle_0), \quad \text{Re}(\mathcal{A}_2) = \frac{G_F}{\sqrt{2}} V_{ud}V_{us}^* z_+\langle\mathcal{O}_+\rangle_2. \tag{5}$$

Assuming this dominance of SM dynamics in CP-conserving data, our determination of the contributions of $(V-A)\times(V-A)$ operators to $\varepsilon'/\varepsilon$ is basically independent of the non-perturbative approach used. In this way a more accurate prediction for $\varepsilon'/\varepsilon$ can be made than currently possible with direct lattice-QCD simulations:

$$\begin{aligned}\left(\frac{\text{Im}(\mathcal{A}_0)}{\text{Re}(\mathcal{A}_0)}\right)_{V-A} &= \text{Im}(\tau)\frac{2y_4}{(1+q)z_-} + \mathcal{O}(p_3)\\ \left(\frac{\text{Im}(\mathcal{A}_2)}{\text{Re}(\mathcal{A}_2)}\right)_{V-A} &= \text{Im}(\tau)\frac{3(y_9+y_{10})}{2z_+}.\end{aligned} \tag{6}$$

In the above expression the factor $q$ is defined in terms of the current-current Wilson coefficients and operators, $q \equiv (z_+(\mu)\langle\mathcal{O}_+(\mu)\rangle_0)/(z_-(\mu)\langle\mathcal{O}_-(\mu)\rangle_0)$. In addition, $\mathcal{O}(p_3)$ encodes subleading order contributions. Notice here that the $(V-A)\times(V-A)$ terms are dominated by short distance (Wilson coefficients).

The remaining contribution coming from the $(V-A)\times(V+A)$ operators depends on the two hadronic parameters $B_6^{(1/2)}$ and $B_8^{(3/2)}$ and comprises the biggest uncertainty in $\varepsilon'/\varepsilon$:

$$\begin{aligned}\left(\frac{\text{Im}(\mathcal{A}_0)}{\text{Re}(\mathcal{A}_0)}\right)_{V+A} &= -\frac{G_F}{\sqrt{2}}\text{Im}(\lambda_\tau)y_6\frac{\langle\mathcal{O}_6\rangle_0}{\text{Re}(\mathcal{A}_0)} + \mathcal{O}(p_5)\\ \left(\frac{\text{Im}(\mathcal{A}_2)}{\text{Re}(\mathcal{A}_2)}\right)_{V+A} &= -\frac{G_F}{\sqrt{2}}\text{Im}(\lambda_\tau)y_8^{\text{eff}}\frac{\langle\mathcal{O}_8\rangle_2}{\text{Re}(\mathcal{A}_2)}.\end{aligned} \tag{7}$$

It is easy to see that the terms with $(V-A)\times(V+A)$ chirality are controlled by long-distance (matrix elements).

To summarise the main ideas mentioned until now: in the SM the amplitudes Re$(\mathcal{A})_{0,2}$ are mostly governed by the $\mathcal{O}_{1,2}$ current-current operators and $\varepsilon'/\varepsilon$ by the QCD penguin $\mathcal{O}_6$ and electroweak penguin $\mathcal{O}_8$ operators.

In the coming years, the RBC-UKQCD collaboration could reduce the statistical uncertainty of the non-perturbative sector. Consequently, the perturbative side would become the main source of uncertainty for $\varepsilon'/\varepsilon$.

## 4. Higher order QCD corrections

When a quark is integrated out as a dynamical degree of freedom its effects have to be taken into account. These threshold corrections are determined through a matching of the effective theories with $n_f$ and $n_f+1$ flavors. The respective calculation requires the equality of the Green's functions in the two theories at the matching scale $\mu_q = \mathcal{O}(m_q)$, where $m_q$ is the quark mass.

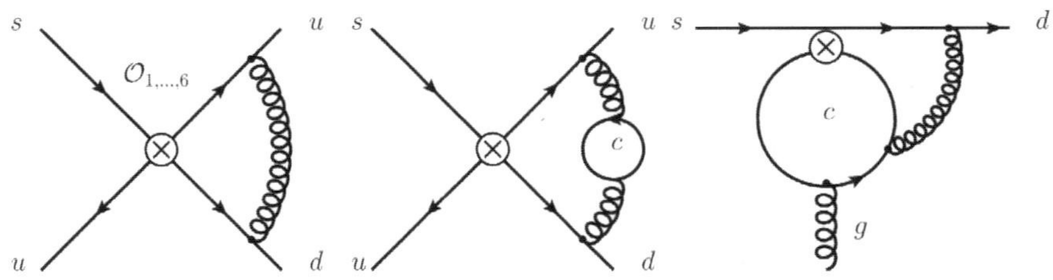


**Figure 1.** Feynman diagrams relevant for the threshold corrections at the charm quark scale. The one-loop current-current diagrams are identical in the four- and three-flavor theories, whereas at NNLO they receive non-trivial corrections from virtual charm quarks.

In order to improve on the present NLO calculation, one needs to include higher order terms in the strong coupling expansion. The completion of this NNLO computation comprises the content of our project [19]. This task is more sophisticated than in the NLO case. In particular the number of diagrams and structures that appear in the study, increases. In addition, there are diverse sources of matching corrections. At one loop the penguin operators are already affected since they explicitly depend on light-quark fields. At NNLO, the matching of the current-current is also non-trivial. Two-loop matrix elements get extra contributions from virtual light quarks. Some relevant diagrams for this calculation are shown in Fig(1). In addition, the strong coupling constant and the light-quark mass are discontinuous beyond leading order (LO).

An important piece in the calculation of the two-loop QCD corrections within the SM is the matrix elements. To simplify this computation we use the so called "modern basis" [18]. This solves the problems arising from the $\gamma_5$ matrix appearing in closed fermion loops in the framework of dimensional regularisation. In our calculations we expand the external momenta up to $\mathcal{O}(k^2)$ and compute the corresponding renormalisation. To simplify the computation we set the mass of the light quarks to zero. This introduces Infrared Divergences (Spurious) in the $n_f + 1$ theory amplitude which have to be cancelled by the Ultra-Violet divergences in the $n_f$ flavor theory. The matching results in finite threshold corrections for the physical operators.

## 5. Scale cancellation

The aim here it is to show the log cancellation that occurs in this procedure and comment about the residual $\mu_q$ scale dependence. For a more exhaustive explanation refer to [14].

The calculation consists of several steps. First, the initial conditions for the Wilson coefficients at the electroweak scale are computed, $\overline{\mathcal{C}}(M_W)$. For this purpose we match the Standard Model Green's functions to those in the five-flavor theory, where the heavy particles have been integrated out. Subsequently, the Wilson coefficients are evolved down to the bottom-quark scale using the renormalisation group equations (RGE),

$$\overline{\mathcal{C}}(\mu_b) = \hat{U}(\mu_b, \mu_W)\overline{\mathcal{C}}(\mu_W), \tag{8}$$

where $\hat{U}(\mu_b, \mu_W)$ describes pure QCD evolution. Afterwards, we compute the threshold corrections at $\mu_b = \mathcal{O}(m_b)$, $\hat{M}(\mu_b)$. For this aim, we match the five-flavor theory onto an effective theory where also the bottom quark has been removed as a dynamical degree of freedom. The resulting Wilson coefficient $\overline{\mathcal{C}}(\mu_b)$ is then evolved down to the charm-quark scale using the RGE,

$$\overline{\mathcal{C}}(\mu_c) = \hat{U}(\mu_c, \mu_b)\hat{M}(\mu_b)\overline{\mathcal{C}}(\mu_b). \tag{9}$$

At this scale, $\mu_c = \mathcal{O}(m_c)$, the matching equation for the charm quark is now evaluated. Here, we match the four-flavor theory onto the three-flavor theory and find the new threshold correction

$\hat{M}(\mu_c)$. Finally, we incorporate these results and perform the renormalisation group evolution down to the scales where the hadronic matrix elements are computed:

$$\overline{C}(\mu_{\text{Lattice}}) = \hat{U}(\mu_{\text{Lattice}}, \mu_c)\hat{M}(\mu_c)\overline{C}(\mu_c). \tag{10}$$

The lines above provide us the general picture without giving much detail. However, to understand how the logarithms are cancelled we have to extend this formalism. In a particular, the calculation can be organised in such a manner that the log cancellations occur separately at each scale. Following the notation used in [14] the evolution matrix is given by

$$\hat{U}(\mu_b, \mu_W) = \hat{K}(\mu_b)\hat{U}^{(0)}(\mu_b, \mu_W)\hat{K}^{-1}(\mu_W) \tag{11}$$

where,

$$\hat{U}^{(0)}(\mu, \mu_0) = \hat{V}\text{diag}\left(\frac{\alpha_s(\mu_0)}{\alpha_s(\mu)}\right)^{a_i}\hat{V}^{-1} \tag{12}$$

stands for the LO term. The matrix $\hat{V}$ and the magic numbers $a_i$ are obtained from the diagonalisation of the LO anomalous dimension matrix (ADM), $\gamma^{(0)T}$:

$$\left(\hat{V}^{-1}\gamma^{(0)T}\hat{V}\right)_{ij} = 2\beta_0 a_i \delta_{ij}. \tag{13}$$

The evolution matrix $\hat{U}(\mu, \mu_0)$ depends on two different scales, with $\mu \ll \mu_0$. Therefore, we split it into two parts:

$$\begin{aligned} \hat{U}^{1/2}(\mu) &= \hat{K}(\mu)\hat{V}\text{diag}\left(\alpha_s(\mu)\right)^{-a_i}\hat{V}^{-1} \\ \hat{U}^{-1/2}(\mu_0) &= \hat{V}\text{diag}\left(\alpha_s(\mu_0)\right)^{a_i}\hat{V}^{-1}\hat{K}^{-1}(\mu_0), \end{aligned} \tag{14}$$

with

$$\hat{K}(\mu) = \hat{1} + \frac{\alpha_s(\mu)}{4\pi}\hat{J}^{(1)} + \mathcal{O}(\alpha_s^2), \quad \hat{K}^{-1}(\mu_0) = \hat{1} - \frac{\alpha_s(\mu_0)}{4\pi}\hat{J}^{(1)} + \mathcal{O}(\alpha_s^2). \tag{15}$$

From this factorisation we find that the following object appears in the intermediate states in our calculation:

$$\left[\hat{V}\text{diag}\left(\alpha_s(\mu)\right)^{a_i}\hat{V}^{-1}\hat{K}^{-1}(\mu)\right]_{n_f}\left[\hat{M}(\mu)\right]_{n_f,n_{f+1}}\left[\hat{K}(\mu)\hat{V}\text{diag}\left(\alpha_s(\mu)\right)^{-a_i}\hat{V}^{-1}\right]_{n_{f+1}}. \tag{16}$$

Note that the matching matrix depends on both $\alpha_s^{n_f}(\mu)$ and on $\alpha_s^{n_f+1}(\mu)$. Expressing $\alpha_s^{n+1}$ in terms of $\alpha_s^{n_f}$ and expanding the mass terms we observe that the scale dependence appearing in the evolution down to the $n_f$-flavor theory (third term in Eq.(16)) and the one emerging in the evolution up to this theory (first term in Eq.(16)) are cancelled by the $\log(\mu_q)$ from the light-quark matching, $\hat{M}(\mu_q)$. Therefore this object is individually scale- and scheme-independent. The log cancellation works order by order.

Once can then estimate higher order effects by varying the matching scale $\mu_q$ while not expanding the expression in $\alpha_s$. This would result in a residual scale dependence that is of the size of higher order corrections.

## 6. Conclusions

The experimental data for $\varepsilon'/\varepsilon$ is not well described by the SM. However, before establishing whether NP is present here many questions should be addressed. In the near future, the RBC-UKQCD collaboration could reduce the uncertainties of the non-perturbative calculations. Therefore, the perturbation side would become the main source of uncertainty for $\varepsilon'/\varepsilon$. A missing piece, in the last theoretical study [6], is the one due to the unknown matching corrections at $\mu = m_c$ which could be sizeable since the strong coupling is growing rapidly in this region. Our work is focused on covering this calculation at NNLO and checking the consistency of the perturbation theory. At present we have completed the matching for the current-current and QCD penguin operators. The new result obtained essentially removes the QCD penguin contribution to the perturbative error which was estimated at $1.5 \times 10^{-4}$ in [6]. This leaves the new perturbative error dominated by the missing NNLO-electroweak penguin contribution and roughly half the size of that in [6]. These new calculations will help to improve the SM predictions and potentially motivate to look into New Physics scenarios.

## Acknowledgments

I would like to thank the organisers of BEACH 2016 conference for this wonderful experience.

## References

[1] NA48 Collaboration, J. Batley et al., *A Precision measurement of direct CP violation in the decay of neutral kaons into two pions, Phys. Lett.* **B544** (2002) 97-112, [hep-ex/0208009].
[2] KTeV Collaboration, A. Alavi-Harati et al., *Measurements of direct CP violation, CPT symmetry, and other parameters in the neutral kaon system, Phys. Rev.* **D67** (2003) 012005, [hep-ex/0208007].
[3] KTeV Collaboration, E. Worcester, *The Final Measurement of $\varepsilon'/\varepsilon$ from KTeV,* [arXiv:0909.2555].
[4] W. A. Bardeen, A. J. Buras, J.-M. Gérard, *The $K \to \pi\pi$ Decays in the Large-N Limit: Quark Evolution, Nucl. Phys.* **B293** (1987) 787.
[5] A. J. Buras, J.-M. Gérard, W. A. Bardeen, *Large-N Approach to Kaon Decays and Mixing 28 years Later: $\Delta I = 1/2$ Rule, $\hat{B}_K$ and $\Delta M_K$, Eur. Phys. J.* **C74** (2014), no. 5 2871, [arXiv:1401.1385].
[6] A. J. Buras, M. Gorbahn, S. Jäger, and M. Jamin, *Improved anatomy of $\varepsilon'/\varepsilon$ in the Standard Model,* [arXiv:1507.06345].
[7] A. J. Buras, M. Jamin, M. Lautenbacher, and P.H. Weisz, *Effective Hamiltonians for $\Delta S = 1$ and $\Delta B = 1$ non-leptonic decays beyond the leading logarithmic approximation, Nucl. Phys.* **B370** (1992) 69-104.
[8] A. J. Buras, M. Jamin, M. E. Lautenbacher, and P.H. Weisz, *Two loop anomalous dimension matrix for $\Delta S = 1$ weak non-leptonic decays. 1. $\mathcal{O}(\alpha_s^2)$, Nucl. Phys.* **B400** (1993) 37-74, [hep-ph/9211304].
[9] A. J. Buras, M. Jamin, M. E. Lautenbacher, *Two loop anomalous dimension matrix for $\Delta S = 1$ weak non-leptonic decays. 2. $\mathcal{O}(\alpha\alpha_s^2)$, Nucl. Phys.* **B400** (1993) 75-102, [hep-ph/9211321].
[10] M. Ciuchini, E. Franco, G. Martinelli and L. Reina, *$\varepsilon'/\varepsilon$ at the next-to-leading order in QCD and QED, Phys. Lett.* **B301** (1993) 263-271, [hep-ph/9212203].
[11] A. J. Buras, M. Jamin, M. E. Lautenbacher, *The anatomy of $\varepsilon'/\varepsilon$ beyond leading logarithms with improved hadronic matrix elements, Nucl. Phys.* **B408** (1993) 209-285, [hep-ph/9303284].
[12] M. Ciuchini, E. Franco, G. Martinelli and L. Reina, *The $\Delta S = 1$ effective Hamiltonian including next-to-leading order QCD and QED corrections, Nucl. Phys.* **B415** (1994) 403-462, [hep-ph/9304257].
[13] A. J. Buras, P. Gambino, and U. A. Haisch, *Electroweak penguin contributions to non-leptonic $\Delta F = 1$ decays at NNLO, Nucl. Phys.* **B570** (2000) 117-154, [hep-ph/9911250].
[14] M. Gorbahn and U. Haisch, *Effective Hamiltonian for non-leptonic $|\Delta F| = 1$ decays at NNLO in QCD, Nucl. Phys.* **B713** (2005) 291-332, [hep-ph/0411071].
[15] J. Brod and M. Gorbahn, *$\epsilon_K$ at Next-to-Next-to-Leading Order: The Charm-Top-Quark Contribution, Phys. Rev.* **D82** (2010) 094026, [arXiv: 1007.0684].
[16] T.Blum, P. Boyle, N.Christ, J.Frison, N. Garron, et al., *$K \to \pi\pi \Delta I = 3/2$ decay amplitude in the continuum limit,* [arXiv: 1502.00263].
[17] Z. Bai,T.Blum, P. Boyle, N.Christ, J.Frison, et al., *Standard-model prediction for direct CP violation in $K \to \pi\pi$ decay,* [arXiv: 1505.07863].
[18] K.G. Chetyrkin, M. Misiak and M. Munz, *$|\Delta F| = 1$ nonleptonic effective Hamiltonian in a simpler scheme, Nucl. Phys.* **B520** (1998) 279, [aXiv:hep-ph/9711280].
[19] M. Cerdà -Sevilla, M. Gorbahn, S. Jäger and A. Kokulu, in preparation.

# Measurement of the CKM Angle $\gamma$

**M. Whitehead[1] on behalf of the LHCb collaboration**
[1] CERN, CH-1211 Geneva 23, Switzerland

E-mail: mwhitehe@cern.ch

**Abstract.** A summary of constraints on the CKM angle $\gamma$ from recent LHCb analyses are discussed. The results of a combination of several LHCb measurements of $\gamma$ are presented, giving

$$\gamma = (70.9^{+7.1}_{-8.5})^{\circ},$$

where the single uncertainty is a combination of statistical and systematic sources. This represents the most precise measurement of the angle $\gamma$.

## 1. Introduction

Measurements of $CP$ violation are important to understand the matter-antimatter asymmetry of the universe. In the quark sector, $CP$ violation is included in the Standard Model (SM) of particle physics through a complex phase in the Cabibbo-Kobayashi-Maskawa (CKM) matrix [1, 2]. The parameter $\gamma$ is the least precisely measured angle of the CKM unitarity triangle, where the area of the triangle is proportional to the amount of CP violation in the SM [3]. Theoretically $\gamma$ is very clean ($\Delta\gamma/\gamma \approx 10^{-7}$), because it can be measured through tree-level processes only [4]. Discrepancies between the value of $\gamma$ from loop dominated processes and a precise direct measurement would infer the presence of New Physics effects.

Several methods are available to measure $\gamma$ using decays such as $B^{\pm} \to DK^{\pm}$, where $D$ is either a $D^0$ or $\bar{D}^0$ meson. The GLW method [5, 6] considers decays of the $D$ meson to $CP$ eigenstates, such as the $CP$ even final states $K^+K^-$ and $\pi^+\pi^-$. The ADS approach [7, 8] requires favoured and doubly Cabibbo suppressed (DCS) decays, for example $D \to K^{\pm}\pi^{\mp}$. A third method, known as GGSZ [9], uses $D$ decays to self conjugate final states such as $K^0_S\pi^+\pi^-$.

Three recent LHCb analyses are discussed; determining $\gamma$ with $B^{\pm} \to DK^{\pm}$ with $D \to K^{\pm}\pi^{\mp}$, $K^+K^-$, $\pi^+\pi^-$, $\pi^+\pi^-\pi^+\pi^-$ and $K^{\pm}\pi^{\mp}\pi^+\pi^-$ decays [10] (Sec. 2), measuring $CP$ violation using $B^0 \to DK^*(892)^0$ decays where $D \to K^0_S\pi^+\pi^-$ [11] (Sec. 3) and constraining $\gamma$ using $B^0 \to DK\pi$ decays with $D \to K^+K^-$ and $D \to \pi^+\pi^-$ [12] (Sec. 4). The results of a combination of LHCb $\gamma$ measurements is discussed in Sec. 5 [13].

## 2. Analysis of $B^{\pm} \to DK^{\pm}$ Decays

The GLW/ADS analysis of the decay $B^{\pm} \to DK^{\pm}$, using 2- and 4-body $D$ decays is documented in Ref. [10]. Candidate $B^{\pm} \to D\pi^{\pm}$ decays are used as a high statistic control channel, although results are also given for these channels. Data samples are selected using multivariate analyses to remove background candidates.

Figure 1 shows the $B$ candidate invariant mass distribution for the 2-body DCS $D$ meson decays, where $CP$ violation is clearly visible by eye in the $B^{\pm} \to DK^{\pm}$ channel, and has a

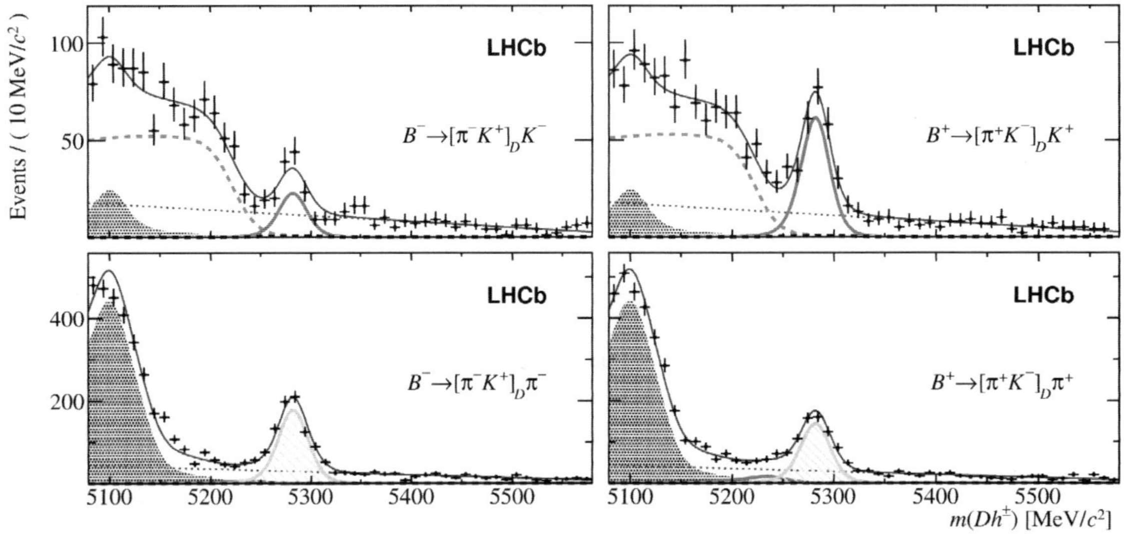


**Figure 1.** Invariant mass distribution of $B^{\pm} \to DK^{\pm}$ decays with the DCS decay $D \to \pi^{\pm} K^{\mp}$. Figure taken from Ref. [10], where the fit model is described.

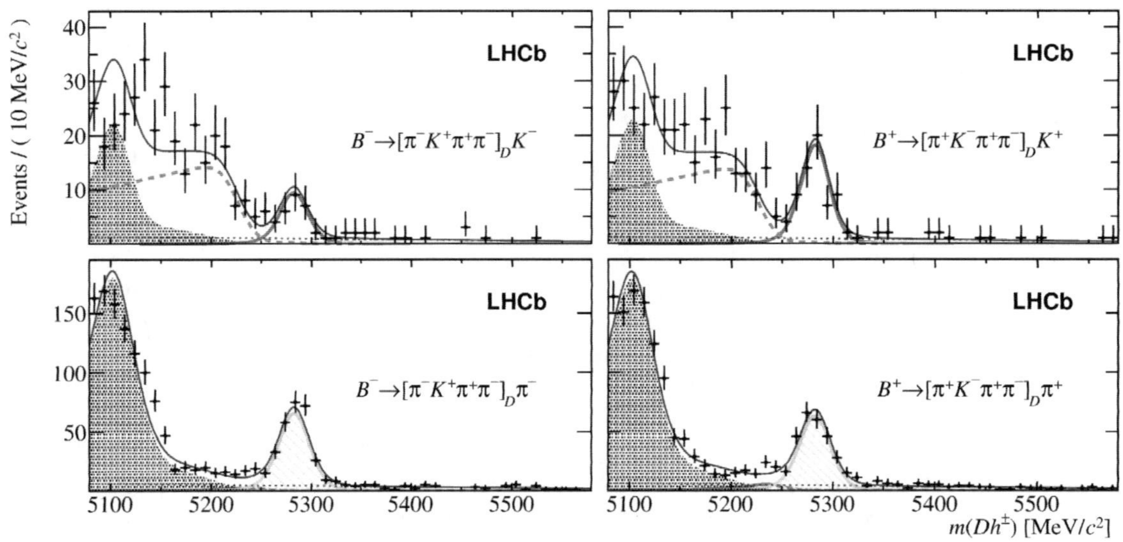


**Figure 2.** Invariant mass distribution of $B^{\pm} \to DK^{\pm}$ decays with the DCS decay $D \to \pi^{\pm} K^{\mp} \pi^+ \pi^-$. Figure taken from Ref. [10], where the fit model is described.

significance of about $8\sigma$. Similarly, Fig. 2 shows similar distributions for the 4-body DCS decays of the $D$ meson. In this case the $CP$ violation effect is visible, but not significant, with the current statistics. In addition, $CP$ violation is observed at the $5\sigma$ level using the $D \to K^+ K^-$ and $D \to \pi^+ \pi^-$ modes combined. The 4-body decay $\pi^+ \pi^- \pi^+ \pi^-$ is also studied for the first time. There are 22 observables that are determined from the yields of the $B^+$ and $B^-$ decays for each $D$ decay final state, which are used as an input to the $\gamma$ combination described in Sec. 5.

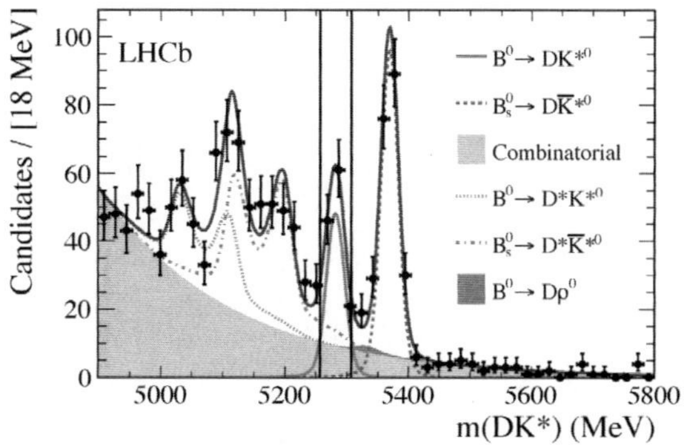


**Figure 3.** Invariant mass distribution of $B^0 \to DK^*(892)^0$ decays where $D \to K^0_S\pi^+\pi^-$, the vertical lines show the signal mass window and components are as described in the legend. Figure taken from Ref. [11], where the fit model is described.

## 3. Study of $B^0 \to DK^*(892)^0$ Decays

A model dependent GGSZ analysis of $B^0 \to DK^{*0}$ decays with $D \to K^0_S\pi^+\pi^-$ by LHCb is detailed in Ref. [11]. Note that a model independent analysis has also been performed [14], but is not discussed in the following. Candidates are selected using a boosted decision tree to reject background candidates. A fit is performed to the $B$ candidate invariant mass, as shown in Fig. 3, to obtain the signal and background yields in the mass window $\pm 25\text{MeV}/c^2$ around the $B^0$ mass.

For candidates in the signal mass window, an amplitude fit is performed to the $D \to K^0_S\pi^+\pi^-$ Dalitz plot distribution for $B^0$ and $\bar{B}^0$ candidates. The amplitude model is taken from a BaBar analysis, described in Ref. [15]. The observables $x_\pm \equiv r_B\cos(\delta_B \pm \gamma)$ and $y_\pm \equiv r_B\sin(\delta_B \pm \gamma)$ are determined to be

$$\begin{aligned}
x_- &= -0.15 \pm 0.14 \pm 0.03 \pm 0.01, \\
y_- &= \phantom{-}0.25 \pm 0.15 \pm 0.06 \pm 0.01, \\
x_+ &= \phantom{-}0.05 \pm 0.24 \pm 0.04 \pm 0.01, \\
y_+ &= -0.65 \ {}^{+0.24}_{-0.23} \ \pm 0.08 \pm 0.01.
\end{aligned}$$

Here the first uncertainty is statistical, the second experimental systematic and the third model dependent systematic. The hadronic parameters $r_B$ and $\delta_B$ are the ratio and strong phase difference between the favoured and suppressed $B$ decay amplitudes. All observables are consistent with zero at roughly $2\sigma$, and are therefore consistent with $CP$ conservation.

## 4. Dalitz Plot Analysis of $B^0 \to DK^+\pi^-$ Decays

Reference [12] describes the GLW-Dalitz analysis of the $B^0 \to DK^+\pi^-$ channel where $D \to K^+K^-$ and $D \to \pi^+\pi^-$. In addition, the favoured $D \to K\pi$ decay is used as a high statistics control channel. Signal candidates are separated from backgrounds using an artificial neural network, and the data are binned in the output variable of the network. This approach preserves the maximum signal yield without sacrificing purity in each bin. Projections of the fits to the $B$ candidate invariant mass distributions are shown in Fig. 5 for the three $D$ decay modes.

Signal mass windows are defined to be 5246.6–5309.9MeV/$c^2$, 5246.9–5310.5MeV/$c^2$ and 5243.1–5312.3MeV/$c^2$ for $D \to K^+\pi^-$, $K^+K^-$ and $\pi^+\pi^-$ samples, respectively. For candidates in the signal windows, a simultaneous Dalitz plot fit is performed to the three $D$ decay samples, following the method described in Refs. [16, 17]. Terms for $CP$ violation are included in the

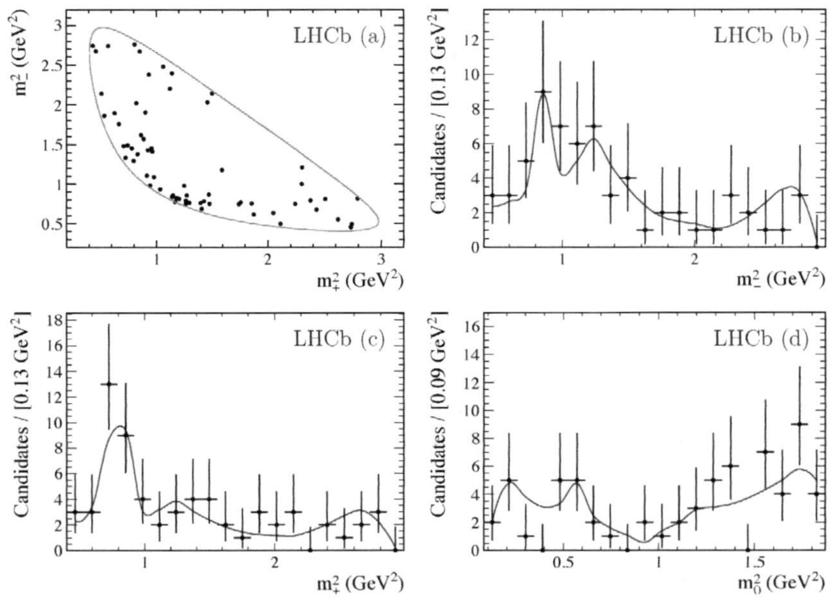


**Figure 4.** Dalitz plot (a) and projections of the amplitude fit on to the invariant mass squares (b, c and d) for $D \to K^0_S\pi^+\pi^-$ decays from the $B^0 \to DK^*(892)^0$ sample. The labels $m_+$, $m_-$ and $m_0$ represent $m(K^0_S\pi^+)$, $m(K^0_S\pi^-)$ and $m(\pi^+\pi^-)$ respectively. Figure taken from Ref. [11], where the amplitude model is described.

$K^*(892)^0$ amplitudes for the $CP$ even $D \to K^+K^-$ and $D \to \pi^+\pi^-$ modes. The observables determined from the amplitude fit, as defined above, are found to be

$$\begin{aligned} x_- &= -0.02 \pm 0.13 \pm 0.14, \\ y_- &= -0.35 \pm 0.26 \pm 0.41, \\ x_+ &= \phantom{-}0.04 \pm 0.16 \pm 0.11, \\ y_+ &= -0.47 \pm 0.28 \pm 0.22, \end{aligned}$$

where the first uncertainty is statistical and the second systematic. These results are consistent with no $CP$ violation.

## 5. Combination of LHCb $\gamma$ Measurements

The combination of several LHCb measurements using a frequentist approach is described in Ref. [13], which is an update of the previous combination [18], including the new results described in Secs. 2–4. The combination takes input from $B \to DK$-like decay modes only, using a total of 71 observables from 10 LHCb analyses. The result of the one dimensional `PLUGIN` scan [19] for $\gamma$ is shown in Fig. 7 (left), the (right) plot shows the breakdown of the result in terms of the different species of $B$ mesons. It is clear that the $B^\pm$ decay modes currently dominate the sensitivity, though the other contributions are non-negligible. The confidence intervals $\gamma \in [62.4,\ 78.0]^\circ$ at 68% CL and $\gamma \in [62.4,\ 78.0]^\circ$ at 95% CL are set and $\gamma$ is measured to be

$$\gamma = (70.9^{+7.1}_{-8.5})^\circ,$$

where the uncertainty includes both statistical and systematic effects. This is currently the most precise measurement of $\gamma$, improving on the previous LHCb combination by approximately two degrees.

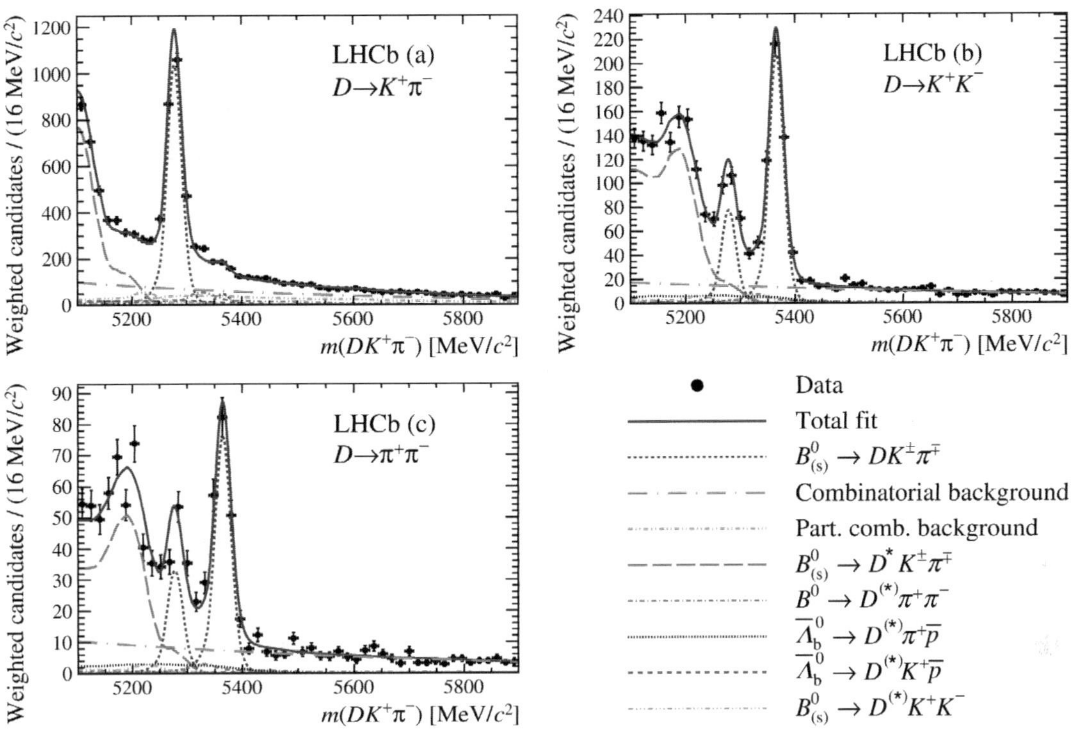


**Figure 5.** Invariant mass distribution of $B^0 \to DK^+\pi^-$ decays with (a) $D \to K^+\pi^-$, (b) $D \to K^+K^-$ and (c) $D \to \pi^+\pi^-$. The fit components are as described in the legend and the figure is from Ref. [12].

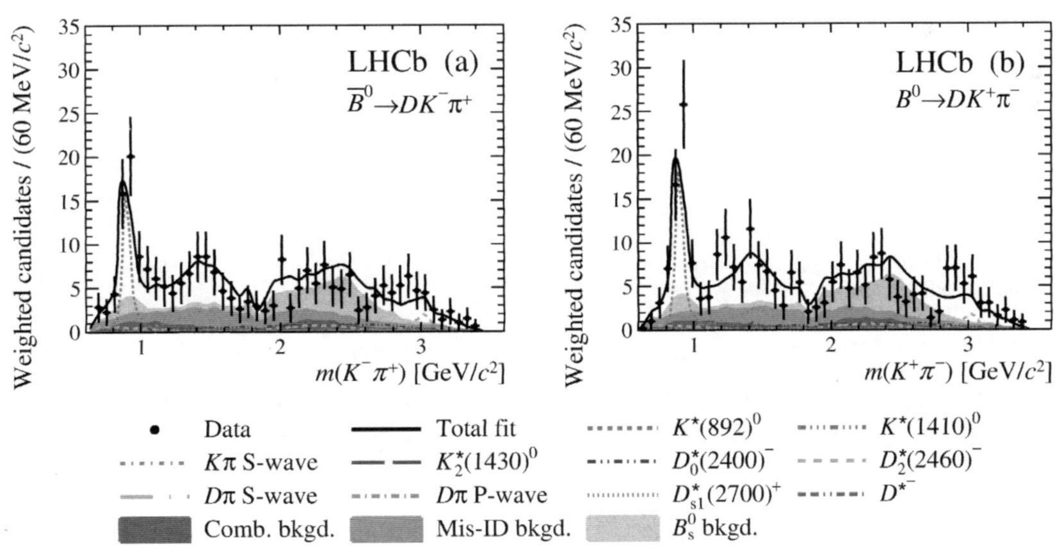


**Figure 6.** Projections of the Dalitz plot fit on $m(K\pi)$ for (a) $\bar{B}^0$ and (b) $B^0$ candidates. The fit components are as described in the legend. Figure taken from Ref. [12].

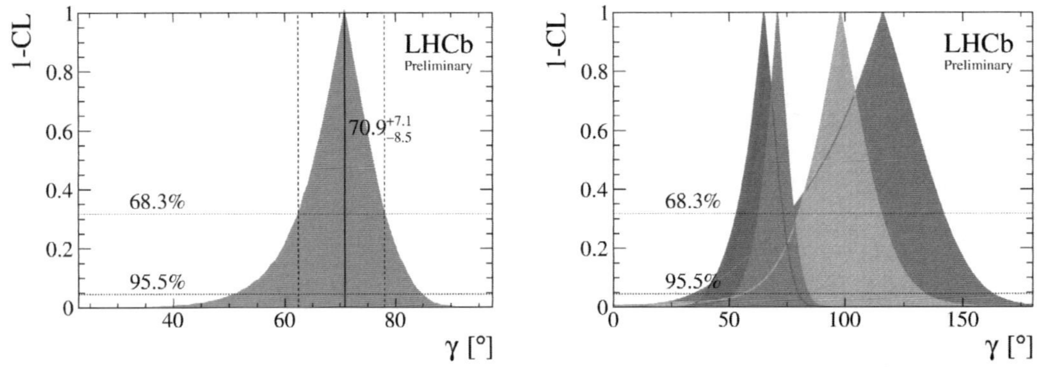


**Figure 7.** (Left) 1-CL curve for the angle $\gamma$ obtained using the `PLUGIN` method, with the central value (solid vertical line), $1\sigma$ uncertainties (dashed vertical lines), and 68.3% and 95.5% CLs (horizontal dotted lines) shown. (Right) a breakdown to show the contributions of the different $B$ meson species; $B^+$ (blue), $B^0$ (yellow), $B_s^0$ (orange) and the full combination (green). Figure taken from Ref. [13].

## 6. Summary

The latest measurements of the CKM angle $\gamma$ from LHCb are presented. The combination of results is consistent with the expected LHC Run 1 sensitivity. To improve the precision of the measurement, new decay modes can be added and updates can be included to existing measurements using Run 1 data. For example, information from $B \to D\pi$-like decays can be included. In the longer term the full Run 2 data sample will provide more than double the current available statistics, with an aim to reduce the uncertainties to around 4°.

## 7. Acknowledgements

I would like to thank my LHCb colleagues for their help in preparing the talk and this document.

## 8. References

[1] Cabibbo N 1963 *Phys. Rev. Lett.* **10** 531–533
[2] Kobayashi M and Maskawa T 1973 *Prog. Theor. Phys.* **49** 652–657
[3] Jarlskog C 1985 *Phys. Rev. Lett.* **55** 1039
[4] Brod J and Zupan J 2014 *JHEP* **01** 051 (*Preprint* `1308.5663`)
[5] Gronau M and Wyler D 1991 *Phys. Lett.* **B265** 172–176
[6] Gronau M and London D 1991 *Phys. Lett.* **B253** 483–488
[7] Atwood D, Dunietz I and Soni A 1997 *Phys. Rev. Lett.* **78** 3257–3260 (*Preprint* `hep-ph/9612433`)
[8] Atwood D, Dunietz I and Soni A 2001 *Phys. Rev.* **D63** 036005 (*Preprint* `hep-ph/0008090`)
[9] Giri A, Grossman Y, Soffer A and Zupan J 2003 *Phys. Rev.* **D68** 054018 (*Preprint* `hep-ph/0303187`)
[10] Aaij R *et al.* (LHCb) 2016 *Phys. Lett.* **B760** 117–131 (*Preprint* `1603.08993`)
[11] Aaij R *et al.* (LHCb collaboration) 2016 submitted to JHEP (*Preprint* `1605.01082`)
[12] Aaij R *et al.* (LHCb) 2016 *Phys. Rev.* **D93** 112018 (*Preprint* `1602.03455`)
[13] Aaij R *et al.* (LHCb collaboration) 2016 *LHCb-CONF-2016-001*
[14] Aaij R *et al.* (LHCb collaboration) 2016 *JHEP* **06** 131 (*Preprint* `1604.01525`)
[15] del Amo Sanchez P *et al.* (BaBar) 2010 *Phys. Rev. Lett.* **105** 121801 (*Preprint* `1005.1096`)
[16] Gershon T 2009 *Phys. Rev.* **D79** 051301 (*Preprint* `0810.2706`)
[17] Gershon T and Williams M 2009 *Phys. Rev.* **D80** 092002 (*Preprint* `0909.1495`)
[18] Aaij R *et al.* (LHCb collaboration) 2014 *LHCb-CONF-2014-004*
[19] Bodhisattva S, Walker M and Woodroofe M 2009 *Statist. Sinica* **19** 301–314

Journal of Physics: Conference Series **770** (2016) 012029 doi:10.1088/1742-6596/770/1/012029

# Hadron Spectroscopy, exotics and $B_c^+$ physics at LHCb

**Biplab Dey**

Sezione INFN di Milano, Milano, Italy

E-mail: biplab.dey@cern.ch

**Abstract.** The LHCb experiment is designed to study properties and decays of heavy flavored hadrons produced from $pp$ collisions at the LHC. During Run 1, it has recorded the world's largest data sample of beauty and charm hadrons, enabling precision spectroscopy studies of such particles. Several important results obtained by LHCb, such as the discovery of the first pentaquark states and the first unambiguous determination of the $Z_c(4430)^-$ as an exotic state, have dramatically increased the interest in spectroscopy of heavy hadrons. An overview of the latest LHCb results on the subject, including the discovery of four strange exotic states decaying as $X \to J/\psi\phi$, is presented. LHCb has also made significant contributions to the field of $B_c^+$ physics, the lowest bound state of the heavy flavor $\bar{b}$ and $c$ quarks. A synopsis of the latest results is given.

## 1. Introduction

The quark model enunciated by Gell-Mann *et al.* [1] predicts, besides conventional $q\bar{q}$ mesons and $qqq$ baryons, any other SU(3) color-neutral combination of quarks and gluons such as $gg$ glueballs, $q\bar{q}g$ hybrids, $q\bar{q}q\bar{q}$ tetraquarks, $qqqq\bar{q}$ pentaquarks, and other exotics. Experimentally, such "exotic" states have been hard to find, especially in the light quark sector. The situation changed in 2003, after the discovery of the narrow exotic tetraquark state, $X(3872)$, by the Belle collaboration [2]. Since then, several other states with exotic configurations have been seen, interestingly, all involving heavy quark systems. With the first unambiguous spin-parity assignments of the $X(3872)$ [3] and $Z_c(4430)^-$ [4] states, the discovery of two pentaquark states [5], among other results, the LHCb experiment has been contributing heavily towards our understanding of these particles. In this talk we touch upon some recent results from LHCb in exotic hadron spectroscopy, as well as in the rapidly evolving field of the doubly heavy-flavored $B_c^+$ meson.

## 2. $J/\psi\phi$ exotic states in $B^+ \to J/\psi\phi K^+$

The $X(4140)$ state, first claimed by the CDF collaboration in 2008 [6] as a narrow, near-threshold peak in the $J/\psi\phi$ invariant mass in the decay $B^+ \to J/\psi\phi K^+$, has generated both considerable interest as well as confusion. Results from several experiments [6, 7, 8, 9, 10, 11] have seen disagreements on the properties of the $X(4140)$ and higher lying states. Theoretical interpretations range from tetraquark, molecular or hybrid states, or re-scattering phenomena, including the so-called "cusp effect" [12] at the $D_s^{\pm}D_s^{*\mp}$ production threshold. An earlier LHCb analysis based on a data sample corresponding to 0.37 fb$^{-1}$ of integrated luminosity at

$\sqrt{s} = 7$ TeV found no evidence of the $X(4140)$ state. With a larger data sample corresponding to the entire Run I dataset, the latest LHCb results [13, 14] incorporate an amplitude analysis of the complete decay chain $B^+ \to J/\psi(\to \mu^+\mu^-)\phi(\to K^+K^-)K^+$. The estimated signal yield is $N_{\rm sig} = 4289 \pm 151$, and the estimated background fraction in the signal region is $(23 \pm 6)\%$. This constitutes the largest and cleanest world dataset for this decay mode.

For the amplitude analysis, there are three relevant production amplitudes that can contribute: $B^+ \to J/\psi K^{*+}(\to \phi K^+)$, $B^+ \to X(\to J/\psi\phi)K^+$, and $B^+ \to Z_c^+(\to J/\psi K^+)\phi$. Each of the three decay amplitudes is constructed in its corresponding helicity basis. The spin-projections of all intermediate states are summed over coherently, while the spin-projections of the final state muons are summed incoherently. The lineshapes of resonant structures are taken as relativistic Breit-Wigner distributions with mass-dependent widths, folded with the corresponding angular-momentum barrier factor. For non-resonant components, the lineshape is taken just as the angular-momentum barrier factor. The transition matrix element depends on six kinematic variables describing the full decay chain, and the angular analysis fits to the joint six-dimensional differential rate.

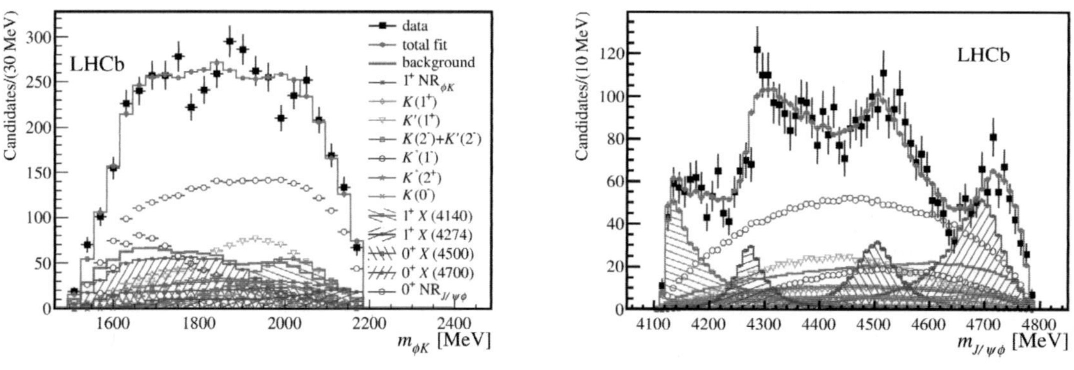


(a) (b)

**Figure 1.** Results of the amplitude fit projected on to variables (a) $m(\phi K)$ and (b) $m(J/\psi\phi)$.

Figure 1 shows the results of the default amplitude fit, projected onto the "$K^*$" and "$X$" invariant mass variables (the exotic $Z_c$ contributions are found not to dominate). Although the overall $m(\phi K)$ distribution in Fig. 1a appears flat, a rich spectrum of $K^{*+} \to \phi K^+$ resonances is found in the angular distributions. The near-threshold $J/\psi\phi$ structure in Fig. 1b is found to be consistent with previous world data on $X(4140)$ in the mass, while the width is found to be larger. A nearby second state consistent with the $X(4272)$ found by the CDF collaboration [15] is seen, and both states prefer spin-parity assignments $J^P = 1^+$. The higher $m(J/\psi\phi)$ region is also found to require two $J^P = 0^+$ resonances, $X(4500)$ and $X(4700)$. The significances of all four exotic states are found to exceed $5\sigma$.

## 3. Reconfirmation of the pentaquarks

### 3.1. *Pentaquarks in* $\Lambda_b^0 \to J/\psi p\pi^-$

After the 2015 discovery of two pentaquark states $P_c(4380)^+$ and $P_c(4450)^+$ in the mode $\Lambda_b^0 \to P_c^+(\to J/\psi p)K^-$ by the LHCb collaboration [5], one question that naturally arises is if these states occur in other decay modes as well. Analysing the same Run I dataset, but now in the Cabibbo suppressed mode, $\Lambda_b^0 \to J/\psi p\pi^-$, the data is found to be consistent with the two pentaquarks coupling via the $\Lambda_b^0 \to P_c^+\pi^-$ mode [16]. Figures 2a and 2b show the Feynman diagrams in $\Lambda_b^0 \to J/\psi p\pi^-$, analogous to those in the $\Lambda_b^0 \to J/\psi pK^-$ [5], but with the $s$-quark now replaced by a $d$-quark. In addition, for the $p\pi^-$ mode, the third diagram in Fig. 2c contributes as well, which can potentially result in the ratio of the branching fractions $R_{\pi/K} = \mathcal{B}(\Lambda_b^0 \to P_c^+\pi^-)/\mathcal{B}(\Lambda_b^0 \to P_c^+K^-)$ to deviate from that expected from Cabibbo suppression [17].

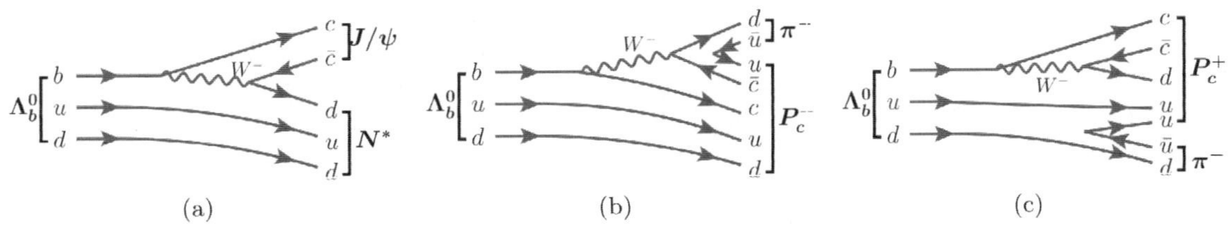


**Figure 2.** Feynman diagrams for the decay $\Lambda_b^0 \to J/\psi p\pi^-$: (a) $N^*$ resonances (b) "external" $W^{*-}$ exchange, and (c) "internal" $W^{*-}$ exchange.

Since $p\pi^-$ is a Cabibbo-suppressed mode, compared to $pK^-$, the statistics is approximately 15 times smaller ($N_{\rm sig} = 1885 \pm 50$), with thrice the percentage of background. The dominant contributions to the decay amplitude come from $\Lambda_b^0 \to J/\psi N^*(\to p\pi^-)$. $N^*$ resonances are modeled as relativistic Breit-Wigner distributions, except the $N(1535)$, which is modeled by a Flatte lineshape, since, in addition to coupling to $p\pi^-$, it couples to the near-threshold $n\eta$ mode as well. The last component contributing is $\Lambda_b^0 \to Z_c^-(\to J/\psi\pi^-)p$, where $Z_c^-$ is an exotic contribution, such as the $Z_c(4200)^-$ tetraquark candidate claimed by the Belle collaboration [18]. Since the Cabibbo suppressed dataset is limited in statistics, the properties of the $P_c$ and $Z_c$ exotic candidates are kept as in previous world data and a full amplitude fit is performed, following the formalism in the original pentaquark analysis [5]. Two different models of the $N^*$ contributions are included: "reduced" (RM), and "extended" (EM), the latter including a wider set of resonances and allowed couplings. Figure 3 shows the background-subtracted data and fit results projected on to the variables $m(p\pi^-)$ and $m(J/\psi p)$. The pentaquark contributions are more prominent at higher $m(p\pi^-)$ masses, as visible in Fig. 3b.

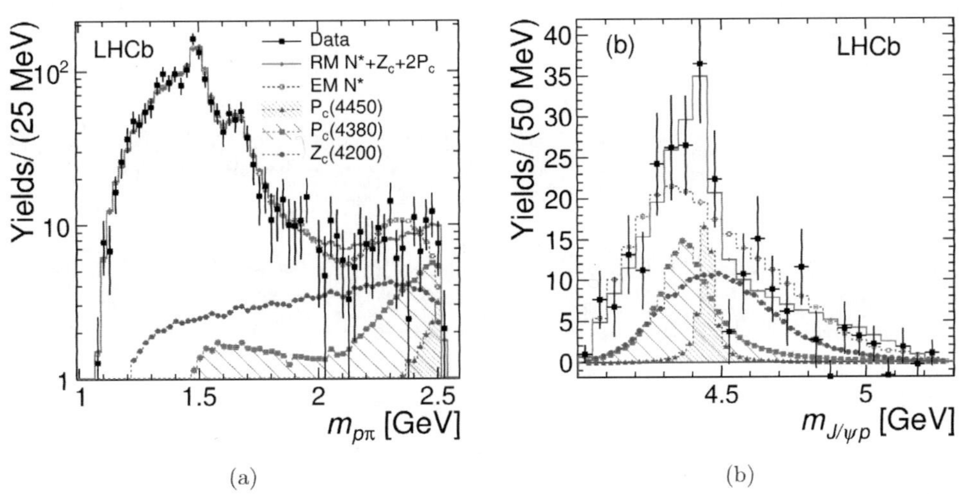


**Figure 3.** Projections of the background-subtracted data and the amplitude fit results in $\Lambda_b^0 \to J/\psi p\pi^-$ projected on to: (a) $m(p\pi^-)$, and (b) $m(J/\psi p)$ with $m(p\pi^-) > 1.8$ GeV.

Journal of Physics: Conference Series **770** (2016) 012029 doi:10.1088/1742-6596/770/1/012029

The significance for exotic contributions (the $P_c$'s and the $Z_c(4200)^-$) is found to be 3.1 $\sigma$ and $R_{\pi/K} = 0.05$, and 0.033, for $P_c(4380)^+$ and $P_c(4450)^+$, respectively, consistent with expectations from Cabibbo suppression [17], pointing to a negligible contribution from the diagram in Fig. 2c.

### *3.2. Model-independent confirmation of exotic activity in $\Lambda_b^0 \to J/\psi pK^-$*

One of the important sources of systematic uncertainties in the exotic states searches is the often poorly known spectrum of conventional hadronic resonances. For pentaquarks in $\Lambda_b^0 \to J/\psi pK^-$ [5], these comprise the excited $\Lambda_J^* \to pK^-$ states of spin-$J$. To circumvent this problem, the LHCb adopted a model-independent method [19] first proposed by the BaBar collaboration in the context of $Z_c(4430)^-$ searches [20]. The underlying principle is that if the highest spin of the contributing $\Lambda_J^*$ states is $J_{\rm max}$, then the maximal power of $x = \cos\theta_{pK}$ in the differential decay rate is given by the Legendre polynomial, $P_{l_{\rm max}}(x)$, of order $l_{\rm max} = 2J_{\rm max}+1$. Here $\theta_{pK}$ is the helicity angle of the daughter kaon in the mother $\Lambda_J^*$ rest-frame. Typically, physical $\Lambda_J^*$ resonances occurring at a given $m(pK)$ mass will have a limited range of $l_{\rm max}$ as depicted in Fig. 4a: resonances of higher spins occur at higher masses, as demarcated by the red and blue lines.

On the other hand, possible exotic states in the $J/\psi\, p$ or $J/\psi\, K^-$ systems "reflect" on the entire spectrum of spin-$J$ $pK^-$ states in the 3-body Dalitz plane $\Lambda_b^0 \to J/\psi pK^-$. At a given $m(pK^-)$ invariant mass, the presence of such unphysically large spin-$J$ components, corresponding to large $l_{\rm max}$, beyond the limits in Fig. 4a, signals the present of exotic activity. Further, since the Legendre polynomials constitute a complete basis of orthonormal functions, the $l_{\rm max}$ component in the $\cos\theta_{pK}$ distribution can be extracted via a counting experiment, weighting each event by the corresponding $P_{l_{\rm max}}(\cos\theta_{pK})$ function. The analysis uses the same Run I dataset and selection as in the earlier LHCb pentaquark paper. [5]. Figure 4b shows the efficiency-corrected and background-subtracted distribution in $m(J/\psi\, p)$, compared with predictions from an "exotic" model with large $l_{\rm max} = 31$ (shown by the broken lines) that can accommodate reflections from exotics, compared to a non-exotic "$\Lambda_J^*$-only" model (shown by the blue curve), with $l_{\rm max}$ as a function of $m(pK^-)$ given by Fig. 4a. The "$\Lambda_J^*$-only" model clearly does not describe the data, which is a strong pointer towards the presence of exotic activity. Based on a likelihood ratio study using pseudoexperiments, between the "$\Lambda_J^*$-only" and "exotic" models, LHCb was also able to place a numeric estimate of the exotic significance at above 9$\sigma$.

We underscore the point that the efficacy of this method is its independence from requiring any detailed knowledge of the complicated $\Lambda_J^*$ spectrum.

## 4. Non-confirmation of the $X(5568)$ tetraquark

The D0 collaboration has recently claimed a 5.1 $\sigma$ evidence for a novel "4-flavored" ($\bar{b}su\bar{d}$) exotic tetraquark state $X(5568)^\pm \to B_s^0\pi^\pm$, using $p\bar{p}$ collisions at $\sqrt{s} = 1.97$ TeV. Further, the relative production rate between the $X(5568)$ and $B_s^0$, multiplied by the $X \to B_s^0\pi^\pm$ branching fraction has been claimed to be $\rho_X^{\rm D0} \sim 8.6\%$.

LHCb has searched for the $X(5568)$ decaying to $B_s^0\pi^\pm$ with the full Run I dataset [21], employing the modes $B_s^0 \to D_s^-\pi^+$ and $B_s^0 \to J/\psi\,\phi$, with $D_s^- \to K^+K^-\pi^-$, $J/\psi \to \mu^+\mu^-$ and $\phi \to K^+K^-$. The overall signal yield is around 110,000 $B_s$ decays, roughly twenty times larger than in D0, and with a higher purity. The selection requirements for the $B_s^0$ and the companion $\pi^\pm$ follow those in previous well-understood LHCb analyses. To facilitate a clean extraction of the $B_s^0$ candidates, its transverse momentum, $p_{\rm T}$, is required to be greater than 5, 10, or 15 GeV. No significant $X(5568)$ is seen for either of the three $p_{\rm T}(B_s^0)$ choices, as shown in Fig. 5 and upper limits of the order $\rho_X^{\rm LHCb} \sim 2\%$ at the 95% confidence level, are placed.

We also note that preliminary results from the CMS collaboration [22] also show no indication of a resonant-like structure around the purported $X(5568)$ mass. Results from the ATLAS and CDF collaborations are still awaited.

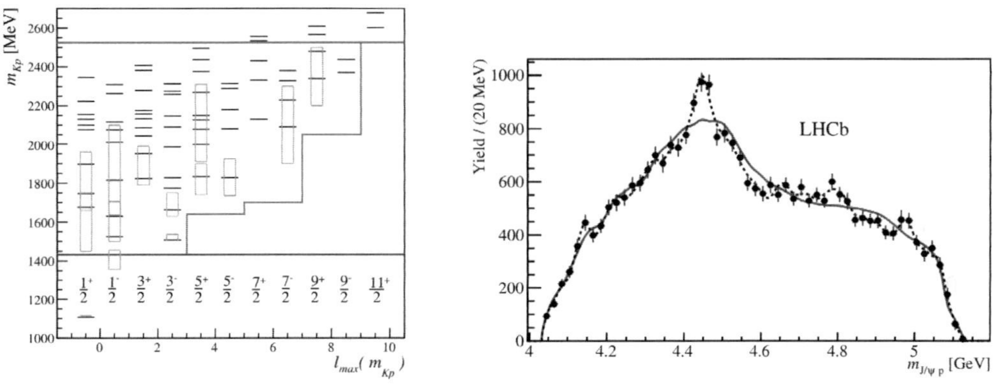


(a) (b)

**Figure 4.** Model-independent exotic evidence in $\Lambda_b^0 \to J/\psi p K^-$: (a) at a given $m(pK^-)$, the range of expected spin-$J$ physical $\Lambda_J^*$ resonances, where $l_{\max} = 2J_{\max} + 1$ and the green bands are the experimentally observed states; (b) comparison between the background subtracted and efficiency corrected data (black markers), predictions from a non-exotic "$\Lambda_J^*$-only" model (blue curve) and exotic "$l_{\max} = 31$" predictions (broken lines).

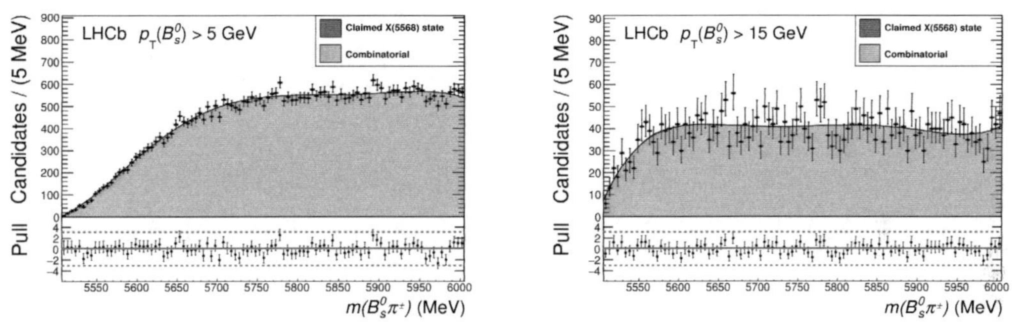


(a) (b)

**Figure 5.** Fits to the LHCb $B_s^0\pi^\pm$ spectrum [21] showing no significant excess at the $X(5568)$ mass for $p_T(B_s^0)$ greater than (a) 5 GeV and (b) 15 GeV.

## 5. $B_c^+$ physics at LHCb

The $B_c^+$ $(\bar{b}c)$ meson is unique in being the only established hadronic state composed of two different heavy flavor quarks. Unlike the $b\bar{b}$ or the $c\bar{c}$ onia states, due to its flavor quantum numbers $B = -C = \pm 1$, the $B_c^+$ can not decay strongly, but only weakly, either via $\bar{b} \to \bar{c}W^+$, $c \to sW^+$, or the so-called weak-annihilation (WA) process $\bar{b}c \to W^+$. The latter can potentially receive contributions from New Physics, such as charged Higgs exchange at the tree-level. Further, compared to $B^+$ decays, the WA process in $B_c^+$ decays is Cabibbo favored by the factor $|V_{cb}/V_{ub}|^2 \sim 100$. LHCb has performed several measurements involving the $B_c^+$, including its decay modes with charmonium final states $J/\psi 3\pi$, $J/\psi K^+$, $\psi(2S)\pi^+$, $J/\psi D_s^{(*)+}$, $J/\psi K^+K^-\pi^+$, $J/\psi 3\pi^+ 2\pi^-$ that derive from the Cabibbo suppressed $\bar{b} \to \bar{c}$ transition, as well $B_c^+ \to B_s^0\pi^+$, involving the Cabibbo favored $c \to s$ transition. The most precise measurements of its mass [23] and lifetime [24] come from LHCb as well and are consistent with expectations from Lattice

Journal of Physics: Conference Series **770** (2016) 012029 doi:10.1088/1742-6596/770/1/012029

QCD and heavy quark effective theory.

More recently, there has been emphases on WA studies via charmless $B_c^+ \to h^+h^-h'$ decays such as $B_c^+ \to \{KKK, \pi\pi\pi, KK\pi, p\bar{p}K, p\bar{p}\pi\}$ [25, 26]. Theory predictions for the WA processes exist only for the quasi two-body modes, with branching fractions ranging in $10^{-6}$ to $10^{-8}$ [27]. Since the $B_c^+$ has a mass of around 6274 MeV, decays to three light hadrons result in large available phase-space, and the formation of intermediate $D^0$, $B^0_{(s)}$ and $\psi^{(\prime)}$ resonances also contribute. Using the full Run I dataset, LHCb has searched for the rare decay $B_c^+ \to p\bar{p}\pi^+$ [25] and also measured $R_p \equiv \frac{f_c}{f_u} \times \mathcal{B}(B_c^+ \to p\bar{p}\pi^+)$, where $f_c(f_u)$ is the fragmentation fraction of a $b$ quark into $B_c^+$ $(B^+)$. In the $J/\psi$ veto region, $m(p\bar{p}) < 2.85$ GeV, no signal was detected and an upper limit at the 95% confidence level is set as $R_p < 3.6 \times 10^{-8}$. The latest LHCb results [26] on the $B_c^+ \to K^+K^-\pi^+$ mode hint at WA, but more data is needed.

## 6. Conclusions

During the Run I phase of LHC running, the LHCb experiment has made several important strides in exotic hadron spectroscopy searches and $B_c^+$ physics. The original pentaquark discovery has been reconfirmed by two further measurements and evidence for four possible new strange exotics have been seen in the $J/\psi\phi$ spectrum, while the $X(5558)$ tetraquark state claimed by the DO collaboration stands unconfirmed at the LHC, as yet. With significant production rates of $B_c^+$ mesons and acceptance inside the LHCb detector, several precision or first measurements of its mass, lifetime and decay modes have been made. The interesting problem of whether the weak annihilation process $\bar{b}c \to W^+$ in $B_c^+$ 3-body charmless decays are enhanced from New Physics contributions has also being probed. Studies are also ongoing on the spectroscopy of $B_c^{+*}$ states. With a five fold increase in statistics expected before the ongoing Run II data-taking period ends in 2018, further detailed studies of these questions are expected.

## References


[1] Gell-Mann M 1964 *Phys. Lett.* **8** 214–215
[2] Choi S K *et al.* (Belle) 2003 *Phys. Rev. Lett.* **91** 262001 (*Preprint* hep-ex/0309032)
[3] Aaij R *et al.* (LHCb) 2013 *Phys. Rev. Lett.* **110** 222001 (*Preprint* 1302.6269)
[4] Aaij R *et al.* (LHCb) 2014 *Phys. Rev. Lett.* **112** 222002 (*Preprint* 1404.1903)
[5] Aaij R *et al.* (LHCb) 2015 *Phys. Rev. Lett.* **115** 072001 (*Preprint* 1507.03414)
[6] Aaltonen T *et al.* (CDF) 2009 *Phys. Rev. Lett.* **102** 242002 (*Preprint* 0903.2229)
[7] Shen C P *et al.* (Belle) 2010 *Phys. Rev. Lett.* **104** 112004 (*Preprint* 0912.2383)
[8] Aaij R *et al.* (LHCb) 2012 *Phys. Rev.* **D85** 091103 (*Preprint* 1202.5087)
[9] Chatrchyan S *et al.* (CMS) 2014 *Phys. Lett.* **B734** 261–281 (*Preprint* 1309.6920)
[10] Abazov V M *et al.* (D0) 2014 *Phys. Rev.* **D89** 012004 (*Preprint* 1309.6580)
[11] Lees J P *et al.* (BaBar) 2015 *Phys. Rev.* **D91** 012003 (*Preprint* 1407.7244)
[12] Swanson E S 2016 *Int. J. Mod. Phys.* **E25** 1642010 (*Preprint* 1504.07952)
[13] Aaij R *et al.* (LHCb) 2016 (*Preprint* 1606.07898)
[14] Aaij R *et al.* (LHCb) 2016 (*Preprint* 1606.07895)
[15] Aaltonen T *et al.* (CDF) 2011 (*Preprint* 1101.6058)
[16] Aaij R *et al.* (LHCb) 2016 *Phys. Rev. Lett.* **117** 082003 (*Preprint* 1606.06999)
[17] Cheng H Y and Chua C K 2015 *Phys. Rev.* **D92** 096009 (*Preprint* 1509.03708)
[18] Chilikin K *et al.* (Belle) 2014 *Phys. Rev.* **D90** 112009 (*Preprint* 1408.6457)
[19] Aaij R *et al.* (LHCb) 2016 *Phys. Rev. Lett.* **117** 082002 (*Preprint* 1604.05708)
[20] Aubert B *et al.* (BaBar) 2009 *Phys. Rev.* **D79** 112001 (*Preprint* 0811.0564)
[21] Aaij R *et al.* (LHCb) 2016 *accepted by Phys. Rev. Lett.* (*Preprint* 1608.00435)
[22] CMS Collaboration (CMS) 2016
[23] Aaij R *et al.* (LHCb) 2014 *Phys. Rev. Lett.* **113** 152003 (*Preprint* 1408.0971)
[24] Aaij R *et al.* (LHCb) 2015 *Phys. Lett.* **B742** 29–37 (*Preprint* 1411.6899)
[25] Aaij R *et al.* (LHCb) 2016 *Phys. Lett.* **B759** 313–321 (*Preprint* 1603.07037)
[26] Aaij R *et al.* (LHCb) 2016 (*Preprint* 1607.06134)
[27] Descotes-Genon S, He J, Kou E and Robbe P 2009 *Phys. Rev.* **D80** 114031 (*Preprint* 0907.2256)

Journal of Physics: Conference Series **770** (2016) 012030 doi:10.1088/1742-6596/770/1/012030

# Modeling new XYZ states at JPAC

**A Pilloni**

Theory Center, Thomas Jefferson National Accelerator Facility,
12000 Jefferson Avenue, Newport News, VA 23606, USA

E-mail: pillaus@jlab.org

**Abstract.** The observation of the unexpected $XYZP$ resonances has challenged the usual heavy quarkonium framework. One of the most studied exotic states, the $X(3872)$, happens to be copiously produced in high-energy hadron collisions. We discuss how this large prompt production cross-section, together with the comparison with light nuclei production data, disfavors a loosely-bound molecule interpretation, and calls for a new interpretation for the exotic hadron resonances. We also present the research of the Joint Physics Analysis Center in Hadron Spectroscopy.

## 1. Introduction

The last decade witnessed the observation of many unexpected $XYZP$ resonances in the heavy quarkonium sector. Their production and decay rates are not compatible with a standard charmonium interpretation [1–3]. The most popular phenomenological interpretation for many of these states is the so-called *hadron molecule*, *i.e.* a loosely bound state of two mesons, interacting via long-range light meson exchange. The main competing model is the *compact tetraquark*, which is dominated by short-range color interaction [4–6]. Also, the possibility for some of these states to be mere kinematical effects has been discussed [7–9].

## 2. Production of $X(3872)$ at hadron colliders

The $X(3872)$ is known to have a large prompt production cross section at hadron colliders, similar to the one of the ordinary $\psi(2S)$ charmonium state. The closeness of the $X(3872)$ to the $\bar{D}^0 D^{*0}$ threshold might be a $\bar{D}^0 D^{*0}$ molecule [1], with a binding mechanism provided by some inter-hadron potential (one-pion exchange). If this is the case, one can derive a model-independent relation between the partial width of $X \to \bar{D}^0 D^{*0}$ and the binding energy, which can be tested in future high-statistics experiments [10]. Nevertheless, it is hard to explain the formation of a loosely bound molecule, with a binding energy $< 192$ keV, at the vertex of hard collisions at energies of some TeVs. This issue was first raised in [11], which used MC simulations to find a cross section $\sim 300$ times smaller than the experimental value, if one considered all the $\bar{D}^0 D^{*0}$ pairs with a maximum relative momentum $k_0^{\max} \sim 50$ MeV to be molecular candidates.

Ref. [12] explained the large experimental cross section by including the final state interactions between the two mesons, in the Migdal-Watson framework. The authors argued that the very presence of strong rescattering allows for a relative momentum between the constituents of the order of the mass of the mediator (pion). The cross section scales as $(k_0^{\max})^3$, and reaches rapidly

[1] The charge-conjugate mode is understood

the experimental value. This approach was criticised in [13], by noticing that the present Migdal-Watson approach requires the two rescattering hadrons to be separated in phase space from the debris of the reaction, whereas in hadron collisions a huge number of pions interferes with the $\bar{D}^0 D^{*0}$ rescattering. The controversy remained somehow unsolved [14]. Similar approaches appeared in the literature to estimate the production for several exotic hadrons [15–17]; however, in the absence of precise final state interactions calculations, the uncertainty in $k_0^{\max}$ reflects in a $> O(10)$ uncertainty in the cross sections, which makes any estimate of sizeable cross sections unreliable.

Refs. [18, 19] proposed a more mechanistic way to take into account final state interactions: it was considered that some of the large number of pions produced might elastically interact with the would-be-molecule constituents, thus changing the relative momentum in the center of mass of the pair. If this interaction reduces the relative momentum of even a small part of the many large-$k_0$ pairs, there could be a significant effect of feed-down of pairs towards lower bins, even in the very low energy region below 50 MeV. Populating that region means increasing the formation probability of the loosely bound $X$. The simulations show that this effect indeed occurs, but is not large enough to justify the large cross section.

## 3. Comparison of $X(3872)$ with light nuclei production

If the $X(3872)$ is a real molecule, and if long-range final state interactions are indeed responsible for such a large cross section, one expects the cross sections of light nuclei to have a similar behavior, especially at high values of $p_\perp$. In Ref. [20], we used the Glauber theory to rescale the ALICE data on helium-3 ($^3$He) and hypertriton ($^3_\Lambda$H) production in Pb-Pb collisions to $pp$ collisions, and we extrapolated these and the deuteron data to $p_\perp \gtrsim 15$ GeV. The comparison in Figure 1 shows that the extrapolated hypertriton differential cross section in $pp$ collisions would fall short by about 2–3 orders of magnitude with respect to the $X$ production, and much more according to a hydrodinamical model (blast wave) in the right panel. The drop of the deuteron cross section, which is directly measured in $pp$ collisions, appears definitely faster. One might assume the very opposite point of view, *i.e.* that at high $p_T$ the production is dominated by the short-range nature of the state, and expect the slope of the differential cross section to depend on the quark counting only. This would explain why the $X(3872)$ (4-quark state, either compact or molecule-like) is produced more than the deuteron (6-quark state), but is at odds with the steeper descent of the deuteron with respect to hypertriton (9-quarks).

The main problem for the production of loosely bound molecular states in proton-proton collisions is the difficulty in producing the constituents close enough in phase space. It is well known that the interaction of elementary partons with the collective hot dense medium causes relevant energy loss of the partons themselves. This effect is usually quantified by the nuclear modification factor $R_{AA}$, which compares the particle yield in Pb-Pb collisions with that in $pp$. While for ordinary hadrons medium effects generally lead to a suppression of the particle yield, conversely molecular states with small binding energy are expected to be enhanced. This would favor their coalescence into the final bound state by reducing their relative momenta directly at the parton level. A naïve estimate based on available ALICE data suggests for values $R_{AA} \sim 5$ at $p_\perp = 5$ GeV. This confirms the enhancement for the production of hadron molecules. One naturally expects such an enhancement to be even more relevant for 3-body nuclei like $^3$He and the hypertriton. Its role would be to further *decrease* the extrapolated cross section in prompt $pp$ collisions. Even though qualitative conclusions can already be drawn, a quantitative analysis substantiated by data at higher $p_\perp$ is necessary for a definitive comparison with the $X$ case.

## 4. Hybridized tetraquarks

In Ref. [21] we proposed a new interpretation for some of the $XZ$ states: they would result from an hybridization between the discrete levels of the tetraquark potential and the levels of

Journal of Physics: Conference Series **770** (2016) 012030 doi:10.1088/1742-6596/770/1/012030

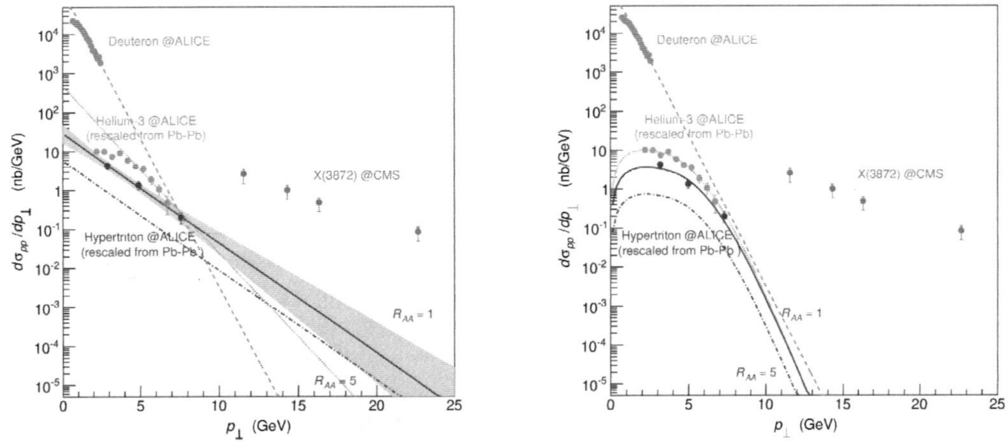


**Figure 1.** Comparison between the prompt production cross section in $pp$ collisions of $X(3872)$, deuteron, $^3$He, and hypertriton [20]. The hypertriton and $^3$He data are fitted with an exponential curve (left panel), or with a hydrodinamical model (right panel).

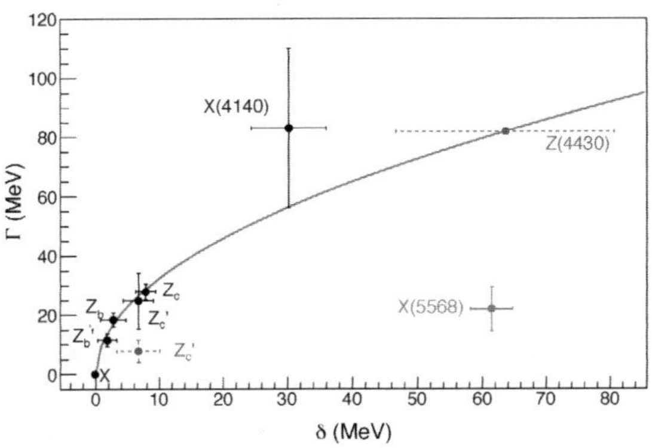


| | Thr. | $\delta$ (MeV) | $A\sqrt{\delta}$ (MeV) | $\Gamma$ (MeV) |
|---|---|---|---|---|
| $X(3872)$ | $D^0 D^{*0}$ | 0 | 0 | 0 |
| $Z_c(3900)$ | $\bar{D}^0 D^{*+}$ | 7.8 | 27.9 | 27.9 |
| $Z_c'(4020)$ | $\bar{D}^{*0} D^{*+}$ | 6.7 | 25.9 | 24.8 |
| $X(4140)$ | $J/\psi\ \phi$ | 30.1 | 54.7 | 83.0 |
| $Z_b(10610)$ | $\bar{B}^0 B^{*+}$ | 2.7 | 16.6 | 18.4 |
| $Z_b'(10650)$ | $\bar{B}^{*0} B^{*+}$ | 1.8 | 13.4 | 11.5 |
| $X(5568)$ | $B_s^0 \pi^+$ | 61.4 | 78.4 | 21.9 |
| $X_{bs}$ | $B^+ \bar{K}^0$ | 5.8 | 24.1 | — |

**Figure 2.** Width of the observed exotic mesons as a function of their detuning, $\delta$ to the closest, from below, two-meson thresholds. The red point corresponds to the $X(5568)$ state whose observation has been claimed by D0. We also show the prediction for the $Z(4430)$ width, which underestimates the total width as expected. We show the value of the width of the $Z_c'(4020)$ measured in the $\bar{D}^{*0} D^{*+}$ channel, which is $2\sigma$ away from the one measured in the $h_c\pi$ channel. The only constant fitted from data is $A = (10.3 \pm 1.3)\,\mathrm{MeV}^{1/2}$.

the continuous spectrum of the two-meson potential. Consider a diquark-antidiquark state. An effective hamiltonian can give a rough estimate for the energy of these levels [4–6]. The wave function can be Fierz rearranged in a number of color singlet pairs which can be of the form hidden-flavor + light meson or two open flavor mesons, having quantum numbers compatible with the initial tetraquark state. For these meson pairs to have nonnegligible interaction and being able to rescatter into the diquarkonium, the relative kinetic energy cannot exceed an $E_{\rm max}$ of few tens of MeV. A level in the continuous spectrum of the two-body system and the near discrete level of the compact tetraquark can match, if their energy difference is smaller than $E_{\rm max}$. If this happens, a sort of 'hybridization' of the two-meson state into the compact structure occurs. The possibility of having a true, shallow bound state enhancing the meson-meson scattering length is unfavored because of the prompt production issue. The Feshbach formalism provides an alternative mechanism (see some related discussions in [19, 22–24]). The scattering length $a$ in the meson-meson channel is given by

$$a \sim a_P - C\frac{|\langle\Psi_n|H_{QP}|\Psi_\alpha\rangle|^2}{E_n - E_\alpha + i\epsilon} \equiv \left(1 - \frac{\varkappa}{\delta - E + i\epsilon}\right) a_P \tag{1}$$

with $a_P$ the original small scattering length in the open channel, $C > 0$ a positive constant (depending on the reduced mass of the would-be molecule), $H_{PQ}$ the hamiltonian which connects the two channels, and $\kappa$ an effective coupling. It is clear that the *detuning* $\delta$, *i.e.* the distance in energy between the expected tetraquark discrete level and the onset of the continuous spectrum starting from the closest molecular threshold is positive and small. Conversely, $\delta < 0$ can suggest either the formation of a true bound state (unfavored by prompt production), or a repulsion in the meson-meson channel, which is incompatible with hybridization. This might be the reason which forbids the charged partners of the $X(3872)$ and provides isospin violation. The width of the state can thus be calculated, $d\Gamma \sim \delta(E-\delta)\ |\varkappa\, a_P|\ (2m)^{3/2}\sqrt{E}\, dE/m$ which can be effectively estimated to be $\Gamma \sim A\sqrt{E}$. The square root dependence is given by the phase space. However, if the detuning $\delta$ falls outside the $[0, E_{\max}]$ interval, the integral is essentially zero and there is no significant contribution to the width: this contradicts the pure phase space prediction, according to which the further the threshold, the more it contributes to the decay width.

We observe that the widths and detunings in a broad class of observed resonances strictly obey this law with a common value for the $A$ parameter — this can be appreciated by the very good fit in Figure 2. The fact that all data can be fitted with the same proportionality constant $A$, strongly supports the described states as sharing the same nature. This does not straightforwardly generalize to excited tetraquarks, where the closed channel is itself not stable against a de-excitation into its allowed tetraquark ground state. Similarly, we do not extend the analysis to pentaquarks. Tetraquarks in the form of diquarkonia might be difficult to be formed once a generic four-quark system is placed in a small region of space. That means that diquarkonia might not be formed independently of this resonant mechanism. More precisely, the hadronization state could contain diquarkonium in its mixture with such a small probability that, if not enhanced by some other mechanism, pure diquarkonium levels are not accessible in present experimental conditions. This might provide a set of 'dynamical selection rules': very little probability of being formed is translated into the experimental evidence of the fact that a particular state is not observed.

## 5. The Joint Physics Analysis Center

The discovery of exotic $XYZ$ states brought a renewed interest in hadron spectroscopy. The Joint Physics Analysis Center (JPAC) was set up to develop theoretical and phenomenological analysis methods to support hadron physics experiments, and recently started working on $XYZ$ states. The project started in 2013 as a joint venture between Indiana University, George Washington University and Jefferson Lab. Currently it has expanded to 20 researchers distributed among the three founding institutions plus the Universities of Bonn and Mainz in Germany, IFIC-University of Valencia in Spain, Universiteit Gent in Belgium, and UNAM in Mexico. The JPAC researchers aim to apply $S$-matrix theory principles, *i.e.* analyticity, crossing symmetry and unitarity [25], to develop scattering amplitudes for several hadron reactions of theoretical and experimental interest The work is performed in close collaboration with experimentalists (from BESIII, COMPASS, CLAS, GlueX, KLOE and LHCb), to implement the amplitudes in the existing data-analysis software. The methods to write the amplitudes include the $K$-matrix and $N/D$ parametrizations [26], complemented with chiral constraints or Regge asymptotics if needed. The amplitudes can be analytically continued from the real axis into the complex energy plane and the unphysical Riemann sheets, to unravel the existing resonances contributing to the physical reactions. We are focused on hadron spectroscopy delivering amplitudes for three-body meson decays [27–31], meson-baryon scattering [32, 33], Regge phenomenology to probe hadron structure [34], and meson photoproduction [35–37].

I briefly review the most recent worked published by our group. The $\eta \to 3\pi$ decay is of particular interest because it is isospin-breaking and provides insight on the light-quark mass difference. In [30] we performed a simultaneous global fit to the KLOE-2 and WASA-at-COSY data using the Khuri-Treiman equations, determining $Q^2 = \frac{m_s^2 - \bar{m}_{u,d}^2}{m_d^2 - m_u^2} = (21.6 \pm 0.4)^2$ [30]. The LHCb collaboration reported two pentaquarks in the $J/\psi p$ channel, but the nature of these states is still to be understood. In [37] we built up a full model for the $\gamma p \to J/\psi p$ reaction to discuss the pentaquark photoproduction. We analyzed the scarce available experimental data. The $P_c(4450)$ was introduced as a Breit-Wigner amplitude and the background was modeled with a Pomeron exchange. We found that data allow for the existence of a pentaquark whose $J/\psi\, p$ branching ratio has to be $< 30\%$ (17%) at 95% confidence level, if $J^P = 3/2^-$ $(5/2^+)$. Jefferson Lab has just approved an experiment to measure this reaction [38]. The analysis of the reactions where the $Z_c(3900)$ state has been seen is currently ongoing. To ease the exchange of information between JPAC and other theorists and experimentalists, we created a public interactive website [39]. The JPAC web page hosts downloadable versions of the codes to compute several of the reactions analyzed by the group [40]. The codes can also be run on the web page and the outputs downloaded.

## Acknowledgments
I wish to thank A. Esposito, A.D. Polosa, and A. Szczepaniak for invaluable comments and discussions. My participation to the conference was supported by the U.S. Department of Energy, Office of Science, Office of Nuclear Physics under contract DE-AC05-06OR23177.

## References
[1] Esposito A, Guerrieri A L, Piccinini F, Pilloni A and Polosa A D 2014 *Int.J.Mod.Phys.* **A30** 1530002 (*Preprint* 1411.5997)

[2] Faccini R, Pilloni A and Polosa A D 2012 *Mod.Phys.Lett.* **A27** 1230025 (*Preprint* 1209.0107)

[3] Chen H X, Chen W, Liu X and Zhu S L 2016 *Phys.Rept.* **639** 1–121 (*Preprint* 1601.02092)

[4] Maiani L, Piccinini F, Polosa A D and Riquer V 2005 *Phys.Rev.* **D71** 014028 (*Preprint* hep-ph/0412098)

[5] Maiani L, Riquer V, Faccini R, Piccinini F, Pilloni A *et al.* 2013 *Phys.Rev.* **D87** 111102 (*Preprint* 1303.6857)

[6] Maiani L, Piccinini F, Polosa A and Riquer V 2014 *Phys.Rev.* **D89** 114010 (*Preprint* 1405.1551)

[7] Szczepaniak A P 2015 *Phys.Lett.* **B747** 410–416 (*Preprint* 1501.01691)

[8] Szczepaniak A P 2016 *Phys.Lett.* **B757** 61–64 (*Preprint* 1510.01789)

[9] Guo F K, Meißner U G, Nieves J and Yang Z 2016 *Preprint* 1605.05113)

[10] Polosa A 2015 *Phys.Lett.* **B746** 248–250 (*Preprint* 1505.03083)

[11] Bignamini C, Grinstein B, Piccinini F, Polosa A and Sabelli C 2009 *Phys.Rev.Lett.* **103** 162001 (*Preprint* 0906.0882)

[12] Artoisenet P and Braaten E 2010 *Phys.Rev.* **D81** 114018 (*Preprint* 0911.2016)

[13] Bignamini C, Grinstein B, Piccinini F, Polosa A, Riquer V *et al.* 2010 *Phys.Lett.* **B684** 228–230 (*Preprint* 0912.5064)

[14] Artoisenet P and Braaten E 2011 *Phys.Rev.* **D83** 014019 (*Preprint* 1007.2868)

[15] Guo F K, Meißner U G and Wang W 2014 *Commun.Theor.Phys.* **61** 354–358 (*Preprint* 1308.0193)

[16] Guo F K, Meißner U G, Wang W and Yang Z 2014 *Eur.Phys.J.* **C74** 3063 (*Preprint* 1402.6236)

[17] Guo F K, Meißner U G, Wang W and Yang Z 2014 *JHEP* **05** 138 (*Preprint* 1403.4032)

[18] Esposito A, Piccinini F, Pilloni A and Polosa A 2013 *J.Mod.Phys.* **4** 1569–1573 (*Preprint* 1305.0527)

[19] Guerrieri A, Piccinini F, Pilloni A and Polosa A 2014 *Phys.Rev.* **D90** 034003 (*Preprint* 1405.7929)

[20] Esposito A, Guerrieri A L, Maiani L, Piccinini F, Pilloni A, Polosa A D and Riquer V 2015 *Phys.Rev.* **D92** 034028 (*Preprint* 1508.00295)

[21] Esposito A, Pilloni A and Polosa A D 2016 *Phys.Lett.* **B758** 292–295 (*Preprint* 1603.07667)

[22] Papinutto M, Piccinini F, Pilloni A, Polosa A D and Tantalo N 2013 *Preprint* 1311.7374

[23] Braaten E and Kusunoki M 2004 *Phys.Rev.* **D69** 074005 (*Preprint* hep-ph/0311147)

[24] Blitz S H and Lebed R F 2015 *Phys.Rev.* **D91** 094025 (*Preprint* 1503.04802)

[25] Eden R J, Landshoff P V, Olive D I and Polkinghorne J C 1966 *The analytic S-matrix* (Cambridge: Cambridge Univ. Press)

[26] Bjorken J D 1960 *Phys.Rev.Lett.* **4**, 473

[27] Guo P, Danilkin I V and Szczepaniak A P 2015 *Eur.Phys.J.* **A51** 135 (*Preprint* 1409.8652)

[28] Szczepaniak A P and Pennington M R 2014 *Phys.Lett.* **B737** 283–288 (*Preprint* 1403.5782)

[29] Guo P, Danilkin I V, Schott D, Fernández-Ramírez C, Mathieu V and Szczepaniak A P 2015 *Phys.Rev.* **D92** 054016 (*Preprint* 1505.01715)

[30] Guo P, Danilkin I V, Fernández-Ramírez C, Mathieu V and Szczepaniak A P 2016 *Preprint* 1608.01447

[31] Danilkin I V, Fernández-Ramírez C, Guo P, Mathieu V, Schott D, Shi M and Szczepaniak A P 2015 *Phys.Rev.* **D91** 094029 (*Preprint* 1409.7708)

[32] Fernández-Ramírez C, Danilkin I V, Manley D M, Mathieu V and Szczepaniak A P 2016 *Phys.Rev.* **D93** 034029 (*Preprint* 1510.07065)

[33] Mathieu V, Danilkin I V, Fernández-Ramírez C, Pennington M R, Schott D, Szczepaniak A P and Fox G 2015 *Phys.Rev.* **D92** 074004 (*Preprint* 1506.01764)

[34] Fernández-Ramírez C, Danilkin I V, Mathieu V and Szczepaniak A P 2016 *Phys.Rev.* **D93** 074015 (*Preprint* 1512.03136)

[35] Shi M, Danilkin I V, Fernández-Ramírez C, Mathieu V, Pennington M R, Schott D and Szczepaniak A P 2015 *Phys.Rev.* **D91** 034007 (*Preprint* 1411.6237)

[36] Mathieu V, Fox G and Szczepaniak A P 2015 *Phys.Rev.* **D92** 074013 (*Preprint* 1505.02321)

[37] Hiller Blin A N, Fernández-Ramírez C, Jackura A, Mathieu V, Mokeev V I, Pilloni A and Szczepaniak A P 2016 *Phys.Rev.* **D94** 034002 (*Preprint* 1606.08912)

[38] Meziani Z E *et al.* 2016 *Preprint* 1609.00676

[39] Mathieu V 2016 *AIP Conf.Proc.* **1735** 070004 (*Preprint* 1601.01751)

[40] http://www.indiana.edu/~jpac/

Journal of Physics: Conference Series **770** (2016) 012031 doi:10.1088/1742-6596/770/1/012031

# Implications of Lepton Nonuniversality in the Beauty Sector

**A Giri[a*], R Mohanta[b] and S Sahoo[b]**

[a]Physics Department, IIT Hyderabad, Kandi - 502285, India
[b]School of Physics, University of Hyderabad, Hyderabad - 500046, India

E-mail: * giria@iith.ac.in

**Abstract.**
The phenomenon of CP violation in the standard model (SM) framework and the decay dynamics have been established from the data obtained from the B factories and so far we have not seen anything new. Nevertheless, there have been instances of deviations in many measured observables in the flavor sector, as far as the data and predictions are concerned. Here we will mention some deviations obtained in measurements related to lepton universality, as seen from the data, and try to understand their implications. To accommodate the observed data we will consider a leptoquark model, which seems to be one interesting model beyond the SM.

## 1. Introduction

The course of high energy physics is going through an interesting and exciting phase. We have in one hand many experiments, conducted in the last few decades, measured various observables which are in excellent agreement with the model proposed by Glashow, Salam and Weinberg, which is also known as the standard model in the literature. One exception being that of the observation of neutrino oscillation. At the same time, there are many fundamental unsolved questions like the hierarchy problem, dark matter and baryon asymmetry of the Universe etc., which make ourselves believe that there is something beyond that of the SM. In fact, the resounding success of the SM has led us with no option but to believe that whatever may be the form of the new physics the low energy limit of the same is the SM. The $V - A$ current structure of the weak interactions has been established long ago and, in fact, in the past has been very instrumental in providing many interesting and accurate results. It is interesting to note that we have observed, in the last few years, some kind of unusual results involving leptons, in particular the heavy $\tau$ (or the third generation leptons). The belief is that the couplings of third generation fermions to the electroweak symmetry breaking is comparatively stronger due to their large masses and therefore, sensitive to new physics that modifies the $V - A$ structure of the SM. From this point of view the study of $B^{(*)} \to \tau\bar{\nu}$ and $B \to D^{(*)}\tau\bar{\nu}$ charge current processes are really interesting. In recent measurements, BaBar [1] and Belle [2] have reported $3.5\sigma$ deviation in the ratio of branching fractions of $\bar{B} \to \bar{D}^{(*)}\tau\bar{\nu}$ over $\bar{B} \to \bar{D}^{(*)}l\bar{\nu}$, where $l = e, \mu$,

$$R_D = \frac{\mathrm{Br}\left(\bar{B} \to D\tau\bar{\nu}\right)}{\mathrm{Br}\left(\bar{B} \to Dl\bar{\nu}\right)} = 0.421 \pm 0.058, \quad R_{D^*} = \frac{\mathrm{Br}\left(\bar{B} \to D^*\tau\bar{\nu}\right)}{\mathrm{Br}\left(\bar{B} \to D^*l\bar{\nu}\right)} = 0.337 \pm 0.025,$$

from their corresponding SM predictions,

$$R_D^{\rm SM} = 0.305 \pm 0.012, \qquad R_{D^*}^{\rm SM} = 0.252 \pm 0.004. \tag{1}$$

The above results might indicate the violation of lepton universality. It should be noted here that the observables measured, as mentioned above, are ratio of two processes, where in the numerator and denominator the initial state is the same, whereas the final sates differ depending upon the kind of leptons involved. This actually helps us to reduce the theoretical uncertainties since most of the contributing terms are same and uncertainties actually cancel out. Since these decays occur at the tree level in the SM, the general expectation is that models with masses of the new particles near the TeV scale are required to explain the anomaly. The branching ratios of semileptonic $b \to cl\bar{\nu}$ processes can be computed precisely due to the light leptons mass, thus the deviation in $R_{D^{(*)}}$ is obviously from the new physics affecting the $\bar{B} \to \bar{D}^{(*)}\tau\bar{\nu}$ process. The branching ratios and $R_{D^{(*)}}$ anomaly in the SM and in various new physics models have been investigated in the literature. Similarly, another interesting observable, reported by LHCb [3], is the lepton non universality in $B \to Kl^+l^-$ process,

$$R_K = \frac{{\rm Br}\left(\bar{B} \to K\mu^+\mu^-\right)}{{\rm Br}\left(\bar{B} \to Ke^+e^-\right)} = 0.745^{+0.090}_{-0.074} \pm 0.036, \tag{2}$$

which has $2.6\sigma$ deviation from the SM value, $R_K = 1.0003 \pm 0.0001$, in the dilepton invariant mass squared bin $\left(1 \le q^2 \le 6\right)$ GeV$^2$. In the semileptonic decay, the rate of $B \to K^*\mu^+\mu^-$ and the angular observables $P_5'$ [4] have $3\sigma$ deviations from the SM predictions. The discrepancy of $3\sigma$ is also found in the decay rate of the $B_s \to \phi\mu^+\mu^-$ process [5].

In this article, we pursue the analysis of rare semileptonic decays of $B$ meson to leptons of second and third generations and we extend the SM by an additional leptoquark model which is built based on the $SU(3)_c \times SU(2)_L \times U(1)_Y$ gauge symmetries. The study of famous $R_{D^{(*)}}$ anomaly and the lepton non-universality in the $b \to c\tau\bar{\nu}$ decay processes are the main interests of this work. We calculate the branching ratios, forward-backward asymmetries and the $\tau$-polarisations of $B \to D^{(*)}\tau\bar{\nu}$ processes in the leptoquark model. We estimate the branching ratio of rare leptonic $B^*_{u,c} \to \tau\bar{\nu}$ decay process of $B^*_{u,c}$ vector meson. In the leptoqaurk model, we also explore the possibility of lepton nonuniversality parameters in the $B^* \to \tau\bar{\nu}$ and $\Lambda_b \to \Lambda_c\tau\bar{\nu}$ processes. Leptoquarks can couple (decay) to a quark and lepton of the same generation simultaneously and carry both lepton and baryon number. They can have spin 0 (scalar leptoquarks) or spin 1 (vector leptoquarks) and can be characterized by their fractional electric charge and Fermion numbers $F = 3B + L$, where $B$ and $L$ are the baryon number and lepton number respectively. Such leptoquarks exist in some extended SM theories [6] such as grand unified theories based on $SU(5)$, $SU(10)$, Pati-Salam model, technicolor model and composite model etc. To avoid rapid proton decay, we consider that the leptoquark does not couples to diquarks and therefore conserve baryon and lepton numbers. The leptoquark model in the context of $B$-physics anomalies has been taken up in the literature.

The outline of this paper is follows. In section 2, we describe the effective Hamiltonian involving $b \to c\tau\bar{\nu}$ quark level transition in the SM. In section 3 we discuss the new physics contributions coming from vector leptoquarks and show how they can explain the observed anomalies in $b$-sector. Our results are presented in Section 4.

## 2. Effective Hamiltonian for $b \to c\tau\bar{\nu}_l$ and $b \to sl^+l^-$ processes

In this section we write the relevant effective Hamiltonian in the SM as given by [7]

$$\mathcal{H}_{eff} = \frac{4G_F}{\sqrt{2}} V_{cb}\Big[\left(\delta_{l\tau} + C^l_{V1}\right)\mathcal{O}^l_{V1} + C^l_{V2}\mathcal{O}^l_{V2} + C^l_{S1}\mathcal{O}^l_{S1} + C^l_{S2}\mathcal{O}^l_{S2} + C^l_T\mathcal{O}^l_T\Big], \tag{3}$$

where $G_F$ is the Fermi constant, $V_{cb}$ is the Cabibbo-Kobayashi-Maskawa (CKM) matrix element and the index $l$ stands for neutrino flavour, $l = e, \mu, \tau$. The $C_X^l$'s, where $X = V_{1,2}, S_{1,2}, T$ are the Wilson coefficients and the corresponding operators are

$$\begin{aligned}
\mathcal{O}_{V1}^l &= \left(\bar{C}_L \gamma^\mu b_L\right)\left(\bar{\tau}_L \gamma_\mu \nu_{lL}\right), \\
\mathcal{O}_{V2}^l &= \left(\bar{C}_R \gamma^\mu b_R\right)\left(\bar{\tau}_L \gamma_\mu \nu_{lL}\right), \\
\mathcal{O}_{S1}^l &= \left(\bar{C}_L b_R\right)\left(\bar{\tau}_R \nu_{lL}\right), \\
\mathcal{O}_{S2}^l &= \left(\bar{C}_R b_L\right)\left(\bar{\tau}_R \nu_{lL}\right), \\
\mathcal{O}_{T}^l &= \left(\bar{C}_R \sigma^{\mu\nu} b_L\right)\left(\bar{\tau}_R \sigma_{\mu\nu} \nu_{lL}\right),
\end{aligned} \tag{4}$$

where $L(R) = (1 \mp \gamma_5)/2$ are the projection operators. Since the flavour of neutrino is not observed at $B$-factories, all generations of neutrinos can be taken into the account to reveal the signature of new physics. In the SM, the contribution to the $b \to c l \bar{\nu}_l$ process is indicated as $\delta_{l\tau}$ and the Wilson coefficients ($C_X^l$) are zero, which can only be generated by new physics models. These new couplings can be bound experimentally, so that the effects of the new operators can be scrutinized in physical observables.

Similarly, the effective Hamiltonian describing the processes induced by $b \to s l^+ l^-$ transitions in the SM is given by [8]

$$\begin{aligned}
\mathcal{H}_{eff} \quad = \quad & -\frac{4G_F}{\sqrt{2}} V_{tb} V_{ts}^* \left[ \sum_{i=1}^{6} C_i(\mu) O_i + C_7 \frac{e}{16\pi^2} \Big( \bar{s} \sigma_{\mu\nu} (m_s P_L + m_b P_R) b \Big) F^{\mu\nu} \right. \\
& \left. + C_9^{eff} \frac{\alpha}{4\pi} (\bar{s} \gamma^\mu P_L b) \bar{l} \gamma_\mu l + C_{10} \frac{\alpha}{4\pi} (\bar{s} \gamma^\mu P_L b) \bar{l} \gamma_\mu \gamma_5 l \right],
\end{aligned} \tag{5}$$

where $\alpha$ is the fine structure constant, $V_{tb} V_{ts}^*$ is the product of CKM matrix element and $C_i$'s are the Wilson coefficients evaluated at the renormalization scale $\mu = m_b$ [9]. In the following subsections we will explain the possible leptoquarks relevant for the $b \to c l \bar{\nu}_l$ and $b \to s l^+ l^-$ quark level transitions.

## 3. Vector leptoquarks and new Physics

Here we consider the new physics model in which the new particle interacts both with quarks and leptons simultaneously, called leptoquark, and carries both the baryon and lepton numbers. Leptoquarks have ten different multiplets [10] under the $SU(3)_c \times SU(2)_L \times U(1)_Y$ SM gauge symmetries with flavour non-diagonal couplings. Half of them have scalar nature and other halves have vector nature under the Lorentz transformation. The scalar (vector) leptoquarks have spin 0 (1) and could potentially contribute to the FCNC processes involving the quark level transitions $b \to s l^+ l^-$ and $b \to c l^- \bar{\nu}$. Out of all possible leptoquark multiplets, six scalar and vector leptoquark bosons are relevant for the $b \to c l \bar{\nu}$ processes. Here $S_{1,3}$ and $R_2$ are the scalar leptoquark bosons, $U_{1,3}^\mu$ and $V_2^\mu$ are the vector leptoquark bosons. The vector leptoquarks with charge= 2/3 and with 0 fermion noumber can mediate both $b \to s l^+ l^-$ and $b \to c l^- \bar{\nu}$ quark level transitions. Therefore, $U_1^\mu = (3, 3, 2/3)$ and $U_3^\mu = (3, 1, 2/3)$ (where the numbers in the parenthesis represent the respective quantum numbers under the standard model gauge group, $SU(3)_c \times SU(2)_L \times U(1)_Y$) are only valid vector leptoquarks to study both $R_{K^{(*)}}$ and $R_{D^{(*)}}$ anomaly. In this work, we investigate the $U_1^\mu = (3, 3, 2/3)$ and $U_3^\mu = (3, 1, 2/3)$ vector leptoquarks, which have charge=2/3, fermion no =0 and can mediate both $b \to s l^+ l^-$ and $b \to c l^- \bar{\nu}$ quark level transitions. In order to avoid rapid proton decay, we do not consider diquark interactions, as the presence of both leptoquark and diquark interactions will violate baryon and lepton number.

The interaction Lagrangian of $U^{\mu}_{1,3}$ leptoquarks with the SM fermion bilinear is given as [7, 10]

$$\mathcal{L}^{LQ} = \left(h^{ij}_{1L}\bar{Q}_{iL}\gamma^{\mu}L_{jL} + h^{ij}_{1R}\bar{d}_{iR}\gamma^{\mu}l_{jR}\right)U_{1\mu} + h^{ij}_{3L}\bar{Q}_{iL}\boldsymbol{\sigma}\gamma^{\mu}L_{jL}U_{3\mu}, \tag{6}$$

where $Q_L(L_L)$ is the left handed quark (lepton) doublet, $u_R(d_R)$ and $l_R$ are the right-handed up (down) quark and lepton singlet respectively and $\psi^c = C\bar{\psi}^T = C\gamma^0\psi^*$ is the charge-conjugated fermion field of $\psi$. The leptoquark couplings are represented by $h^{ij}$, where $i, j$ are the generation indices of quarks and leptons respectively.

Here the fermions are stated in the gauge eigen basis in which Yukawa couplings of the up type quarks and the charged leptons are diagonal, whereas the down-type quark fields are rotated into the mass eigenstate basis by the CKM matrix. Now performing the Fierz transformations, we obtain the additional Wilson coefficients to the $b \to c\tau\bar{\nu}_l$ process as [10],

$$C^l_{V_1} = \frac{1}{2\sqrt{2}G_F V_{cb}}\sum_{k=1}^{3} V_{k3}\left[\frac{h^{2l}_{1L}h^{k3\,*}_{1L}}{2M^2_{U_1^{2/3}}} - \frac{h^{2l}_{3L}h^{k3\,*}_{3L}}{2M^2_{U_3^{2/3}}}\right], \tag{7a}$$

$$C^l_{V_2} = 0, \tag{7b}$$

$$C^l_{S_1} = -\frac{1}{2\sqrt{2}G_F V_{cb}}\sum_{k=1}^{3} V_{k3}\frac{2h^{2l}_{1L}h^{k3\,*}_{1R}}{M^2_{U_1^{2/3}}}, \tag{7c}$$

where $V_{k3}$ denotes the CKM matrix element and $M_{U^{2/3}_{1(3)}}$ is the mass of the leptoquark.

After expanding the $SU(2)$ indices of Eqn. (6), one can notice that $U_{1,3}$ vector leptoquarks also contributes additional Wilson coefficients to the $b \to sl^+l^-$ processes as

$$C_9^{NP} = -C_{10}^{NP} = \frac{1}{2\sqrt{2}G_F V_{tb}V^*_{ts}}\left[\frac{h^{2l}_{1L}h^{k3\,*}_{1L}}{M^2_{U_1^{2/3}}} + \frac{h^{2l}_{3L}h^{k3\,*}_{3L}}{M^2_{U_3^{-1/3}}}\right]. \tag{8a}$$

$$C_9^{\prime NP} = C_{10}^{\prime NP} = \frac{1}{2\sqrt{2}G_F V_{tb}V^*_{ts}}\frac{h^{2l}_{1R}h^{k3\,*}_{1R}}{M^2_{U_1^{2/3}}}, \tag{8b}$$

$$-C_P^{NP} = C_S^{NP} = \frac{1}{\sqrt{2}G_F V_{tb}V^*_{ts}}\frac{h^{2l}_{1L}h^{k3\,*}_{1R}}{M^2_{U_1^{2/3}}}, \tag{8c}$$

$$C_P^{\prime NP} = C_S^{\prime NP} = \frac{1}{\sqrt{2}G_F V_{tb}V^*_{ts}}\frac{h^{2l}_{1R}h^{k3\,*}_{1L}}{M^2_{U_1^{2/3}}}. \tag{8d}$$

## 4. Results and Discussion

In order to use the scenario of vector leptoquarks we constrain the parameter space in terms of the couplings from the existing data and thereafter use the same values to explain the anomalies, as mentioned above [11]. Looking at the figures it may be concluded that the possibility of vector type of leptoquarks can be thought of as an alternative option at least in the context of the subject matter discussed here.

The deviations in the observables $R_D$ and $R_{D^*}$, in comparison to that of the SM predictions, have been reported sometime ago both by Belle and BaBar. With the announcement of new result from the LHCb the situation has not changed anyhow and as a matter of fact the combined deviation from all three experiments is still more than $3\sigma$ away from the SM expectation. At the same time we have also noticed the lepton nonuniversality in the form of $R_K$, where both first

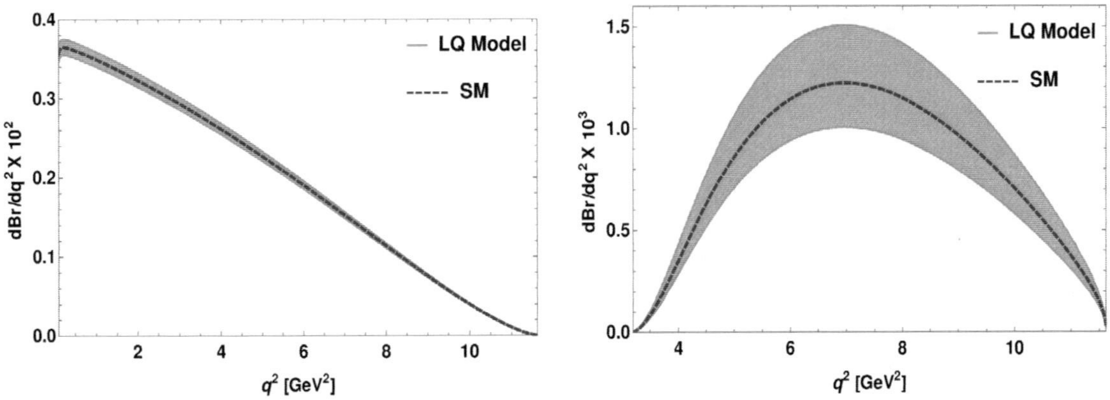


**Figure 1.** The variation of branching ratio of $B \to D\mu\bar{\nu}_l$ process (left panel) and $B \to D\tau\bar{\nu}_l$ process (right panel) with respect to $q^2$ in the leptoquark model. Here dashed lines are for SM and bands represent leptoquark model.

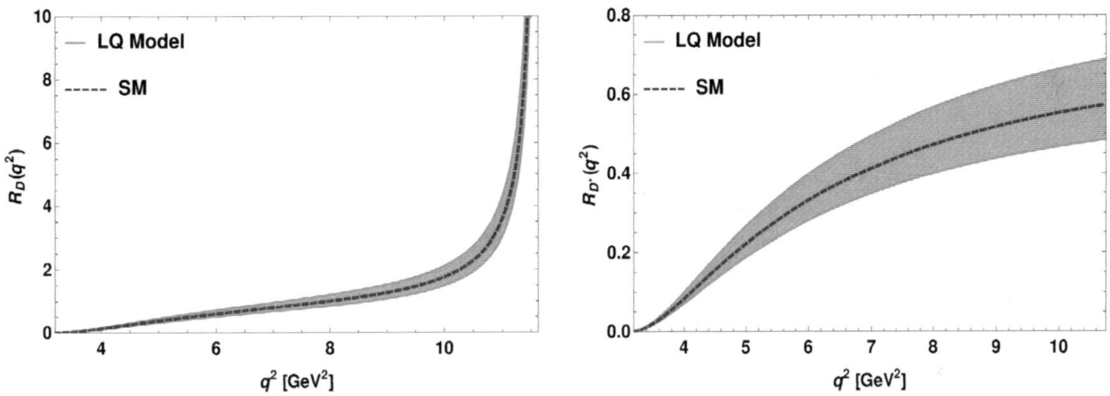


**Figure 2.** The $q^2$ variation of lepton non-universality $R_D(q^2)$ (left panel) and $R_{D^*}(q^2)$ (right panel) in leptoquark model.

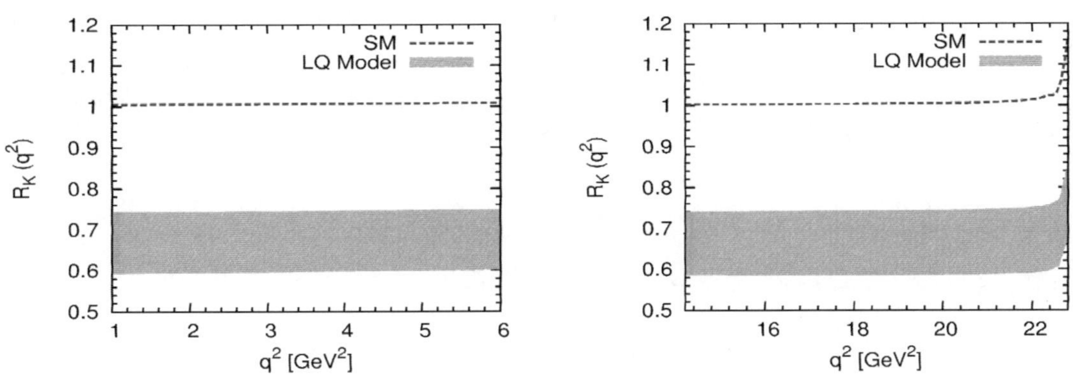


**Figure 3.** $R_K^{\mu e}(q^2)$ in low $q^2$( left panel)and high $q^2$ (right panel) in the leptoquark model.

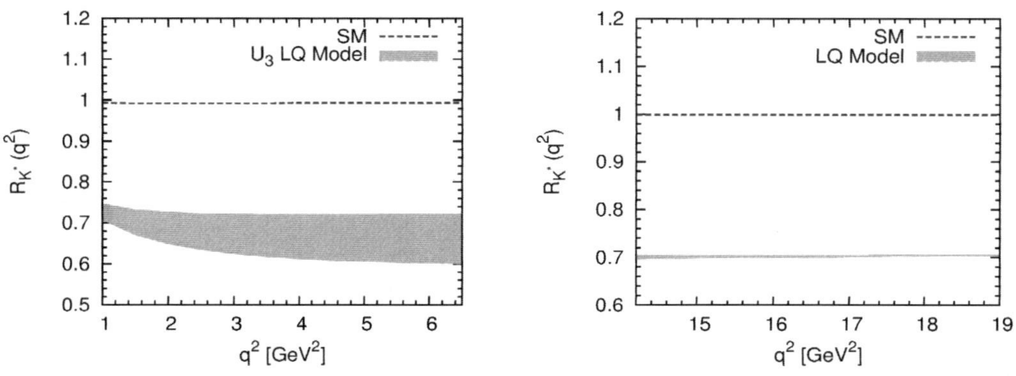


**Figure 4.** $R^{\mu e}_{K^*}(q^2)$ in low $q^2$ (left panel) and high $q^2$ ( right panel) in the leptoquark model.

and second generation of leptons are involved. Past studies in the literature have indicated many scenarios for the possible reason behind such discrepancies. In a model independent analysis it was shown that new physics in the form vector type of couplings could be a possible candidate option for such a discrepancy. In this report we have considered the vector leptoquark model to explain the discrepancies obtained in the so-called $R_D$ and $R_{D^*}$ problems and also the lepton nonuniversality observable $R_K$. It is interesting to note that we can simultaneously explain both these anomalies, one in the tree level decay and another one in the loop suppressed process, using the vector leptoquarks. Therefore, using the scenario of vector leptoquarks, as the possible new physics candidature, one can attempt to understand the current discrepancies in the beauty sector related to lepton non-universality and we hope more refined measurements will resolve these puzzles in the next few years (using additional data obtained from LHCb and Belle II) or else will give some smoking gun signal for physics beyond the SM.

## Acknowledgments

The authors would like to thank Science and Engineering Research Board (SERB), Government of India for financial support through grant No. SB/S2/HEP-017/2013.

## References

[1] Lees J *et al.* [BaBar Collaboration] 2012 Phys. Rev. Lett. **109** 101802; Lees J *et al.* [BaBar Collaboration] 2013 Phys. Rev. D **88** 072012
[2] Bozek A *et al.* [Belle Collaboration] 2010 Phys. Rev. D **82** 072005
[3] Aaij R *et al.* [LHCb Collaboration] 2014 Phys. Rev. Lett. **113** 151601
[4] Aaij R *et al.* [LHCb Collaboration] 2013 Phys. Rev. Lett. **111** 191801
[5] Aaij R *et al.* [LHCb Collaboration] 2013 JHEP **1307** 084
[6] Georgi H and Glashow S L 1974, Phys. Rev. Lett. **32** 438; Pati J C and Salam A 1974, Phys. Rev. D **10** 275; Langacker P 1981 Phys. Rep. **72** 185; Schrempp B and Shrempp F 1985 Phys. Lett.B **153** 101; Kaplan D B 1991 Nucl. Phys.B **365** 259
[7] Sakaki Y Watanabe R Tanaka M and Tayduganov A (2013), Phys. Rev. D **88** 094012
[8] Buras A J and Munz M 1995, Phys. Rev. D **52** 186
[9] Hou W S Kohda M and Xu F 2014 Phys. Rev. D **90** 013002
[10] Buchmuller W Ruckl R and Wyler D 1987, Phys. Lett. B **191** 442; Erratum-ibid. 1999 B **448** 320
[11] Sahoo S Mohanta R and Giri A 2016, arxiv: 1609.04367

# Search for Lepton Number Violation and resonances in the $K^{\pm} \rightarrow \pi\mu\mu$ at the NA48/2 experiment

**C Lazzeroni on ehalf of the A 8/ Colla oration**[1]
School of Physics and Astronomy, University of Birmingham, UK
E-mail: c.lazzeroni@bham.ac.uk

**Abstract.** Thanks to a large sample of kaon decays collected in 2003-2004, the NA48/2 experiment at CERN is able to set an upper limit on the rate of the lepton number violating decay $BR(K^{\pm} \rightarrow \pi^{\mp}\mu^{\pm}\mu^{\pm}) < 8.6 \times 10^{-11}$ at 90%CL, and to set limits at the $10^{-10} - 10^{-9}$ level on the appearance of resonances (including heavy neutral leptons $N_4$ and inflatons $\chi$) in the $K^{\pm} \rightarrow \pi\mu\mu$ decays as functions of the resonance mass and lifetime.

## 1 ntroduction

The observation of neutrino oscillations is so far the only indication of lepton flavour non-conversation; however models of physics beyond the Standard Model (SM) often include more general phenomena of Lepton Flavour Violation (LFV) and Lepton Number Violation (LNV). Besides, neutrino oscillations have demonstrated the massive nature of neutrinos; hence it is natural to include right-handed neutrino states in SM extensions. One of these extensions is the Neutrino Minimal Standard Model ($\nu$MSM)[1], in which three massive right-handed neutrinos are introduced to explain simultaneously neutrino oscillations, dark matter and baryon asymmetry of the Universe. The $\nu$MSM can be further extended by adding a real scalar field, to incorporate inflation and provide a common source for electroweak symmetry breaking and for right-handed neutrino masses[2]. These SM extensions predict new particles, such as heavy neutral leptons and inflatons, which could be produced in $K^{\pm} \rightarrow \pi\mu\mu$ decays: the Lepton Number Violating (LNV) $K^{\pm} \rightarrow \pi^{\mp}\mu^{\pm}\mu^{\pm}$ decay, forbidden in the SM, could instead proceed via the production of on-shell Majorana neutrinos, while inflatons $\chi$ could be produced in

[1] G. Anzivino, R. Arcidiacono, W. Baldini, S. Balev, J.R. Batley, M. Behler, S. Bifani, C. Biino, A. Bizzeti, B. Bloch-Devaux, G. Bocquet, N. Cabibbo, M. Calvetti, N. Cartiglia, A. Ceccucci, P. Cenci, C. Cerri, C. Cheshkov, J.B. Chèze, M. Clemencic, G. Collazuol, F. Costantini, A. Cotta Ramusino, D. Coward, D. Cundy, A. Dabrowski, P. Dalpiaz, C. Damiani, M. De Beer, J. Derré, H. Dibon, L. DiLella, N. Doble, K. Eppard, V. Falaleev, R. Fantechi, M. Fidecaro, L. Fiorini, M. Fiorini, T. Fonseca Martin, P.L. Frabetti, L. Gatignon, E. Gersabeck, A. Gianoli, S. Giudici, A. Gonidec, E. Goudzovski, S. Goy Lopez, M. Holder, P. Hristov, E. Iacopini, E. Imbergamo, M. Jeitler, G. Kalmus, V. Kekelidze, K. Kleinknecht, V. Kozhuharov, W. Kubischta, G. Lamanna, C. Lazzeroni, M. Lenti, L. Litov, D. Madigozhin, A. Maier, I. Mannelli, F. Marchetto, G. Marel, M. Markytan, P. Marouelli, M. Martini, L. Masetti, K. Massri, E. Mazzucato, A. Michetti, I. Mikulec, N. Molokanova, E. Monnier, U. Moosbrugger, C. Morales Morales, D.J. Munday, A. Nappi, G. Neuhofer, A. Norton, M. Patel, M. Pepe, A. Peters, F. Petrucci, M.C. Petrucci, B. Peyaud, M. Piccini, G. Pierazzini, I. Polenkevich, Yu. Potrebenikov, M. Raggi, B. Renk, P. Rubin, G. Ruggiero, M. Savrié, M. Scarpa, M. Shieh, M.W. Slater, M. Sozzi, S. Stoynev, E. Swallow, M. Szleper, M. Valdata-Nappi, B. Vallage, M. Velasco, M. Veltri, S. Venditti, M. Wache, H. Wahl, A. Walker, R. Wanke, L. Widhalm, A. Winhart, R. Winston, M.D. Wood, S.A. Wotton, A. Zinchenko, M. Ziolkowski.

$K^{\pm} \to \pi^{\pm}\chi$ decays, followed by $\chi \to \mu^{+}\mu^{-}$ [3]. In the following, the decays of interest $K^{\pm} \to \pi^{\mp}\mu^{\pm}\mu^{\pm}$ are denoted as KLNV and $K^{\pm} \to \pi^{\pm}\mu^{+}\mu^{-}$ are denoted as KLNC.

**Experimental apparatu , data takin and election of data ample**

The NA48/2 experiment, located at the CERN SPS, was a multi-purpose experiment which collected data in 2003-2004, with simultaneous and collinear $K^{+}$ and $K^{-}$ beams of the same momentum $(60 \pm 3.7)$ GeV/$c$ produced by the 400 GeV/$c$ SPS primary proton beam impinging on a Beryllium target. The kaon beams were steered into a 114 m long decay region, contained in a vacuum cylindrical tank sealed downstream by a convex Kevlar window; such window separated the vacuum from a magnetic spectrometer hosted in helium at atmospheric pressure. The spectrometer was formed of 4 drift chambers and a dipole magnet providing a horizontal momentum kick of 120 MeV/$c$. The spatial resolution of each chamber was $\sigma_x = \sigma_y =$ 90$\mu$m, while the momentum resolution of the spectrometer was $\sigma(p)/p = (1.02 \oplus 0.044p)\%$, where the momentum $p$ is measured in GeV/$c$. A hodoscope was placed downstream of the spectrometer and provided fast signals for trigger purposes, as well as time measurements for charged particles with a resolution of about 150 ps. The hodoscope was followed by a liquid Kripton electromagnetic calorimeter (LKr) electromagnetic calorimeter with a depth of 127 cm, corresponding to 27 radiation lengths. The LKr calorimeter energy resolution was $\sigma(E)/E = (3.2/\sqrt{E} \oplus 9.0/E \oplus 0.42)\%$, where $E$ is the energy expressed in GeV. Downstream of the LKr, a muon detector consisted of three $2.7 \times 2.7$ m$^2$ planes of plastic scintillator strips, each preceded by a 80 cm thick iron wall and alternately aligned horizontally and vertically. The strips were 2.7 m long and 2 cm thick, and they were read out by photomultipliers at both ends. The widths of the strips were 25 cm in the first two planes, and 45 cm in the third plane. A detailed description of the NA48/2 beam line and the detector layout can be found in [4, 5].

The event selection for KLNV and KLNC decays and for resonance searches is based on the reconstruction of a three-track vertex, since the resolution of 50 cm on the vertex longitudinal position makes decays mediated by short-lived resonance particles (lifetime less than 10 ps) indistinguishable from a genuine three-track decay. The size of the selected $K \to \pi\mu\mu$ samples (denoted $K_{\pi\mu\mu}$ below) is normalised relative to the $K^{\pm} \to \pi^{\pm}\pi^{+}\pi^{-}$ decay channel (denoted $K_{3\pi}$ below). The advantage of this choice of normalisation channel is that both samples are collected concurrently using the same trigger logic; and that, given that the muon and pion masses are close, both final states have similar topologies leading to first order cancellation of the systematic effects induced by imperfect kaon beam description, local detector inefficiencies, and trigger inefficiency. There are no restrictions applied to additional energy depositions in the LKr calorimeter, or to extra tracks not belonging to the three-track vertex, to decrease the sensitivity to accidental activity.

The selections of both modes includes the requirement of a vertex satisfying the following criteria: total three-tracks charge $= \pm 1$; vertex longitudinal position within the 98 m long fiducial decay volume; vertex tracks with momentum within the range (5;55) GeV/$c$; total momentum of the three track consistent with the beam nominal range (55;65) GeV/$c$; total transverse momentum of the three tracks with respect to the beam axis direction less than 0.010 GeV/$c$. If several vertices satisfy the above conditions, the one with the lowest fit $\chi^2$ is considered. The vertex tracks are required to be consistent in time and to be in the detector geometric acceptances. Track separations are required to exceed 2 cm in the first chamber plane to suppress photon conversions, and 20 cm in the LKr and muon detector front planes to minimise particle misidentification due to shower overlaps and to multiple Coulomb scattering.

In order to select specifically the KLNV ( KLNC ) candidates, a vertex composed of one $\pi^{\pm}$ candidate and a pair of identically (oppositely) charged muon candidates is required, with the following additional criteria: for the pion, the ratio of energy $E$ in the LKr calorimeter to momentum $p$ measured in the spectrometer is $E/p < 0.95$ to suppress electrons; the pion has no in-time associated hits in the muon detector; the muon candidates have $E/p < 0.2$ and associated hits in the first two planes of the muon detector; the pion candidate is required to have momentum above 15 GeV/$c$ to ensure high muon rejection efficiency (in these conditions, the muon identification efficiency has been measured to be above 99% for $p > 15$ GeV/$c$). In addition the invariant mass of the three tracks in the $\pi^\pm\mu^\pm\mu^\pm$ ($\pi^\pm\mu^+\mu^-$) hypothesis must satisfy the requirement $|M_{\pi\mu\mu} - M_K| < 5$ MeV/$c^2$ ($|M_{\pi\mu\mu} - M_K| < 8$ MeV/$c^2$), where $M_K$ is the nominal charged kaon mass, and the mass resolution is $\pm 2.5$ MeV/$c^2$. An additional requirement is applied to the $K_{\pi\mu\mu}$ samples, when searching for resonances: the invariant mass $M_{ij}$ of the $ij$ pair ($ij = \pi^\pm\mu^\mp, \mu^+\mu^-$) must satisfy the condition $|M_{ij} - M_X| < 2\sigma(M_ij)$, where $M_X$ is the assumed resonance mass and $\sigma(M_{ij})$ is the resolution on the invariant masses $M_{ij}$.

Independently, the following criteria are applied to select the $K_{3\pi}$ sample: pion identification criterion as above is applied to the odd-sign pion only, to symmetrize the selection of the signal and normalisation modes; the invariant mass of the three tracks is in the range $|M_{3\pi} - M_K| < 5$ MeV/$c^2$, corresponding to about $\pm 3$ times its resolution. The number of charged kaon decays in the 98 m long fiducial decay region is computed as $N_K = N_{3\pi}D/Br(K_{3\pi})A(K_{3\pi}) = (1.64 \pm 0.01) \times 10^{11}$, where $N_{3\pi} = 1.37 \times 10^7$ is the number of data candidates reconstructed within the $K_{3\pi}$ selection, $D = 100$ is the downscaling factor, $BR(K_{3\pi})$ is the nominal branching ratio of the $K_{3\pi}$ decay mode, and $A(K_{3\pi}) = (14.955 \pm 0.004)\%$ is the acceptance of the $K_{3\pi}$ selection evaluated with MC simulations.

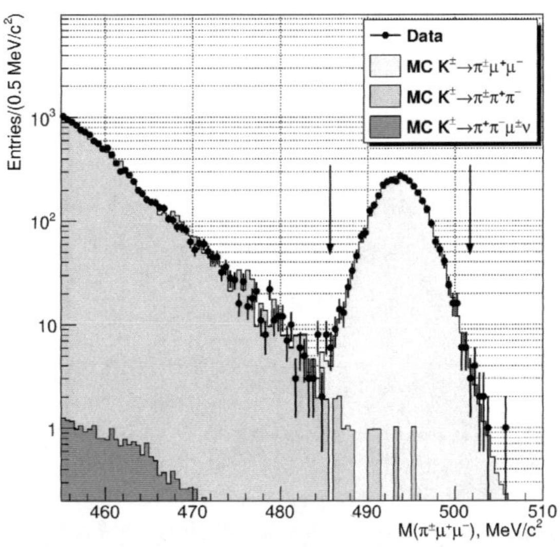


**i ure 1** Invariant mass distribution of data and MC events for passing the KLNC selection.The signal mass region is indicated with vertical arrows.

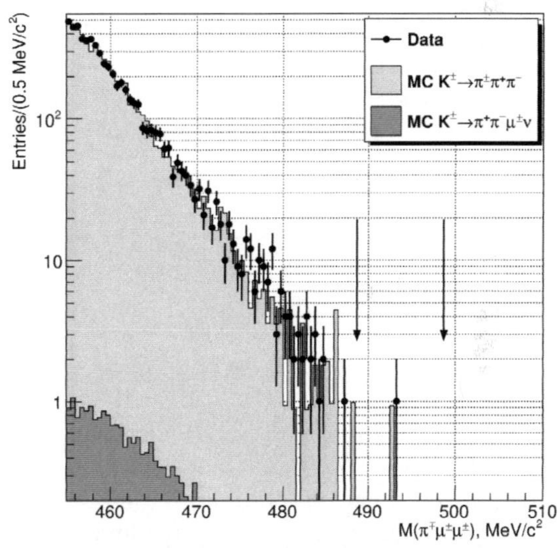


**i ure** Invariant mass distribution of data and MC events for passing the KLNV selection.The signal mass region is indicated with vertical arrows.

## 3 e ult

The statistical analysis to obtain the results shown here has been performed using a quasi-Newton minimisation algorithm to find numerically the 90% confidence intervals for the case of a Poisson process in presence of unknown backgrounds, by applying an extension of the Rolke-Lopez method [6].

The invariant mass distributions of data and MC events passing the KLNV and KLNC selections are shown in Fig. 1 and 2. One event is observed in the signal mass region (indicated by vertical arrows) after applying the KLNV selection, while 3489 candidates are selected with the KLNC selection.

The main source of background in the KLNV sample is the $K_{3\pi}$ decay with two subsequent pion decays; the number of expected background events is $1.163 \pm 0.867_{stat} \pm 0.021_{ext} \pm 0.116_{syst}$. The corresponding upper limit at 90% CL on the number of $K^{\pm} \rightarrow \pi^{\mp}\mu^{\pm}\mu^{\pm}$ signal events in the KLNV sample is $N < 2.92$; using the values of the signal acceptance $(20.62 \pm 0.01)\%$ estimated with MC simulations and the number of kaon decays in the fiducial volume, this leads to a constraint on the signal branching ratio $BR(K^{\pm} \rightarrow \pi^{\mp}\mu^{\pm}\mu^{\pm}) < 8.6 \times 10^{-11}$ at 90% CL. The total systematic uncertainty on the quoted upper limit is 1.5%; the largest source is the limited accuracy of the MC simulations (1.0%), followed by uncertainties on the branching ratios of the normalisation and background channels.

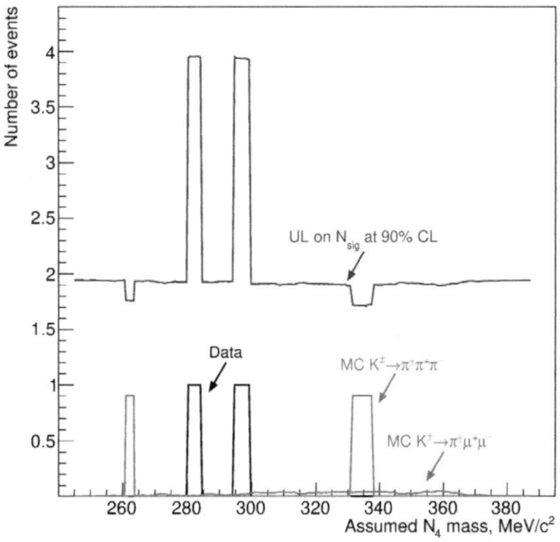


**i ure 3** Numbers of observed data events (black) and expected background events ($K_{3\pi}$ in green, $K^{\pm} \rightarrow \pi^{\pm}\mu^{+}\mu^{-}$ in red) passing the $M_{\pi\mu}$ cut with the KLNV selection. The obtained upper limits at 90% CL on the numbers of signal candidates are shown for each resonance mass value (light blue).

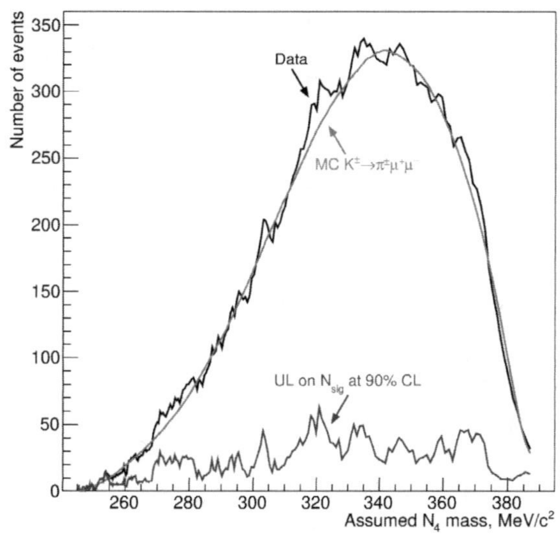


**i ure** Numbers of observed data events (black) and expected background events ($K^{\pm} \rightarrow \pi^{\pm}\mu^{+}\mu^{-}$ in red) passing the $M_{\pi\mu}$ cut with the KLNC selection. The obtained upper limits at 90% CL on the numbers of signal candidates are shown for each resonance mass value (light blue).

The estimated $K_{3\pi}$ background contamination in the KLNC sample is $(0.36 \pm 0.10)\%$, based on MC simulations; this allows to consider the KLNC decay as the only background for the resonance searches in the collected KLNC sample. The acceptances for $K^{\pm} \rightarrow \mu^{\pm}X$ ($K^{\pm} \rightarrow \pi^{\pm}X$) decays with the subsequent $X \rightarrow \pi^{\pm}\mu^{\mp}$ ($X \rightarrow \mu^{+}\mu^{-}$) decay as a function of the resonance mass and lifetime has been evaluated with dedicated MC simulations; due to the required three-track topology of the selected events, the acceptances scale as $1/\tau$ for resonance lifetimes $\tau > 1$ ns. A peak search assuming different mass hypotheses has been performed over the distributions of the invariant masses $M_{ij}$ (as defined above) of the selected $K_{\pi\mu\mu}$ samples. The mass steps of the resonance search is taken to be equal to $\sigma(M_{ij})/2$ and the half-width of the signal mass windows around the assumed mass is $2\sigma(M_{ij})$, where $\sigma(M_{ij})$ is the resolution on the invariant masses $M_{ij}$. It is worth mentioning that, as a consequence, the results obtained in the neighbouring mass hypotheses are highly correlated, since the signal mass window is about 8 times wider than the mass step of the resonance scan. In total, 284 (267) and 280 mass

Journal of Physics: Conference Series **770** (2016) 012032 doi:10.1088/1742-6596/770/1/012032

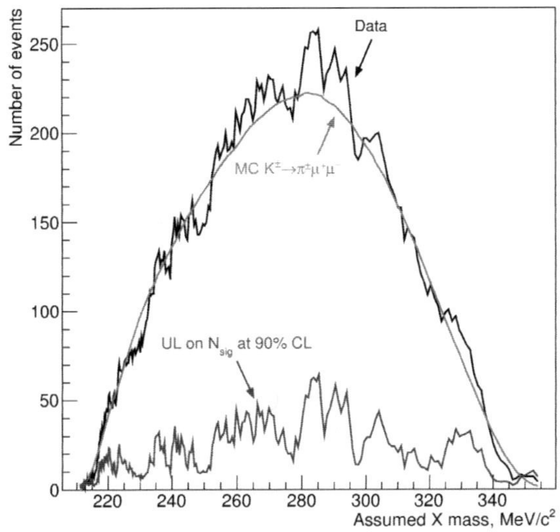


**i ure** Numbers of observed data events (black) and expected background events ($K^{\pm} \rightarrow \pi^{\pm}\mu^{+}\mu^{-}$ in red) passing the $M_{\mu\mu}$ cut with the KLNC selection. The obtained upper limits at 90% CL on the numbers of signal candidates are shown for each resonance mass value (light blue).

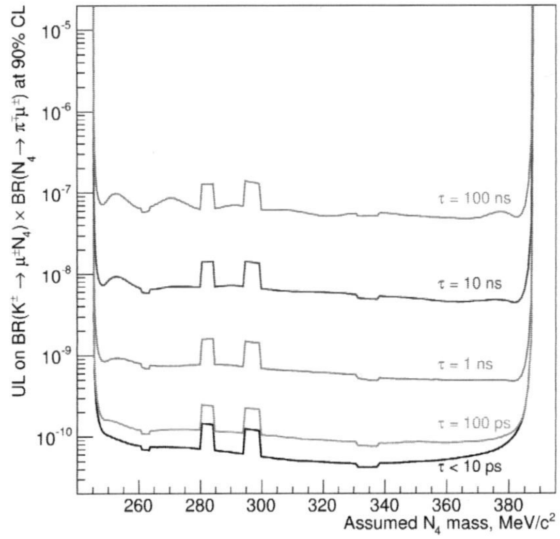


**i ure** Obtained upper limits at 90% CL on the products of branching ratios as functions of the resonance mass and lifetime for $Br(K^{\pm} \rightarrow \mu^{\pm}N_4)Br(N_4 \rightarrow \pi^{\mp}\mu^{\pm})$.

hypotheses have been tested in the search of resonances in the $M_{\pi\mu}$ distribution of the KLNV (KLNC) candidates and in the $M_{\mu\mu}$ distribution of the KLNC candidates respectively. Here the statistical analysis has been performed in each mass window. The values of the observed data events in the signal mass window; the number of MC background events observed in the signal mass window and normalised to the size, with respect to the data volume, of the MC sample used to evaluate it; and the upper limit at 90% confidence level on the number of signal events obtained are shown for each mass hypothesis of the resonance searches in Fig. 3, 4, 5. For each of the three resonance searches performed, the local significance $z$ of the signal has been evaluated for each mass hypothesis as $z = (N_{obs} - N_{exp})/\sqrt{(\delta N_{obs}^2 + \delta N_{exp}^2)}$, where $N_{obs}$ is the number of observed events, $N_{exp}$ is the number of expected background events, and $\delta N_{obs}$ and $\delta N_{exp}$ is the statistical uncertainty of $N_{obs}$ and $N_{exp}$ respectively. No signal is observed, as the local significances never exceed 3 standard deviations.

In absence of a signal, upper limits on the product $BR(K^{\pm} \rightarrow p_1 X)BR(X \rightarrow p_2 p_3)$ ($p_1p_2p_3 = \mu^{\pm}\pi^{\mp}\mu^{\pm}, \mu^{\pm}\pi^{\pm}\mu^{\mp}, \pi^{\pm}\mu^{+}\mu^{-}$) as a function of the resonance lifetime are obtained for each mass hypothesis, by using the values of the acceptances and the upper limits on the number of signal events for such mass hypothesis; the upper limits as a function of the resonance mass, for several values of the resonance lifetime, are shown in Fig. 6, 7, 8 .

## Conclu ion

The NA48/2 experiment at CERN has established an upper limit of $8.6 \times 10^{-11}$ for the branching ratio of the Lepton Number Violating $K^{\pm} \rightarrow \pi^{\mp}\mu^{\pm}\mu^{\pm}$ decay, improving the best previous limit [7] by more than one order of magnitude. In addition, upper limits have been derived on the products of branching ratios $BR(K^{\pm} \rightarrow \mu^{\pm}N_4)BR(N_4 \rightarrow \pi\mu)$ and $BR(K^{\pm} \rightarrow \pi^{\pm}\chi)BR(\chi \rightarrow \mu^{+}\mu^{-})$ as functions of the resonance mass and lifetime; such limits

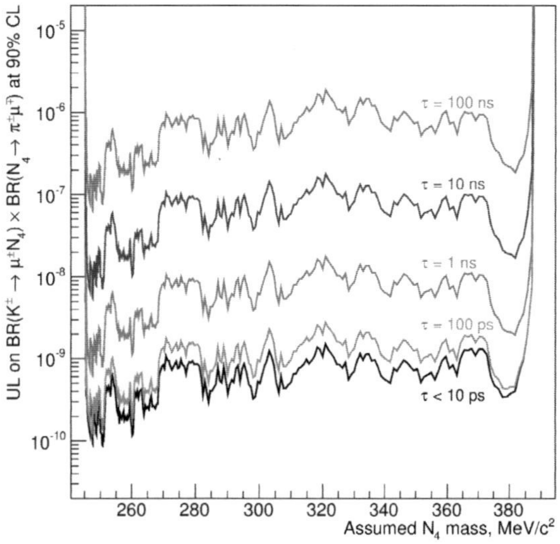


**i ure 7** Obtained upper limits at 90% CL on the products of branching ratios as functions of the resonance mass and lifetime for $Br(K^{\pm} \to \mu^{\pm} N_4) Br(N_4 \to \pi^{\pm}\mu^{\mp})$.

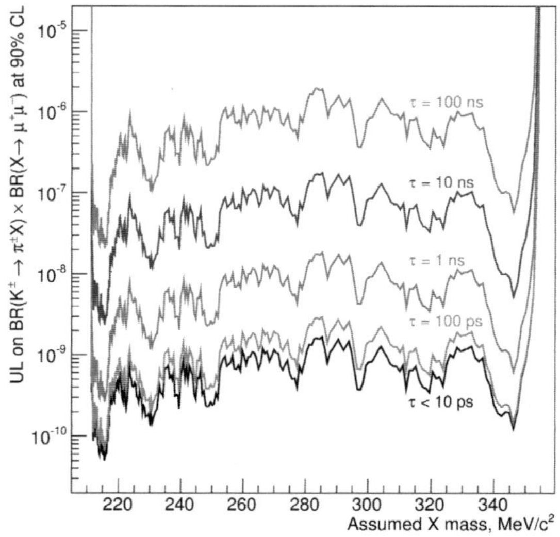


**i ure 8** Obtained upper limits at 90% CL on the products of branching ratios as functions of the resonance mass and lifetime for $Br(K^{\pm} \to \pi^{\pm}\chi) Br(\chi \to \mu^{+}\mu^{-})$.

are in the $10^{-10} - 10^{-9}$ range for resonance lifetimes below 100 ps.

**eference**

[1] T. Asaka and M. Shaposhnikov, Phys. Lett. B 620, 17 (2005).
[2] M. Shaposhnikov and I. Tkachev, Phys. Lett. B 639, 414 (2006).
[3] F. Bezrukov and D. Gorbunov, Phys. Lett. B 736, 494 (2014).
[4] J.R. Batley et al. (NA48/2 collaboration), Eur. Phys. J. C 52, 875 (2007).
[5] V. Fanti et al. (NA48 collaboration), Nucl. Instrum. Methods A 574, 433 (2007).
[6] W.A. Rolke and A.M. Lopez, Nucl. Instrum. Methods A 458, 745 (2001).
[7] J.R. Batley et al. (NA48/2 collaboration), Phys. Lett. B 697, 107 (2011).

Journal of Physics: Conference Series **770** (2016) 012033 doi:10.1088/1742-6596/770/1/012033

# On the Standard Model prediction for $R_K$ and $R_{K^*}$

**A Pattori** [1,2]

[1] Physik-Institut, Universität Zürich, CH-8057 Zürich, Switzerland
[2] Dipartimento di Fisica e Astronomia 'G. Galilei', Università di Padova, Italy

E-mail: pattori@phyisik.uzh.ch

**Abstract.** In this article a recent work is reviewed, where we evaluated the impact of radiative corrections in $R_K$ and $R_{K^*}$. We find that, employing the cuts presently applied by the LHCb Collaboration, such corrections do not exceed a few percent. Moreover, their effect is well described (and corrected) by existing Monte Carlo codes. Our analysis reinforces the interest of these observables as clean probe of physics beyond the Standard Model.

## 1. Introduction

The Lepton Flavor Universality (LFU) ratios

$$R_M[q^2_{\rm min},\ q^2_{\rm max}] = \frac{\int_{q^2_{\rm min}}^{q^2_{\rm max}} \frac{{\rm d}\Gamma(B\to M\mu^+\mu^-)}{{\rm d}q^2}{\rm d}q^2}{\int_{q^2_{\rm min}}^{q^2_{\rm max}} \frac{{\rm d}\Gamma(B\to Me^+e^-)}{{\rm d}q^2}{\rm d}q^2}\ , \qquad (1)$$

where $q^2 = m^2_{\ell\ell}$, are very clean probes of physics beyond the Standard Model (SM): they have small theoretical uncertainties and are sensitive to possible new interactions that couple in a non-universal way to electrons and muons [1]. A strong interest in $R_K$ has recently been raised by the LHCb result [2]

$$R_K\left[1\ {\rm GeV}^2,\ 6\ {\rm GeV}^2\right] = 0.745^{+0.090}_{-0.074} \pm 0.036\ , \qquad (2)$$

that differs from the naïve expectation $R^{\rm (SM)}_{K^{(*)}} = 1$ by about $2.6\sigma$. The interest in this anomaly is further raised by its combination with other $b \to s\ell^+\ell^-$ observables [3, 4], and by the independent hints of violations of LFU observed in $B \to D^{(*)}\tau\nu_\ell$ decays [5–7], presenting an overall $3.9\sigma$ tension with the SM predictions [8, 9].

While perturbative and non-perturbative QCD contributions cancel in $R_{K^{(*)}}$ (beside trivial kinematical factors), this is not necessarily the case for QED corrections. In particular, QED colinear singularities induce corrections of order $(\alpha/\pi)\log^2(m_B/m_\ell)$ to $b \to s\ell^+\ell^-$ transitions [10–12] that could easily imply 10% effects in $R_{K^{(*)}}$. In a recent paper [13] we have estimated these corrections to precisely quantify up to which level a deviation of $R_K$ or $R_{K^*}$ from 1 can be considered a clean signal of physics beyond the SM.

Journal of Physics: Conference Series **770** (2016) 012033 doi:10.1088/1742-6596/770/1/012033

## 2. factorizable QED corrections in $R_M$

While discussing the $B \to M\ell^+\ell^-$ decay, the key observable we are interested in is the differential lepton-pair invariant-mass distribution

$$\mathcal{F}_M^\ell(q^2) = \frac{\mathrm{d}\Gamma(B \to M\ell^+\ell^-)}{\mathrm{d}q^2} , \tag{3}$$

since this is the observable involved in the $R_M$ definition.

A complete evaluation of QED corrections to $\mathcal{F}_M^\ell(q^2)$ is a non-trivial task, due to the interplay of perturbative and non-perturbative dynamics (see e.g. [14]). However, the problem is drastically simplified if we are only interested in the low dilepton invariant mass region, and if interested in possible deviations from $R_{K^{(*)}}^{(\mathrm{SM})} = 1$ exceeding 1%. In this case the problem is reduced to evaluating $\log(m_\ell)$ enhanced terms, whose origin can be unambiguously traced to soft and colinear photon emission. The latter represents a universal correction factor [15, 16] that can be implemented, by means of appropriate convolution functions, irrespective of the specific short-distance structure of the amplitude.

Following the above observation, the $\log(m_\ell)$ enhanced terms in $\mathcal{F}_M^\ell(q^2)$ can be factorized and are independent from the spin of the meson $M$. In formulas:

$$\mathcal{F}_M^\ell(q^2) = \int_{q^2}^{q^2_{0,\max}} \frac{\mathrm{d}q_0^2}{q_0^2}\, \mathcal{F}_M^{(0)}(q_0^2)\, \omega\left(\frac{q^2}{q_0^2}, \frac{2m_\ell^2}{q_0^2}\right) . \tag{4}$$

$\mathcal{F}_M^{(0)}(q_0^2)$ is the differential distribution at tree level. $q^2_{0,\max}$ is the maximum value for the initial dilepton invariant mass squared (pre bremsstrahlung) compatible with the cut applied in the experimental analysis, namely that the reconstructed $B$-meson mass (from the measurement of leptons and hadron momenta) is above a minimum value $m_B^{\mathrm{rec}}$. The function $\omega(x, x_\ell)$, which represents the probability density function that a dilepton system retains a fraction $\sqrt{x}$ of its original invariant mass after bremsstrahlung, includes both real and virtual QED corrections. Explicit expressions for $q^2_{0,\max}$ and $\omega(x, x_\ell)$ can be found in [13].

## 3. Numerical results

The relative impact of radiative corrections in $B \to K^+\ell^+\ell^-$, namely a plot of the ratio

$$\mathcal{R}_K^\ell(q^2) = \frac{\mathcal{F}_K^\ell(q^2)}{\mathcal{F}_K^{(0)}(q^2)} , \tag{5}$$

is shown in Figure 1 in the region $q^2 \in [1, 9]$ GeV$^2$. The different colors correspond to different lepton masses (red for the electron and blue for the muon). Dashed and full lines correspond to different choices of $m_B^{\mathrm{rec}}$. We have choosen for the latter the two values used in Ref. [2] for the analysis of electron modes ($m_B^{\mathrm{rec}} = 4.880$ GeV, full lines) and muon modes ($m_B^{\mathrm{rec}} = 5.175$ GeV, dashed lines).

The first point to be noted in Figure 1 is that $\mathcal{R}_K^\ell(q^2)$ is a smooth function for sufficiently low values of $q^2$, while a sudden rise appear close to the resonance region. The latter is a manifestation of the radiative return from the $J/\Psi$ peak. The position where the $J/\Psi$ contamination appears depends only from the cut imposed on $m_B^{\mathrm{rec}}$. Even for the looser cut applied in the electron case the region $q^2 \in [1, 6]$ GeV$^2$ is free from the $J/\Psi$ contamination and can be estimated with good theoretical accuracy.

The second point to be noted is that in the regular region of the spectrum radiative corrections reach (or even exceed) the 10% level for the electrons (as naively expected); however, the net effect in $R_K$ is significantly smaller. Indeed the magnitude of the corrections is larger for electron

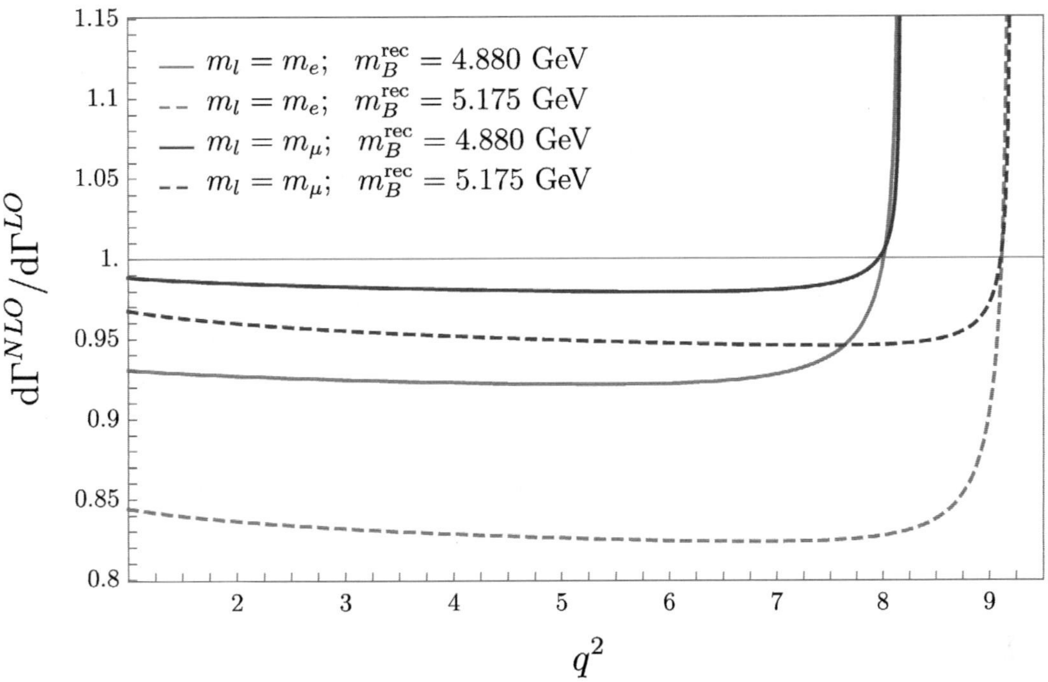


**Figure 1.** Relative impact of radiative correction in $B \to K^+\ell^+\ell^-$ decays for $q^2 \in [1, 9.5]$ GeV$^2$, with different cuts on the reconstructed mass and different lepton masses.

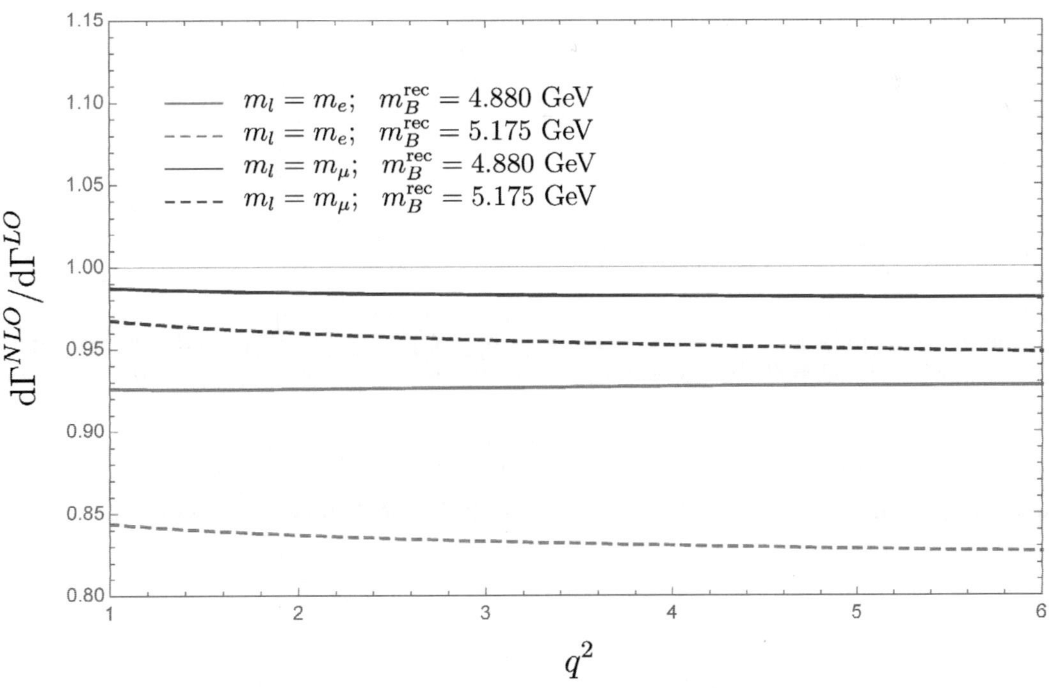


**Figure 2.** Relative impact of radiative correction in $B \to K^*\ell^+\ell^-$ for $q^2 \in [1, 6]$ GeV$^2$, with different cuts on the reconstructed mass and different lepton masses.

**Table 1.** Relative impact on $R_K$ (left) and $R_{K^{(*)}}$ (right) of radiative corrections for $q^2 \in [1,6]$ GeV$^2$, with different cuts on the reconstructed mass and different lepton masses.

| $B \to K\ell^+\ell^-$ | $\ell = e$ (%) | $\ell = \mu$ (%) |
|---|---|---|
| $m_B^{\rm rec} = 4.880$ GeV | $-$ 7.6 | $-1.8$ |
| $m_B^{\rm rec} = 5.175$ GeV | $-16.9$ | $-4.6$ |

| $B \to K^*\ell^+\ell^-$ | $\ell = e$ (%) | $\ell = \mu$ (%) |
|---|---|---|
| $m_B^{\rm rec} = 4.880$ GeV | $-$ 7.3 | $-1.7$ |
| $m_B^{\rm rec} = 5.175$ GeV | $-16.7$ | $-4.5$ |

**Table 2.** Relative contribution of radiative corrections due emission from the meson leg, in the $B^+ \to K^+\ell^+\ell^-$ case, for $q^2 \in [1,6]$ GeV$^2$.

| | |
|---|---|
| $m_B^{\rm rec} = 4.880$ GeV | $-0.02\%$ |
| $m_B^{\rm rec} = 5.175$ GeV | $-0.18\%$ |

vs. muons, but it increases for $m_B^{\rm rec} \to m_B$. This implies that the specific choice of $m_B^{\rm rec}$ cuts applied by the LHCb collaboration, i.e. a loose cut for the electrons and a tighter cut for the muons, gives rise to a natural compensation of the QED corrections to $R_K$.

The integrated corrections that quantify the modifications to $R_K$ are reported in Table 1. Given the choice of $m_B^{\rm rec}$ applied in Ref. [2], we estimate that radiative corrections induce a *positive* shift of the central value of $R_K$ of about $\Delta R_K = +3\%$. This effect is taken into account by the LHCb collaboration, who estimated the impact of radiative corrections with PHOTOS [17], and properly corrected for in the result reported. We have explicitly checked that our estimate of $\Delta R_K$ is in agreement with that obtained with PHOTOS up to differences of $\pm 1\%$.

In order to check the smallness of the non-log($m_\ell$) enhanced terms, in Table 2 we report the effect of the radiation from the meson leg, that is IR divergent but has no colinear singularities. We evaluated these terms developing the corresponding radiator function (see Ref. [18]), whose implementation depends only on $m_B^{\rm rec}$. As can be seen from Table 2, the results are well below the 1% level.

The impact of radiative corrections in the $B \to K^*\ell^+\ell^-$ decays is shown in Figure 2 and summarized by the integrated values reported in Table 1. The situation is very similar to the $B^+ \to K^+\ell^+\ell^-$ : employing the same $m_B^{\rm rec}$ cuts for electron and muon modes in Ref. [2], we find that the net impact of radiative corrections is $\Delta R_{K^*} = +2.8\%$. Also in this case the effect is well described by PHOTOS and therefore can be properly corrected in future experimental analyses.

## 4. Conclusions

The experimental result in Eq. (2) has stimulated a lot of theoretical activity. In view of this result and, especially, in view of possible future experimental improvements in the determination of $R_K$ or $R_{K^*}$, we have re-examined the SM predictions of these LFU ratios.

As we have shown, log($m_\ell$)-enhanced QED corrections may induce sizable deviations from $R_{K^{(*)}}^{\rm (SM)} = 1$, even up to 10%, depending on the specific cuts applied to define physical observables. In particular, a key role is played by the cuts on $q^2 = m_{\ell\ell}^2$ and on the reconstructed $B$-meson mass. The former is important to avoid rapidly varying regions in the dilepton spectrum (where the theoretical tools to compute QED corrections become unreliable), while the latter defines the physical IR cut-off of the rates. Employing the cuts presently applied by the LHCb

Collaboration, the corrections in $R_K$ do not exceed 3%. Moreover, their effect is well described (and corrected for them in the experimental analysis) by existing Monte Carlo codes.

According to our analysis, a deviation from $R_{K^{(*)}}^{(\mathrm{SM})} = 1$ exceeding the 1% level, performed along the lines of Ref. [2] in the region $1\ \mathrm{GeV}^2 < q^2 < 6\ \mathrm{GeV}^2$, would be a clear signal of physics beyond the Standard Model.

## Acknowledgments

The author would like to thank the organisers of the BEACH2016 conference for the invitation to give his talk. This research was supported in part by the Swiss National Science Foundation (SNF) under contract 200021-159720.

## References

[1] Hiller G and Kruger F 2004 *Phys. Rev.* **D69** 074020 (*Preprint* hep-ph/0310219)
[2] Aaij R *et al.* (LHCb) 2014 *Phys. Rev. Lett.* **113** 151601 (*Preprint* 1406.6482)
[3] Altmannshofer W and Straub D M 2015 *Proceedings, 50th Rencontres de Moriond Electroweak Interactions and Unified Theories: La Thuile, Italy, March 14-21, 2015* pp 333–338 (*Preprint* 1503.06199) URL https://inspirehep.net/record/1353682/files/arXiv:1503.06199.pdf
[4] Descotes-Genon S, Hofer L, Matias J and Virto J 2016 *JHEP* **06** 092 (*Preprint* 1510.04239)
[5] Lees J P *et al.* (BaBar) 2012 *Phys. Rev. Lett.* **109** 101802 (*Preprint* 1205.5442)
[6] Huschle M *et al.* (Belle) 2015 *Phys. Rev.* **D92** 072014 (*Preprint* 1507.03233)
[7] Aaij R *et al.* (LHCb) 2015 *Phys. Rev. Lett.* **115** 111803 [Addendum: Phys. Rev. Lett.115,no.15,159901(2015)] (*Preprint* 1506.08614)
[8] Fajfer S, Kamenik J F and Nisandzic I 2012 *Phys. Rev.* **D85** 094025 (*Preprint* 1203.2654)
[9] Na H, Bouchard C M, Lepage G P, Monahan C and Shigemitsu J (HPQCD) 2015 *Phys. Rev.* **D92** 054510 [Erratum: Phys. Rev.D93,no.11,119906(2016)] (*Preprint* 1505.03925)
[10] Huber T, Hurth T and Lunghi E 2015 *JHEP* **06** 176 (*Preprint* 1503.04849)
[11] Huber T, Hurth T and Lunghi E 2008 *Nucl. Phys.* **B802** 40–62 (*Preprint* 0712.3009)
[12] Huber T, Lunghi E, Misiak M and Wyler D 2006 *Nucl. Phys.* **B740** 105–137 (*Preprint* hep-ph/0512066)
[13] Bordone M, Isidori G and Pattori A 2016 (*Preprint* 1605.07633)
[14] Gratrex J, Hopfer M and Zwicky R 2016 *Phys. Rev.* **D93** 054008 (*Preprint* 1506.03970)
[15] Yennie D R, Frautschi S C and Suura H 1961 *Annals Phys.* **13** 379–452
[16] Weinberg S 1965 *Phys. Rev.* **140** B516–B524
[17] Davidson N, Przedzinski T and Was Z 2016 *Comput. Phys. Commun.* **199** 86–101 (*Preprint* 1011.0937)
[18] Isidori G 2008 *Eur. Phys. J.* **C53** 567–571 (*Preprint* 0709.2439)

# Kaonic atoms and strangeness in nuclei: SIDDHARTA-2 and AMADEUS experiments

M Iliescu[1], M Bazzi[1], G Belloti[3], C Berucci[1,6], A.M Bragadireanu[5], D Bosnar[12], M Cargnelli[6], C.Curceanu[1], A D Butt[3,4], R Del Grande[1], A D'Uffizi[1], L Fabietti[10], C Fiorini[3,4], F Ghio[7], C Guaraldo[1], R S Hayano[8], T Ishiwatari[6], M Iwasaki[9], P Levi Sandri[1], J Marton[6], M Miliucci[1], S Okada[9], D Pietreanu[5], K Piscicchia[1, 14], A Romero Vidal[11], A Scordo[1], H Shi[8], D L Sirghi[1,5], F Sirghi[1,5], H Tatsuno[9], I Tucakovic[1], O Vazquez Doce[10], E Widmann[6] S Wycech[13] and J Zmeskal[6]

[1] INFN, Laboratori Nazionali di Frascati, Via Enrico Fermi 40, I-00044, Frascati (Roma), Italy

[2] Department of Physics and Astronomy, University of Victoria, Victoria, BC, Canada

[3] Politecnico di Milano, Dip.di Elettronica, Informazione e Bioingegneria, Milano, Italy

[4] INFN Sezione di Milano, Milano, Italy

[5] Horia Hulubei National Institute of Physics and Nuclear Engineering, Magurele, Romania

[6] Stefan-Meyer-Institut f¨ur subatomare Physik, Vienna, Austria

[7] INFN Sezione di Roma I and Istituto Superiore di Sanit`a, Roma, Italy

[8] The University of Tokyo, Tokyo, Japan

[9] RIKEN, The Institute of Physics and Chemical Research, Saitama, Japan

[10] Technische Universit¨at München, Physik Department, Garching, Germany

[11] University of Santiago de Compostela, Santiago de Compostela, Spain

[12] University of Zagreb, Zagreb, Croatia

[13] National Centre for Nuclear Studies, 00-681 Hoza 69 , Warsaw, Poland

[14] Museo Storico della Fisca e Centro Studi e Ricerche ``Enrico Fermi", Piazza del Viminale 1-00184 Roma, Italy

**(SIDDHARTA, SIDDHARTA-2 and AMADEUS Collaborations)**

**E-mail:** Mihai.Iliescu@lnf.infn.it

Journal of Physics: Conference Series **770** (2016) 012034 doi:10.1088/1742-6596/770/1/012034

**Abstract**. The dynamics of the strong interaction processes in the non-perturbative regime is currently approached by lattice calculations and effective field theories (ChPT), still lacking several experimental results, fundamental for a good understanding of the strangeness sector. Among these, the information provided by the low-energy kaon nucleon/nuclei interaction, accessible through the study of kaonic atoms and kaonic nuclear processes, plays a key-role. The lightest atomic systems, the kaonic hydrogen and the kaonic deuterium, deliver, in a model-independent way, the isospin-dependent kaon-nucleon scattering lengths, through the X-ray spectroscopy of the exotic atoms de-exciting to the fundamental level. The most precise kaonic hydrogen measurement to-date, together with an exploratory measurement of kaonic deuterium, were carried out in 2009 at the DAΦNE collider, by the SIDDHARTA collaboration. Nowadays, an upgraded setup was built, for a precise measurement of kaonic deuterium and, eventually, of heavier exotic atoms. A correlated study of the kaon-nuclei interaction at momenta below 130 MeV/c is carried out by the AMADEUS collaboration, using the KLOE detector and dedicated targets inserted near the collider interaction point. Preliminary results of the study of charged antikaons interacting with nuclei are shown, including a discussion of the still controversial Λ(1405).

## 1. Low energy QCD in the strange sector

The KN interaction at threshold provides important information on the relationship between spontaneous and explicit chiral symmetry breaking in low-energy QCD. Though the use of kaonic atoms as a perfect tool to extract without extrapolation the scattering lengths was predicted in the late 70s, reliable experimental data was available only from the 1990s, after the measurement of the strong-interaction energy shift $\varepsilon$ and width $\Gamma$ of kaonic-hydrogen 1s atomic state, at KEK-PS in Japan [1] and afterwards, at DAΦNE in Italy [2]. From the obtained values, the $K^-p$ scattering length was extracted using a Deser-Trueman-like formula [3], lately improved to account for the isospin-breaking corrections [4]. The measurement permitted a consistent interpretation of the scattering data [5] and the calculation of chiral SU(3) meson-baryon effective Lagrangian. Nevertheless, the amplitudes below the threshold were extrapolated with large uncertainties due to the experimental precision. Successively, the SIDDHARTA collaboration obtained more accurate values for both $\varepsilon$ and $\Gamma$ for kaonic hydrogen [9], thus providing stronger constraints on the theoretical models. The progress motivated new calculations regarding the structure of Λ(1405) and investigations on the existence of kaonic nuclear clusters [6], both now under experimental study within the AMADEUS collaboration. In this context, the importance of kaonic deuterium ($K^-D$) X-ray spectroscopy has been well recognized, since the isospin-dependent scattering lengths $a_0$ and $a_1$, mediated in the $K^-p$ process, cannot be determined without the measurement of a system containing a neutron. As the neutron alone cannot form atoms, the $K^-D$ transitions on the 1s level will provide the missing information. Up to now, no experimental results have been achieved, due to the extremely low yield ($\sim 10^{-3}$) of the $K^-D$ transitions to the fundamental state, suppressed by the Stark effect coupled to a large contribution of the strong interaction, inducing $K^-$ nuclear absorption from higher levels.

## 2. Current experimental status

SIDDHARTA technique and results are fully described in [9], so only a short account will be given, to facilitate the understanding of the upgrade proposed by the SIDDHARTA-2 collaboration for reaching a significant measurement of the $K^-D$ transition parameters.

The SIDDHARTA experiment was performed at the DAΦNE $e^+e^-$ collider in Frascati, Italy. The two beams energy is tuned to create Φ mesons, which decay in $K^+K^-$ pairs with a BR of 49.1%. The monochromatic, low-energy charged kaons are degraded, then stopped in a cryogenic gaseous target, producing kaonic atoms. The target is a critical item, the yields of kaonic atom X-rays decreasing with the gas density, due to Stark mixing. On the other hand, a too low density does not permit efficient stopping and increases the kaons in-flight decay. For the SIDDHARTA optimized working point, the yield was ~1.2% when filled with hydrogen and will presumably drop to ~0.1% for deuterium. The trigger was given by $K^-K^+$ coincident hits on fast scintillators. The signal was detected with 144 silicon drift detectors (SDDs), 1 $cm^2$ each, surrounding the target. The SDDs, developed by the collaboration, have an energy resolution of 150 eV FWHM at 6 keV, while their drift time is below 800 ns FWHM. The trigger signal, correlated with the fast X-ray detectors pulse, conferred a high background rejection ($10^4$), most of the last consisting in electromagnetic (EM) showers from the beam losses, uncorrelated with the Φ production (asynchronous background).

## 3. SIDDHARTA-2 experiment

Having acquired a good knowledge of the problems imposed by rare kaonic atom transition measurements and an accurate understanding of the background sources generating EM and hadronic cascades, the SIDDHARTA group has developed a set of methods aiming to increase the signal over background ratio (S/B) by a factor of 14 or better, thus allowing the kaonic deuterium measurement. The changes and add-ons required for reaching the above objective are briefly described, as follows:

-larger area, faster SDD detector array; the solution to improve the SDD time resolution consists in the reduction of the single element size (from 10 to 8 mm) and the replacement of the integrated J-FET (thermally limited), with a newly-developed amplifier on the ceramics, able to operate at very low temperatures (below 50 K). The shorter path and the higher carrier mobility consent a faster charge packet drift to the anode and therefore, a reduced time window (350 ns instead of 900 ns) for each trigger, with consequent suppression of the asynchronous background.

-a gas-moderation detector [19] measures the prompt time of the secondaries from $K^-$ absorption on nuclei. The functioning is based on the kaon moderation time in a high-density gas, longer than the corresponding one in solids. Moreover, a fraction of the stopped $K^+$ (~50%) decays after $K^-$ absorption, so its contribution to background is also removed. The system consists in scintillators surrounding the vacuum chamber, read at both ends by PMs coupled to mirrors and light-guides (to cope with the narrow space between the setup and the shielding against beam background).

-a new cryogenic target in reinforced kapton (13 cm diameter, 7 cm height), operating few hundred mK above the liquid point (~25 K) at a pressure of 4 bar (5% LHD), for more efficient kaon stopping.

-a veto system, consisting in scintillators read by SiPMs, placed behind each SDD array, to reject the hadronic background coming from border hits of MIPs, depositing energy in the X-ray range.

-a kaon trigger with geometric acceptance optimized to match the kaon gas stopping distribution.

-a new $K^+$ - $K^-$ discrimination detector, to further reduce the background from the $K^+$ mesons.

-an improved X-ray calibration system, providing low-rate in-situ calibration, as well as high rate calibration between physics runs, to compensate very small fluctuations in each single SDD response.

-mechanical and cryogenic improvements of the vacuum chamber, necessary to add more cooling power to the SDDs and to the cryogenic target.

All described items were optimized by GEANT4 simulation, considered reliable after having reproduced, in the same framework, the SIDDHARTA result, both in terms of signal and background, to 7% accuracy. A sketch of the SIDDHARTA-2 setup is shown in the Fig.1 (top), while the Monte Carlo configuration and the expected spectrum of the $K^-D$ transition, for an acquired luminosity of 800 $pb^{-1}$ and assuming a yield of 0.1%, are plot on the bottom. The fit on the simulated signal indicates the $K^-D$ 1s level, modified by the strong interaction, can be determined with a precision of 30 eV for the shift and 70 eV for the width, corresponding to a statistical significance of 20 σ for the Kα line.

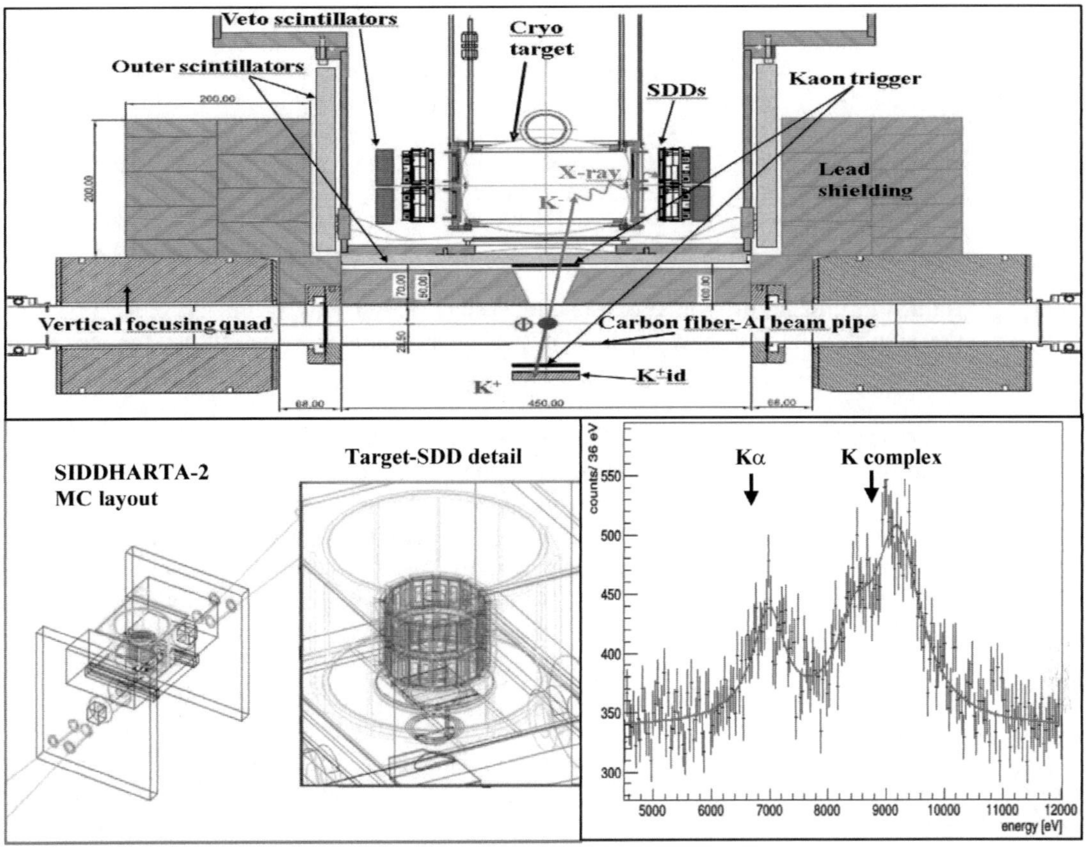


Figure 1: The SIDDHARTA-2 layout (top), with the Monte Carlo configuration (bottom left) and the expected K-D spectrum (bottom right).

The estimated values are comparable with the precision obtained for kaonic hydrogen in SIDDHARTA and will allow to determine the antikaon-nucleon isospin-dependent scattering lengths, required by the non-perturbative QCD models dealing with strangeness

## 4. AMADEUS

A research line complementary to that of exotic atoms, the AMADEUS experiment at DAΦNE, deals with the investigation of the low-energy kaon-nucleon/nuclei hadronic interaction. For that purpose, signals from $K^-$ absorption at-rest and in-flight (with momenta below 127 MeV/c), are investigated using a sample corresponding to a total integrated luminosity of 2 $fb^{-1}$ collected by KLOE [12] during the 2004 and 2005 runs. In a first phase the KLOE detector was used as an active target, the nuclear interaction of negatively charged kaons in the gas filling the KLOE Drift Chamber (DC) [13] (mainly 4He) and the DC entrance wall [14] (mainly 12C) was investigated. In the period November/December 2012 a successful data taking was performed with a dedicated pure carbon solid target implemented in the central region of KLOE. A total integrated luminosity of 90 $pb^{-1}$ was collected providing a high statistic sample of purely at-rest $K^-$ nuclear interactions [15]. The ongoing analyses are mainly concerned with the investigation of the $K^-$ multi-nucleon absorption processes (related the possible formation of $K^-$ multi-nucleon bound states [16]) and the $\Sigma^*/\Lambda^*$ properties and behavior in nuclear environment.

| | yield / $K^-_{stop} \cdot 10^{-2}$ | $\sigma_{stat} \cdot 10^{-2}$ | $\sigma_{syst} \cdot 10^{-2}$ |
|---|---|---|---|
| 2NA-QF | 0.124 | ± 0.019 | +0.004 -0.008 |
| 2NA-FSI | 0.265 | ± 0.027 | +0.021 -0.022 |
| Tot 2NA | 0.366 | ± 0.032 | +0.022 -0.031 |
| 3NA | 0.267 | ± 0.067 | +0.043 -0.020 |
| Tot 3body | 0.532 | ± 0.072 | +0.047 -0.032 |
| 4NA + Uncorr. bkg. | 0.753 | ± 0.052 | +0.024 -0.074 |

Table 1: Production probability of the $\Sigma^0$p final state for different intermediate processes normalized to the number of $K^-$ stopped in the DC wall. The statistical and systematic errors are shown as well.

In [17], a clean sample of $\Sigma^0$p events, from $K^-$ captures in the KLOE DC wall, was reconstructed. $\Sigma^0$p is (together with Λp) an expected decay channel of the pp$K^-$ cluster, free from the Σ conversion processes, which strongly affect the uncorrelated Λp production. A simultaneous fit of several kinematic variables was performed by a simulated cocktail containing $K^-$ absorptions on two or more nucleons (together with an uncorrelated background source). The study allowed isolating, with good precision, the $K^-$ absorption on two nucleons (2NA) free from any final state interaction (QF) of the produced $\Sigma^0$p pairs (2NA-QF). More difficult is to distinguish between 2NA and 3NA when 2NA is followed by final state interactions of the hyperon or nucleon with the residual nucleus. The obtained results are summarized in Table 1. The second step of the analysis consisted in the search of the pp$K^-$ bound state produced in $K^-$ interactions with $^{12}$C, decaying into a $\Sigma^0$p ppair. The pp$K^-$ was simulated similarly to the 2NA-QF process but sampling the mass of the pp$K^-$ state with a distribution, rather than the Fermi momenta of the two nucleons in the initial state. A second fit of the experimental data was carried out including a pp$K^-$ component decaying into $\Sigma^0$p. A systematic scan of possible binding energies and widths, varying within 15-75 MeV and 30-70 MeV respectively, was done. The best fit resulted in a binding energy of 45 MeV and a width of 30 MeV. The corresponding pp$K^-$ yield extracted from the fit is $(0.043 \pm 0.009 \text{ stat } ^{+0.004}_{-0.005} \text{ syst}) * 10^{-2}$. In order to test the significance of the bound state with respect to a statistical fluctuation an f-test has been performed. The contribution of the pp$K^-$ component was found to be only significant at the level of 1σ. Although the measured spectra are compatible with the hypothesis of a pp$K^-$ contribution, the significance of the result is not sufficient to claim the discovery of the state.

The possible production of kaonic bound states is strongly influenced by the parameters of the Λ(1405) state. We are presently investigating the $\Sigma^0\pi^0$ decay of the Λ(1405), produced in $K^-$ absorptions in-flight on $^4$He and $^{12}$C. For kaon momenta of about 100 MeV the $K^-$-N sub-threshold region is accessible, thus allowing to explore the Λ(1405) properties in the region 1415-1432 MeV. In order to extract the Λ(1405) shape it is crucial to disentangle the $\Sigma^0\pi^0$ resonant and non-resonant production. An analysis in this direction is ongoing, aiming to extract the yield of the non-resonant transition amplitude in the $\Lambda\pi^-$ isospin 1 channel. The $\Lambda\pi^-$ momentum spectra, including all the involved processes in $K^-$ capture on 4He and possible final state interactions, were calculated in [18]. The spectra are expressed in terms of two $K^-n \rightarrow \Lambda\pi^-$ transition amplitudes: the isospin I=1 S-wave amplitude ($|f^s|$) and the resonant I=1 P-wave amplitude, dominated by the Σ(1385). $|f^s|$ below threshold can be inferred given the well-known Σ(1385) amplitude. The goal is to get information on the corresponding I=0 sub-threshold amplitude in the Σπ channel.

**Acknowledgments**
We thank C. Capoccia, G. Corradi, B. Dulach, and D. Tagnani from LNF-INFN; and H. Schneider, L. Stohwasser, and D. Stuükler from Stefan-Meyer-Institut, for their fundamental contribution in designing and building the SIDDHARTA setup. We thank as well the DAΦNE staff for the excellent working conditions and permanent support. Part of this work was supported by the European Community-Research Infrastructure Integrating Activity "Study of Strongly Interacting Matter"

(HadronPhysics2, Grant Agreement No. 227431, and HadronPhysics3 (HP3) Contract No. 283286) under the Seventh Framework Programme of EU; HadronPhysics I3 FP6 European Community program, Contract No. RII3-CT-2004- 506078; Austrian Science Fund (FWF) (P24756-N20); Austrian Federal Ministry of Science and Research BMBWK 650962/0001 VI/2/2009; Romanian National Authority for Scientific Research, Contract No. 2-CeX 06-11-11/2006; the Grant-in-Aid for Specially Promoted Research (20002003), MEXT, Japan; and the Croatian Science Foundation, under project HRZZ 1680.

**References**

[1] M Iwasaki et al. Observation of the kaonic hydrogen $K_\alpha$ x-ray *Phys. Rev. Lett.* **78**, 3067-3069 (1997) T M Ito et al. Observation of kaonic hydrogen atom x rays *Phys. Rev. C* **58** 2366-2382 (1998)

[2] G Beer et al. Measurement of the kaonic hydrogen X-ray spectrum DEAR Collaboration, *Phys. Rev. Lett.* **94** 212302 (2005)

[3] S Deser et al. Energy level displacements in pi mesonic atoms *Phys. Rev.* **96** 774-776 (1954) T L Trueman Energy level shifts in atomic states of strongly-interacting particles *Nucl. Phys.* **26**, 57-67 (1961) A Deloff S Matrix Approach to Hadronic Atom Level Shifts *Phys. Rev. C* **13**, 730-736 (1976)

[4] U G Meißner, U Raha, A Rusetsky Spectrum and decays of kaonic hydrogen *Eur. Phys. J. C* **35**, 349-357 (2004) A Ciepl´y, J Smejkal Kaonic hydrogen versus the K-p low-energy data *Eur. Phys. J. A* **34**, 237-241 (2007)

[5] A D Martin: Kaon-Nucleon Parameters *Nucl. Phys. B* **179**, 33-48 (1981) D N Tovee et al. Some properties of the charged sigma hyperons *Nucl. Phys. B* **33**, 493-504 (1971) R J Nowak et al. Charged Sigma Hyperon Production by $K^-$ Meson Interactions at Rest *Nucl. Phys. B* **139**, 61-71 (1978)

[6] A Cieplý et al. Chirally motivated K− nuclear potentials *Phys. Lett. B* **702**, 402-407 (2011) Y Ikeda, T Hyodo, W Wise Improved constraints on chiral SU(3) dynamics from kaonic hydrogen *Phys. Lett. B* **706**, 63-67 (2011) Y Ikeda, T Hyodo, W Wise Chiral SU(3) theory of antikaon-nucleon interactions with improved threshold constraints *Nucl. Phys. A* **881**, 98-114 (2012)

[7] M Boscolo et al. Description and performances of luminosity and background detectors at the upgraded e+e-DAFNE collider *Nucl. Instr. Meth. Phys. Res. A* **621**, 157-170 (2010)

[8] T Koike, T Harada, Y Akaishi Cascade calculation of K–p and K–d atoms *Phys. Rev. C* **53**, 79-87 (1996)

[9] M Bazzi et al. A new measurement of kaonic hydrogen X-rays SIDDHARTA Collaboration, *Phys. Lett. B* **704**, 113-117 (2011) M Bazzi et al. Kaonic hydrogen X-ray measurement in SIDDHARTA SIDDHARTA Collaboration, *Nucl. Phys. A* **881**, 88-97 (2012).

[10] B Borasoy, R Nißler, W Weise Kaonic Hydrogen and K-p Scattering *Phys. Rev. Lett.* **94**, 213401 (2005) B Borasoy, U.-G Meißner, R Nißler : K-p scattering length from scattering experiments *Phys. Rev. C* **74**, 055201 (2006)

[11] SIDDHARTA-2 Collaboration, Proposal of Laboratori Nazionali di Frascati of INFN (2010)

[12] F Bossi, E De Lucia, J Lee-Franzini, S Miscetti, M Palutan and KLOE coll. *Riv. Nuovo Cim.* **31** (2008) 531-623

[13] M Adinolfi et al., [KLOE Collaboration] *Nucl. Inst. Meth. A* **488**, (2002) 51

[14] M Adinolfi et al. [KLOE Collaboration] *Nucl. Inst. Meth. A* **482**, (2002) 368

[15] C Curceanu, K. Piscicchia et al. *Acta Phys. Polon. B* **46** 1, 203-215, (2015)

[16] Y Akaishi and T Yamazaki Emission spectra and invariant masses of Λ and p in the stopped-K−NN absorption process in $^4$He and $^6$Li, *Nucl. Phys. A* **792**, 229 (2007)

[17] O Vazquez Doce et al. *Phys. Lett. B* **758**, 134 (2016).

[18] K Piscicchia, S Wycech, C Curceanu *Nucl. Phys. A* **954** (2016) 75-93

[19] M Bazzi et al. *JINST* **8**, (2013) T11003

Journal of Physics: Conference Series **770** (2016) 012035 doi:10.1088/1742-6596/770/1/012035

# The neutral pion form factor at NA62

**Patrizia Cenci**[1]

Istituto Nazionale di Fisica Nucleare Sezione di Perugia, Via Alessandro Pascoli, 06123 Perugia (I)

E-mail: patrizia.cenci@pg.infn.it

**Abstract**. In 2007 the NA62 experiment at CERN collected a large sample of charged kaon decays with a highly efficient trigger selecting events with electrons in the final state. The kaon beam represents a rich source of tagged neutral pion decays in vacuum. The electromagnetic transition form factor slope of the $\pi^0$ in the time-like region has been measured from about $10^6$ fully reconstructed $\pi^0$ Dalitz decays collected in 2007. The preliminary result $a = (3.68 \pm 0.51_{stat} \pm 0.25_{syst}) \times 10^{-2}$ is the most precise to date. This value is compatible with theoretical expectations and consistent with the previous measurements.

[1] On behalf of the NA62 Collaboration: G. Aglieri Rinella, R. Aliberti, F. Ambrosino, R. Ammendola, B. Angelucci, A. Antonelli, G. Anzivino, R. Arcidiacono, I. Azhinenko, S. Balev, M. Barbanera, J. Bendotti, A. Biagioni, L. Bician, C. Biino, A. Bizzeti, T. Blazek, A. Blik, B. Bloch-Devaux, V. Bolotov, V. Bonaiuto, M. Boretto, M. Bragadireanu, D. Britton, G. Britvich, M.B. Brunetti, D. Bryman, F. Bucci, F. Butin, E. Capitolo, C. Capoccia, T. Capussela, A. Cassese, A. Catinaccio, A. Cecchetti, A. Ceccucci, P. Cenci, V. Cerny, C. Cerri, B. Checcucci, O. Chikilev, S. Chiozzi, R. Ciaranfi, G. Collazuol, A. Conovaloff, P. Cooke, P. Cooper, G. Corradi, E. Cortina Gil, F. Costantini, F. Cotorobai, A. Cotta Ramusino, D. Coward, G. D'Agostini, J. Dainton, P. Dalpiaz, H. Danielsson, J. Degrange, N. De Simone, D. Di Filippo, L. Di Lella, S. Di Lorenzo, N. Dixon, N. Doble, B. Dobrich, V. Duk, V. Elsha, J. Engelfried, T. Enik, N. Estrada, V. Falaleev, R. Fantechi, V. Fascianelli, L. Federici, S. Fedotov, M. Fiorini, J. Fry, J. Fu, A. Fucci, L. Fulton, S. Gallorini, S. Galeotti, E. Gamberini, L. Gatignon, G. Georgiev, A. Gianoli, M. Giorgi, S. Giudici, L. Glonti, A. Goncalves Martins, F. Gonnella, E. Goudzovski, R. Guida, E. Gushchin, F. Hahn, B. Hallgren, H. Heath, F. Herman, T. Husek, O. Hutanu, D. Hutchcroft, L. Iacobuzio, E. Iacopini, E. Imbergamo, O. Jamet, P. Jarron, E. Jones, T. Jones K. Kampf, J. Kaplon, V. Kekelidze, S. Kholodenko, G. Khoriauli, A. Khotyantsev, A. Khudyakov, Yu. Kiryushin, A. Kleimenova, K. Kleinknecht, A. Kluge, M. Koval, V. Kozhuharov, M. Krivda, Z. Kucerova, Yu. Kudenko, J. Kunze, G. Lamanna, G. Latino, C. Lazzeroni, G. Lehmann-Miotto, R. Lenci, M. Lenti, E. Leonardi, P. Lichard, R. Lietava, L. Litov, R. Lollini, D. Lomidze, A. Lonardo, M. Lupi, N. Lurkin, K. McCormick, D. Madigozhin, G. Maire, C. Mandeiro, I. Mannelli, G. Mannocchi, A. Mapelli, F. Marchetto, R. Marchevski, S. Martellotti, P. Massarotti, K. Massri, P. Matak, E. Maurice, A. Mefodev, E. Menichetti, E. Minucci, M. Mirra, M. Misheva, N. Molokanova, J. Morant, M. Morel, M. Moulson, S. Movchan, D. Munday, M. Napolitano, I. Neri, F. Newson, A. Norton, M. Noy, G. Nuessle, T. Numao, V. Obraztsov, A. Ostankov, S. Padolski, R. Page, V. Palladino, G. Paoluzzi, C. Parkinson, E. Pedreschi, M. Pepe, F. Perez Gomez, M. Perrin-Terrin, L. Peruzzo, P. Petrov, F. Petrucci, R. Piandani, M. Piccini, D. Pietreanu, J. Pinzino, I. Polenkevich, L. Pontisso, Yu. Potrebenikov, D. Protopopescu, F. Raffaelli, M. Raggi, P. Riedler, A. Romano, P. Rubin, G. Ruggiero, V. Russo, V. Ryjov, A. Salamon, G. Salina, V. Samsonov, C. Santoni, G. Saracino, F. Sargeni, V. Semenov, A. Sergi, M. Serra, A. Shaikhiev, S. Shkarovskiy, I. Skillicorn, D. Soldi, A. Sotnikov, V. Sugonyaev, M. Sozzi, T. Spadaro, F. Spinella, R. Staley, A. Sturgess, P. Sutclie, N. Szilasi, D. Tagnani, S. Trilov, M. Valdata-Nappi, P. Valente, M. Vasile, T. Vassilieva, B. Velghe, M. Veltri, S. Venditti, P. Vicini, R. Volpe, M. Vormstein, H. Wahl, R. Wanke, P. Wertelaers, A. Winhart, R. Winston, B. Wrona, O. Yushchenko, M. Zamkovsky, A. Zinchenko.

## 1. Introduction

Hadron-photon couplings are fundamental observables in particle physics. They can be described in terms of form factors merging effects due to the underlying structure of hadrons. For this reason, precision measurements of Transition Form Factors (TFF) are very significant probes to investigate electromagnetic interactions and the internal structure of hadrons. Improved measurements have been recently made available from studies of the radiative and Dalitz meson decays as well as from meson production in photon-photon processes, allowing to set new stricter constraints to theoretical models.

Pions are the lighest mesons, well suited to probe low energy hadron dynamics. Since pions are produced in most of the kaon decays, kaon experiments are ideal environments for precision studies of the pion properties.

The NA62 experiment [1] is the last generation kaon experiment at the CERN SPS, exploiting an intense beam of charged kaons. Its main goal is the measurement of the Branching Ratio (BR) of the $K^+ \to \pi^+ \nu\bar{\nu}$ ultra-rare decay with 10% precision. This process is a golden decay in flavour physics, highly suppressed in the Standard Model (SM). The BR prediction is of the order of $10^{-10}$ and it is calculated with minimal theoretical uncertainty. Only one measurement is available so far, compatible with the SM within a large uncertainty. Due to that, the precision measurement of BR($K^+ \to \pi^+ \nu\bar{\nu}$) is a remarkable achievement and a very sensitive probe of the flavour sector of the SM. NA62 is currently running and will take data until the year 2018.

At an early stage of the NA62 experimental program, in 2007, an improved setup of NA48/2 [2], the former CERN experiment with charged kaons, has been exploited to test lepton flavour universality by measuring the ratio $R_K = \Gamma(K^\pm \to e\nu_e) / \Gamma(K^\pm \to \mu\nu_\mu)$ between the leptonic decay rates of the charged kaons into electrons and muons. NA62 has collected about $2 \times 10^{10}$ $K^\pm$ decays in 2007 and the ratio $R_K$ has been measured with a 0.4% accuracy, achieving the most precise result to date [3].

Neutral pions are produced in four of the six main decay modes of the charged kaons. For this reason, the NA62 experiment can also be considered as a $\pi^0$ factory. Based on the 2007 data sample, the slope $a$ of the $\pi^0$ electromagnetic TFF has been measured in the time-like region with an unprecedented level of precision. The preliminary results of the measurement are presented.

## 2. Measurement of the $\pi^0$ electromagnetic Transition Form Factor slope parameter in NA62

The $\pi^0$ meson mostly decays electromagnetically as $\pi^0 \to \gamma\gamma$ ($\pi^0_{2\gamma}$). When one of the photons has an off-shell mass above 2 electron masses, a $\pi^0$ Dalitz decay $\pi^0 \to \gamma e^+ e^-$ ($\pi^0_D$) occurs, with BR = (1.174 ± 0.035)% [4]. The leading order Feynman diagram of the $\pi^0_D$ decay is represented in figure 1 (left).

The measurement of the $\pi^0$ TFF slope paramete is exploited in NA62 by studying the decay $K^\pm \to \pi^\pm \pi^0$ ($K_{2\pi}$), which has BR = (20.67 ± 0.08 )% [4], followed by the $\pi^0_D$ decay. A clean sample of $\pi^0$ mesons is produced in $K_{2\pi}$ decays: the charged pion is used to tag the production of the $\pi^0$ and, since no neutrinos are present in the final state, background processes can be suppressed by applying stringent kinematic constraints on the reconstructed particle quantities.

The kinematics of the $\pi^0_D$ decay width can be expressed in terms of the $x$ and $y$ Dalitz variables:

$$x = \left( \frac{M_{e^+e^-}}{m^2_{\pi^0}} \right)^2, \quad y = \frac{2 p_{\pi^0}(p_{e^+} - p_{e^-})}{m^2_{\pi^0}(1-x)} \tag{1}$$

The variable limits are $r^2 \leq x \leq 1$ and $-\beta \leq y \leq \beta$, with $r = 2m_e / m_{\pi^0}$ and $\beta = (1-r^2/x)^{1/2}$.

The differential $\pi^0_D$ decay width normalized to the leading $\pi^0_{2\gamma}$ decay mode can be factorized in terms of a point-like part and two further ingredients: the $\pi^0$ electromagnetic TFF, $F(x)$, which describes the photon-pion coupling including effects due to the inner particle structure, and the radiative corrections $\delta(x, y)$ to the process. The formula is:

$$\frac{1}{\Gamma_0(\pi^0_{2\gamma})}\frac{\partial^2\Gamma(\pi^0_D)}{\partial x\partial y}=\frac{\alpha}{4\pi}\frac{(1-x)^3}{x}\left(1+y^2+\frac{r^2}{x}\right)\times\left|F(x)\right|^2\times[1+\delta(x,y)] \tag{2}$$

Small variations of $F(x)$ are expected in the allowed kinematic region, hence it can be parametrized by a linear expression as a function of $x$: $F(x) = (1 + ax)$, where $a$ is the $\pi^0$ electromagnetic TFF slope parameter. The value of the TFF slope parameter is obtained by adjusting the MC simulation to the data in order to get the best comparison for the $\pi^0_D$ decay $x$-spectrum obtained from the normalized differential decay width (2) integrated over y:

$$\frac{1}{\Gamma_0(\pi^0_{2\gamma})}\frac{d\Gamma(\pi^0_D)}{dx}=\frac{2\alpha}{3\pi}\frac{(1-x)^3}{x}\left(1+\frac{r^2}{2x}\right)\sqrt{1-\frac{r^2}{x}}\times(1+ax)^2\times[1+\delta(x,y)] \tag{3}$$

The first theoretical investigation of the $\pi^0$ electromagnetic TFF [5] estimated $F(x)$ to be dominated by two resonances, the $\rho$ and $\Omega$ mesons. In this modelling approach, called Vector Meson Dominance (VMD), $a$ is defined in terms of the $\pi^0$, $\rho$ and $\Omega$ masses as: $a \approx m(\pi^0)^2\ [m(\rho)^{-2} - m(\Omega)^{-2}]\ /\ 2 \approx 0.03$.

In general, the $\pi^0$ TFF slope parameter is expected to be in agreement with the VMD model. Since then, the $\pi^0$ TFF has been thoroughly studied theoretically [6–10] and measured experimentally in the time-like [11-14] as well as in the space-like [15-18] regions.

Radiative corrections to the $\pi^0_D$ differential decay rate are fundamental for the TFF measurement, as their contribution to the Leading Order $x$-spectrum is comparable to the effect of the TFF. They have been first studied using the soft-photon approximation [19], later extended by adding virtual photon contributions and photon bremsstrahlung [20] and recently revised to include the 1-loop 1-photon irreducible contributions [21] (see diagrams in figure 1 (center)). The latter calculations have been implemented in NA62 in the event generator of the $\pi^0_D$ decay MC simulation.

Figure 1 (right) shows a comparison of the Leading Order $\pi^0_D$ decay $x$-spectrum, the Leading Order $x$-spectrum including TFF effects with slope a = 0.3, as expected in the VMD model, and the Leading Order $x$-spectrum corrected for radiative effects. The three curves are very close each other and barely visible, since the $\pi^0_D$ TFF effects to the Leading Order decay $x$-spectrum are tiny and comparable to the size of radiative corrections. The $\pi^0$ electromagnetic TFF also enters the prediction of significant observable quantities: it influences the rate of the rare decay $\pi^0\to e^+e^-$ and the hadronic light-by-light scattering contribution to the magnetic moment of the muon $(g\text{-}2)_\mu$ [22]. For those reasons, precise model independent measurements of the $\pi^0$ TFF slope parameter are essential to test the theoretical models.

### *2.1. The NA62 experimental setup in 2007*

An improved setup of the NA48/2 experiment has been exploited to address the NA62 experimental program in 2007. The detailed description of the experimental apparatus can be found in [2].

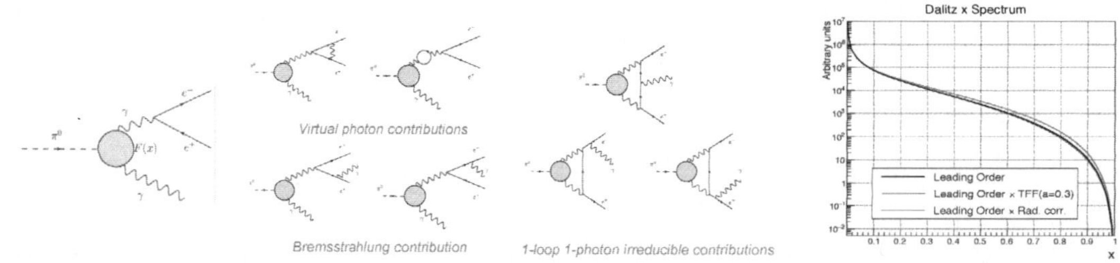


**Figure 1.** (left) Leading Order Feynmann diagram of the $\pi^0_D$ decay. (center) Diagrams of the $\pi^0_D$ radiative corrections. (right) Predictions of the $\pi^0_D$ decay $x$-spectrum: Leading Order, Leading Order including $\pi^0$ TFF contribution with VMD slope $a = 0.3$, Leading Order corrected for radiative effects.

Primary protons with 400 GeV/c momentum, extracted from the CERN SPS, were sent onto a beryllium target producing charged particles, among which the 6% were kaons. The NA48/2 beam line was designed to deliver simultaneous $K^+$ and $K^-$ beams. In 2007, $K^+$ and $K^-$ beams with (74±1.4) GeV/c momentum have been alternatively produced. The momentum of the charged particles from $K^\pm$ decays was measured by a magnetic spectrometer consisting of four drift chambers (DCH) and a dipole magnet. The spectrometer was located in a tank filled with helium at atmospheric pressure. A hodoscope made of horizontal and vertical scintillator slabs provided a fast trigger for charged particles with sub-nanosecond resolution. A quasi-homogeneous liquid Krypton electromagnetic calorimeter (LKr), segmented into 2×2 $cm^2$ projective cells, was used to precisely measure the energy deposited by photons and electrons. A hadron calorimeter and a muon veto system, essential to distinguish muons from pions, were located further downstream.

In 2007 the data taking was carefully tuned for the $R_K$ measurement and the trigger system was optimized to collect events with energy depositions in the LKr calorimeter. The beam intensity was lowered by a factor 10 with respect to NA48/2 to get an effective reduction of background due to accidental events; a minimum bias trigger was exploited at the lower trigger level to ensure an efficient data collection; an optimized electron selection was applied at the higher trigger level.

In 2007 the NA62 experiment collected about $2\times10^{10}$ kaon decays in the fiducial decay region. This corresponds to about $5\times10^9\,\pi^0$ mesons generated from $K_{2\pi}$ events.

### 2.2. Event Selection

A sample of pure Dalitz decays $\pi^0_D$ has been identified by fully reconstructing the kinematics of the $K^\pm\rightarrow\pi^\pm\pi^0_D$ ($K_{2\pi D}$) decay chain. Being the $\pi^0_D$ decay an almost prompt process, with few μm mean free path, final states with two same-sign tracks, one opposite-sign track and a photon have been selected by applying the following criteria.

Three tracks in the magnetic spectrometer, originated from the same vertex inside the fiducial decay region, have been required. Each track must be within the DCH acceptance and have a reconstructed momentum between 2 and 74 GeV/c. No additional track is allowed. The photon candidate is identified by a single energy deposit in LKr, in time with the tracks, separated in space from the track impact points at the calorimeter surface. The total reconstructed momentum of photon and tracks must be consistent with the nominal beam momentum and direction within the experimental resolution. The total reconstructed transverse momentum with respect to the nominal beam axis must be compatible with zero, since no neutrino is present in the final state.

The kinematic properties of the $K^\pm\rightarrow\pi^\pm\pi^0_D$ decay chain are exploited for the mass assignment to the tracks. The opposite-sign track is assigned an electron mass. Two hypotheses are built for the possible mass assignments of the same-sign tracks. For each hypothesis, the reconstructed invariant masses $M_{\gamma e^+e^-}$ and $M_{\pi\pi^0_D}$, and the $x$ and $y$ variables are computed. The following conditions are required: $M_{\gamma e^+e^-}$ between 115 and 145 MeV/$c^2$, i.e. consistent with the $\pi^0$ mass; $M_{\pi\pi^0_D}$ between 465 and 510 MeV/$c^2$, i.e. consistent with the $K^\pm$ mass; $x$ and $|y| < 1$. Solely and exclusively those events with only one valid hypothesis are selected. The trigger selection is also reproduced, with tighter conditions, to eliminate edge effects due to different resolution between online and offline analysis. The signal region is defined as $x > 0.01$ since in the simulation the event acceptance at low $x$ value is not well reproduced. This choice does not affect the final result since the $\pi^0_D$ $x$-spectrum is not sensitive to TFF effects for $x$ close to null value (see figure 1 (right)).

The final sample amounts to $1.05\times10^6$ fully reconstructed $\pi^0_D$ decays, with a small contribution given by $K^\pm\rightarrow\pi^0_D\,\mu^\pm\nu$ ($K_{\mu3D}$) events. The distributions of the reconstructed invariant masses $M_{\gamma e^+e^-}$ and $M_{\pi\pi^0_D}$ are illustrated in figure 2 (left and center), with cut arrows. Figure 2 (right) also shows the $x$-spectrum of the selected $\pi^0_D$ candidates. In this figure, the distribution of the final $\pi^0_D$ data sample (solid red line) is compared with the MC predictions showing the individual contributions from $K_{2\pi D}$ and $K_{\mu 3D}$ decays. The acceptances have been evaluated with MC simulations and amount, respectively, to 1.81% for $K_{2\pi D}$ decays and to 0.02% for $K_{\mu 3D}$ decays.

### *2.3. Preliminary Results*

A $\chi^2$ fit of the $x$-spectrum of data and MC simulation with different slopes is used to extract the TFF slope value using an equipopulous binning. The different hypotheses are tested by reweighing the MC events with a known slope $a_{sim}$ = 0.032 to obtain simulated distributions with different values of the slope from the same MC sample. The measurement uncertainties are listed in table 1. The main systematics come from the simulation of the beam spectrum and from the calibration of the spectrometer global momentum scale. The preliminary result, with $\chi^2/ndf = 52.5/49$, is:

$$a = (3.70 \pm 0{:}53_{stat} \pm 0.36_{syst}) \times 10^{-2} = (3.70 \pm 0.64) \times 10^{-2}.$$

This measurement represents the first 5.8σ observation of a positive $\pi^0$ electromagnetic TFF slope in the time-like region of the momentum transfer. The best fit result is shown in figure 3 together with the comparison with previous measurements from $\pi^0_D$ Dalitz decays. The NA62 result improves the precision of the previous measurement by an order of magnitude, and is consistent with theoretical predictions. The final result requires to finalize systematic uncertainty studies and is in preparation.

**Table 1.** Uncertainties of the $\pi^0$ TFF slope measurement

| Source | $\delta a$ ($\times 10^2$) |
|---|---|
| Statistical – data | 0.49 |
| Statistical – MC | 0.20 |
| Beam momentum simulation | 0.30 |
| Spectrometer momentum scale | 0.15 |
| Spectrometer resolution | 0.05 |
| LKr non-linearity and energy scale | 0.04 |
| Particle mis-identification | 0.08 |
| Accidental background | 0.08 |
| Neglected $\pi^0_D$ sources in MC | 0.01 |

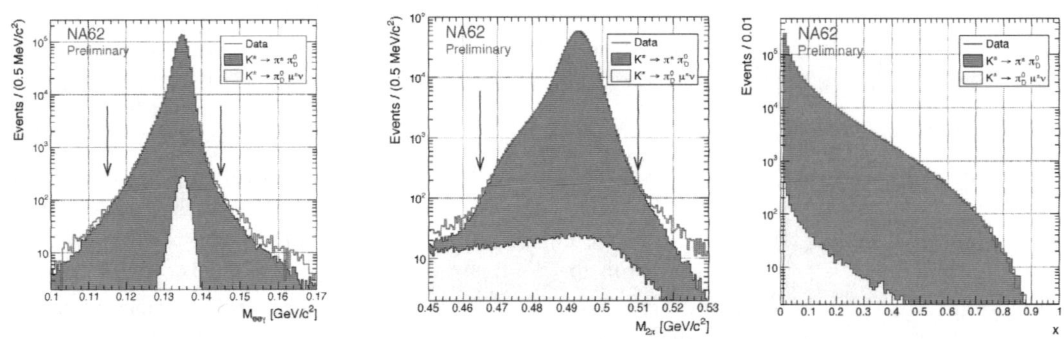


**Figure 2.** The final $\pi^0_D$ data sample (solid red line) compared with MC prediction showing individual contributions from $K^{\pm} \to \pi^{\pm} \pi^0_D$ and $K^{\pm} \to \pi^0_D \mu^+ \nu$ decays. Log-scale distributions of the reconstructed invariant mass $M_{\gamma e^+ e^-}$ of $\gamma e^+ e^-$ events (left), peaking at $M_{\pi^0}$, and of $2\pi$ events $M_{2\pi}$ (center), peaking at $M_K$, with cut arrows. (right) Log-scale distribution of the reconstructed x Dalitz variable.

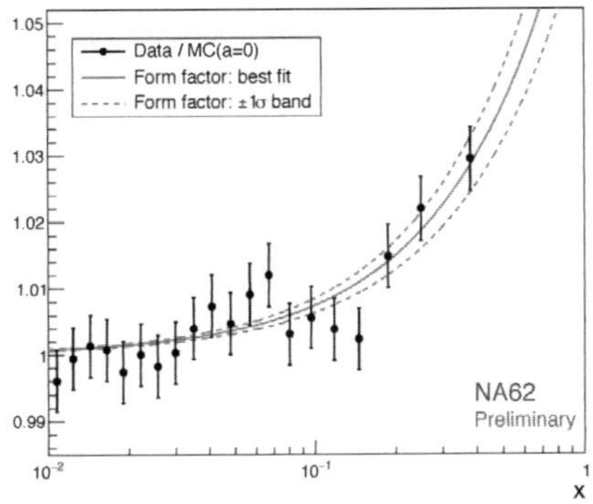


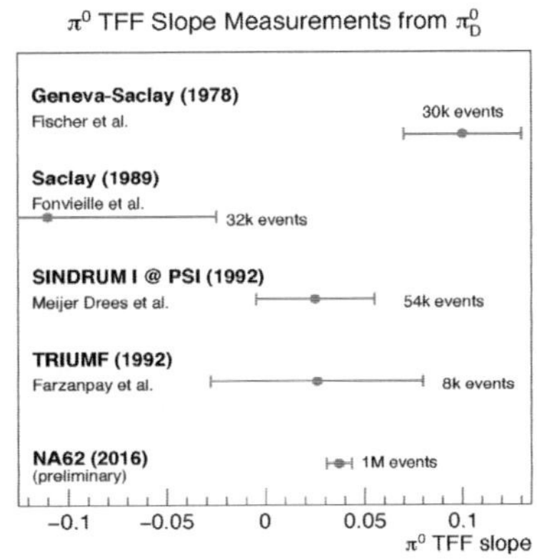


**Figure 3.** (left) Results of the TFF slope parameter fit showing the data/MC ratio as a function of $x$ with the MC sample weighted to obtain a flat TFF (a = 0). The events are divided into 20 equipopulous bins. The solid line represents the squared TFF function with a slope equal to the central value of the fit. The dashed lines indicate the 1σ band. (right) The NA62 measurement of the TFF slope is compared with those of experiments exploiting the same method.

## 3. Conclusions

The NA62 experiment has performed a preliminary measurement of the $\pi^0$ TFF in the time-like region of momentum transfer from a sample of about one million fully reconstructed $\pi_D^0$ decays collected in 2007. The preliminary result $a = (3.70 \pm 0.64) \times 10^{-2}$ is the most precise to date. This result is compatible with theoretical expectations and consistent with the previous measurements.

## References

[1] Anelli G *et al.* (NA62 Collaboration) *Proposal to measure the rare decay* $K^+ \to \pi^+ \nu\bar{\nu}$ *at the CERN SPS* 2005 CERN-SPSC-2005-013 SPSC-P-326
[2] Fanti V *et al.* (NA48 Collaboration) 2007 *Nucl. Instrum. Methods* A **574** 433
[3] Lazzeroni C *et al.* (NA62 Collaboration) 2013 *Phys. Lett.* B **719** 326–336
[4] Patrignani C *et al.* (Particle Data Group) 2016 *Chin. Phys.* C **40** 100001
[5] Gell-Mann M and Zachariasen F 1961 *Phys. Rev.* **124** 953
[6] Lichard P 2011 *Phys. Rev.* D **83** 037503
[7] Kampf K *et al.* 2006 *Eur. Phys. J.* C **46** 191
[8] Husek T and Leupold S 2015 *Eur. Phys. J.* C **75** 586
[9] Masjuan P 2012 *Phys. Rev.* D **86** 094021
[10] Hoferichter M *et al.* 2014 *Eur. Phys. J.* C **74** 3180
[11] Fisher J *et al.* (Geneva-Saclay Collaboration) 1978 *Phys. Lett.* B **73** 359
[12] Fonvieille H *et al.* 1989 *Phys. Lett.* B **233** 65
[13] Forzanpay H *et al.* 1992 *Phys. Lett.* B **278** 413
[14] Meijer Drees R *et al.* (SINDRUM Collaboration) 2014 *Phys. Rev.* D **45** 1439
[15] Berend H J *et al.* (CELLO Collaboration) 1991 *Zeit. Phys.* C **49** 401
[16] Gronberg J *et al.* (CLEO Collaboration) 1998 *Phys. Rev.* D **57** 33
[17] Aubert B *et al.* (BaBar Collaboration) 2009 *Phys. Rev.* D **80** 052002
[18] Uehara S *et al.* (Belle Collaboration) 2012 *Phys. Rev.* D **86** 092007
[19] Lautrup B E and Smith J 1971 *Phys. Rev.* D **3** 1122
[20] Mikaelian K O and Smith J 1972 *Phys. Rev.* D **5** 1763
[21] Husek T *et al.* 2015 *Phys. Rev.* D **92** 054027
[22] Nyffeler A 2016 *Phys. Rev.* D **94** 053006 *Preprint* arXiv:1602.03398 [hep-ph]

Journal of Physics: Conference Series **770** (2016) 012036 doi:10.1088/1742-6596/770/1/012036

# Measurement of the pion form factor at BESIII in the rho-peak region relevant for $(g-2)_\mu$

**C F Redmer for the BESIII Collaboration**

Institut für Kernphysik, Johannes Gutenberg-Universität Mainz, J.-J.-Becher-Weg 45, 55128 Mainz, Germany

E-mail: redmer@uni-mainz.de

**Abstract.** Measurements of hadronic cross sections in $e^+e^-$ collisions are an important input for the Standard Model calculations of the hadronic contribution to the anomalous magnetic moment of the muon, $(g-2)_\mu$. We present a new measurement of the cross section $e^+e^- \to \pi^+\pi^-$, the pion form factor, which dominates the hadronic contribution. Using the method of initial state radiation, 2.93 fb$^{-1}$ of data taken with the BESIII detector at the $\psi(3770)$ peak have been evaluated to determine the cross section in the energy range between 600 and 900 MeV with a systematic uncertainty of 0.9%. The new BESIII data are compared to previous measurements by KLOE and BaBar, and their impact on $(g-2)_\mu$ is discussed.

## 1. Introduction

The anomalous magnetic moment of the muon $a_\mu = \frac{1}{2}(g-2)_\mu$ is one of the most precisely known observables in the Standard Model. It has been determined experimentally since 1961 with ever increasing accuracy. The latest high precision direct measurement has been performed in Brookhaven and resulted in $a_\mu^{exp} = (11659208.9 \pm 6.3) \times 10^{-10}$ [1]. The theoretical prediction of $a_\mu$ within the Standard Model achieved a competitive accuracy, however, there remains a discrepancy of three to four standard deviations from the experimental value. In order to clarify the situation new direct measurements are planned at Fermilab, Chicago, USA and J-PARC, Tokai, Japan, which aim at a fourfold improvement of the experimental precision [2, 3].

The Standard Model calculations involve contributions from QED, $a_\mu^{QED}$, the weak interaction, $a_\mu^{weak}$, and the strong interaction, $a_\mu^{hadr}$. The QED contribution accounts for the largest part of the absolute value of $a_\mu$. It has been determined up to the $10^{th}$ order in perturbation theory [4] and its uncertainty is well under control. The weak contribution is small and also well under control by means of perturbation theory [5]. The hadronic contributions are comparably small, however, their uncertainties completely dominate the total uncertainty of the Standard Model prediction of $a_\mu$ [6]. Due to the running of the strong coupling constant, the hadronic contributions cannot be treated perturbatively in the relevant energy regime. However, experimental input can be exploited to improve the prediction.

The hadronic vacuum polarization causes the largest hadronic contribution $a_\mu^{hVP}$. It can be related by the optical theorem to hadronic cross sections measured in $e^+e^-$ annihilations. These are an important input to the Standard Model calculations of the anomalous magnetic moment of the muon $a_\mu$. The accuracy of the hadronic contribution can be systematically improved by the measurement of $\sigma(e^+e^- \to \text{hadrons})$ at $e^+e^-$ colliders with high precision. Due to the energy dependence of the cross sections and of the kernel function in the dispersion integral, which is used to calculate the contribution of individual cross sections to the leading order of $a_\mu^{hVP}$, hadronic cross sections measured below 1 GeV center-of-mass energy are the most relevant input. The energy region is dominated by the $\rho$-resonance, and it turns out that $a_\mu^{hVP}$ can be determined to more than 70% by the measurement of the cross section $\sigma(e^+e^- \to \pi^+\pi^-)$ [6].

High precision measurements of the two pion cross section have already been performed by BaBar [7, 8] and KLOE [9, 10, 11]. Both experiments claim a precision below one percent, however, their results differ by more than 3% in the $\rho$-peak region. This discrepancy reflects directly in the uncertainty of the leading order contribution $a_\mu^{hVP,LO}$ of the hadronic vacuum polarization. Another high precision measurement of $\sigma(e^+e^- \to \pi^+\pi^-)$ with a precision of better than one percent is needed to clarify the situation and reduce the uncertainty of $a_\mu^{hadr}$.

## 2. The BESIII Detector

The measurement has been done with the BESIII detector [12], which is operated at the symmetric BEPCII $e^+e^-$ collider. The cylindrical detector setup covers 93% of the total solid angle. It consist of several subsystems arranged in layers around the interaction region, which is made from Beryllium. The innermost detector component is a helium-based drift chamber, which is immersed in the 1.0 T magnetic field of a superconducting solenoid. It measures momenta of charged particles with a resolution of 0.5% at 1 GeV, and provides $\frac{dE}{dx}$ information for particle identification. The chamber is surrounded by a time-of-fight system, made from two layers of plastic scintillator, which is used for particle identification. It provides time information for charged particles with a resolution of 90 ps in the barrel part and 110 ps in the end caps. The third detector component located inside the bore of the solenoid is an electromagnetic calorimeter made from 6240 CsI(Tl) crystals. It measures energies of photons with a resolution of 2.5% at 1 GeV. The outermost detector component is the muon counter. It is built into the flux-return yoke of the solenoid and is instrumented with resistive plate counters.

The accelerator allows for data taking at center-of-mass energies between 2.0 GeV and 4.6 GeV, which is routinely performed since 2009. BESIII has collected $1.25 \times 10^9$ $e^+e^- \to J/\psi$ events, more than $500 \times 10^6$ $e^+e^- \to \psi(2S)$ events, 2.93 fb$^{-1}$ at the $\psi(3770)$ peak, and more than 5 fb$^{-1}$ in the center-of-mass region above 4 GeV, which are devoted to studies of the charmonia and charmonium-like states [13, 14, 15]. These samples are among the world's largest data sets of $J/\psi$, $\psi(2S)$, and $\psi(3770)$ mesons. Additional data have been acquired for a $\tau$-mass scan and a high statistics R-scan. The BESIII collaboration uses these data to follow a physics program [16], which focuses on charmonium spectroscopy, charm physics, light hadron spectroscopy, $\tau$ physics, and R-measurements.

## 3. Measurement of the cross section $e^+e^- \to \pi^+\pi^-$

The BESIII measurement of the hadronic cross section $\sigma(e^+e^- \to \pi^+\pi^-)$ is based on the 2.93 fb$^{-1}$ sample taken at the center-of-mass energy of $\sqrt{s} = 3.773$ GeV/c$^2$ [14]. A detailed description of the analysis can be found in Ref. [17].

Events with the final state $\pi^+\pi^-\gamma$ are selected to exploit the method of Initial State Radiation (ISR). The radiated photon reduced the effective center-of-mass energy, so that the energy region of the $\rho$-peak, 600 MeV/c$^2$ – 900 MeV/c$^2$, can be studied. The radiative cross section $\sigma(e^+e^- \to \pi^+\pi^-\gamma_{ISR})$ is related to the non-radiative cross section of interest by the radiator function [18], which describes the probability of emitting a photon depending on the center-of-mass energy, momentum fraction of the radiating particle, and the emission angle of the photon.

The dominating background in the event selection comes from radiative muon pair production $e^+e^- \to \mu^+\mu^-\gamma_{ISR}$. Due to the similarity of pion and muon masses, this background cannot be suppressed based on the event kinematics alone. An artificial neural network has been applied, which combines kinematic information with information from different subdetectors, like shower shapes in the calorimeter crystals and the penetration depth in the muon counter, to successfully differentiate pions from muons.

The network has been trained with distributions from Monte Carlo simulations of the signal $\pi^+\pi^-\gamma_{ISR}$ and the background $\mu^+\mu^-\gamma_{ISR}$ processes. Possible discrepancies between the simulation and experimental distributions have been studied carefully, and corresponding corrections are applied in the analysis.

The validity of the systematic corrections and the reliability of the neural network have been tested by using the network to select a clean sample of radiative muons. The result is compared to the QED prediction of muon production, which is taken from the PHOKARA event generator [19, 20]. The accuracy of the prediction implemented in the generator is claimed to be better than 0.5%. Excellent agreement is found between data and Monte Carlo on the full mass range accessible with the ISR method, but also in the mass range relevant for the pion cross section measurement, 600 MeV – 900 MeV, which, due to the peak of the $\rho$-resonance, makes up for 70% of the total two-pion contribution to $a_\mu^{hVP}$.

In order to calculate the contribution to the anomalous magnetic moment of the muon, the bare cross section of the process $e^+e^- \to \pi^+\pi^-$ has to be determined. There are two ways to extract this cross section. In the first method, the measured number of $\pi^+\pi^-\gamma_{ISR}$ events is normalized to the detection efficiency and the integrated luminosity of the data set to obtain the radiative cross section. After correcting it for vacuum polarization and FSR effects, the radiative cross section is finally divided by the radiator function, in order to obtain the bare, non-radiative cross section $\sigma(e^+e^- \to \pi^+\pi^-)$.

In the second method, the number of efficiency corrected $\pi^+\pi^-\gamma_{ISR}$ events is normalized to the ratio of detected $\mu^+\mu^-\gamma_{ISR}$ events in the same data set and the exact QED prediction for muon production. The advantage compared to the first method is that the integrated luminosity and the vacuum polarization cancel in the ratio. Thus, they do not contribute to the systematic uncertainty. However, this method is limited by the number of detected muons. The results provided by both methods agree well within their errors. Due to the statistical limitation in the number of $\mu^+\mu^-\gamma_{ISR}$ events in data, the first method provides the more precise result and is used to evaluate the contribution to $a_\mu^{hVP}$. A total uncertainty of 0.9% is achieved. The individual contributions to the systematic uncertainty are listed in Table 1.

**Table 1.** Contributions to the systematic uncertainty. The numbers are taken from Ref.[17]

| Source | Contribution |
|---|---|
| Photon efficiency correction | 0.2 |
| Pion tracking efficiency correction | 0.3 |
| Pion ANN efficiency correction | 0.2 |
| Pion e-PID efficiency correction | 0.2 |
| ANN | negl. |
| Angular acceptance | 0.1 |
| Background subtraction | 0.1 |
| Unfolding | 0.2 |
| FSR correction $\delta_{FSR}$ | 0.2 |
| Vacuum polarization correction $\delta_{vac}$ | 0.2 |
| Radiator function | 0.5 |
| Luminosity $\mathcal{L}$ | 0.5 |
| **Sum** | **0.9** |

Taking into account vacuum polarization effects and correcting for FSR, the pion form factor can be calculated from the measured cross section. The Gounaris-Sakurai parametrization of the $\rho - \omega$ interference is fitted to the form factor. Masses and widths of the $\rho$ and $\omega$, as well as the phase and the strength of the interference are free parameters of the fit. The pion form factor and the result of the fit are shown in Figure 1.

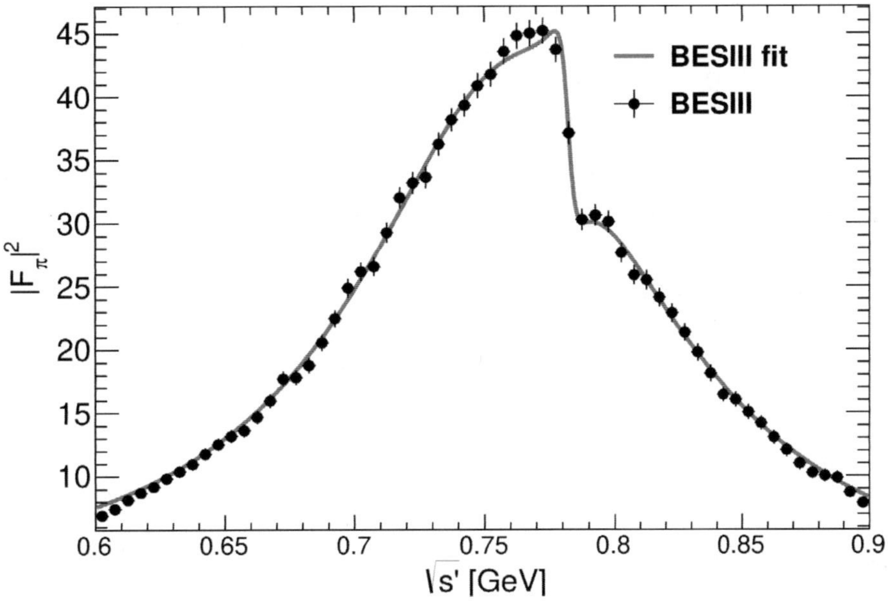


**Figure 1.** Pion form factor measured by BESIII and fitted with the Gounaris-Sakurai parametrization (red line). Figure is taken from Ref. [17].

The parametrization provides an excellent description of the data. The masses and widths obtained from the fit parameters are in good agreement with the existing values from PDG [21]. The fit is used to compare this measurement with the existing measurements from KLOE [9, 10, 11] and BaBar [7, 8].

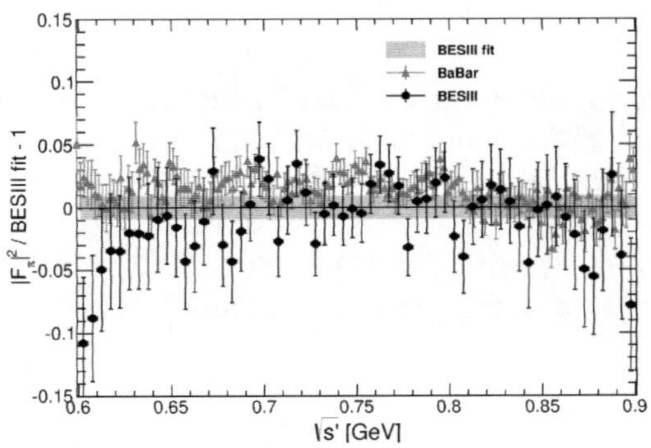


**Figure 2.** Comparison of the fit to the pion form factor of BESIII and the respective data points (black), as well as the BaBar result [7, 8] (red). Figure is taken from Ref. [17].

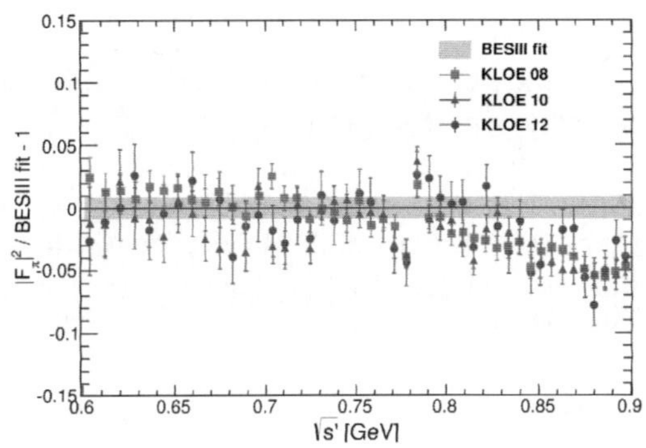


**Figure 3.** Comparison of the fitted pion form factor of BESIII and the KLOE [9, 10, 11] data points. Figure is taken from Ref. [17].

Figure 2 shows the relative difference between the fit result, the BESIII data points (black) and the BaBar data points (red). Within the error bars both measurements agree. Relative to the fit result a step can be observed in the BaBar data, which is approximately located at the $\rho-\omega$ interference. Above this point, the BaBar data points coincide with the BESIII fit result. However, at lower energies there seems to be an offset relative to the BESIII fit.

The comparison with the KLOE measurements in Figure 3 reveals a close to opposite picture. The data points of the three measurements coincide with the BESIII fit below the $\rho-\omega$ interference. Above the interference region all KLOE measurements show a clear systematic offset. Among the individual KLOE results, the KLOE10 [10] measurement seems to have a slight negative offset on the full mass range.

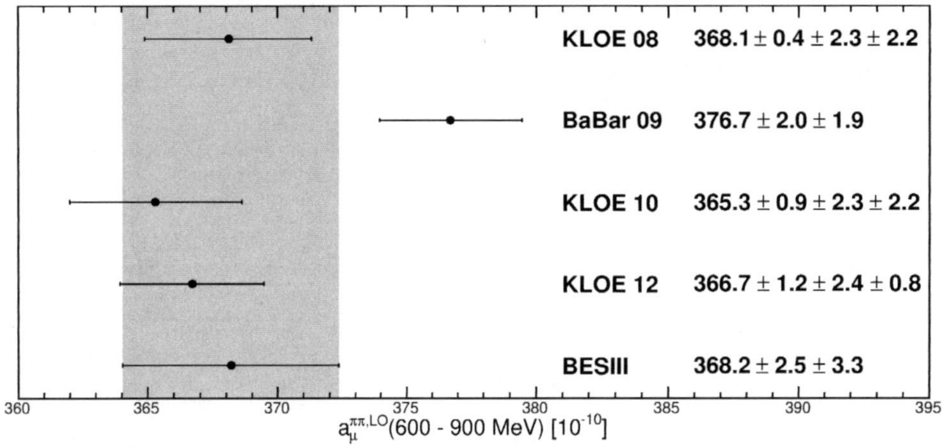


**Figure 4.** Comparison of $a_\mu^{\pi\pi,LO}(600-900\text{MeV})$ from the measurements of KLOE, BaBar, and BESIII. The error bars include statistical and systematical errors. Figure is taken from Ref. [17].

The leading order contribution of the two-pion cross section $a_\mu^{\pi\pi,LO}$ to the total hadronic vacuum polarization contribution to $a_\mu$ is calculated from the cross section $\sigma(e^+e^- \to \pi^+\pi^-)$, which has been corrected for vacuum polarization effects and includes FSR effects. The result of the BESIII measurement is $a_\mu^{\pi\pi,LO}(600-900\text{ MeV}) = 368.2 \pm 2.5_{\text{stat}} \pm 3.3_{\text{syst}} \cdot 10^{-10}$ [17].

Figure 4 shows the comparison of this result with the contributions calculated from the results of BaBar and KLOE. The yellow band indicates the total uncertainty of the BESIII result and demonstrates the competitiveness of the precision with the previous measurements. The three KLOE results agree well with the result of BESIII. Their values of $a_\mu^{\pi\pi,LO}(600-900\text{ MeV})$ reside within the BESIII error band. The result of the BaBar collaboration is shifted towards larger values of $a_\mu^{\pi\pi,LO}(600-900\text{ MeV})$. These deviations coincide with the observed shifts between the different results of the measured cross sections for invariant masses of the pions below the $\rho-\omega$ interference region. Due to the peak of the $\rho$ resonance this mass region dominates the statistics.

The BESIII result confirms the deviation between the experimental value of $a_\mu$ and its Standard Model prediction of three to four standard deviations. The BaBar result would hint at a deviation of only 2.4 standard deviations [22].

## References

[1] Bennett G W *et al.* (Muon g-2) 2006 *Phys. Rev.* **D73** 072003 (*Preprint* hep-ex/0602035)
[2] Venanzoni G (Muon g-2) 2012 *Nucl. Phys. Proc. Suppl.* **225-227** 277–281
[3] Mibe T (J-PARC g-2) 2011 *Nucl. Phys. Proc. Suppl.* **218** 242–246
[4] Aoyama T, Hayakawa M, Kinoshita T and Nio M 2012 *Phys. Rev. Lett.* **109** 111808 (*Preprint* 1205.5370)
[5] Gnendiger C, Stöckinger D and Stöckinger-Kim H 2013 *Phys. Rev.* **D88** 053005 (*Preprint* 1306.5546)
[6] Davier M, Hoecker A, Malaescu B and Zhang Z 2011 *Eur. Phys. J.* **C71** 1515 [Erratum: Eur. Phys. J.C72,1874(2012)] (*Preprint* 1010.4180)
[7] Aubert B *et al.* (BaBar) 2009 *Phys. Rev. Lett.* **103** 231801 (*Preprint* 0908.3589)
[8] Lees J *et al.* (BaBar) 2012 *Phys. Rev.* **D86** 032013 (*Preprint* 1205.2228)
[9] Ambrosino F *et al.* (KLOE) 2009 *Phys. Lett.* **B670** 285–291 (*Preprint* 0809.3950)
[10] Ambrosino F *et al.* (KLOE) 2011 *Phys. Lett.* **B700** 102–110 (*Preprint* 1006.5313)
[11] Babusci D *et al.* (KLOE) 2013 *Phys. Lett.* **B720** 336–343 (*Preprint* 1212.4524)
[12] Ablikim M *et al.* (BESIII) 2010 *Nucl. Instrum. Meth.* **A614** 345–399 (*Preprint* 0911.4960)
[13] Ablikim M *et al.* (BESIII) 2012 *Chin. Phys.* **C36** 915–925 (*Preprint* 1207.2865)
[14] Ablikim M *et al.* (BESIII) 2013 *Chin. Phys.* **C37** 123001 (*Preprint* 1307.2022)
[15] Ablikim M *et al.* (BESIII) 2015 *Chin. Phys.* **C39** 093001 (*Preprint* 1503.03408)
[16] Asner D M *et al.* 2009 *Int. J. Mod. Phys.* **A24** S1–794 (*Preprint* 0809.1869)
[17] Ablikim M *et al.* (BESIII) 2016 *Phys. Lett.* **B753** 629–638 (*Preprint* 1507.08188)
[18] Druzhinin V P, Eidelman S I, Serednyakov S I and Solodov E P *Rev. Mod. Phys.*
[19] Rodrigo G, Czyz H, Kuhn J H and Szopa M 2002 *Eur. Phys. J.* **C24** 71–82 (*Preprint* hep-ph/0112184)
[20] Czyz H, Kuhn J H and Wapienik A 2008 *Phys. Rev.* **D77** 114005 (*Preprint* 0804.0359)
[21] Olive K *et al.* (Particle Data Group) 2014 *Chin.Phys.* **C38** 090001
[22] Davier M, Hoecker A, Malaescu B, Yuan C Z and Zhang Z 2010 *Eur. Phys. J.* **C66** 1–9 (*Preprint* 0908.4300)

# Low-energy $e^+e^-$ Hadronic Cross-Section Measurements and Implications for (g-2) of the Muon

**Janis McKenna (on behalf of the *BABAR* Collaboration)**
Department of Physics and Astronomy, University of British Columbia, Vancouver, BC, Canada
E-mail: janis@physics.ubc.ca

**Abstract.** The *BABAR* Collaboration has performed an intensive program studying low energy hadronic cross sections in $e^+e^-$ annihilations via initial-state radiation. These measurements are crucial to improve the precision of the Standard Model prediction of the muon anomalous magnetic moment, $a_\mu = \frac{1}{2}(g_\mu - 2)$, and could help shed light on the $> 3\sigma$ discrepancy between the Standard Model predicted value for $a_\mu$ and the experimental measurement of $a_\mu$ performed at Brookhaven National Laboratory. We have published results on a number of initial state radiative processes with two to six hadrons in the final state, at effective center-of-mass energies ranging from ~1 to 4.5 GeV. We report here on our most recent results obtained using the entire *BABAR* dataset.

## 1. Introduction and Motivation
We have entered an era of precision tests of the Standard Model of particle physics – any disagreement between the Standard Model and precise experimental measurements may be a hint of new physics beyond the Standard Model.

There is currently a discrepancy of more than $3\sigma$ between the theoretical Standard Model prediction[1] of the anomalous magnetic moment of the muon and the precise experimental measurement performed at the E-821 experiment at Brookhaven National Laboratory[2].

### *1.1. Anomalous Magnetic Dipole Moment of Muon*
The magnetic moment of the muon depends upon its mass $m$, charge $e$, spin $\vec{S}$, and gyromagnetic ratio $g_\mu$:

$$\vec{\mu} = g_\mu \frac{e\hbar}{2m_\mu c}\vec{S} \tag{1}$$

The gyromagnetic ratio for point-like Dirac particles is exactly $g_\mu = 2$ . Deviations from 2 are a result of radiative corrections to the lepton-photon vertex. The anomalous magnetic moment of the muon, $a_\mu$, is the deviation of the gyromagnetic ratio from the value $g_\mu = 2$, so that $a_\mu \equiv (g_\mu - 2)/2$.

The anomalous magnetic moment of the muon, $a_\mu$, has been measured experimentally at Brookhaven's E-821 to a precision of better than 1 part in a million[2], and is precisely predicted theoretically within the framework of the Standard Model.

The theoretical Standard Model prediction of $a_\mu$ has QED, weak and hadronic contributions:

$$a_\mu^{SM} = a_\mu^{QED} + a_\mu^{weak} + a_\mu^{hadLO,VP} + a_\mu^{hadHO,VP} + a_\mu^{had,LBL} \tag{2}$$

The theoretical prediction of $a_\mu$ is limited in precision by the hadronic contributions, and while they are small, they dominate the uncertainty in $a_\mu$. The QED, weak and hadronic contributions to $a_\mu$ are depicted in Figure 1 and their predicted values are listed in Table 1. The QED contribution, $a_\mu^{QED}$, is the dominant contribution to $a_\mu^{SM}$. It has been calculated to 10th order in perturbation theory[3] and is very precisely known with a tiny uncertainty. The weak correction, $a_\mu^{weak}$, is very small and precisely calculable, also with a small uncertainty. The hadronic corrections due to leading order hadronic vacuum polarization, $a_\mu^{hadLO,VP}$, higher order hadronic vacuum polarization, $a_\mu^{hadHO,VP}$, and hadronic light-by-light scattering, $a_\mu^{had,LBL}$, cannot be precisely calculated perturbatively at low energy. The leading order hadronic vacuum polarization, $a_\mu^{hadHO,VP}$ dominates the theoretical uncertainty in $a_\mu^{SM}$, as seen in Table 1. In the future, lattice QCD may make precise predictions of this term, and in Section 3 we describe an alternate method using experimentally measured hadronic cross-sections to decrease the uncertainty in the hadronic vacuum polarization term.

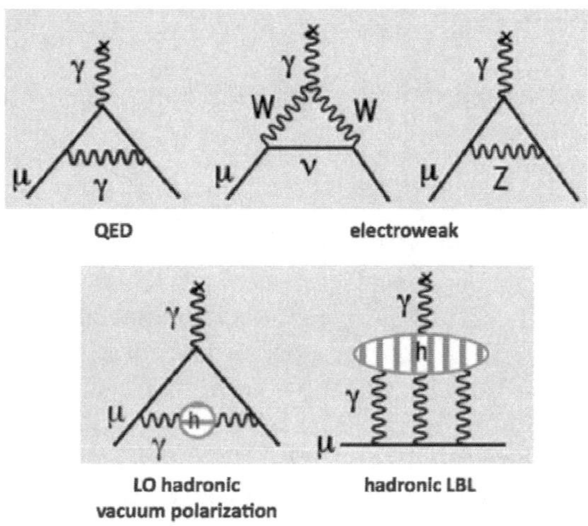


**Figure 1.** QED, weak, and hadronic contributions to $a_\mu^{SM}$. The hadronic vacuum polarization contribution is the dominant contribution to the uncertainty in $a_\mu^{SM}$.

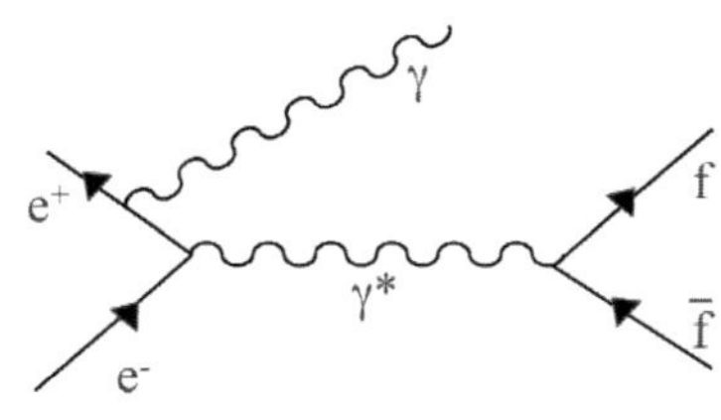

**Figure 2.** The Feynman diagram for $e^+e^- \to f\bar{f}\gamma$ with a photon radiated from the initial state is the basis for the ISR method to study $e^+e^- \to f\bar{f}$ at lower effective center-of-mass energies.

Any discrepancy between the theoretical Standard Model prediction of $a_\mu$ and experimental measurement would be an indication of new physics beyond the Standard Model.

## 2. Direct Measurement of the Anomalous Magnetic Moment of the Muon

The most precise direct measurement of $a_\mu$ was performed at experiment E-821[2] at Brookhaven National Laboratory. The spin precessions of 100 billion $\mu^+$ and $\mu^-$ were measured in a uniform magnetic field, circulating in a storage ring, yielding a 3.6 $\sigma$ discrepancy with the SM prediction:

$$a_\mu = (116\ 592\ 091 \pm 54_{stat} \pm 33_{syst}) \times 10^{-11} \quad \text{(experimentally measured)} \tag{3}$$
$$a_\mu^{SM} = (116\ 591\ 803 \pm 49) \times 10^{-11} \quad \text{(SM theoretical prediction)}$$

The contributions and uncertainties in the Standard Model prediction of $a_\mu$ are listed in Table 1, along with the BNL E-821 measurement[2].

**Table 1.** Contributions to the Standard Model prediction[4] for $a_\mu$ ($\times 10^{-11}$) and comparison with experimental results .

| | | | |
|---|---|---|---|
| QED | 11658478.95 | $\pm$ | 0.08 |
| Weak | 154. | $\pm$ | 1. |
| **Leading Order Hadronic vacuum polarization** | **6923 .** | $\pm$ | **42 .** |
| Higher Order Hadronic vacuum polarization | -98. | $\pm$ | 1. |
| Hadronic light-by-light scattering | 105. | $\pm$ | 26. |
| Theory - Standard Model | 116591803. | $\pm$ | 49. |
| Experiment - E-821 BNL | 116592091. | $\pm$ | 54. $\pm$ 33. |
| Discrepancy (Theory-Experiment) | 288. | $\pm$ | 80. |

## 3. Initial State Radiation and Hadronic Cross-Section Measurements

The dominant contribution to the uncertainly in the Standard Model prediction of $a_\mu^{SM}$ arises from the leading order hadronic vacuum polarization term, $a_\mu^{hadLO,VP}$, which cannot be calculated precisely at low energies.

Fortunately, the precision on the prediction of the hadronic vacuum polarization term may be improved using measured low-energy $e^+e^-$ hadronic cross-sections over a range of effective $e^+e^-$ center-of-mass energies, $\sqrt{s'}$ in a dispersion relation[5]:

$$a_\mu^{hadLO,VP} = \frac{1}{4\pi^3}\int_0^\infty \sigma(e^+e^- \to \gamma^* \to hadrons)(s)\, K(s')\, ds' \tag{4}$$

The kernel $K(s')$[6] and $\sigma(e^+e^- \to \gamma^* \to hadrons)(s')$ each have a $1/s'$ dependance, so low-energy contributions to $a_\mu^{hadLO,VP}$ will dominate. The high-energy part of the integral, above 1.8 GeV, is computed using perturbative QCD, while the lower energy piece, up to 1.8 GeV, is obtained using experimental cross-section measurements.

As seen in Equation 4, the hadronic vacuum polarization term requires integration over a range of $e^+e^-$ center-of-mass energies. A convenient method to effectively vary the center-of-mass energy is the initial state radiation(ISR) method. An ISR event $e^+e^- \to f\overline{f}\gamma$ is shown in Figure 2. The process $e^+e^- \to f\overline{f}\gamma$ with an initial state radiative photon allows us to measure the $e^+e^- \to f\overline{f}$ cross-sections over a range of lower effective $e^+e^-$center-of-mass energies. The ISR process effectively gives a variable center-of-mass energy, without ever having to vary the beam energies of the accelerator. The virtual photon vertex in Figure 2 has contributions from QED, weak, and hadronic processes, as illustrated in Figure 1.

New physics beyond the Standard Model could make an additional vertex contribution to $a_\mu$, and hence result in a discrepancy between the SM prediction and experimental measurement of $a_\mu$. The $(g_\mu - 2)$ of the muon is much more sensitive to new physics than the electron's $(g_e - 2)$; the sensitivity $\propto (m_e/m_\mu)$, so $(g_\mu - 2)$ is a very good probe for new physics.

## 4. *BaBar* Hadronic Cross-Section Measurements Using the ISR Method at *BaBar*

The *BaBar* detector at the Stanford Linear Accelerator Center (SLAC), on the PEP-II $e^+e^-$storage ring, collides 9 GeV electrons on 3.1 GeV positrons. The *BaBar* detector is described elsewhere[7]. Using the *BaBar* detector, we have performed ISR hadronic cross-section measurements and hence critical tests of the muon's $(g_\mu - 2)$ in the Standard Model. Until a few years ago, low energy hadronic cross-section data was dominated by the work at the VEPP machines, in the 1-2 GeV center-of-mass energy region, at Budker Institute for Nuclear Physics.

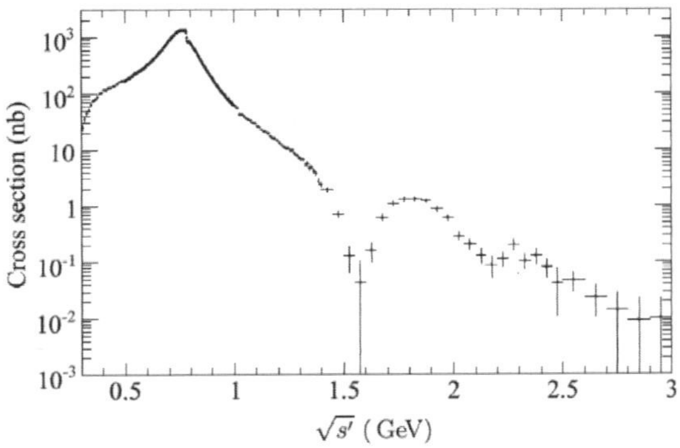


**Figure 3.** The measured cross section for $e^+e^- \to \pi^+\pi^-(\gamma)$ for effective center-of-mass energies, $\sqrt{s'}$, ranging from 0.3 to 3 GeV.[8]

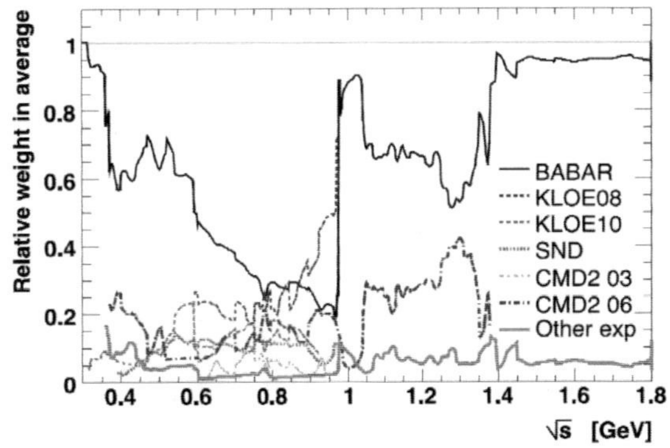


**Figure 4.** Relative local averaging weight per experiment versus effective center-of-mass energy for $e^+e^- \to \pi^+\pi^-$ [8][10][11][12]

*BaBar* contributes significantly to a large number of exclusive ISR cross-section measurements at and just above the VEPP energies, from threshold up to 3-5 GeV, depending on backgrounds of each exclusive hadronic channel. *BaBar* hadronic cross-section measurements are used to investigate topics ranging from spectroscopy to form-factors, in addition to these $(g_\mu - 2)$ studies.

In *BaBar* ISR cross-section analyses, the initial state radiated photon is detected at a large angle, within the sensitive volume of the detector, with energy $\geq 3$ GeV. The topology is such that the ISR photon is roughly back-to-back with the produced hadrons in the event, resulting in high acceptance and good particle identification. Kinematic fits reduce the backgrounds and give good energy resolution. Control samples are used to determine backgrounds, corrections and efficiencies. ISR event reconstruction and selection may be found in [8].

### *4.1. The $e^+e^- \to \pi^+\pi^-(\gamma)$ Channel*

The $\pi\pi$ ISR channels account for approximately 75% of the contributions to $a_\mu^{had}$. The ISR method is used at *BaBar* to measure the $\pi^+\pi^-$ cross-section from the $m_{\pi\pi}$ threshold up to 3 GeV, allowing the possibility of an additional radiated photon, denoted $(\gamma)$. The $\pi^+\pi^-(\gamma)$ cross-section is shown in Figure 3. The $\rho$ resonance, $\rho - \omega$ interference and other structures are evident. The *BaBar* $\pi^+\pi^-$ analysis yields a cross-section measurement which is used to compute the contribution of the $\pi^+\pi^-$ mode to the theoretical prediction of $a_\mu^{SM}$ using the dispersion integral in Equation 4 to obtain:

$$a_\mu^{\pi^+\pi^-(\gamma),LO}[\text{from threshold to 1.8 GeV}] = (514.1 \pm 2.2_{stat} \pm 3.1_{syst}) \times 10^{-10}[9]. \quad (5)$$

The contribution to $a_\mu^{\pi^+\pi^-}$ due to final state radiation is $a_\mu^{\pi^+\pi^-,FSR} = (0.26 \pm 0.12) \times 10^{-10}$. The weights of different experiments in combining results for $\pi^+\pi^-$ contributions to $a_\mu^{\pi^+\pi^-}$ are shown in Figure 4. *BaBar* dominates at all energies shown, except in the 0.75-0.93 GeV region where the KLOE results dominate.[1]. A summary of the leading order hadronic $a_\mu^{\pi\pi}$ contributions are in Figure 5, determined using $e^+e^-$cross-section measurements, as well as results using $\tau$ data.

### *4.2. The $e^+e^- \to K^+K^-(\gamma)$ Channel*

Using the same selection and methods as for the $\pi^+\pi^-$ channel, and subtracting the $J/\psi$ and $\psi(2S)$ resonances, we measured the $K^+K^-(\gamma)$ cross-section for effective center of mass energies

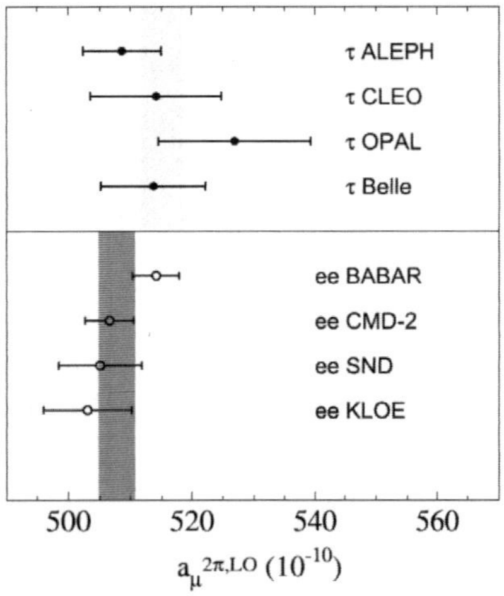


**Figure 5.** Leading order hadronic vacuum polarization $2\pi$ contributions to $a_\mu$, evaluated from threshold to 1.8 GeV, for $\tau$ and $e^+e^-$data [8]

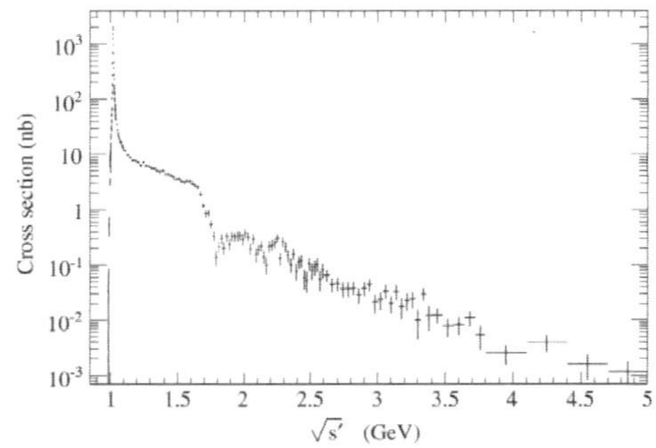


**Figure 6.** The measured cross section for $e^+e^-\to$ $K^+K^-(\gamma)$ for effective center of mass energies, $\sqrt{s'}$, including final state radiation.[13]

ranging from threshold to 5 GeV[13], shown in Figure 6. Final state radiation is measured and included, but is negligible. The *BaBar* $K^+K^-$ analysis yields a cross-section which can be used to compute the contribution of the $K^+K^-$ mode to the theoretical prediction of $a_\mu^{SM}$. The result of the dispersion integral (Equation 4) from threshold to 1.8 GeV is:

$$a_\mu^{K^+K^-(\gamma),LO}[\text{from threshold to 1.8 GeV}] = (22.93\pm0.18_{stat}\pm0.22_{syst}\pm0.03_{VP})\times10^{-10}[13] \quad (6)$$

### *4.3. Cross-sections of Other Channels Contributing to* $a_\mu$

The *BaBar* collaboration has measured many other low-energy single channel cross-sections using the ISR method, all relevant for the determination of the hadronic corrections to the $(g_\mu - 2)$ anomaly. These include recent measurements of $e^+e^-\to\pi^+\pi^-$, $K^+K^-$, $\pi^+\pi^-\pi^+\pi^-$, $K^+K^-K^+K^-$, $p\bar{p}$, $K^0_SK^0_L$, $K^0_SK^0_L\pi^+\pi^-$, $K^0_SK^0_S\pi^+\pi^-$, $K^0_SK^0_SK^+K^-$, $\pi^+\pi^-\pi^0\pi^0$, as well as other channels, as shown in Figure 7.[8, 13, 14, 15, 16, 17, 18, 19, 20, 21, 22]

## 5. Summary and Outlook

Hadronic cross-sections measurements using the ISR method have been performed at *BaBar* for many $e^+e^- \to hadrons$ channels, providing important input for the theoretical prediction of the anomalous magnetic moment of the muon. Measurements performed at *BaBar* have reduced the uncertainty in the hadronic contributions to the predicted $a_\mu$, but a $3\sigma$ discrepancy between the experimental measurements at BNL E-821 and the Standard Model prediction for $a_\mu$ persists. With new $(g_\mu - 2)$ experiments coming online at FNAL (see talk by J. Holzbauer, in these proceedings) and JPARC, improvements in the precision of Standard Model prediction for $(g_\mu - 2)$ is crucially important. BESIII also has new results on this topic (see talk by C. Redmer in these proceedings)

## References

[1] M. Davier, A. Hoecker, B. Malaescu and Z. Zhang, *Eur. Phys. J.* **C71** (2011) 1515. [*Erratum - Eur. Phys. J.* C **72** (2012) 1874].
[2] G.W.Bennett *et al.*(Muon g-2 Collaboration), *Phys. Rev.* **D73** 072003 (2006).

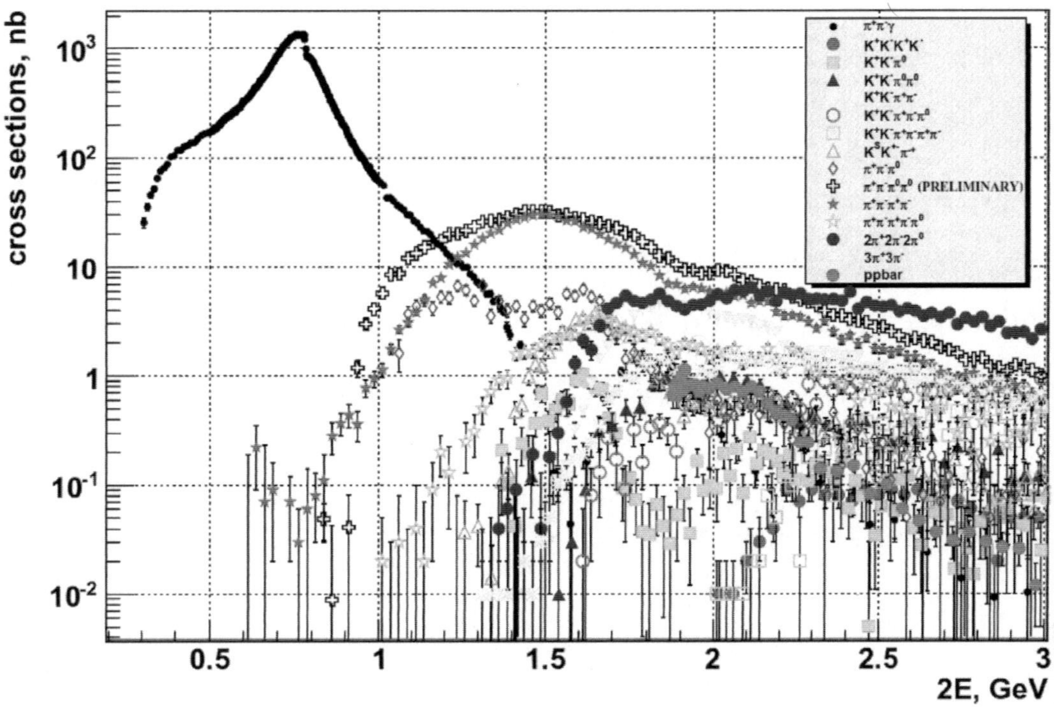


**Figure 7.** $e^+e^- \to$ hadronic cross-sections measured with the *BaBar* detector via ISR.[22]

[3] T.Aoyama, M.Hayakawa, T.Kinoshita, M.Nio, *Phys. Rev. Lett* **109**111807 (2012).
[4] A.Hoecker and W.J. Marciano, in K.A. Olive et al. (Particle Data Group), *Chin. Phys.* **C38** 090001 (2014) (URL: http://pdg.lbl.gov).
[5] F.J.M Farley and E. Picasso, *Ann. Rev. Nucl. and Part. Sci.* **29** 243-282 (1979).
[6] Stanley J. Brodsky and Eduardo de Rafael, *Phys. Rev.* **168**, 1620 (1968).
[7] B.Aubert *et al.* (The *BABAR* Collaboration), *Nucl. Inst. Meth.* **A729** 615-701 (2013).
[8] B. Aubert *et al.* (*BABAR* Collaboration), *Phys. Rev. Lett.* **103** 231801 (2009); J. P. Lees *et al.* (*BABAR* Collaboration), *Phys Rev.***D86** 032013 (2012).
[9] J.P.Lees *et al.* (*BABAR* Collaboration), *Phys Rev.* **D92** 072015 (2015).
[10] F. Ambrosino *et al.* (KLOE Collaboration), *Phys. Lett.* **B 670** 285(2009); F. Ambrosino *et al.* KLOE Collaboration), *Phys. Lett.* **B700** 102 (2011).
[11] V.M. Aulchenko *et al.* (CMD2 Collaboration), *JETP Lett.* **82** 743 (2005); R.R. Akhmetshin *et al.* (CMD2 Collaboration), *JETP Lett.* **84** 413 (2006); R.R. Akhmetshin *et al.* (CMD2 Collaboration), *Phys. Lett.* **B 648** 28 (2007).
[12] M.N. Achasov *et al.* (SND Collaboration), *JETP Lett.* **103** 380 (2006).
[13] J.P.Lees *et al.* (*BABAR* Collaboration), *Phys Rev.* **D88** 032013 (2013).
[14] B.Aubert *et al.* (*BABAR* Collaboration), *Phys Rev.* **D74** 091103 (2006); B.Aubert *et al.* (*BABAR* Collab.), *Phys Rev.* **D76** 012008 (2007).
[15] B.Aubert *et al.* (*BABAR* Collaboration), *Phys Rev.* **D70** 072004 (2004).
[16] B.Aubert *et al.* (*BABAR* Collaboration), *Phys Rev.* **D77** 092002 (2008).
[17] B.Aubert *et al.* (*BABAR* Collaboration), *Phys Rev.* **D71** 052001 (2005); B.Aubert *et al.* (*BABAR* Collaboration), *Phys Rev.* **D76** 012008 (2007); J.P.Lees *et al.* (*BABAR* Collaboration), *Phys Rev.* **D85** 112009 (2012).
[18] B.Aubert *et al.* (*BABAR* Collaboration), *Phys Rev.* **D76** 0092005 (2007).
[19] B.Aubert *et al.* (*BABAR* Collaboration), *Phys Rev.* **D73** 052003 (2006).
[20] B.Aubert *et al.* (*BABAR* Collaboration), *Phys Rev.* **D73** 012005 (2006).
[21] B.Aubert *et al.* (*BABAR* Collaboration), *Phys Rev.* **D76** 0092006 (2007).
[22] V.P. Druzhinin *et al.* , *Rev. Mod. Phys.* **83** 1545 (2011).

Journal of Physics: Conference Series **770** (2016) 012038 doi:10.1088/1742-6596/770/1/012038

# The Muon g-2 Experiment Overview and Status as of June 2016

**J Holzbauer**

University of Mississippi, University, Mississippi 38677, USA

E-mail: jholzbau@fnal.gov

**Abstract.** The Muon g-2 Experiment at Fermilab will measure the anomalous magnetic moment of the muon to a precision of 140 parts per billion, which is a factor of four improvement over the previous E821 measurement at Brookhaven. The experiment will also extend the search for the electric dipole moment (EDM) of the muon by approximately two orders of magnitude, with a sensitivity down to $10^{-21}$ e.cm. Both of these measurements are made by combining a precise measurement of the 1.45T storage ring magnetic field with an analysis of the modulation of the decay rate of higher-energy positrons (from anti-muons), recorded by 24 calorimeters and 3 straw tracking detectors. The recent progress in the alignment of the electrostatic quadrapole plates and the trolley rails inside the vacuum chambers, and in establishing the uniform storage ring magnetic field will be described.

## 1. Introduction

In 2006, an experiment at BNL called E821 reported a final, precisely measured value of g-2 that was approximately 3 sigma away from the theoretical value [1]. In recent years, this tension has remained despite improvements in the theoretical calculation [2]. To help resolve this tension, two new experiments are being built or planned. One is E34, an ultra-cold muon beam experiment in Japan [3, 4]. The other is an improved version of E821, the Muon g-2 Experiment [5, 2], which has reference number E989 and is located at Fermilab in the US. In this proceedings, we will discuss the physics of g-2 and the current standard model (SM) value, the Muon g-2 Experiment procedure and setup, and progress in assembling the new experiment.

## 2. Standard Model and Current Experimental Value

The quantity g is a constant which relates particle spin and magnetic moment. In Dirac theory, the value of g is exactly 2 [6]. However, higher order effects will alter this value. For example, new physics which interacts weakly, such as a new heavy W, could contribute in this way in SUSY models [7]. The muon is studied rather than the electron in this experiment because the muon is much heavier and thus is more sensitive to higher order corrections but still long lived enough to study. The experiment specifically studies what is known as the anomaly, which is $a_\mu$ =(g-2)/2.

The standard model theory calculation breaks down into four main components: quantum electrodynamics (QED) [8], electroweak (EW) [9], hadronic vacuum polarization (HVP) [10, 11] and hadronic light-by-light (HLbL) [12]. The first two, the QED and EW components are well known with relatively small uncertainties. As can be seen in Table 1, the QED component is

Journal of Physics: Conference Series **770** (2016) 012038 doi:10.1088/1742-6596/770/1/012038

the largest contribution to the standard model value but has the smallest uncertainty, while the hadronic terms are less well known. The contribution from hadronic vacuum polarization can be obtained from e+e- to hadrons or tau to hadrons measurements. The hadronic light-by-light term cannot be obtained directly from data, though data can be used indirectly to constrain the contribution. Rather, the value is obtained diagram or lattice calculations.

**Table 1.** Standard model components of the anomaly, taken directly from [2]. Two values are shown for HVP to reflect two recent estimates. The terms lo and ho indicate lower order and higher order, respectively. Other terms are defined in the text.

| | Values in $10^{-11}$ units |
|---|---|
| QED ($\gamma + l$) | $116584718.951 \pm 0.009 \pm 0.019 \pm 0.007 \pm 0.077$ |
| HVP(lo) [10] | $6923 \pm 42$ |
| HVP(lo) [11] | $6949 \pm 43$ |
| HVP(ho) [11] | $-98.4 \pm 0.7$ |
| HLbL | $105 \pm 26$ |
| EW | $153.6 \pm 1.0$ |
| Total SM [10] | $116591802 \pm 42_{\mathrm{H-LO}} \pm 26_{\mathrm{H-HO}} \pm 2_{\mathrm{other}}$ ( $\pm 49_{\mathrm{tot}}$) |
| Total SM [11] | $116591828 \pm 43_{\mathrm{H-LO}} \pm 26_{\mathrm{H-HO}} \pm 2_{\mathrm{other}}$ ( $\pm 50_{\mathrm{tot}}$) |

The value of the anomaly from the BNL experiment [1], corrected for updated constants [13] and given in [2], is $a_\mu^{E821} = 116592089 \pm 63 \times 10^{-11}$ (54 ppm). The new experiment is expected to reduce this uncertainty by a factor of four to 0.14 ppm, which would give a 5 sigma deviation from the standard model calculation, assuming the central value remains the same for both the standard model calculation and the measurement. New data inputs and improvements to the lattice calculations are expected to reduce the hadronic contribution uncertainties (particularly HVP lo), which are the largest sources of theoretical uncertainty. If the standard model uncertainty improves in addition to the experimental uncertainty, the deviation could be around 8 sigma under these same central value assumptions [2].

## 3. The Muon g-2 Experiment Plans and Status

The Muon g-2 Experiment is located at Fermilab. It reuses the BNL storage ring, which was moved carefully from BNL to Fermilab. Most parts, like vacuum chambers, could be shipped by normal means, but the 15 ton cryostat ring was trickier. It is large, heavy and the superconducting coils cannot flex more than approximately 3 mm. This portion was moved by the Emmert Corportation, which used a barge and a large truck. In 2015, the magnet and related cryo-systems were tested, and the magnet was cooled and powered at Fermilab. The required 1.45 Tesla field was achieved and the transportation was deemed a success.

The muons for the storage ring are produced by the Fermilab accelerator complex. The source protons are produced and moved by reused Tevatron equipment, and they collide with a lithium target to produce pions. These pions enter a long delivery ring where they decay to muons and neutrinos. The polarized, positively charged muon beam is then injected into the storage ring. The long delivery ring helps to reduce proton/pion contamination, which was an issue at BNL. The new experiment expects a factor of 20 increase in muon statistics over the BNL experiment and which will reduce the statistical error to 0.1 ppm.

In brief, the experiment consists of a cryo-system surrounding dipole magnets, which contain vacuum chambers with electrostatic quadrapole plates. The magnets and quadrapole plates

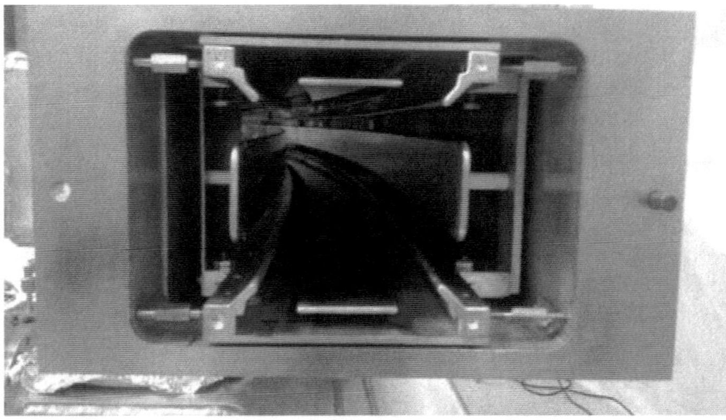

**Figure 1.** View of the cage with trolley rails and (most interior) four electrostatic quadrapole plates, from a vacuum chamber end flange.

work in tandem to keep the muons in the proper location. There are other beam related parts like the inflector, kickers, collimators and a beam monitoring system, as well as a trolley that rides around inside of the ring to measure the field. Detector systems include both straw trackers and lead crystal calorimeters. Trackers were not used in the BNL experiment and should add valuable information to the new g-2 data. Additionally, the calorimeters now have multiple read out channels to reduce pileup.

The measurement consists of two quantities, $\omega_a$ and $\omega_p$. The term $\omega_a$ is the precession frequency measured with high energy decay positrons and $\omega_p$ is the magnetic field (B) normalized to the proton lamour frequency. The spin and cyclotron frequencies can be combined to give the anomaly in terms of $\omega_a$, B, and the charge over mass ratio (e/m). This can be rewritten in terms of $\omega_a/\omega_p$ and the muon, proton magnetic moment ratio from hyperfine splitting. There is an additional electric field term which is removed by choosing the appropriate momenta of the muons, 3.09 GeV/c, known colloquially as the magic momentum. Not all muons will have exactly this momentum, which introduces some uncertainties and corrections into the analysis. Alignment efforts to ensure conformity of the muons are particularly important.

### *3.1. Status of the Ring Installation*

The vacuum chambers contain the electrostatic quadrapole plates, as mentioned earlier. These plates are mounted to a cage structure which contains rails the B field measurement trolley will ride on, see Figure 1. The quadrapole plates are aligned to the design value to within $\pm 0.5$ mm (top/bottom) or $\pm 0.75$ mm (sides) over the length of the chamber, with larger $\pm 2.0$ mm deviations allowed on shorter length scales. The cage rails are aligned to within $\pm 0.5$ mm vertically to allow the trolley to ride smoothly, and the cage (and plates) is positioned radially to be centered on the magic radius, 7112 mm, the location where muons will have the magic momentum value. The vertical alignment is done using a Hamar laser system along with a retroreflector, which allows a final measurement of the vertical alignment in vacuum, through a clear plexiglass flange. Digital calipers and micrometers are used for quadrapole plate alignment, for absolute vertical alignment before the use of the relative laser system, and for the radial alignment.

Journal of Physics: Conference Series **770** (2016) 012038 doi:10.1088/1742-6596/770/1/012038

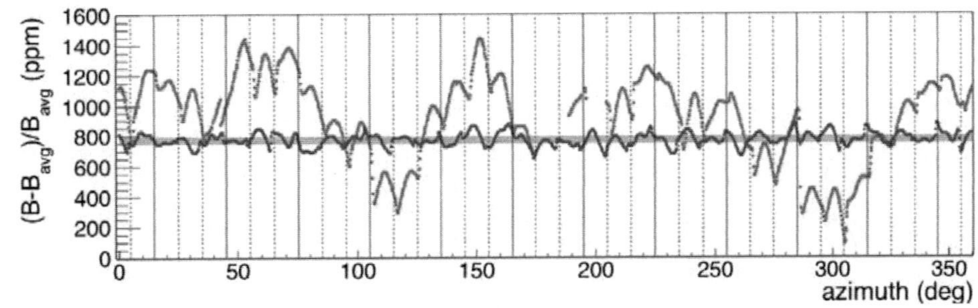


**Figure 2.** Magnetic field variation versus azimuth for the first survey (red) and a more recent June 2016 survey (blue) with the variation goal shown in a magenta band. Additional improvements to the field uniformity will continue through August 2016.

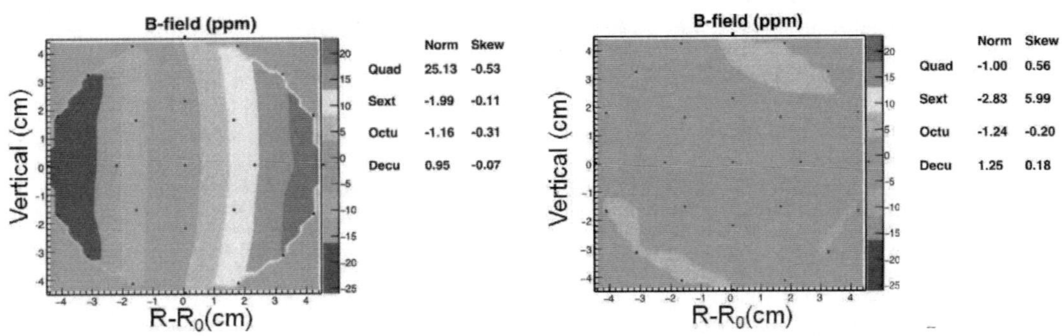


**Figure 3.** Azmuthally averaged magnetic field variation for the first survey (left) and a more recent June 2016 survey (right). Additional improvements to the field uniformity will continue through August 2016.

### *3.2. Status of the Magnetic Field Uniformity*

To produce a uniform magnetic field, iron shims are strategically added throughout the ring. The field is measured and shims are added iteratively. Measurements use a shimming cart containing 25 nuclear magnetic resonance (NMR) probes and 4 capacitative gap sensors as well as 4 position sensors to give the cart location in r, $\theta$, and z.

The initial field measurements when the magnet was first powered up were similar to those at BNL when it was first powered up. A variation was found of $\pm700$ ppm in the field vs azimuth and $\pm25$ ppm in the azimuthally averaged field. As of June 2016, the variation of the field has been reduced to $\pm200$ ppm (RMS of about 40 ppm), as shown in Figure 2 and the azimuthally averaged field variation is reduced to $\pm6$ ppm, as shown in Figure 3. These are much closer to the target field variations of $\pm25$ ppm and $\pm < 1$ ppm, even before the planned laminated shims are applied. This shim application is expected to finish in August of 2016.

## 4. Summary

The Muon g-2 Experiment at Fermilab is well under way in terms of its construction and assembly. It is expected that the new experiment will improve the uncertainty on the measured value versus the previous measurement by a factor of four. The collaboration is very active and working hard to install the various ring components to ensure a uniform B field and correct muon location. Detector systems will be installed soon as well and data taking is expected in 2017, with the potential to resolve the tension of the previous result and the standard model calculation.

## References

[1] Bennett G *et al.* (Muon G-2 Collaboration) 2006 *Phys.Rev.* **D73** 072003 (*Preprint* hep-ex/0602035)
[2] Grange J *et al.* (Muon g-2) 2015 (*Preprint* 1501.06858v1)
[3] Aoki M *et al.* 2011 *JPARC E34 conceptual design report* URL http://g-2.kek.jp/portal/documents.html
[4] Otani M (E34) 2015 *JPS Conf. Proc.* **8** 025008
[5] Carey R, Lynch K, Miller J, Roberts B, Morse W *et al.* 2009
[6] Dirac P 1928 *Proc. R. Soc. (London)* **A117, 610 and A118, 351**
[7] Albrecht J *et al.* (Intensity Frontier Charged Lepton Working Group) 2013 Working Group Report: Charged Leptons *Proceedings, Community Summer Study 2013: Snowmass on the Mississippi (CSS2013): Minneapolis, MN, USA, July 29-August 6, 2013* (*Preprint* 1311.5278) URL https://inspirehep.net/record/1265506/files/arXiv:1311.5278.pdf
[8] Aoyama T, Hayakawa M, Kinoshita T and Nio M 2012 *Phys. Rev. Lett.* **109**(11) 111807 URL http://link.aps.org/doi/10.1103/PhysRevLett.109.111807
[9] Gnendiger C, Stckinger D and Stckinger-Kim H 2013 *Phys. Rev.* **D88** 053005 (*Preprint* 1306.5546)
[10] Davier M, Hoecker A, Malaescu B and Zhang Z 2011 *Eur. Phys. J.* **C71** 1515 [Erratum: Eur. Phys. J.C72,1874(2012)] (*Preprint* 1010.4180)
[11] Hagiwara K, Liao R, Martin A D, Nomura D and Teubner T 2011 *J. Phys.* **G38** 085003 (*Preprint* 1105.3149)
[12] Prades J, de Rafael E and Vainshtein A 2009 *Adv. Ser. Direct. High Energy Phys.* **20** 303–317 (*Preprint* 0901.0306)
[13] Mohr P J, Taylor B N and Newell D B 2008 *Rev. Mod. Phys.* **80**(2) 633–730 URL http://link.aps.org/doi/10.1103/RevModPhys.80.633

# The PADME experiment at INFN LNF

**S Fiore for the PADME Collaboration**[1]
INFN, Sezione di Roma and ENEA, Department of Fusion and Technology for Nuclear Safety and Security, Rome, Italy

E-mail: salvatore.fiore@roma1.infn.it

**Abstract.** At the end of 2015, INFN formally approved a new experiment, PADME, to search for invisible decays of the $A'$ dark photon at the DAFNE Linac in Frascati. The experiment is designed to detect the $A'$ produced in positron-on-fixed target annihilation, by measuring the final state missing mass. The collaboration aims to complete the design and construction of the experiment by the end of 2017 and to collect $\sim 10^{13}$ positrons on target by the end of 2018, thus reaching a sensitivity on the $\varepsilon$ coupling constant of $\sim 10^{-3}$ and on the dark photon mass up to $\sim$ 23.7 MeV/c$^2$

## 1. Introduction

The long standing problem of reconciling the cosmological evidence of the existence of dark matter with the lack of any clear experimental observation of it has recently revived the idea that these new particles are not directly connected with the Standard Model (SM) gauge fields, but only through mediator fields or portals, connecting our world with new secluded or hidden sectors. One of the simplest models just adds an additional U(1) symmetry, with its corresponding vector boson $A'$. All SM particles will be neutral under this symmetry, while the new field will couple to the charged particles of the SM with an effective charge $\varepsilon e$. For this reason the $A'$ is also called Dark Photon (DP)[1, 2]. Additional interest arises from the observation that $A'$ in the mass range 1 MeV/c$^2$ to 100 MeV/c$^2$ and coupling $\varepsilon \sim 10^{-3}$, would explain the discrepancy between theory and observation for the muon anomalous magnetic moment, $(g-2)_\mu$ [3].

If there are no particles in the hidden sector with mass smaller than one half that of the $A'$ boson, the dark photon can only have SM decays (visible decays). Currently, the region of the $\varepsilon$, $m_{A'}$ plane favored by the $(g-2)_\mu$ discrepancy is excluded for an $A'$ boson decaying into SM particles (see Fig.1 left). In the most general case the $A'$ can decay into Dark Matter (DM) invisible decays. In this scenario, there are still unexplored regions in the $(g-2)_\mu$ favored band, as shown in Fig.1 right.

A comprehensive overview of the experimental programs in this field is presented in [6].

[1] The PADME collaboration is: G. Chiodini (INFN Lecce), A.P. Caricato, M. De Feudis, M. Martino, G. Maruccio, A. Monteduro, S.Spagnolo (INFN Lecce e Dip. di Matematica e Fisica, Università del Salento), R. Bedogni, B. Buonomo, F. Bossi, R. De Sangro, G. Finocchiaro, L.G. Foggetta, A. Ghigo, M. Palutan, G. Piperno, B. Sciascia, T. Spadaro (INFN Laboratori Nazionali di Frascati), G.V. Georgiev, V. Kozhuharov (University of Sofia e INFN Laboratori Nazionali di Frascati), U. Dosselli (INFN Padova), F. Ferrarotto, E. Leonardi, P. Valente (INFN Roma1), S. Fiore (INFN Roma1 e ENEA), G.C.Organtini, M.Raggi (INFN Roma1 e Dip. di Fisica, Università di Roma La Sapienza" )

Journal of Physics: Conference Series **770** (2016) 012039 doi:10.1088/1742-6596/770/1/012039

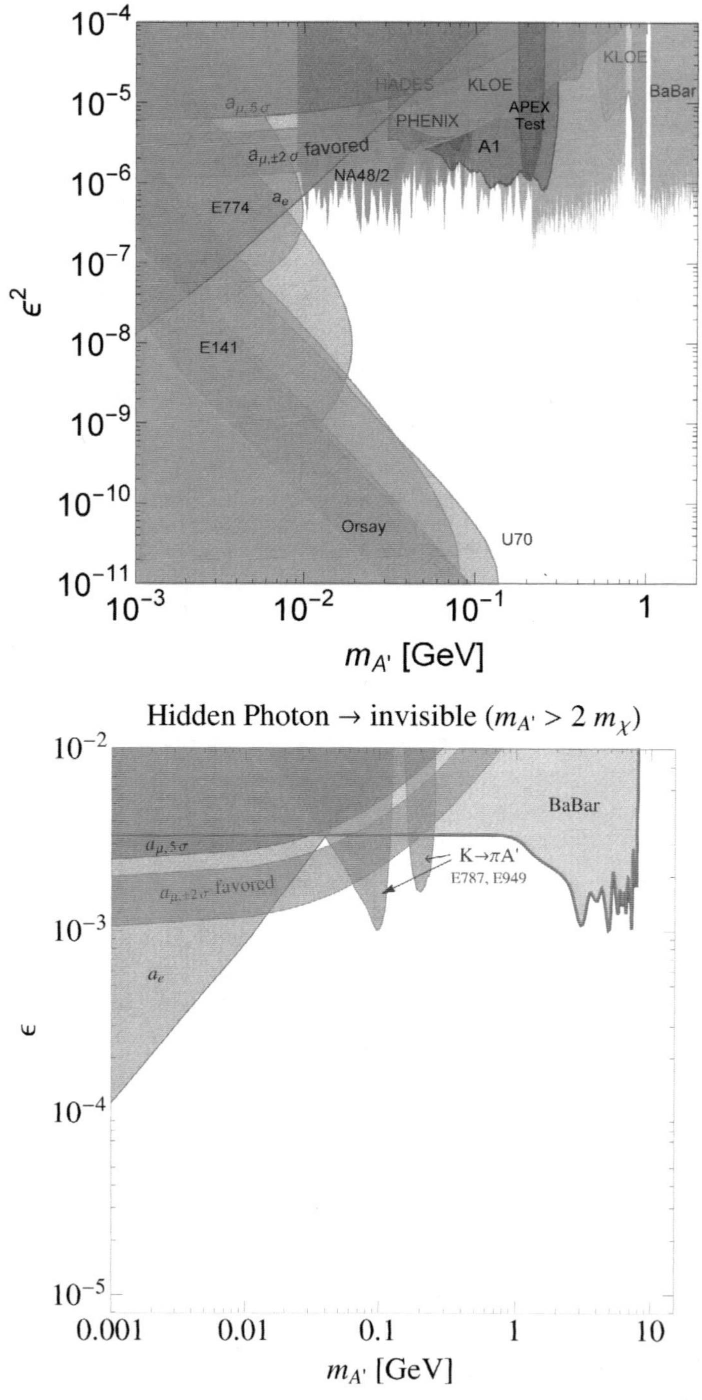


**Figure 1.** Current DP search status: for visible decays (top, adapted from [4]), for the invisible ones (bottom, adapted from [5]). Typical DP exclusion plots have the $A'$ mass on the x-axis and the coupling constant (squared) on the y-axis. In both cases the $2\sigma$ anomalous muon magnetic moment favored band is indicated.

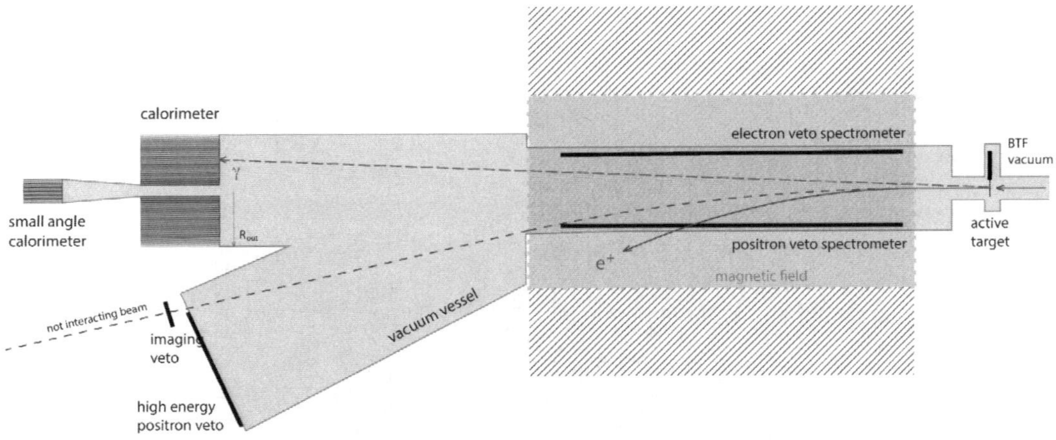


**Figure 2.** PADME detector layout. From right to left: the active target, the $e^+e^-$ vetoes inside the magnetic dipole, the high energy $e^+$ veto near the non-interacting beam exit, the ECAL and the SAC. The distance between the ECAL and the target is 3 m.

## 2. The PADME experiment

At the end of 2015, INFN formally approved a new experiment, PADME, to search for invisible decays of the $A'$ at the DAΦNE Linac of the Laboratori Nazionali di Frascati (LNF). PADME (Positron Annihilation into Dark Mediator Experiment) is designed to detect invisible decaying DPs produced in the process $e^+ e^- \to A' \gamma$, where the $e^+$ are accelerated by the DAΦNE LINAC to 550 MeV and interact with $e^-$ in a fixed target [7, 8].

## 3. The Frascati Beam Test Facility

PADME will be built in the Beam Test Facility (BTF) experimental hall, hosting a transfer-line from the DAΦNE LINAC [9]. The BTF is able to provide up to 50 bunches/s with a maximum energy of 550 and 800 MeV for positrons and electrons, respectively, and with duration (at constant intensity) ranging from 1.5 to 40 ns. The energy spread is 0.5%, while the beam spot size can vary by orders of magnitude: $[0.5, 25]$ mm (vertical) $\times$ $[0.6, 55]$ mm (horizontal). The number of particles that can be provided per bunch ranges from 1 to $10^{10}$. The BTF hall will be rearranged in order to host the PADME experiment.

## 4. The detector

The detector will identify events with a single photon generated in the $e^+e^-$ annihilation, measuring also the missing squared invariant mass of the final state. This will be evaluated by exploiting energy-momentum conservation and the fully constrained initial state of an $e^+$ beam, with known momentum and position, on an active fixed target. The $A'$ boson squared invariant mass $M^2_{miss}$ can be estimated as:

$$M^2_{miss} = \left(\vec{P}_{e^-} + \vec{P}_{beam} - \vec{P}_{\gamma}\right)^2,$$

$\vec{P}_{e^-} = \vec{0}$ and $P_{\vec{beam}} = 550$ MeV along the initial beam direction are the respective $e^-$ and $e^+$ momenta, and $\vec{P}_{\gamma}$ is the photon final state.

The detector, which is shown in Fig. 2, consists of [7]:

- a diamond active target, which allows to measure the beam intensity and position (precision of $\approx 5$ mm) by means of perpendicular graphite strips. The low Z of the diamond helps to

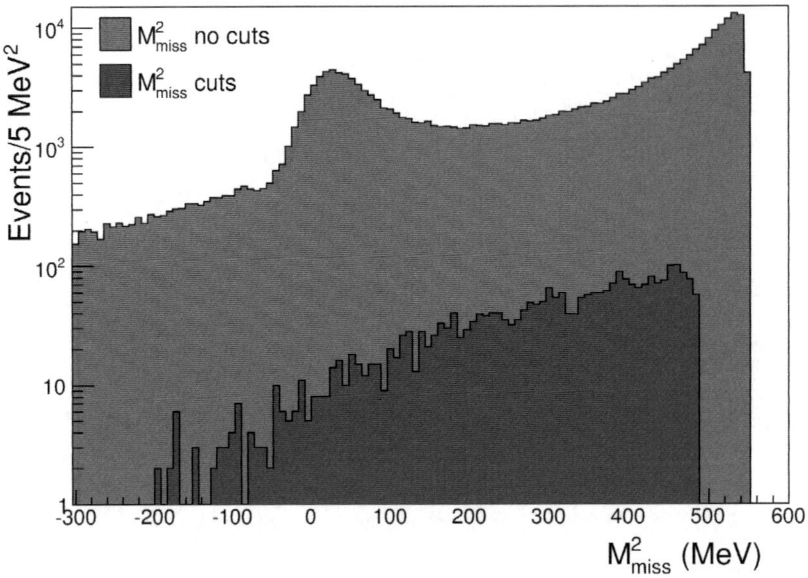


**Figure 3.** Background before (red) and after (blue) event selection [7].

reduce the occurrence of bremsstrahlung processes. The area is $2 \times 2\,\mathrm{cm}^2$ and the small thickness ($50\,\mu$m or $100\,\mu$m) reduces the probability of $e^+$ multiple interactions.

- a dipole magnet, located 20 cm after the target, designed to deflect non-interacting beam particles out of the detector and to direct the positrons that lost part of their energy (mainly through bremsstrahlung) toward the vetoes. The field is 0.5 T over a gap of 23 cm for 1 m length.
- positrons/electrons veto, divided in two parts: one located inside the dipole for positrons and electrons, and another one, near the beam exit, for high energy positrons that lost only a small part of their energy, mainly through bremsstrahlung processes. It is composed of $1 \times 1 \times 16\,\mathrm{cm}^3$ bars of plastic scintillators. The arrays inside the magnet are $\approx 1$ m long, while the high energy positron one is $\approx 0.5$ m long.
- an electromagnetic calorimeter (ECAL), made of 616 BGO crystals $2 \times 2 \times 22\,\mathrm{cm}^3$, placed at 3 m from the target. Energy resolution is expected to be $\sim \frac{(1-2)\%}{\sqrt{E}}$. The ECAL has cylindrical shape (30 cm radius) with a central square hole of 10 cm side to allow the bremsstrahlung radiation to pass through and reach the Small Angle Calorimeter. This is necessary because of the long BGO decay time of 300 ns: the ECAL would be continuously "blinded" by the bremsstrahlung photons rate. The angular coverage is $(20, 93)$ mrad.
- a small Angle Calorimeter (SAC), made of 49 $2 \times 2 \times 20\,\mathrm{cm}^3$ lead glass (SF57) crystals, located along the line of flight of the low-energy photons coming from the target and passing through the ECAL central hole. Its main task is to veto $3\gamma$ events. The lead glass Cerenkov response time makes it a good candidate for this task, being fast enough for the expected photon rate. The angular coverage is $(0, 20)$ mrad.

The DP signature is a single $\gamma$ in the ECAL with no other particles detected in the vetoes. Since $E_{beam} = 550$ MeV, the largest $A'$ boson mass produced in annihilation processes is 23.7 MeV.

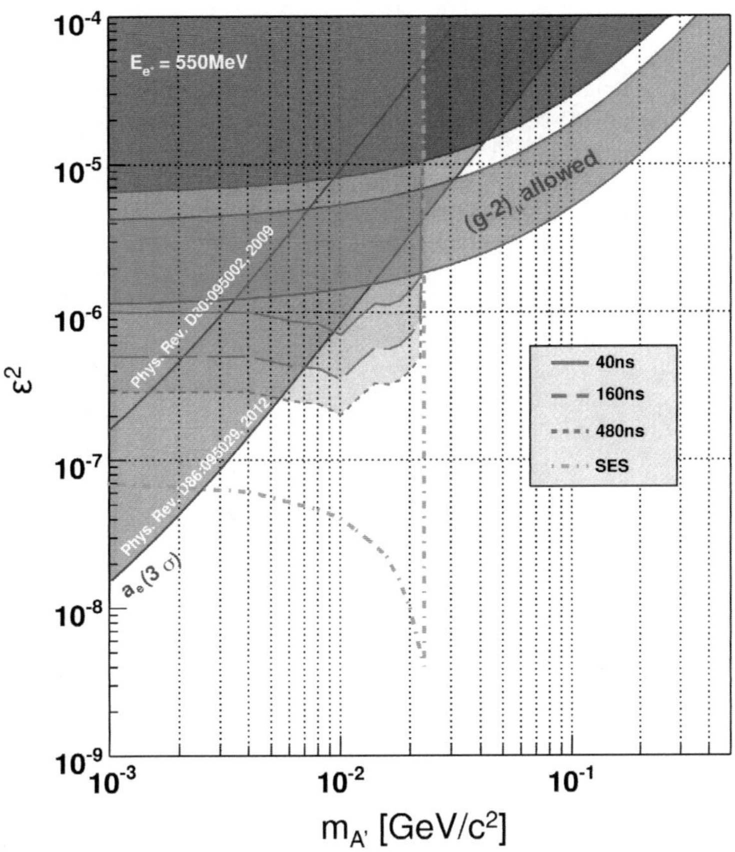


**Figure 4.** PADME sensitivity to $A' \to invisible$. Smaller values of the coupling constant $\varepsilon$ can be explored by increasing the bunch length. SES refers to single event sensitivity.

## 5. Backgrounds and sensitivity

The SM physical processes expected to take place in the $e^+$ beam interaction with the target are bremsstrahlung and $e^+e^- \to \gamma\gamma(\gamma)$ [7]. The probability that their kinematics will mimic a DP production event in our detector can be reduced through an optimization of the ECAL geometry and granularity and of the veto system. The beam intensity plays a crucial role for the pileup probability: clusters cannot be resolved in time by the calorimeter if they are temporally too close to each other [7]. Fig. 3 shows the background reduction obtained requiring only one cluster in the ECAL, with no hits in the vetoes, no photons with energy $> 50\,\mathrm{MeV}$ in the SAC, and cluster energy in a range optimized depending on $m_{A'}$.

The DP sensitivity evaluation is based on $2 \times 5 \cdot 10^{10}$ GEANT4 simulated 550 MeV positrons on target extrapolated to $10^{13}$ $e^+$. This number of particles can be obtained by running PADME for 2 years at 60% efficiency with 5000 $e^+$ per bunch (40 ns) at a repetition rate of 50 Hz. The obtained result for a DP decaying to invisible particles is shown in Fig. 4 for different bunch durations: the favored $(g-2)_\mu$ region can be explored in a model independent way (the only hypothesis on the DP is the coupling to leptons) up to masses of 23.7 MeV [7]. Single Event Sensitivity (SES) refers to the sensitivity in absence of background.

Journal of Physics: Conference Series **770** (2016) 012039 doi:10.1088/1742-6596/770/1/012039

## 6. Conclusions

Theoretical models with a Dark Photon can provide a solution to the Dark Matter puzzle. Additionally, a DP with mass in the $[1, 100]$MeV range and coupling constant $\varepsilon \sim 10^{-3}$ would explain the muon anomalous magnetic moment discrepancy.

PADME will perform a model independent search for an invisible decaying DP, using the accelerator complex at the INFN LNF. The collaboration aims to reaching a sensitivity on $\varepsilon$ of $\sim 10^{-3}$ for DP with masses up to 23.7 MeV.

## References

[1] B. Holdom, Phys. Lett. B **166**, 196 (1986).
[2] P. Galison and A. Manohar, Phys. Lett. B **136**,
[3] M. Pospelov, Phys. Rev. D **80**, 095002 (2009).
[4] B. Echenard, R. Essig and Y. M. Zhong, JHEP **1501**, 113 (2015).
[5] R. Essig *et al.*, JHEP **1311**, 167 (2013).
[6] M. Raggi and V. Kozhuharov, Riv. Nuovo Cim. **38**, 449 (2015).
[7] M. Raggi and V. Kozhuharov, AdHEP **2014**, 959802 (2014).
[8] M. Raggi, V. Kozhuharov and P. Valente, EPJ Web Conf. **96**, 01025 (2015).
[9] G. Mazzitelli *et al.*, Nucl. Instrum. Meth. A **515**, 524 (2003).

Journal of Physics: Conference Series **770** (2016) 012040 doi:10.1088/1742-6596/770/1/012040

# Dark Sector Searches at Jefferson Laboratory

**A Celentano**

INFN Sezione di Genova, Via Dodecaneso 33, 16146 Genova, IT

E-mail: andrea.celentano@ge.infn.it

**Abstract.**
The Jefferson Laboratory community recently joined the dark sector search program with different, complementary efforts: APEX, HPS, DarkLight, and BDX. These fixed-target, electron-beam experiments are designed to search for Dark Matter in the mass range 10 MeV/c$^2$ - 1 GeV/c$^2$, the so-called "light Dark Matter" region. Each experiment is optimized to look for a specific light Dark Matter signal signature.

In this paper, after a brief introduction to the light Dark Matter hypothesis, I present the four experiments, describing their scientific goal, the measurement setup, and the obtained or expected results.

## 1. Introduction

The overwhelming evidence for the existence of Dark Matter is based on multiple, independent astrophysical and cosmological observations, such as the stellar rotation curves in galaxies, the power spectrum of Cosmic Microwave Background (CMB) fluctuations, the ratios of light element yields from Big Bang Nucleosynthesis (BBN), the morphology of galaxy cluster collisions, and astrophysical mass measurements based on gravitational lensing. All the above measurements consistently indicate that 85% of the total matter and 25% of the total energy of our Universe comprises an electrically neutral, non relativistic population of "Dark Matter" [1].

The most popular mechanism to describe the observed abundance of DM in the Universe ($\Omega_{DM} = 0.242 \pm 0.004$ [2]) is the so-called "freeze-out" process, which explains the present amount of DM as a relic left-over from the primordial era of thermal equilibrium with SM particles, made possible by a proper non-gravitational interaction mechanism. The equilibrium broke when the interaction rate dropped below the Universe expansion rate. The thermal origin sets a lower limit on the DM mass, of the order of 10 keV/c$^2$, if DM has to explain large-scale structures as observed in the Universe [3].

The simplest model deriving from the above paradigm of thermally-generated DM in the early Universe is that of "Weakly Interacting Massive Particles" (WIMPs). In this model, DM particles have mass $m_{DM} = 10 \div 1000$ GeV/c$^2$, and interact with SM particles through the Weak Force, with a thermally-averaged cross-section $\langle\sigma v\rangle$ of about $10^{-26}$ cm$^3$s$^{-1}$. The fact that no additional interactions are required to yield the observed DM abundance is the so-called "WIMP miracle". At present, a large number of running and planned experiments are exploring the heavy DM region [4], but no positive signal has yet been found.

On the other hand, the Light Dark Matter (LDM) hypothesis is, so far, almost unexplored. This scenario foresees DM particles in the mass range $m_{DM} = 10 \div 1000$ MeV/c$^2$, and thus is conceptually distinct from heavy WIMP DM. It requires a new interaction between LDM and

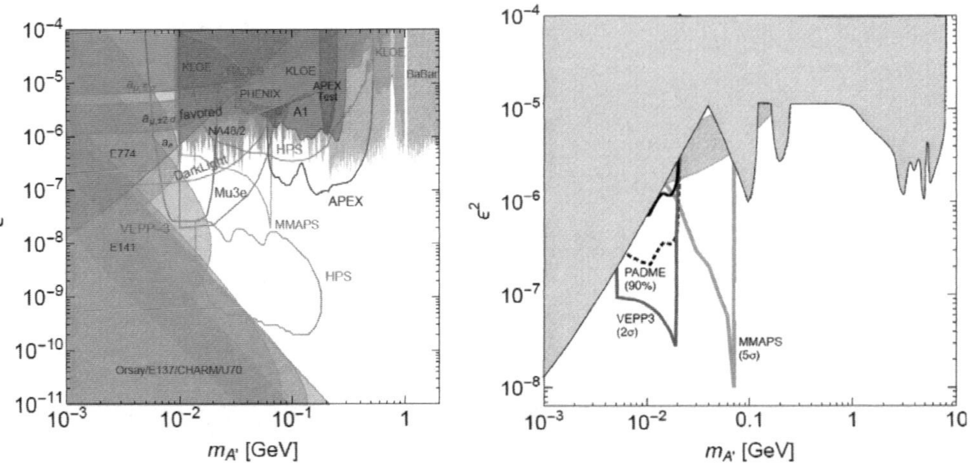


**Figure 1.** Exclusion limits on the dark photon parameter space from past (shaded) and planned (lines) experiments, for visibly (left) and invisibly (right) decaying $A'$. (Figures from [4]).

SM particles, with a correspondingly light, sub-GeV force carrier, in order to be compatible with the DM thermal origin. The simplest mechanism to provide this interaction, the so-called "vector portal", is through a new vector force carrier from an additional $U(1)_D$ gauge group under which LDM is charged, also called "dark photon", or $A'$ (for a complete review of this and other possibilities, see [5]). In this model, the $A'$ couples to ordinary photons through a "kinetic mixing" mechanism [6], i.e. through a loop-order effect, mediated by a GUT-scale massive particle that carries both the standard-model weak hypercharge and the dark-force equivalent. This induces an effective parity-conserving interaction $\varepsilon e A'_\mu J^\mu_{\rm EM}$ of the $A'$ to the SM electromagnetic current, with coupling $\varepsilon e$. The "vector-portal" scenario is regulated by four parameters: the LDM particles mass $m_\chi$, the mediator mass $m'_A$, the $A'$-SM electric charge coupling $\varepsilon$, and the $A'$-LDM coupling $g_D$. The strength of the kinetic mixing parameter is unknown, however, certain ranges can be motivated from quantum-mechanic effects of heavier particles. If the $A'$-SM coupling arises from a single-loop mechanism as described above, it is natural for $\varepsilon$ to be of the order of $10^{-4} \div 10^{-2}$ [6], while higher order effects, foreseen in GUT theories, predict a lower value, of about $10^{-6} \div 10^{-3}$ [7]. Similarly, the LDM thermal origin requirement sets further limits on different combinations of the model parameters, depending on the fine details of the interaction [4].

A comprehensive LDM experimental program has been launched in the last few years, investigating both the existence of $\chi$ particles and dark photons [4]. In particular, dark photon searches are sensitive to different possible $A'$ decay modes, that include both the so-called "visible" decay to SM particles, mainly lepton pairs of opposite charge, and the "invisible decay" to LDM particles. The relative strength of the two processes depends on the actual values of the model parameters. In particular, for an on-shell $A'$, the "visible" decay to $e^+e^-$ pairs is the main reaction channel if $m'_A < 2m_\chi$, while the "invisible" decay dominates in the opposite case. The phenomenology of off-sheel particles is richer [8], but still the distinction between the two regimes is governed by the particle masses. A summary of the results reported from already completed experiements is illustrated in Fig. 1, together with projections from planned or proposed measurements.

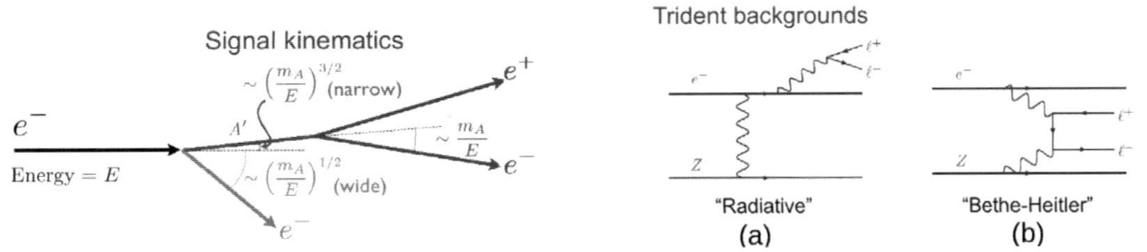


**Figure 2.** Left: the expected kinematics for the $A'$ production by Bremsstrahlung off an incoming electron and the subsequent decay to an $e^+e^-$ pair. Right: sample diagrams of radiative trident ($\gamma^*$) and Bethe-Heitler trident reactions that comprise the primary background to the $A'$ search.

## 2. Dark photon experiments at Jefferson Laboratory

High luminosity, fixed target electron-beam experiments provide a unique way to probe the aforementioned dark photon visible decay mode and the corresponding parameter space [9]. In these experiments, an electron beam impinges on a high Z target, producing heavy photons that subsequently decay to pairs of fermions. The $A'$ is then reconstructed by measuring the decay products momenta. The kinematic of this process, reported in Fig. 2 (left panel), is characterized by a forward emitted $A'$ that carries most of the incident electron beam energy $E$. This results in a very small opening angle of the decay products, of the order of $\theta \simeq M_A/E$. Therefore, the $A'$ decay products are emitted at a very low angle with respect to the beam-line, and carry most of the beam energy. Furthermore, depending on the value of the parameters, long-lived $A'$ are possible in the aforementioned parameter space, depending on the value of $\varepsilon$.

The main background sources associated to this reaction are QED trident processes, i.e. Bethe-Heitler trident reactions and radiative trident production (see Figure 2, right-panel). Although the Bethe-Heitler process has a much larger total cross-section, it can be significantly reduced by exploiting the very different kinematics. Instead, the radiative trident process is an irreducible background, with the same kinematics as the signal. However, it presents a smooth, continuous distribution in the $e^+e^-$ invariant mass distribution, in contrast to the peak at the dark photon for the signal, and occurs promptly at the production target, while the signal can be displaced for small $\varepsilon$ values.

### *2.1. The APEX experiment*

The APEX experiment [10] is a fixed-target experiment in Hall-A, that makes use of the two High Resolution Spectrometers (HRSs) installed there [11] to search for a visibly-decaying dark photon in the $e^+e^-$ final state. A proper dipole septum magnet between the thin Ta target and the HRS aperture allows the detection of scattered leptons down to 5° with respect to the incident beam. Each spectrometer is equipped with two vertical drift chambers, each with two orthogonal tracking planes, for momentum reconstruction, Cerenkov and plastic scintillator counters for particle identification and triggering, and a lead glass calorimeter. A schematic of the APEX experimental setup is illustrated in Fig. 3.

In 2010, the APEX collaboration succesfully completed a test run, at 2.260 GeV electron beam energy, with currents up to 150 $\mu$A. The HRSs detectors were configured for central momenta of 1.13 GeV/c, with a corresponding acceptance of $\pm 4.5\%$. Events were measured by demanding a tight time coincidence (12.5 ns) between the two spectrometers and requiring good quality tracks in both drift chamber systems. Finally, a cut on the energy sum of the two particles, which can't exceed the primary beam energy, has been introduced to further reduce accidentals. The final data-set consits of $\simeq 770k$ events, with $O(7.5\%)$ contamination, mainly from accidentals. The $e^+e^-$ mass resolution was 0.85 - 1.11 MeV/c$^2$, depending on the invariant

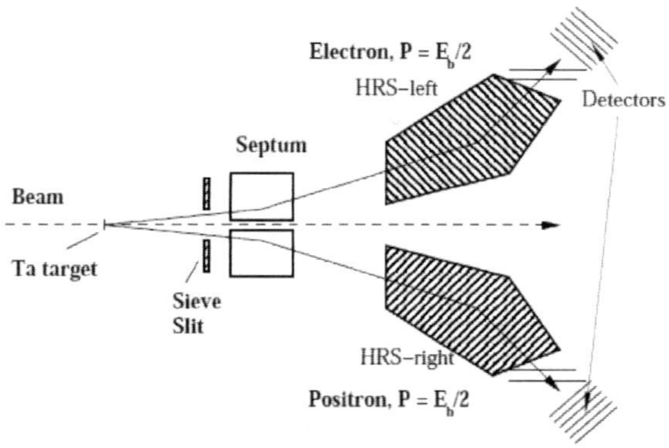


**Figure 3.** Schematic of the APEX experimental setup. Figure from [10].

mass. No excess of events was measured in the $e^+e^-$ invariant mass spectrum, on top of the QED background. Therefore, a sophisticated statistical technique was employed to determine the $2\sigma$ exclusion limit on the kinetic mixing parameter $\varepsilon$, as a function of the dark photon mass (see Fig. 1).

The APEX experiment has been fully approved by JLab, for a 34-days run period, and is scheduled to run in 2017-2018. The run plan foresees different configurations involving different beam energies and spectrometer settings, in order to cover a large portion of parameter space.

### *2.2. The HPS experiment*

The Heavy Photon Search (HPS) experiment [12] is a fixed-target experiment in Hall-B designed to search for dark photons in the mass range between 20 MeV/c$^2$ and 1 GeV/c$^2$ and coupling $\varepsilon^2$ between $10^{-5}$ and $10^{-10}$, through the decay to $e^+e^-$. HPS employes two complementary search techniques: resonance search and detached vertexing. In particular, through the latter, the experiment can explore a unique region in the parameter space corresponding to small cross sections. The foreseen reach is illustrated in Fig. 1.

The HPS detector (see Fig. 4) is a compact spectrometer, made by a $\simeq$ 1 m long silicon tracker (SVT) inside a dipole magnet, to measure charged particle trajectories and production vertices, and a lead-tunsgtate electromagnetic calorimeter (ECal), for particle identification and triggering. HPS employs a three-magnet system with the second dipole serving as an analyzing magnet: the thin Tungsten target ($\simeq 10^{-3}X_0$) is located in front of the analyzing magnet, at about 10 cm from the first SVT layer. To reduce beam-related backgrounds, the target and the SVT are mounted in a vacuum chamber, with the ECal being placed downstream the analyzing magnet. The SVT is made of 6 layers of Si modules, each with two sensors, mounted at a small stereo angle (100 for the first 3 layers and 50 for the others) to provide three-dimensional tracking and vertexing. The SVT is split in two modules, mounted on top and on bottom of the beam plane, leaving a central "dead-region" where the degraded beam can pass through the detector unobstructed. The edge of the sensors have to be placed at 0.5 mm from the electron beam plane to achieve the desired acceptance for low-mass dark photons. The ECal is also split into two halves, each made of 221 $PbWO_4$ crystals arranged in five rows. The light from each crystal is read out by a Large Area Avalanche Photodiode (APD), and corresponding signal is read by a custom amplier before being processed by a FADC-based DAQ system.

In July 2012, a successfull test run [13] demonstrated the feasibility of the measurement and of the detector operation. The experiment received full approval from the laboratory for a 180-day run, with different beam energy configurations, to be completed in different data-taking periods, up through 2020. So far, two data-taking periods have been completed: in 2015, 1.7 days (10 mC) at 1.06 GeV, and in 2016, 5.4 days (92.5 mC) at 2.3 GeV.

IOP Publishing
doi:10.1088/1742-6596/770/1/012040

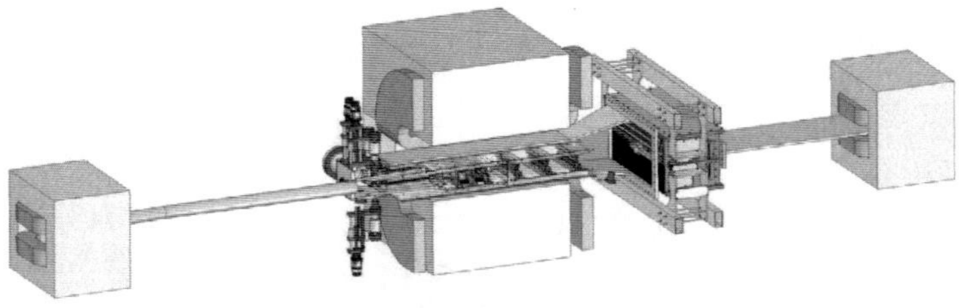

**Figure 4.** Schematic of the HPS detector setup. From left to right: the first dipole magnet, the target system, the silicon vertex tracker (SVT) inside the analyzing dipole magnet, the electromagnetic calorimeter (ECal), the third dipole magnet.

### *2.3. The DarkLight experiment*

The DarkLight experiment [14] is a fixed gas target $A'$ search, proposed at the Jefferson Laboratory Low Energy Recirculating Facility (LERF). In the experiment, the low energy (100 MeV), very high current (5 mA) electron beam will impinge on a $\simeq 10^{19}$ cm$^{-2}$ long hydrogen target, thus reaching a luminosity of $\simeq 3\cdot 10^{35}$ cm$^{-2}$ s$^{-1}$. The experiment aims at reconstructing the complete final state in the reaction $ep \to epe^-e^+$, and, therefore, is sensitive both to the visible and to the invisible dark photon decay, the latter process being measured through the missing mass of the final state $e-p$ system.

The DarkLight detector, cilindrically simmetric around the beam line, is made by a Si detector installed inside the target cell for low-energy recoiling proton detection, a 4-layer MicroMegas system working in conjuction with a 0.5 T solenoidal magnet for $e^+$ and $e^-$ measurement, and an array of plastic scintillators acting as a photon veto. The foreseen $e^+e^-$ invariant mass resolution is $\simeq 1$ MeV/c$^2$.

In 2012, a crucial test demonstrated the feasibility of steering the 0.4 MW electron beam through the 2 mm diameter gas target aperture, with negligible losses [15]. The collaboration has planned a phased approach toward the final experiment: in the first phase, expected in 2016, accelerator studies will be performed, in order to understand the LERF operations in conjuction with the windowless target. At the same time, detector tests will be performed, potentially with a pilot DM search, with only a partial coverage detector. The second phase, currently under design, will consist of the full statistics measurement, with a complete detector.

## 3. The BDX experiment

The "Beam Dump eXperiment" (BDX) [16] is an e$^-$ beam, thick-target search at Jefferson Laboratory for LDM in the mass range $10 \div 1000$ MeV/c$^2$. BDX will detect dark matter particles produced by the primary $e^-$ beam impinging on the JLab Hall-A beam dump, by measuring the scattering on a detector placed about 20 m downstream. The BDX detector will measure both $\chi-$nucleon and $\chi-$electron scattering. These two reactions have a very different signature in the detector: MeV nucleon recoil vs. GeV EM shower. Thus, the sensitivity to both is a powerful handle to control systematics.

The BDX detector is a segmented homogeneus calorimeter, made by CsI(Tl) crystals formerly used in the end cap BaBar Ecal, with an inproved SIPM-based readout. The calorimeter is made of about 800 crystals, resulting in a 0.5 $m^3$ active volume. The crystals are arranged in a sequence of arrays, each $30-cm$ long. In order to suppress cosmogenic and beam-related backgrounds, the calorimeter is surrounded by two active-vetoes, made by plastic scintillator layers read by PMTs and SiPMs. Between the inner and the outer veto, a 5cm thick lead vault shields the calorimeter from low energy photons. The detector will be placed about 8 m undeground, in a new experimental hall, in order to shield it from cosmogenic particles and also to match the depth of the Hall-A beam dump. To shield the detector from beam-related backgrounds, about 10 m of concrete and iron will be placed between the beam dump and the new hall. A schematic of the experimental setup is illustrated in Fig. 5.

The BDX proposal has been submitted to the 2016 JLab Program Advisory Commitee,

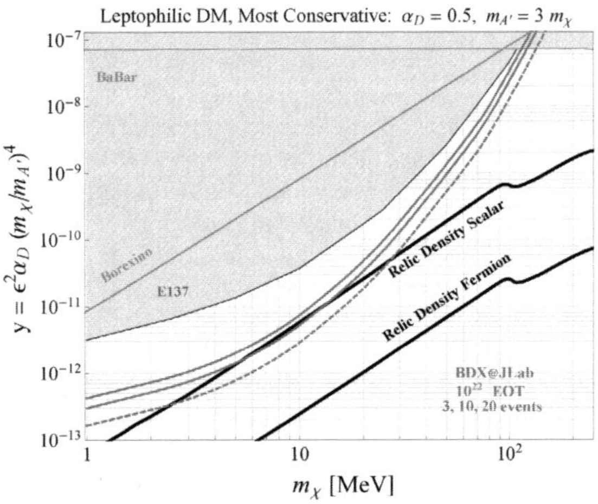


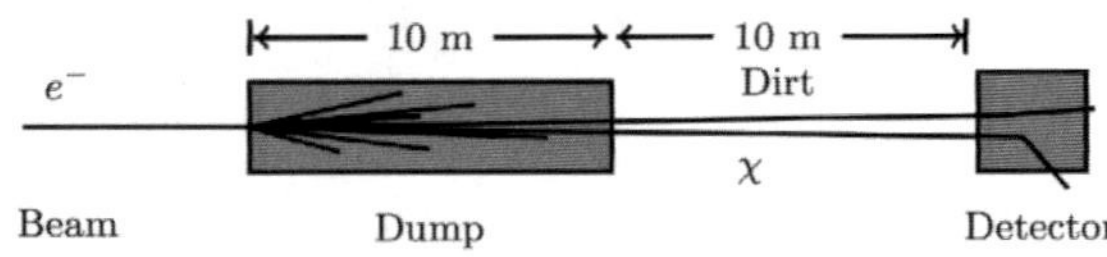


**Figure 5.** Top: schematic of the BDX experiment layout. Left: the foreseen BDX reach, for $10^{22}$ EOT. The $y$ variable on the vertical axis is defined as $y = \varepsilon^2\alpha_D(m_\chi/m'_A)^4$.

receiveing conditional approval. With an electron current up to 60 $\mu$A, and few years of run, BDX will be capable of collecting up to $10^{22}$ electrons on target (EOT), ruling-out a significant portion of the LDM parameter space. The expected BDX reach is shown in Fig. 5, for a leptofilic LDM: a further discussion of the reach, for different LDM scenarios, can be found in [16].

## 4. Conclusions

The LDM scenario (MeV to GeV range) is a compelling possibility, alternative to the "traditional" WIMP paradigm, to explain the observed DM relic density. A new experimental program has been launched in the last few years to explore this so-far unexplored hypothesis. The Jefferson Lab community joined this program with four complementary efforts, searching both for light dark matter particles (BDX experiment) and for the corresponding light mediator with the SM, the so-called "dark photon" (APEX, HPS, DarkLight experiments). The results already obtained by APEX, as well as those soon expected by HPS, cover a large portion of the "dark photon" parameter space, for the visible decay to an $e^+e^-$ pair. The DarkLight and BDX experiments, instead, will be able to explore the invibible decay scenario, and also to directly measure LDM particles in a so-far unexplored mass range.

## References

[1] Olive K A *et al* 2014 *Chin. Phys. C* **38** 090001
[2] Ade A R *et al* 2015 (Planck Collaboration) Planck 2015 results. XIII. Cosmological parameters, *Preprint* arXiv:1502.01589
[3] Peter A 2012 Dark Matter: A Brief Review *Preprint* arXiv:1201.3942
[4] Alexander J *et al* 2016 Dark Sectors 2016 Workshop: Community Report *Preprint* arXiv:1608.08632
[5] Essig R *et al* 2013 Working Group Report: New Light Weakly Coupled Particles *Preprint* arXiv:1311.0029
[6] Holdom B 1986 *Phys. Lett. B* **166** 196
[7] del Aguila F, Coughlan G D and Quiros M 1988 *Nucl. Phys. B* (307) 633
[8] Izaguirre E, Krnjaic G, Schuster P and Toro N 2013 *Phys. Rev. D* **88** 114015
[9] Bjorken J D, Essig R, Schuster P and Toro N 2009 *Phys. Rev. D* **80** 075018
[10] Abrahamyan S *et al.* (APEX Collaboration) 2011 *Phys. Rev. Lett.* **107** 191804
[11] Alcorn J *et al.* 2004 *Nucl. Instrum. Method. A* **522** 294
[12] Celentano A (HPS Collaboration) 2014 *J. Phys.: Conf. Series* **556** 012064
[13] Battaglieri M *et al.* (HPS Collaboration) 2015 *Nucl. Instrum. Meth. A* **777** 91
[14] Balewski J *et al.* (DarkLight Collaboration) 2014 The DarkLight Experiment: A Precision Search for New Physics at Low Energies *Preprint* arXiv:1412.4717
[15] Alarcon R. *et al.* 2013 *Phys. Rev. Lett.* **111** 164801
[16] Battaglieri M *et al.* (BDX Collaboration) 2016 Dark matter search in a Beam-Dump eXperiment (BDX) at Jefferson Lab *Preprint* arXiv:1607.01390

Journal of Physics: Conference Series **770** (2016) 012041 doi:10.1088/1742-6596/770/1/012041

# Physics Opportunity with an Electron-Ion Collider

**Patrizia Rossi**

Jefferson Lab
12000 Jefferson Avenue, Newport News VA 23606, USA

E-mail: rossi@jlab.org

**Abstract.** Understanding the emergence of nucleons and nuclei and their interactions from the properties and dynamics of quarks and gluons in Quantum Chromodynamics (QCD) is a fundamental and compelling goal of nuclear science. A high-energy, high-luminosity polarized electron-ion collider (EIC) will be needed to explore and advance many aspects of QCD studies in the gluon dominated regions in nucleon and nuclei. The federal Nuclear Science Advisory Committee unanimously approved a high-energy electro-ion collider to explore a new frontier in physics research. In fact, the committee calls the collider the country's next "highest priority" in new facility construction, and is one of four main recommendations contained in its 2015 Long Range Plan for Nuclear Science. Two proposals for the EIC are being considered in the U.S.: one each at Jefferson Laboratory (JLab) and at Brookhaven National Laboratory (BNL). An overview of the physics opportunities an EIC presents to the nuclear science community in future decades is presented.

## 1. Introduction

Recently, the last missing piece in the current Standard Model was discovered: the Higgs boson. It plays a unique role by explaining why the other elementary particles, except the photon and gluon, are massive. However, this is not the end of the story. The interactions of quarks and gluons in high-energy experiments are well described by the QCD, but an exact understanding of how QCD works under normal conditions found in the everyday world is quite limited. Gluons, the carriers of the strong force, bind the quarks together inside nucleons and nuclei and generate nearly all of the visible mass in the universe. However, despite their importance, fundamental questions remain about the role of gluons in nucleons and nuclei. For instance, it is not yet understood why parton distribution functions, the most prominent quantities that describe the relation between the basic degrees of freedom of QCD and the observable physical states, reveal a large abundance of soft gluons inside the proton at very small $x$ (the fraction of the proton momentum carried by the struck partons).

Past and present experiments have provided critical information to understand how the properties and structure of nuclear matter emerge from the dynamics encoded in QCD, but many open questions still remain:

- How are the gluons and sea quarks, and their intrinsic spins, distributed in space and momentum inside the nucleon? What is the role of sea quark and gluon orbital motion in building the nucleon spin?
- How do gluon and sea quarks contribute to the nucleon-nucleon force, as manifested in the internal landscape of light nuclei?

- Can one find evidence for saturation of the gluon density?
- How do quarks and gluons propagate in nuclear matter and join together to form hadrons?

These questions can be answered with a powerful new electron-ion collider which will provide unprecedented precision and versatility, and whose realization is enabled by recent advances in accelerator technology. The federal Nuclear Science Advisory Committee unanimously approved it as one of the four main recommendations contained in its 2015 Long Range Plan for Nuclear Science [1]. Two options are under study: at Jefferson Lab and at Brookhaven National Lab. The basic scientific requirements for such a facility are:

- Highly polarized ($\sim$ 70%) electron, proton, and light ion (d,$^3He$) beams
- Ion beams from deuteron to the heaviest nuclei (Uranium or Lead)
- Variable center of mass energies from ~20 - ~100 GeV, upgradable to ~150 GeV
- High collision luminosity $\sim 10^{33-34} cm^{-2}s^{-1}$

The EIC will be unique in colliding polarised electrons off polarised protons and light nuclei, providing the spin degrees of freedom essential to pursue a physics program driven by spin structure, multi-dimensional tomographic images of protons and nuclei, and discovery of the role of collective effects of gluons in nuclei [2]. Such a facility would capitalize on the powerful new experimental techniques for exploring nucleon structure that are being developed for the 12 GeV JLab program and the RHIC program, and apply them to the low $x$ region where the dynamics is dominated by the gluons. Addressing this kinematic regime with high luminosity and fully polarized beams is necessary to complete our understanding of the basic partonic structure of the nucleon.

## 2. The Nucleon Structure

Although the nucleon is the only stable hadron in the Standard Model, its structure is not fully understood. Understanding it from first principles is considered a milestone of hadronic physics and numerous experiments are, and will continue to be, devoted to its study.

### *2.1. The Nucleon Spin*

The decomposition of the proton's overall intrinsic spin into quark and gluon contributions remains a fascinating open question. We know that the spin of quarks and antiquarks is only responsible for $\sim$ 30% of the proton spin. We know also, that the latest more precise RHIC results [3] indicate that the gluons' spin contribution in the currently explored kinematic region is non-zero [4]. However, as the partons' total helicity contribution to the proton spin is very sensitive to the minimum momentum fraction $x$ accessible by the experiments, we can not yet assess whether parton spin alone can account for the overall proton spin, or whether additional contributions are needed from the orbital angular momentum of partons. With the unique capability to reach two orders of magnitude lower in $x$ and to span a wider range of momentum transfer $Q$ than previously achieved, the EIC would be able to quantify precisely how much the intrinsic spin of quarks of various flavors and gluons contributes to the proton spin in just few month of operation. This is shown in the two plots of Fig. 1, for the parton helicity distributions (left) and for the gluon helicity contribution versus the quark helicity contribution (right).

### *2.2. The Three Dimentional Imaging of Quarks and Gluons*

In understanding the microscopic structure of the nucleon, we have so far relied mainly on two types of physical quantities: the spatial distribution of charge and current probed through elastic lepton scattering and described by the elastic form factors, and the longitudinal momentum distributions probed through Deep Inelastic Scattering (DIS) experiments and described by the parton distribution functions. Although form factors and parton distribution functions have

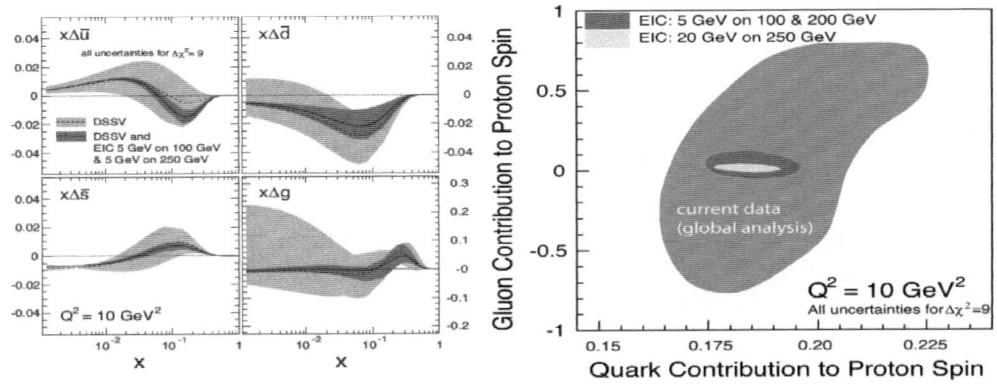


**Figure 1.** Left: Uncertainty bands on helicity parton distributions, in the first DSSV analysis [5, 6] (light bands) and with EIC data (darker bands), using projected inclusive and semi-inclusive EIC data sets. Right: Accuracies for the correlated truncated integrals of $\Delta\Sigma$ and $\Delta g$ over $0.001 \leq x \leq 1$, on the basis of "DSSV+" analysis (outer area) and projected for an EIC (inner area)[7].

provided much to shape our physical picture of the nucleon, they have similar deficiencies. The form factors contain no dynamical information on the constituents, such as their speed and angular momentum, and the momentum distributions provide no knowledge of the spatial locations. A unified picture of the nucleon in momentum and space requires to invoke a new framework provided by the the Wigner distributions [8], a quantum mechanical concept. They have inspired a three-dimensional description of the nucleon through the construction of quantities known as generalized parton distributions (GPDs) and transverse momentum-dependent parton distributions (TMDs). The GPDs encode the correlation between the quark/gluon transverse position in the nucleon and its longitudinal momentum, i.e., they contain information how the transverse spatial shape of the nucleon changes when probing quarks with different longitudinal momentum. In other words, GPDs can be viewed as form factor distributions at different values of the longitudinal momentum of the quark. They can be measured in exclusive electron-nucleon scattering processes at large $Q^2$ where a virtual photon interacts with a single quark in the nucleon. The most prominent processes to access GPDs are Deeply Virtual Compton Scattering (DVCS), which has been recognized as the cleanest process. The TMDs arise by integrating the Wigner distributions over the spatial position of the parton and are functions of both the longitudinal momentum fraction $x$ and transverse motion $k_T$ of partons. They offer a momentum tomography of the nucleon complementary to the spatial tomography of GPD. When adding the spin degree of freedom, TMDs may link the parton spin to the parent proton spin and to the transverse motion, and also the parton transverse momentum to the nucleon spin, and are hence related to the orbital motion of partons inside the proton. At leading order there are eight independent TMDs, each one related to different couplings between spin and transverse motion. One of these is the Sivers function which describes the correlation between the momentum direction of the struck parton and the spin of its parent nucleon. As a result, in a transversely polarized nucleon the quark distribution is azimuthally asymmetric in the transverse momentum space. Fig. 2 demonstrates such deformation for the up quark distribution.

Although there has been tremendous progress in understanding TMDs and GPDs, an EIC with its wide kinematical coverage, high polarization and high luminosity will give an unprecedent insight into the three-dimensional description in both momentum and space of quarks, antiquark and gluons. As an exemple of what can be reached with an EIC, the upper part of Fig. 3 shows the spacial distribution of sea quarks in an unpolarized proton (left) and in a polarized one (right) obtained from a GPD fit to simulate data for the DVCS cross section and spin asymmetry measurements. In the bottom part of the same figure, the corresponding parton density in the transverse plane is reported.

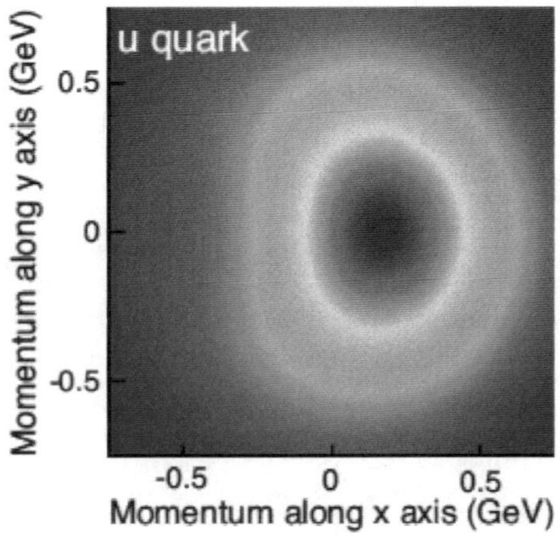


**Figure 2.** The density in the transverse-momentum plane for unpolarized quarks with $x = 0.1$ in a nucleon polarized along the $\hat{y}$ direction. The anisotropy, due to the proton polarization, is described by the Sivers function.

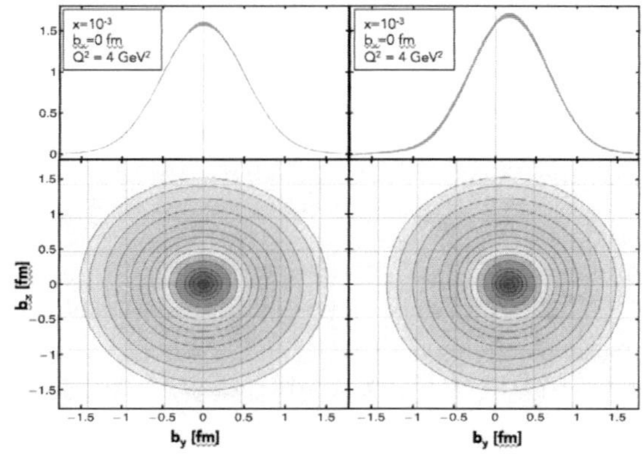

**Figure 3.** Upper: the spatial distribution of sea quarks in an unpolarized proton (left) and in a proton polarized along the positive x axis (right) obtained from a GPD fit data for the DVCS cross section and spin asymmetry measurements with the EIC. Bottom: The corresponding density of partons in the transverse plane.

## 3. The nucleus: a laboratory for QCD

### *3.1. Hadronization and energy loss in cold QCD matter*

The experimental program of the EIC is targeted to answer also fundamental questions concerning the dynamics of quarks and gluons in a nuclear environment, such as:

- How do hadrons emerge from quarks and gluons?
- What is nature telling us about confinement?

Semi-inclusive DIS in e+A collisions provides *i)* a known and stable nuclear medium; *ii)* well-controlled kinematics of hard scattering; *iii)* final state particles with well-known properties. Thus cold QCD matter could be an excellent femtometer-scale detector of the hadronization process from its controllable interaction with the produced quark (or gluon). With an EIC we will reach unprecedent precision and control on parton kinematics by selecting $\nu$ (the virtual photon energy range), and on the length-in-medium by selecting appropriate nuclei. The data will provide essential information for understanding the response of the nuclear medium to a colored, fast-moving quark. For instance, by measuring pion and $D^0$ meson production in both $e + p$ and $e + A$ collisions, the EIC will provide the first measurement of the quark mass dependence of the response of nuclear matter to a fast-moving quark. The dramatic difference between them, shown in Fig. 4, would be easily discernable at the EIC.

*3.2. Gluon saturation*

The nucleon structure in the small-$x$ region is dominated by gluons, whose density increases rapidly with decreasing $x$. At low $x$, the large soft-gluon density growth is quenched by the non-linear process of gluon-gluon recombination. This mechanism necessarily generates a dynamic scale known as the saturation scale, $Q_s$, at which gluon splitting and recombination reach a balance. The existence of such a state of saturated, soft gluon matter, often referred to as the Color Glass Condensate (CGC), is a direct consequence of gluon self-interactions in QCD. It has been conjectured that the CGC has universal properties common to nucleons and all nuclei. HERA, RHIC and LHC have found hints of saturated gluonic matter. The construction of an EIC will allow an exploration of this region in great detail (with luminosities 100x that of HERA). In particular, colliding electrons with a wide variety of nuclei, the saturation region will be more accessible due to the amplification of the saturation scale with nuclear size [9]. By measuring the ratio of diffractive cross-sections over the total DIS cross-sections in both $e + A$ and $e + p$ collisions, as shown in Fig. 5, the EIC would provide the first unambiguous evidence for the novel QCD matter of saturated gluons.

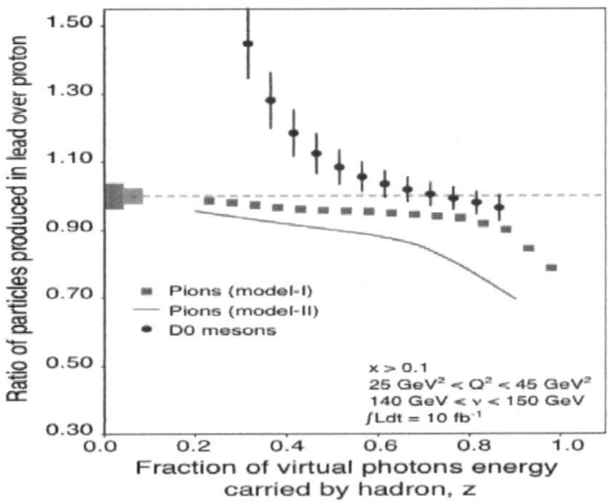


**Figure 4.** The ratio of $\pi$ production in $e + Pb$ collisions over that in $e + d$ collisions (full square symbols) is always below unity, while the ratio of heavy meson ($D^0$) production (full circle symbols) can be less than as well as larger than unity due to the difference in hadronization.

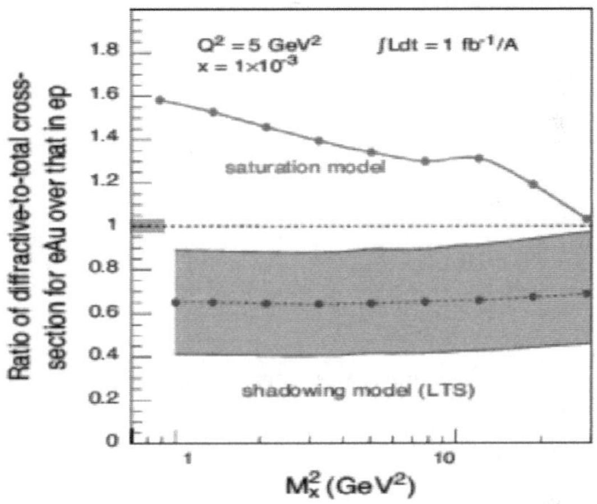


**Figure 5.** The ratio of diffractive over total cross section for DIS on gold normalized to DIS on proton for different values of mass squared of hadrons produced in collisions with and without saturation.

## 4. The EIC Designs and Detectors

Two proposals for the EIC are being considered in the U.S.: one each at JLab (JLEIC) and at BNL (eRHIC) [2]. Both designs build on the DOE's significant investments in infrastructure. eRHIC is based on the existing Relativistic Heavy Ion Collider (RHIC) hadron facility with its two intersecting superconducting rings, each 3.8 km in circumference (Fig. 6, left). JLEIC will utilize the 12 GeV CEBAF as an injector to a collider facility. The storage rings would be in a "figure 8" layout to mitigate the effects of depolarizing resonances and facilitate high beam polarization (Fig. 6, right).

The physics-driven requirements on the EIC accelerator parameters and extreme demands on the kinematic coverage for measurements make integration of the detector into the accelerator a particularly challenging feature of the design. Driven by the demand for high precision on

Journal of Physics: Conference Series **770** (2016) 012041 doi:10.1088/1742-6596/770/1/012041

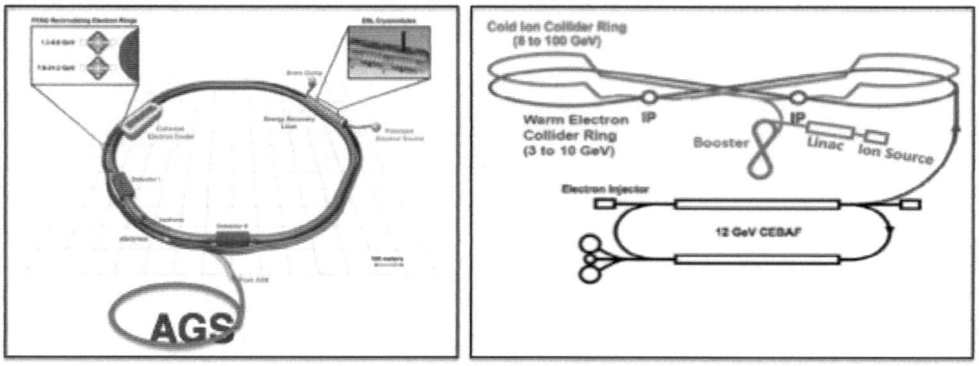


**Figure 6.** Left: The schematic of eRHIC at BNL, which would require construction of an electron beam facility to collide with the RHIC beam at up to three interaction points [2]. Right: The schematic of JLEIC at JLab, which would require construction of an ion linac, and an electron-ion collider ring with at least two interaction points, around the 12 GeV CEBAF [2].

particle detection and identification of final state particles in both $e + p$ and $e + A$ programs, R&D efforts are under way on various novel ideas for modern particle detector systems.

## 5. Conclusions

The EIC is conceived to profoundly impact our understanding of the structure of nucleons and nuclei in terms of quarks, sea quarks and gluons. Its high luminosity, energy, and beam polarization will enable the study of the three-dimensional quark-gluon structure of nucleons in yet unexplored regions of phase spaces in QCD. Il it will also allow color neutralization to be studied, and saturated gluonic matter to be probed and explored in detail. The EIC promises to unite and extend the scientific programs at CEBAF and RHIC in dramatic and fundamentally important ways, and to propel them to the next QCD frontier. Two designs have been proposed by scientists at RHIC and JLab, both capable of delivering a frontier accelerator facility. With the endorsement of the US Nuclear Science Advisory Committee, the US EIC is on track to become a reality in the near future. Its realization will ensure US leadership in nuclear science research.

*Acknowledgements:* This material is based upon work supported by the U.S. Department of Energy, Office of Science, Office of Nuclear Physics under contract DE-AC05-06OR23177.

## References

[1] Aprahamian A *et al.*, 2015 *http : //science.energy.gov/ ∼ /media/np/nsac/pdf/2015LRP/2015_LRPNS_091815.pdf*.
[2] Accardi A *et al.*, 2016 *Eur. Phys. J. A* **52** 268.
[3] Adamczyk L *et al.*, 2015 *Phys. Rev. Lett.* **115** 092002.
[4] de Florian D, Sassot R, Stratmann M and Vogelsang W, 2014 *Phys. Rev. Lett.* **113** 012001.
[5] de Florian D, Sassot R, Stratmann M and Vogelsang W, 2008 *Phys. Rev. Lett.* **101** 072001.
[6] de Florian D, Sassot R, Stratmann M and Vogelsang W, 2009 *Phys. Rev. D* **80** 034030.
[7] Aschenauer E C, Sassot R and Stratmann M, 2012 *Phys. Rev. D* **86** 054020.
[8] Hillery M, O'Connell R, Scully M and Wigner E P, 1984 *Phys. Rept.* **106** 121.
[9] Kowalski H and Teaney D, 2003 *Phys. Rev. D* **68** 114005.

# Heavy Quark Production at an Electron-Ion Collider

**E Chudakov**[1], **D Higinbotham**[1], **Ch Hyde**[2], **S Furletov**[1], **Yu Furletova**[1], **D Nguyen**[1,3], **M Stratmann**[4], **M Strikman**[5], **C Weiss**[1]

[1] Jefferson Lab, Newport News, VA 23606, USA
[2] Old Dominion University, Norfolk, VA 23529, USA
[3] University of Virginia, Charlottesville, VA 22904, USA
[4] Tübingen University, 72076 Tübingen, Germany
[5] Pennsylvania State University, University Park, PA 16802, USA

E-mail: weiss@jlab.org

**Abstract.** An Electron-Ion Collider (EIC) with center-of-mass energies $\sqrt{s_{eN}} \sim$ 20–100 GeV and luminosity $L \sim 10^{34}$ cm$^{-2}$ s$^{-1}$ would offer new opportunities to study heavy quark production in high-energy electron or photon scattering on protons and nuclei. We report about an R&D project exploring the feasibility of direct measurements of nuclear gluon densities at $x >\sim 0.1$ (gluonic EMC effect, antishadowing) using open charm production at EIC. We describe the charm production rates and angle-momentum distributions at large $x$ and discuss methods of charm reconstruction using next-generation detector capabilities ($\pi/K$ identification, vertex reconstruction). The results could be used also for other physics applications of heavy quark production at EIC (fragmentation functions, jets, heavy quark propagation in nuclei).

## 1. Electron-Ion Collider

An Electron-Ion Collider (EIC) is being developed as a next-generation facility for nuclear physics and has been recommended for future construction in the 2015 U.S. Department of Energy's Long-Range Plan [1]. The present EIC designs envisage electron-proton ($ep$) center-of-mass (CM) energies $\sqrt{s}_{ep} \sim$ 20–100 GeV, with possible extensions to higher energies, and aim to deliver luminosities $\sim 10^{34}$ cm$^{-2}$ s$^{-1}$ over the full energy range [2, 3]. Acceleration of a wide variety of nuclear beams would be possible, ranging from the deuteron ($A = 2$) to heavy ions ($A \sim 200$). In electron-nucleus ($eA$) scattering at the collider the CM energy per nucleon is lower compared to $ep$ by a factor $\sqrt{Z/A} \approx 0.7$ (0.6) for light (heavy) nuclei, and the luminosity per nucleon is approximately the same as in $ep$. Such a facility would significantly expand the "energy-luminosity frontier" in electromagnetic scattering (see figure 1), particularly for nuclei, and enable qualitative advances in exploring short-range structure and Quantum Chromodynamics. The EIC physics program includes studies of the nucleon's three-dimensional partonic structure (gluon spin, quark spin flavor decomposition, transverse momentum, spatial structure, correlations), the dynamics of color fields in nuclei (quarks/gluon densities in nuclei, nuclear shadowing, physics of high gluon densities), and the conversion of color charge to hadrons (color transparency, parton propagation in medium, hadronization) [4, 5].

The EIC would vastly expand the capabilities for studying heavy quark production in electromagnetic scattering as compared to present facilities. Measurements of open charm and beauty electro- and photoproduction were performed at the HERA $ep$ collider at $x_{\rm B} <$ 0.01, using various methods of charm/beauty identification; see [6, 7] and references therein.

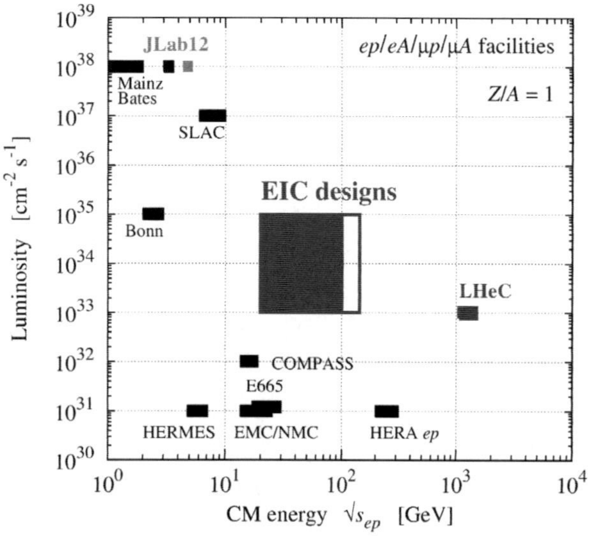


**Figure 1.** CM energies and luminosities of experimental facilities for electromagnetic scattering at multi-GeV energies ($ep/eA/\mu p/\mu A$; past, present, and future). The coverage of the EIC designs is indicated by the blue box ($ep$, $Z/A = 1$).

Open charm production was also observed at the COMPASS $\mu N$ fixed-target experiment [8]. The EIC luminosity is two orders-of-magnitude higher than that of HERA and would permit measurements of open charm/beauty production with much higher rates, extending the kinematic coverage to the region of large $x_B$ ($\sim 0.1$) and rare processes such as high-$p_T$ jets. Heavy quark production with electromagnetic probes could for the first time be measured on nuclear targets and used to study the gluonic structure of nuclei and the propagation of heavy quarks through cold nuclear matter with full control of the initial state. Next-generation detection capabilities at the EIC — tracking, vertex detection, and especially $\pi/K$ identification — would open up new channels for charm/beauty reconstruction compared to HERA and further boost the rate of identified heavy quarks for physics purposes. The study of possible EIC applications to heavy quark production therefore deserves special attention. Here we consider heavy quark production as a new method for measuring the gluon densities in nuclei.

## 2. Heavy quarks as probe of nuclear gluons

Heavy quark production in DIS can serve as a direct probe of the gluon density in the target. At leading order in perturbative QCD the heavy quark pair is produced through photon-gluon fusion (see figure 2a) and samples the gluon density at momentum fractions $x > ax_B$ ($x_B$ is the Bjorken variable, $a = 1 + 4m_h^2/Q^2$, and $m_h$ is the heavy quark mass), at an effective scale $\mu^2 \approx 4m_h^2$; see [9] and references therein. Higher-order QCD corrections are known and theoretical uncertainties have been quantified [10]. The HERA results have shown good agreement with the QCD predictions [6]. With the EIC heavy quark production could thus become a practical tool for measuring the gluon densities in the nucleon and in nuclei.

Of particular interest is the gluon density in nuclei at large $x$. Measurements of inclusive DIS have shown that the valence quark densities in nuclei are suppressed at $x > 0.3$ (EMC effect) and inspired numerous theoretical studies of QCD in nuclei [11]. The nuclear modifications of gluons are largely unknown at present, and basic questions remain to be answered (see figure 2b): Is the nuclear gluon density suppressed at $x > 0.3$ like the valence quarks (gluonic EMC effect)? Is the nuclear gluon density enhanced at $x \sim 0.1$ (gluon antishadowing)? The answers to these questions would offer insight into the change of the nucleon's gluonic structure due to nuclear binding and the QCD structure of nucleon-nucleon interactions. Information on the nuclear gluons at $x > 0.1$ can been obtained indirectly from the $Q^2$ dependence of inclusive nuclear DIS

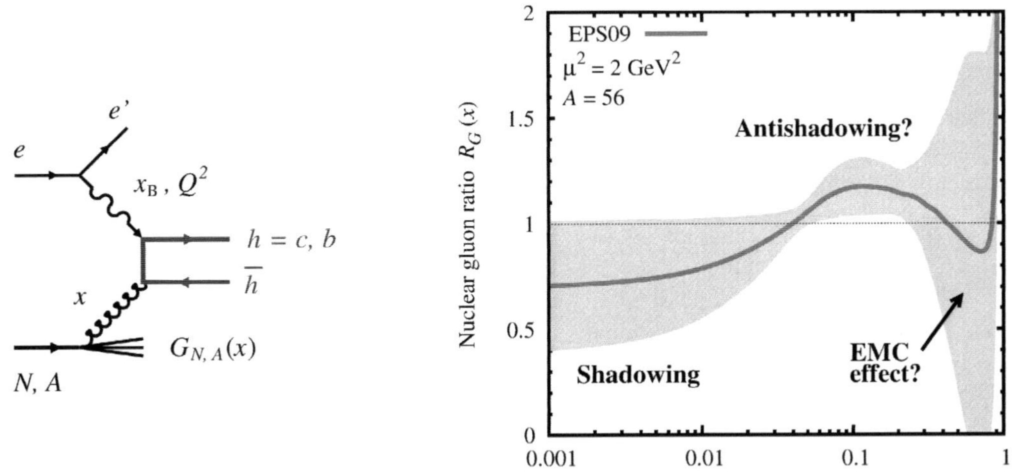


(a) (b)

**Figure 2.** (a) Heavy-quark production in DIS at LO (photon-gluon fusion). (b) The nuclear gluon density ratio $R_G(x,\mu^2) = G_A(x,\mu^2)/[AG_N(x,\mu^2)]$ and its uncertainty at a scale $\mu^2 = 2$ GeV$^2$, obtained from the EPS09 analysis of nuclear PDFs [12].

cross sections (DGLAP evolution), but the reach of the present fixed-target data is very limited. EIC would improve the situation by extending the inclusive measurements over a larger $Q^2$ and $W$ range and separating longitudinal and transverse structure functions. A much more powerful method would be direct measurements of the nuclear gluon density at a fixed scale using heavy quark production.

Here we report about an R&D project studying the feasibility of direct measurements of large-$x$ nuclear gluons using heavy quark production at EIC [13, 14]. The tasks include (a) estimating the charm production rates at EIC and the angle-momentum distributions of the heavy mesons and their decay products; (b) exploring new methods of charm reconstruction appropriate for large $x_B$ using the EIC detector capabilities; (c) quantifying the impact on nuclear gluons and the theoretical uncertainties. We note that a detailed design of the JLab EIC detector has yet to be completed, and that simulations of charm reconstruction methods at this stage are necessarily at the generic level. While the studies of charm/beauty production reported here focus on the specific application to large-$x$ nuclear gluons, many of the results are more general and can be used for other physics studies with heavy quarks at EIC (heavy quark fragmentation functions, jets, propagation and hadronization in nuclei).

## 3. Charm production in DIS at EIC

Charm production rates in DIS at EIC have been estimated using QCD expressions and the HVQDIS code [15] (see figure 3). The charm rates drop rapidly above $x_B \sim 0.1$ due to the decrease of the gluon density. The fraction of DIS events with charm production changes from $\sim 10\%$ at $x_B \sim 0.01$ to $\sim 1\%$ at $x_B \sim 0.1$ (exact numbers depending on $Q^2$). The charm fraction increases with $Q^2$ at fixed $x_B$ as expected for a gluon-dominated process. With an integrated luminosity of 10 fb$^{-1}$ charm production numbers of $\sim 10^6$ ($\sim 10^5$) can be achieved in DIS at $x_B \sim 0.01$ ($\sim 0.1$). Higher charm rates could be achieved by lowering the $Q^2$ cutoff and/or including charm photoproduction. These numbers define the starting point for charm physics analysis. The challenge in gluon measurements at $x_B \sim 0.1$ will be to identify charm events with an efficiency of $\sim$ few %, in the presence of a DIS background that is $\sim 100$ times larger.

Journal of Physics: Conference Series **770** (2016) 012042 doi:10.1088/1742-6596/770/1/012042

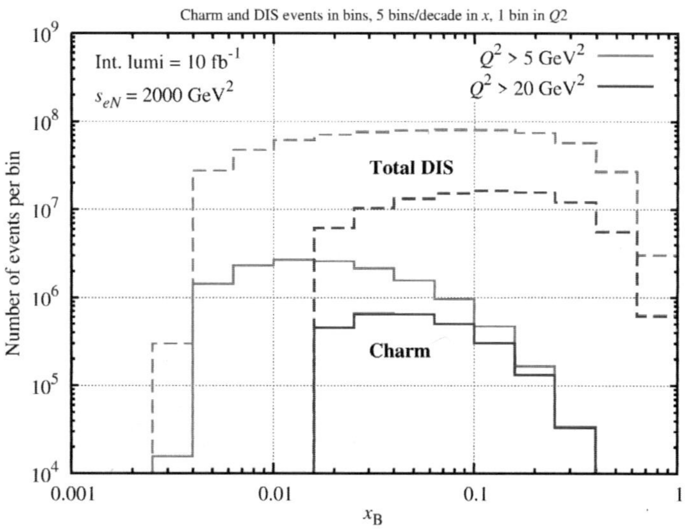


**Figure 3.** Estimated number of DIS events (dashed lines) and charm events (solid lines) in DIS at EIC (CM energy $s_{eN} = 2000$ GeV$^2$, integrated nucleon luminosity 10 fb$^{-1}$). The bins in $x_B$ are 5 per decade as indicated on the plot. $Q^2$ is integrated from the lower value indicated (5 or 20 GeV$^2$) to the kinematic limit at the given $x_B$.

The nucleon momenta in the medium-energy EIC proton/ion beams are $\sim$ few 10 GeV; e.g., collisions of 10 GeV electrons on 50 GeV nucleons at $s_{eN} = 2000$ GeV$^2$. With this setup the charm quarks produced in DIS at $x_B \sim 0.1$ typically emerge with large angles in the lab frame (which approximately coincides with the virtual photon-gluon CM frame) and carry moderate momenta $\sim<$ 10 GeV. (The actual charm angle and momentum distributions exhibit a complex dependence on $x_B$ and $Q^2$ and have to be determined by kinematic transformations.) The charm quark distributions are imparted on the produced charmed hadrons and their final decay hadrons $(\pi, K, p)$. A major advantage of the medium-energy EIC design is that these hadrons emerge at large angles and moderate momenta, where good particle identification (PID), tracking, and vertex detection capabilities can be provided by the central detector. In contrast, with a high-energy collider (HERA) the hadrons from large-$x_B$ charm decays would appear at forward angles and with much larger momenta, rendering their detection more difficult.

## 4. Toward charm reconstruction at EIC

Charm events in DIS have to be identified through the charmed hadrons ($D$ mesons, $\Lambda_c$ baryons) that are produced by charm quark fragmentation and subsequently decay into $\pi/K/p$, using either exclusive decay channels or inclusive modes (jets).

A summary of significant exclusive decays of $D/\Lambda_c$ into charged $\pi/K$ is given in table 1. The theoretical reconstruction efficiency in this method is determined by the product of the fragmentation ratio into the charmed hadron and the branching ratio for the exclusive decay. HERA experiments made extensive use of the $D^{*+}$ channel, which exhibits a distinctive two-step decay $D^{*+} \rightarrow D^0\pi^+_{\text{slow}}$, $D^0 \rightarrow K^-\pi^+$, and can be reconstructed without PID or vertex detection. However, this channel offers an overall reconstruction efficiency of only $\sim 1\%$, which is not sufficient for gluon measurements at $x_B \sim 0.1$. The EIC detector will provide PID capabilities for charged $\pi/K$ separation and permit use of other exclusive channels for charm reconstruction (see the example of $D^0$ reconstruction through the $K^-\pi^+$ decay in figure 4). Combining the charged exclusive decays in table 1 one could achieve a theoretical charm reconstruction efficiency of up to $\sim 10\%$, which would significantly expand the physics reach at large $x_B$. Vertex detection can substantially improve the signal/background ratio in charm reconstruction through exclusive decays (see figure 4), but it reduces the overall reconstruction efficiency, because it rejects events with a short decay length. The optimization of charm reconstruction with exclusive decays for

| $h_c$ | $f$ | Decay | BR |
|---|---|---|---|
| $D^0$ | 59% | $K^-\pi^+$ | 3.9% |
| | | $K^-\pi^+\pi^+\pi^-$ | 8.1% |
| $D^+$ | 23% | $K^-\pi^+\pi^+$ | 9.2% |
| $D^{*+}$ | 23% | $(K^-\pi^+)_{D0}\ \pi^+_{\text{slow}}$ | 2.6% |
| | | $(K^-\pi^+\pi^+\pi^-)_{D0}\ \pi^+_{\text{slow}}$ | 5.5% |
| $D_s^+$ | 9% | $(K^+K^-)_\phi\ \pi^+$ | 2.3% |
| $\Lambda_c^+$ | 8% | $pK^-\pi^+$ | 5.0% |

**Table 1.** Channels for charm reconstruction through exclusive decays into charged hadronic final states. Columns: 1) charmed hadron $h_c$; 2) fragmentation fraction $f(c \to h_c)$; 3) significant charged decay channels; 4) branching ratio. The fragmentation fractions are from ZEUS $\gamma p$ [7]. They do not add up to 100% because $D^0$ is also produced through $D^{*+}$ decays.

the purpose of gluon measurements at $x_B \sim 0.1$ at EIC is the object of on-going R&D [13].

Charm production in DIS can also be identified through inclusive modes, by selecting jet events with a displaced secondary vertex indicating a $D$ meson decay. A selection method based on the decay length significance distribution was used in the last HERA experiments at $x_B < 0.01$ [6]. Such methods can in principle achieve much larger charm reconstruction efficiencies than exclusive channels, especially when combined with PID ($\sim$ 30% was assumed in the EIC simulations in [2]). Their feasibility for gluon measurements at $x_B \sim 0.1$, where the fraction of charm events is only $\sim$ 1% of DIS, needs to be explored further. Note that this assessment requires detailed assumption about the performance of the EIC vertex detector.

Another possible strategy for large-$x$ gluon measurements with charm is to focus on exceptional $c\bar{c}$ pairs with large transverse momenta $p_T \gg 1$ GeV. While they are produced with a small cross section, such configurations represent a very distinctive final state that is practically free from hadronic background.

## 5. Nuclear ratio measurements

The application of charm production at EIC to the study of nuclear gluons at large $x$ also requires analysis of the uncertainties specific to nuclear ratio measurements [14]. This includes (a) controlling the relative nuclear luminosity in measurements with different ion beams through physics processes; (b) separating effects of nuclear final-state interactions on the observed meson spectrum from initial-state modifications of the nuclear gluon density using the different $A$-dependence of the two mechanisms; (c) quantifying the impact of the charm production pseudodata on the nuclear gluon densities. Results will be reported in due course [13].

## 6. Summary

A medium-energy EIC would offer excellent opportunities for measurements of charm/beauty production in $ep/eA$ and $\gamma p/\gamma A$ scattering through a unique combination of energy, luminosity, and next-generation detection capabilities. The charm rates appear sufficient to constrain nuclear gluons at $x > 0.1$, if charm reconstruction could be performed with an overall efficiency of $\sim$ few %. Practical methods to achieve such efficiency are being investigated. Heavy quark production at EIC could of course also be used for other physics purposes (e.g. heavy quark fragmentation functions, jets, propagation and hadronization in nuclei), which may impose less stringent requirements. Simulating such measurements would elucidate other aspects of charm production/reconstruction with EIC and provide further impulses for detector development.

This material is based upon work supported by the U.S. Department of Energy, Laboratory Directed Research and Development funding, under contract DE-AC05-06OR23177. The work of M.S. was supported by the U. S. DoE under award DE-FG02-93ER40771.

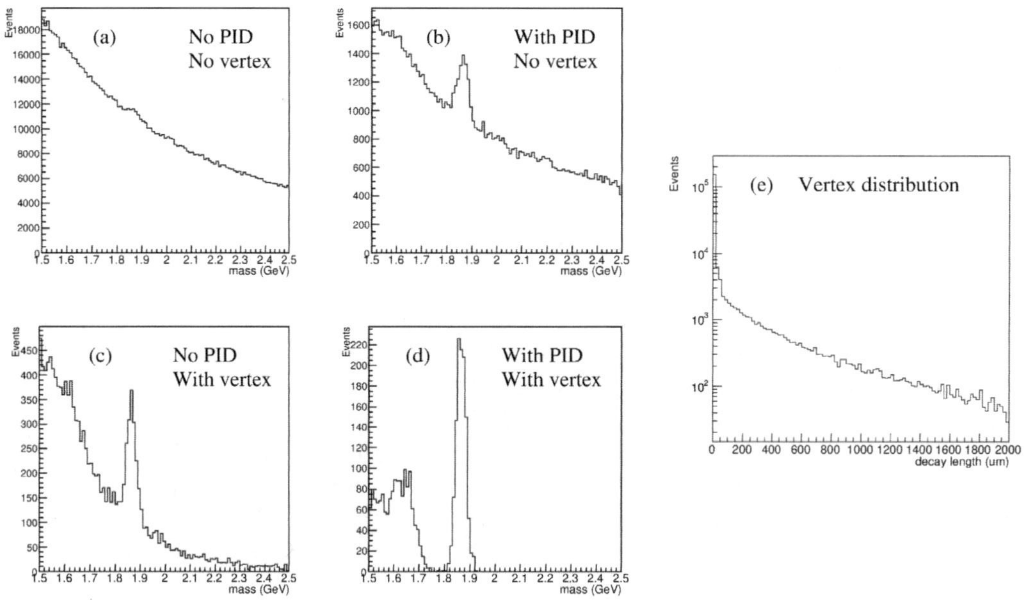


**Figure 4.** Impact of PID and vertex cuts on $D^0$ meson reconstruction from the $K^-\pi^+$ exclusive decay. Plots (a)–(d) show the invariant mass spectrum of two charged tracks/mesons in a sample of charm events with $Q^2 > 10$ GeV$^2$ and $x_B > 0.05$ (PYTHIA 6 simulation, arbitrary normalization of event sample, background from non-charm DIS events not included). (a) No PID (charged tracks), no vertex cut; (b) with PID ($K^-, \pi^+$), no vertex cut; (c) no PID, with vertex cut; (d) with PID and vertex cut. Plot (e) shows the vertex distribution of the $D^0$ decay in the sample. The vertex cut was applied at 100 $\mu$m.

## References

[1] U.S. Department of Energy Office of Science, Reaching for the horizon: The 2015 Long Range Plan for Nuclear Science, available at `http://science.energy.gov/np/nsac`
[2] Accardi A *et al.* 2016 Eur. Phys. J. A **52** 268
[3] For current information on the EIC machine designs, see: `https://eic.jlab.org/wiki/` (JLab) and `https://wiki.bnl.gov/eic/` (BNL).
[4] Accardi A, Guzey V, Prokudin A and Weiss C 2012 Eur. Phys. J. A **48** 92
[5] Boer D *et al.* 2011 Gluons and the quark sea at high energies: Distributions, polarization, tomography, *Preprint* arXiv:1108.1713 [nucl-th].
[6] Aaron F D *et al.* [H1 Collaboration] 2011 Eur. Phys. J. C **71** 1509 Abramowicz H *et al.* [ZEUS Collaboration] 2014 JHEP **1409** 127; Abramowicz H *et al.* [H1 and ZEUS Collaborations] 2013 Eur. Phys. J. C **73** 2311
[7] Abramowicz H *et al.* [ZEUS Collaboration], 2013 JHEP **1309** 058
[8] Adolph C *et al.* [COMPASS Collaboration], 2012 Eur. Phys. J. C **72**, 2253; 2013 Phys. Rev. D **87** 052018 [arXiv:1211.6849 [hep-ex]].
[9] Gluck M, Reya E and Stratmann M, 1994 Nucl. Phys. B **422** 37
[10] Baines J et al. 2006 Heavy quarks (Working Group 3): Summary Report for the HERA-LHC Workshop Proceedings, *Preprint* hep-ph/0601164.
[11] Malace S, Gaskell D, Higinbotham D and Cloet I, 2014 Int. J. Mod. Phys. E **23** 1430013
[12] Eskola K J, Paukkunen H and Salgado C A, 2009 JHEP **0904** 065
[13] Weiss C et al., JLab LDRD Project LD1601: Nuclear gluons with charm at EIC, `https://wiki.jlab.org/nuclear_gluons/`
[14] Chudakov E *et al.* 2016 Probing nuclear gluons with heavy quarks at EIC *Preprint* arXiv:1608.08686 [hep-ph].
[15] Harris B W and Smith J, 1998 Phys. Rev. D **57** 2806 (1998)

# The $\overline{\text{P}}$ANDA Detector at FAIR

**W Ikegami Andersson**
**on behalf of the $\overline{\text{P}}$ANDA collaboration**

Department of Physics and Astronomy, Uppsala University, Box 516, S-751 20 Uppsala, Sweden

E-mail: walter.ikegami_andersson@physics.uu.se

**Abstract.** The future $\overline{\text{P}}$ANDA detector at FAIR is a state-of-the-art internal target detector designed for strong interaction studies. By utilizing an antiproton beam, a rich and unique physics programme is planned. The $\overline{\text{P}}$ANDA experiment, as well as feasibility studies for hyperon and charmonium physics, are discussed.

## 1. Introduction

The accepted theory of the strong interaction is Quantum Chromodynamics (QCD). It describes the properties of quarks and their interactions through gluons, the force mediator of the strong interaction. QCD is very successful in predicting processes at high energies where the coupling constant $\alpha_s$ is small and perturbation theory is applicable. However, at low energies, the theory becomes strongly coupled as $\alpha_s$ grows large. In this non-perturbative regime, it is still hard to make predictions from first principles. This give rise to many questions: Are there effective degrees of freedom which can systematically describe resonances and bound states? Where are the exotic resonances and bound states predicted by QCD? The future $\overline{\text{P}}$ANDA detector is the ideal experiment to answer these questions. A major part of the $\overline{\text{P}}$ANDA physics programme is to collect large statistics and high-quality data and test QCD in the non-perturbative regime. In the following sections, the $\overline{\text{P}}$ANDA experiment is introduced and three aspects of its physics programme will be highlighted.

## 2. The $\overline{\text{P}}$ANDA experiment

The Facility for Antiproton and Ion Research (FAIR) is an accelerator facility under construction at GSI (Gesellschaft für Schwerionenforschung) in Darmstadt, Germany. The $\overline{\text{P}}$ANDA detector will be situated in the high-energy storage ring (HESR) at FAIR. HESR will provide an antiproton beam with momenta from 1.5 up to 15 GeV/c, which corresponds to center-of-mass energies between 2.0 and 5.5 GeV/c$^2$. HESR will support two modes of operation, high resolution mode and high luminosity mode. In the high luminosity mode, a luminosity of $\mathcal{L} \approx 2 \times 10^{32}$ cm$^{-2}$s$^{-1}$ will be provided. Stochastic cooling will be used to achieve a beam momentum resolution of $\Delta p/p = 10^{-4}$. The high resolution mode will provide a luminosity of $\mathcal{L} \approx 2 \times 10^{31}$ cm$^{-2}$s$^{-1}$ and a beam momentum resolution of $\Delta p/p = 5 \times 10^{-5}$ with the use of stochastic cooling. During the startup phase, the accumulator preceding HESR will not be present. At this stage, HESR will provide a luminosity of $\mathcal{L} \approx 10^{31}$ cm$^{-2}$s$^{-1}$.

The $\overline{\text{P}}$ANDA detector, shown in Figure 1, consists of a target spectrometer (TS) and a forward spectrometer (FS) which provide almost full coverage of the solid angle. A cluster jet or pellet target system will be used to provide either a cluster beam of a target gas or frozen hydrogen pellets. Thin foils will be used for $\bar{p}A$ studies. The interaction point is surrounded by the Micro Vertex Detector (MVD) which has a spatial resolution of 50 $\mu$m in x- and y-direction and 100 $\mu$m in the z-direction. In addition to the MVD, the Straw Tube Tracker (STT) and Gas Electron Multiplier (GEM) stations will be used for tracking charged particles ($\Delta p_t/p_t = 1.2\%$). Photons will be reconstructed with the Electromagnetic Calorimeter (EMC). The EMC consists of a barrel, a forward endcap and backward endcap and is made up of 17200 $PbWO_4$ crystals. Particle identification of pions, kaons and protons will utilize information from a Time-of-Flight (ToF) and a barrel Detection of Internally Reflected Cherenkov light (DIRC) detector. A solenoid magnet will provide a homogeneous magnetic field up to 2T in the beam direction. Outside of the solenoid, muon chambers are placed for identification of muons.

The FS covers polar angles below 10° horizontally and 5° vertically. Charged particles will be detected using the Forward Tracking System (FTS), which consists of multiple straw tube layers, in conjunction with a dipole magnet. The Forward Time-of-Flight (FToF) and the aerogel Ring Imaging Cherenkov Counter (FRICH) detectors will provide particle identification. Calorimetry is performed with a Shashlyk-type Calorimeter (FSC). Behind the FSC, the Muon Range System is placed for identification of muons. A detailed description of the $\overline{\text{P}}$ANDA detector and its components are given in Ref. [1]. All simulation studies in the following sections assume the full detector setup.

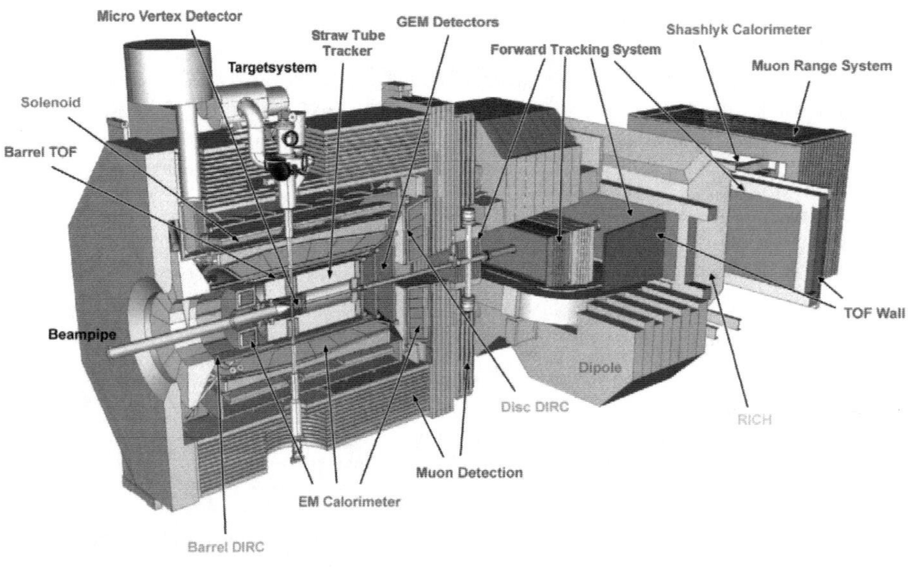


**Figure 1.** The $\overline{\text{P}}$ANDA detector setup.

## 3. Charmonium spectroscopy

One open question within the Standard Model is the existence of exotic states such as glueballs, hybrids and multiquark states, which are expected by QCD. From experiment, candidates for exotics exist, but none of them have been identified unambiguously. The field of charmonium spectroscopy is an exciting field with many new states discovered in the past 13 years. So far, all of the predicted charmonium states with masses below the $D\overline{D}$ threshold have been observed with properties in line with predictions. However, the situation above the $D\overline{D}$ threshold is less clear. Many predicted states have not yet been observed. On the other hand, unexpected states have been measured by various experiments *e.g.* BaBar, Belle, BESIII. The properties of these states are not in agreement with the Quark Model expectations of the charmonium spectra.

One of the most intriguing state is the $X(3872)$. It was first observed by Belle in 2003 in the decay $X(3872) \to J/\psi\pi^+\pi^-$ [2] and has since then been confirmed by several other experiments [3, 4, 5, 6, 7]. Its quantum numbers, $J^{PC} = 1^{++}$, have been determined by LHCb [8]. This makes the $X(3872)$ a candidate for the excited $\chi_{C1}(2P)$ state. However, since its mass is very close to the $D^{*0}\overline{D}^0$ threshold, it could also be a loosely bound $D^{*0}\overline{D}^0$ molecule or a virtual scattering state. In the case of a molecule or virtual state, the line-shape of the $X(3872)$ state is expected to be different from a Breit-Wigner [9, 10]. Thus, a precise measurement of the line-shape is the key to identify its nature. The upper limit for the width of the $X(3872)$ is $\Gamma < 1.2$ MeV, which was measured by Belle [11]. A more precise measurement of the width is unfeasible with currently running experiments as they are limited by their detector resolutions. By producing the $X(3872)$ in formation reactions $A + B \to X(3872)$, the width can be measured with higher precision by the means of a resonance scan. However, no current experiment can produce the quantum numbers $J^{PC} = 1^{++}$ in leading order. $\overline{\text{P}}$ANDA is the ideal experiment as $\overline{p}p$ annihilation can produce all non-exotic $J^{PC}$ quantum numbers directly.

To illustrate the potential of measuring the width of the $X(3872)$, a Monte Carlo study has been performed. The process $X(3872) \to J/\psi\pi^+\pi^- \to l^+l^-\pi^+\pi^-$ along with relevant background channels were generated at 40 scan points around the nominal center-of-mass energy with a step size of $dE \approx 70$ keV. A MC sample equivalent to two days of beam time was simulated for each scan point which corresponds to a total of 80 days of beam time. Both the high luminosity mode and the high resolution mode with the startup luminosity have been tested. In order to get an impression of the expected sensitivity and bias of a width measurement, the toy MC experiment was repeated $N_{\text{MC}} = 300$ times. A Breit-Wigner distribution was fitted to the scan points in the energy dependent yield distribution. The difference between the extracted width $\Gamma_{\text{meas}}$ and the input width $\Gamma_0$ in each simulation was noted. The sensitivity and bias are based on the mean value and the root-mean-square of the $\Gamma_{\text{meas}} - \Gamma_0$ distribution and are defined as

$$\frac{\Delta\Gamma_{\text{meas}}}{\overline{\Gamma}_{\text{meas}}} = \frac{\text{RMS}}{\text{Mean} + \Gamma_0} \pm \frac{\text{RMS}}{\text{Mean} + \Gamma_0}\sqrt{\frac{1}{2(N_{\text{MC}} - 1)} + \frac{\text{RMS}^2}{N_{\text{MC}} \cdot (\text{Mean} + \Gamma_0)^2}} \tag{1}$$

$$\frac{\overline{\Gamma}_{\text{meas}} - \Gamma_0}{\Gamma_0} = \frac{\text{RMS}}{\Gamma_0} \pm \frac{\text{RMS}}{\Gamma_0\sqrt{N_{\text{MC}}}}. \tag{2}$$

The sensitivity and bias are calculated for the following values of the input width $\Gamma_0 \in [50, 70, 100, 130, 180, 250, 500]$ keV. See Figure 2.

The sensitivity is expected to be better than 20% for input widths $\Gamma_0 > 90$ keV for high luminosity mode and $\Gamma_0 > 120$ keV for high resolution mode with the startup luminosity. Furthermore, the experimental bias is negligible within errors. A similar study where a $\overline{D}D^*$ molecule picture is assumed has also been performed. More details about the study, including a sensitivity test of a molecule description, can be found in Ref. [12].

## 4. Baryon spectroscopy

Baryon spectroscopy is another field that explores the dynamics of quarks and gluons confined into hadrons. Baryon spectroscopy has played an integral part in the development of QCD. One of the great successes of baryon spectroscopy is the formulation of the quark model. Starting in the 1950s, many new particles were discovered and soon it became clear that these particles could not all be elementary. Among the newly discovered particles, some seemed to match the representation of a SU(3) symmetry. The Eightfold Way was introduced in 1962 to organize

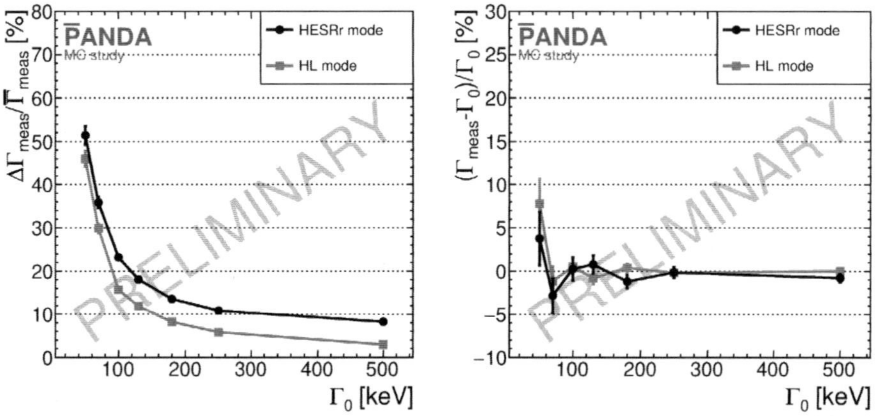


**Figure 2.** The expected sensitivity (left) and bias (right) for the high luminosity mode (red squares) and the high resolution mode (black circles) of HESR.

spin-$\frac{1}{2}$ baryons into an octet and spin-$\frac{3}{2}$ baryons into a decuplet. In particular, the spin-$\frac{3}{2}$ decuplet predicted the existence of a triple-strange particle with a mass of 1680 MeV/c$^2$. Two years later, the $\Omega^-$ was discovered which established the Eightfold way [13]. Following it, the quark model was formulated by Gell-Mann and Zweig [14][15].

Today, there are several experiments active in both light and heavy quark baryon spectroscopy. Efforts within light-baryon spectroscopy are mainly focused on $N^*$ and $\Delta$ resonances. These resonances have been well studied at various experiments in $\pi N$ and $\gamma N$ production channels. In the heavy quark $(c, b)$ sector, B-factories (BaBar and Belle) and LCHb have opened up the possibility of studying baryons with charmed and even bottom quarks. Charmed baryons can be produced in B-meson decays at BaBar and Belle and bottomed baryons can be produced in $Z^0 \to \bar{b}b$ decays at LHCb. However, the strangeness sector is not as active, which is reflected in scarce data of strange baryons. Currently, there is only one well established excited $\Xi^*$ resonance and no excited $\Omega^*$ resonances in the PDG [16].

The prospects of measuring excited multistrange baryons at $\overline{\text{P}}$ANDA have been investigated [17]. The $\overline{\Xi}^+\Xi^-(1820)$ + *c.c.*, where $\Xi^-(1820) \to \Lambda K^-$ and $\overline{\Xi}^+ \to \overline{\Lambda}\pi^+$, process has been simulated. The decay $\Xi^-(1820) \to \Lambda K^-$ was assumed to have a branching ratio of 100%. With a production cross section of $\sigma(\overline{p}p \to \overline{\Xi}^+\Xi^-(1820)) \approx 1\mu$b and the startup phase luminosity, an event rate of 15000 d$^{-1}$ is expected, shown in Table 1.

## 5. Hyperon dynamics

$\overline{\text{P}}$ANDA will produce a large amount of antihyperon-hyperon pairs through the reaction $\overline{p}p \to \overline{Y}Y$. In such reactions, the energy scale is given by the mass of the strange quark $m_s \approx 100$ MeV/c$^2$, which is below the QCD cut-off $\Lambda_{\text{QCD}}$ scale. In this energy region, it is unclear what the relevant degrees of freedom are. It could either be quarks and gluons or hadrons. Since perturbation theory is no longer applicable in this regime, alternative phenomenological models are needed. Descriptions of $\overline{p}p \to \overline{Y}Y$ have been made in both constituent quark-gluon pictures [18, 19, 20, 21, 22] as well as kaon exchange pictures [18, 23, 24, 25, 26]. There have also been efforts to combine the two approaches [27].

Spin observables are very powerful tools to discriminate between different model predictions. The PS185 collaboration has provided a large sample of single-strangeness hyperons produced in the $\overline{p}p \to \overline{Y}Y$ reaction [28]. The measurements of spin observables published by the PS185

Journal of Physics: Conference Series **770** (2016) 012043 doi:10.1088/1742-6596/770/1/012043

**Table 1.** Expected production rates of hyperons at $\overline{\text{P}}$ANDA with luminosity $\mathcal{L} = 10^{31}\text{cm}^{-2}\text{s}^{-1}$. The cross sections $\sigma(\overline{p}p \to \overline{\Lambda}\Sigma^0)$ and $\sigma(\overline{p}p \to \overline{\Xi}^+\Xi^-)$ are extrapolated values. The cross sections $\sigma(\overline{p}p \to \overline{\Omega}^+\Omega^-)$ and $\sigma(\overline{p}p \to \overline{\Lambda}_c\Lambda_c)$ are based on theoretical predictions.

| $p_{\overline{p}}$ (GeV/c) | Reaction | $\sigma$ ($\mu$b) | Eff (%) | Decay | Rate |
|---|---|---|---|---|---|
| 1.64 | $\overline{p}p \to \overline{\Lambda}\Lambda$ | 64 | 10 | $\Lambda \to p\pi^-$ | 28 $\text{s}^{-1}$ |
| 4.0 | $\overline{p}p \to \overline{\Lambda}\Sigma^0$ | ~40 | 30 | $\Sigma^0 \to \gamma$ | 30 $\text{s}^{-1}$ |
| 4.0 | $\overline{p}p \to \overline{\Xi}^+\Xi^-$ | ~2 | 20 | $\Xi^- \to \Lambda\pi^-$ | 2 $\text{s}^{-1}$ |
| 12.0 | $\overline{p}p \to \overline{\Omega}^+\Omega^-$ | ~0.002* | ~30 | $\Omega^- \to \Lambda K^-$ | ~4 $\text{h}^{-1}$ |
| 12.0 | $\overline{p}p \to \overline{\Lambda}_c\Lambda_c$ | ~0.1* | ~30 | $\Lambda_c \to \Lambda\pi^+$ | ~2 $\text{d}^{-1}$ |
| 4.6 | $\overline{p}p \to \overline{\Xi}^+\Xi^-(1820)$ | 1 | 5 | $\Xi^-(1820) \to \Lambda K^-$ | 15000 $\text{d}^{-1}$ |

collaboration have been decisive in testing different polarization predictions and have sparked ambitions of measuring and testing spin observables in $\overline{p}p \to \overline{Y}Y$ reactions with multi-strange as well as charmed hyperons.

The parity violating weak decay of hyperons means that the spin observables are accessible in the angular distributions of their decay particles. For example, when considering the decay $\Lambda \to p\pi^-$, the $\Lambda$ polarization manifests in the angular distribution of the decay proton $I(\cos\theta_p) = \frac{1}{4\pi}(1+\alpha_\Lambda P_n \cos\theta_p)$, where $\alpha_\Lambda = 0.642\pm0.013$ [16] is the decay asymmetry parameter. The polarization is also accessible from hyperons which decay into other hyperons. In the decay $\Xi^- \to \Lambda\pi^-$, where $\Lambda \to p\pi^-$, the polarization as well as the $\beta$ and $\gamma$ decay asymmetry parameters are accessible in the joint angular distribution of the $\Lambda$ hyperon and the proton. Hyperons with spin-$\frac{3}{2}$ such as the $\Omega^-$ have seven polarization parameters. Three polarization parameters are accessible in the angular distribution of the decay $\Omega^- \to \Lambda K^-$. In the case of the decay $\Lambda \to p\pi^-$, the remaining four polarization parameters are obtained in the joint angular distribution of the $\Lambda$ hyperon and the proton.

$\overline{\text{P}}$ANDA will have an unpolarized beam and unpolarized target. In these settings, two spin observables are accessible; the polarization and the spin correlation. Furthermore, with the increased center-of-mass energy achievable, multi-strange as well as single-charmed hyperon-antihyperon pairs can be produced. Monte Carlo studies have been performed to explore the possibilities of measuring spin observables [29, 30, 31]. In Table 1, the expected production rates are given. The rates are calculated assuming a luminosity of $\mathcal{L} = 10^{31}$ $\text{cm}^{-2}\text{s}^{-1}$. The cross section of $\overline{p}p \to \overline{\Lambda}\Lambda$ is well known. The $\overline{\Lambda}\Sigma^0$ and $\overline{\Xi}^+\Xi^-$ cross sections are obtained by extrapolating measurements at other energies. The $\overline{\Omega}^+\Omega^-$ and $\overline{\Lambda}_c^-\Lambda_c^+$ cross sections are based on theoretical predictions [32, 33, 34, 35, 36]. From the expected rates, the spin observables of single-strange hyperon production can be measured with unprecedented precision during the first year of data taking. Furthermore, the spin observables of double-strange hyperons can be measured for the first time. As more data is accumulated the following years, the polarization parameters of the $\Omega^-$ hyperon can be measured as well.

## 6. Other physics topics

$\overline{\text{P}}$ANDA has a broad physics programme, comprising many topics which were not described in these proceedings. The topics include light meson spectroscopy, open charm physics, structure of nucleons, in-medium properties of hadrons and CP-violation studies. The modular design of $\overline{\text{P}}$ANDA also opens up the possibility to produce and study hypernuclei.

## 7. Summary

During the first year of operation, the spin observables in single-strange hyperon production in $\overline{p}p \to \overline{Y}Y$ processes will be measured with higher statistical precision and for the first time, spin observables of double-strange hyperon production will be measured. Furthermore, the full spectrum of single- and double-strange excited hyperons can be measured with unprecedented statistical precision. As the facility matures and more data is collected, the polarization parameters of the $\Omega^-$ can be measured for the first time and the full triple-strange excited hyperon spectrum will be accessible. By producing the $X(3872)$ in direct formation, an energy scan of its line-shape can be performed which will allow to distinguish between the different theoretical interpretations of its nature. With 80 days of beam time, such a measurement is sensitive to widths an order of magnitude smaller than the current upper limit for the width of the $X(3872)$.

## References

[1] Kotulla M *et al.* PANDA Collaboration 2005 *PANDA Technical Progress Report*
[2] Choi S-K *et al.* Belle Collaboration 2003 *Phys. Rev. Lett.* **91**, 262001
[3] Auber B *et al.* Babar Collaboration 2005 *Phys. Rev. D* **71** 071103
[4] Acosta D *et al.* CDF II Collaboration 2004 *Phys. Rev. Lett.* **93** 072001
[5] Abazov V M *et al.* D0 Collaboration 2004 *Phys. Rev. Lett.* **93** 162002
[6] Aaij R *et al.* LHCb Collaboration 2012 *Eur. Phys. J. C* **72** 1972
[7] Chatrchyan S, Khachatryan V *et al.* CMS collaboration 2013 *J. High Energ. Phys.* **1304** 154
[8] Aaij R *et al.* LHCb Collaboration 2013 *Phys. Rev. Lett.* **110** 222001
[9] Hanhart C *et al.* 2007 *Phys. Rev. D* **76** 034007
[10] Braaten E and Lu M 2008 *Phys. Rev. D* **77**, 014029
[11] Choi S-K *et al.* Belle Collaboration 2011 *Phys. Rev. D* **84** 052004
[12] Nerling F 2016 *Talk at the 11th International Workshop on Heavy Quarkonium*
[13] Barnes V E *et al.* 1964 *Phys. Rev. Lett.* **12** 204
[14] Gell-Mann M 1964 *A Schematic Model of Baryons and Mesons Phys. Lett.* **8** 214
[15] Zweig G 1964 *An SU(3) model for strong interaction symmetry and its breaking. Version 1* CERN-TH-401
[16] K A Olive *et al.* 2014 *Particle Data Group Chin. Phys. C* **38** 090001.
[17] Pütz J 2016 *Proceedings of FAIRNESS 2016* Garmisch-Partenkirchen, Germany
[18] Kohno M and Weise W 1986 *Phys. Lett. B* **179** 15
[19] Rubinstein H R and Snellman H 1985 *Phys. Lett. B* **165** 187
[20] Furui S and Faessler A 1987 *Nucl. Phys. A* **468** 669
[21] Burkardt M and Dillig M 1988 *Phys. Rev. C* **37** 1362
[22] Alberg M A *et al.* 1988 *Z. Phys. A* **331** 207
[23] Tabakin F and Eisenstein R A *Phys. Rev. C* **31**, 1857 (1985)
[24] La France P *et al.* 1988 *Phys. Lett. B* **214** 317
[25] Timmermans R G E *et al.* 1992 *Phys. Rev. D* **45** 2288
[26] Haidenbauer J *et al.* 1992 *Phys. Rev. C* **46** 2516
[27] Ortega P G *et al.* 2011 *Phys. Lett. B* **696** 352
[28] Johansson T 2003 *Proceedings of 8th Int. Conf. on Low Energy Antiproton Physics* 95
[29] Erni W *et al.* 2009 PANDA collaboration *Physics Performance Report* arXiv: 0903.3905 [hep-ex]
[30] Thomé E 2012 *Multi-Strange and Charmed Antihyperon-Hyperon Physics for PANDA* (Ph. D. Thesis, Uppsala University)
[31] Grape S 2009 *Studies of PWO Crystals and Simulations of the* $\overline{p}p \to \overline{\Lambda}\Lambda, \overline{\Lambda}\Sigma^0$ *Reactions for the PANDA experiment* (Ph.D. Thesis, Uppsala University)
[32] Kaidalov A B and Volkovitsky P E Z 1994 *Phys. C* **63** 51
[33] Titov A I and Kampfer B 2008 *Phys. Rev. C* **78** 025201
[34] Gornitschnig A T *et al.* 2009 *Eur. Phys. J. A* **42** 43
[35] He J *et al.* 2011 *Phys. Rev. D* **84** 114010
[36] Khodjamirian A *et al.* 2012 *Eur. Phys. J. A* **48** 31

Journal of Physics: Conference Series **770** (2016) 012044 doi:10.1088/1742-6596/770/1/012044

# Belle II Physics Prospects, Status and Schedule

**J Bennett**

Department of Physics, Carnegie Mellon University, 5000 Forbes Ave, Pittsburgh, PA 15213

E-mail: jvbennett@cmu.edu

**Abstract.** The second generation $B$-factory at the SuperKEKB facility in Tsukuba, Japan is beginning to take shape. The highly anticipated Belle II experiment will have a rich physics program at the intensity frontier, in complement to existing experiments in the energy frontier. Accelerator commissioning has been making good progress, as has the construction and installation of the Belle II detector. An overview of the physics prospects at Belle II, as well as the status and schedule of the experiment, is presented.

## 1. Introduction

The Belle II experiment inherits a very successful flavor physics program from its progenitors, the Belle and BaBar experiments. The first generation $B$-factories boasted many important results, including experimental confirmation of the CKM mechanism as the source of CP violation (CPV) in the Standard Model (SM). This decisive evidence lead to the awarding of the 2008 Nobel prize in physics to Makoto Kobayashi and Toshihide Maskawa "for the discovery of the origin of the broken symmetry which predicts the existence of at least three families of quarks in nature" [1].

Together, the Belle and BaBar collaborations recorded a total data sample of over 1.5 $ab^{-1}$, or about $1.25 \times 10^9$ $B\bar{B}$ pairs. Most of this was collected at the center of mass energy of the $\Upsilon(4S)$, which is just above the threshold for $B$-meson pair production. From the time of the first measurement of the CKM parameter $\sin 2\beta$ that established CPV in the B system in 2001 to the state of the art in 2015, global fits to the data relevant to the CKM unitarity triangle have made a remarkable improvement. Current results show excellent agreement between the SM and results from the $B$ factories and LHCb [2]. Even with the improved precision, however, there is still room for contributions from new physics such as flavor-changing neutral currents (FCNC), lepton flavor violation (LFV), and new sources of CPV. A new physics (NP) amplitude in $B_d$ mixing, on the scale of about 20 (2) TeV for tree-level (one-loop) processes, is perfectly compatible with the current data [3]. With the size of the data samples expected at Belle II, it should be possible to improve the current limits on contributions from NP by a factor of five for all CP violating phases.

In contrast to experiments at the LHC, which search for new physics at the energy frontier, Belle II will search for NP in the flavor sector at the intensity frontier, relying on massive data samples and high precision measurements for its physics program. While LHCb, for example, relies on the direct production of new particles and is limited by beam energy, Belle II will search for signatures of new particles or processes through measurements of suppressed flavor physics reactions or in deviations from SM predictions. Any observed discrepancies could then be interpreted in terms of NP models. Studies like these at the "flavor frontier" will enable searches for NP in virtual particle loops on a scale of up to 100 TeV [3]. Of course, the Belle II physics program is not entirely dedicated to CKM measurements. Dark sector searches, LFV, and studies of QCD exotics are just a few more examples of the rich physics program that will be pursued in the next decade at Belle II.

## 2. SuperKEKB and Belle II

In order to reach the design goal to collect a 50 $ab^{-1}$ data sample at Belle II, the KEKB accelerator has undergone a significant upgrade. The new SuperKEKB accelerator makes use of nanobeams to reach the design luminosity of $8 \times 10^{35}$ $cm^{-2}s^{-1}$, an impressive 40 times higher than that of KEKB. This improvement is achieved in part by greatly reducing the beam size. Of course, luminosity of this rate will result in much higher background rates, leading to greater detector occupancy, radiation damage, and other effects that must be considered. The data acquisition system (DAQ) and computing resources have also been upgraded to accommodate the burden of a higher event rate.

The Belle II detector builds upon the success of its predecessor in order to enable a strong physics research program at SuperKEKB. The colliding electron and positron beams are slightly asymmetric, resulting in a boosted center of mass (CM) system to enable time-dependent CPV measurements. With a coverage of greater than 90% of $4\pi$, the hermiticity of the Belle II detector is improved relative to that of Belle due to a reduction in the CM boost. Most of the Belle II data will be collected at the $\Upsilon(4S)$ resonance, just above the threshold for $B$ meson pair production, with a goal of collecting 50 $ab^{-1}$ by the year 2024.

While many components of the Belle II detector are based on the design of the Belle detector, many improvements have been made to cope with the conditions at SuperKEKB and to improve the expected performance [4]. Immediately surrounding the beam-pipe are upgraded silicon vertex detectors, including two pixel layers (PXD) based on DEPFET sensors and four layers of double-sided silicon strip sensors (SVD). A large central drift chamber (CDC) will provide tracking with precise momentum measurements as well as particle identification (PID) through measurements of ionization energy loss. Surrounding the CDC is the upgraded barrel PID system, which consists of an imaging time of propagation counter (TOP). A focusing aerogel ring imaging Cherenkov detector (ARICH) provides PID measurements in the endcap region. The electromagnetic calorimeter (ECL) has been equipped with upgraded electronics to suppress the much higher beam-related backgrounds expected at SuperKEKB. The barrel region of the $K_L$ and muon detector (KLM) consists primarily of resistive plate chambers. The innermost layers of the barrel region and the entirety of the endcap region is made up of plastic scintillators to cope with the high neutron flux both from the interaction point and from the beamline. A schematic view of the Belle II detector is given in Fig. 1.

## 3. Physics prospects

There are several significant advantages of pursuing a physics program at Belle II. For instance, Belle II will operate in a lower background environment relative to hadron machines like LHCb. The flavor tagging efficiency at Belle II is also superior to that at LHCb by about an order of magnitude. Belle II can measure $K_S$ and $K_L$ particles, which has an impact on most time-dependent CPV measurements. A large sample of $\tau$ leptons will enable studies of rare decays and searches for LFV in a low background environment. Interestingly, the systematics of Belle II will be quite different from those of LHCb. This allows for important confirmation by one experiment of any hints of NP observed by the other.

One of the most important advantages of SuperKEKB and Belle II is the ability to collect very clean samples of quantum-correlated $B\bar{B}$ pairs by running at the $\Upsilon(4S)$ resonance. By fully reconstructing one of the $B$ mesons, it is possible to tag the flavor of the other $B$, determine its momentum, and isolate the tracks of its decay particles. In particular, this allows for a study of reactions with final state neutrinos. Due to the low efficiency of full reconstruction tagging (on

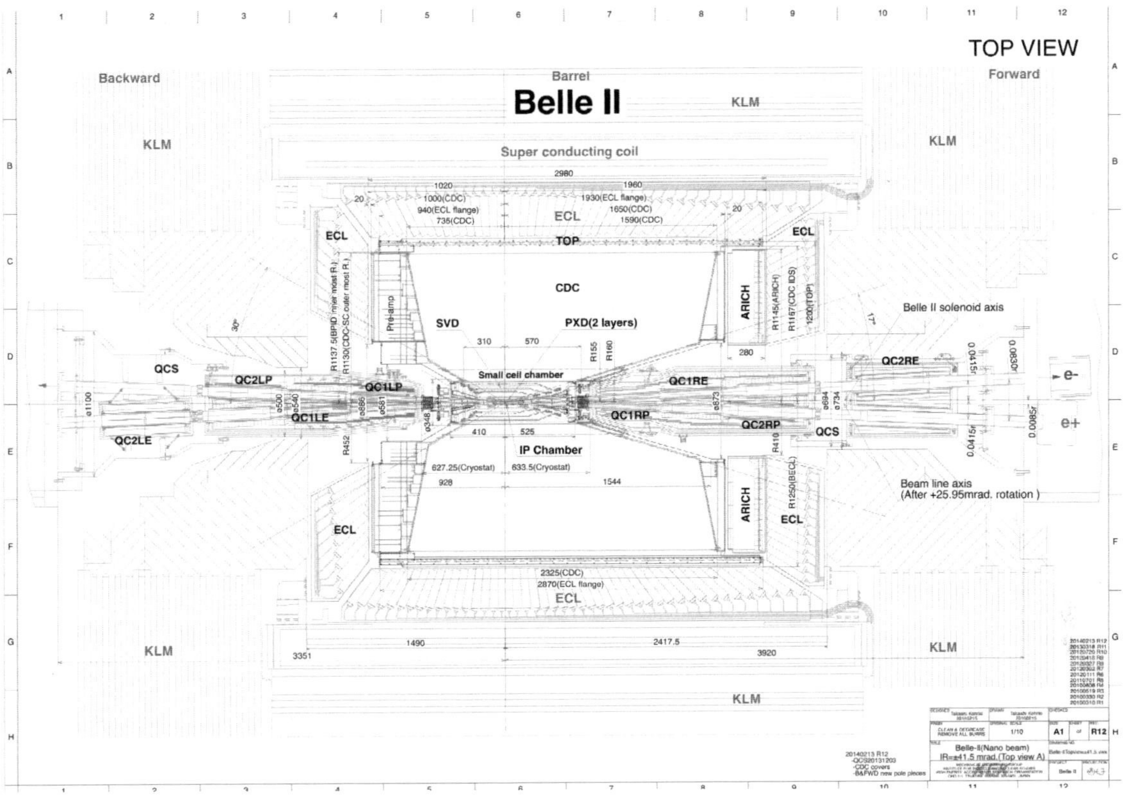


**Figure 1.** A schematic view of the Belle II detector.

the order of 0.1%), the first generation B factories did not make extensive use of this method. However, given the high statistics samples that will be available at Belle II, this analysis method will be used for precision measurements of CKM elements and searches for NP.

Some of the earliest physics results to come out of Belle II could be in the area of bottomonium spectroscopy. There has been considerable progress recently in Lattice QCD [5] and Belle II has the opportunity to search for some of the missing states. The clean environment at Belle II allows for inclusive searches for new states by reconstructing a single resonance and searching the recoil system. This method of study was successful at Belle and BaBar, for example in the study of P-wave spin-singlet states and the discovery of the $Z_b$ states [6, 7]. Indeed XYZ spectroscopy provides an interesting area of research with much still to be done to quantify and confirm the plethora of states that have been observed recently.

One of the biggest questions that Belle II will seek to explore is whether or not there are any new CP-violating phases, which are included in most theories involving NP. Some of these theories allow for large deviations from SM predictions for $B$-meson decays. Additional sources of CPV may be important to understand the origin of the observed matter–antimatter asymmetry of the universe. Therefore, it will be important to search for new sources of CPV by comparing the mixing-induced CP asymmetries in penguin transitions with tree-dominated modes. One example of this is in the study of time-dependent CPV in $B$ decays to $\phi K^0$, $\eta' K^0$, and $K^0 K^0 K^0$, similar to the Belle analysis of the same reactions [8]. With the full Belle II data sample, the uncertainty on these measurements will be on the order of 1-3% [9]. This level of precision will be important to search for any discrepancies with respect to $B$ decays to $J/\psi K^0$. Any such discrepancies would be an important potential source of evidence for NP.

Belle II will also employ other probes for NP, including studies of radiative and electroweak processes like $b \to s\gamma$, $b \to d\gamma$, and $b \to sll$. In helicity-changing NP models like the left-right symmetric model, the helicity-suppressed amplitude could be enhanced in transitions like $B^0 \to K_S^0\pi^0\gamma$ [10], leading to a sizable CP asymmetry. With the improved vertex position resolution and reconstruction efficiency of $K_S^0$ decays at Belle II, the precision of measurements of this reaction could reach about 3.5%.

Leptonic $B$ decays provide an experimentally challenging environment to test some NP models. The final states contain one or more neutrinos and the signal sides include only a single charged track. The average branching fraction measurement of $B$ decays to $\tau\nu$ using hadronic [11, 12] and semileptonic [13, 14] tags at Belle and BaBar is $(1.14 \pm 0.22) \times 10^{-4}$, slightly higher than the SM prediction of $7.58 \times 10^{-5}$ [2]. Decays like these are useful for placing constraints on models that include a charged Higgs. Using fully reconstructed hadronic and semileptonic tags at Belle II, it should be possible to measure the branching fraction of $B$ decays to $\tau\nu$ with a precision of 3-5%. Interestingly, leptonic $B$ decays should also provide a competitive measurement of $|V_{ub}|$ with the 50 ab$^{-1}$ data sample.

Another useful avenue to study the two-Higgs-doublet model, one of the simplest possible extensions of the SM, is in semileptonic $B$ decays, which proceed via first-order electroweak interactions. Reactions of this type involving electrons and muons are less sensitive to non-SM contributions, but can be used to measure the CKM elements $|V_{cb}|$ and $|V_{ub}|$. Experimentally challenging decays involving a $\tau$ lepton are sensitive to additional amplitudes and provide a means to search for NP. The most recent measurements of $\mathrm{R}(D)$ and $\mathrm{R}(D^*)$ pose a combined significance of 4.0 $\sigma$ disagreement with the SM [15]. With only a 5 ab$^{-1}$ sample, Belle II should also be able to confirm the excess in the $\mathrm{R}(D^*)$ flavor anomaly [16, 17]. With the full 50 ab$^{-1}$, Belle II should be able to measure $\mathrm{R}(D)$ and $\mathrm{R}(D^*)$ with a precision of 3.5% and 2%, respectively.

The charm sector provides a unique environment to probe for NP since CP violating effects are expected to be small in this sector. Mixing in heavy flavor systems is dominated by short-distance effects, which are suppressed in the charm system. This makes $D^0 - \bar{D}^0$ mixing a promising avenue for NP searches. However, the sizable effects of hadronic interactions pose a challenge to studies in the charm sector. Belle II will make significant contributions to searches for CPV in charm decays by measuring final states containing neutral particles, which are necessary for isospin analyses to isolate penguin contributions.

Lepton flavor violation is highly suppressed in the SM. Various NP models yield branching ratios for these processes on the order of $10^{-7}$-$10^{-10}$ [18, 19, 20, 21, 22]. The hadron machines are generally not competitive with Belle II for $\tau$ decay measurements due to trigger and other limitations. With the full data sample, Belle II can access LFV decay rates of over 100 times smaller than those of Belle for the cleanest channels.

## 4. SuperKEKB status and schedule

The tentative schedule for SuperKEKB indicates that physics quality data should be available in the fall of 2018. Construction and installation of the Belle II detector is currently ongoing. The simple background-conditioning "BEAST" detector was used for phase 1 commissioning starting in February of 2016. Phase 2 commissioning is scheduled to begin in the fall of 2017 and will consist of a more elaborate inner background commissioning detector and the full Belle II outer detector. Full superconducting final focus will be available, so it is possible that early physics results could be available during phase 2. However, the vertex detector will not be installed, so the phase 2 physics program will be limited. Phase 3 running with the full detector is expected to begin in the fall of 2018, with a sample of about 300 fb$^{-1}$ during the first run. An overview of the SuperKEKB schedule is given in Fig. 2.

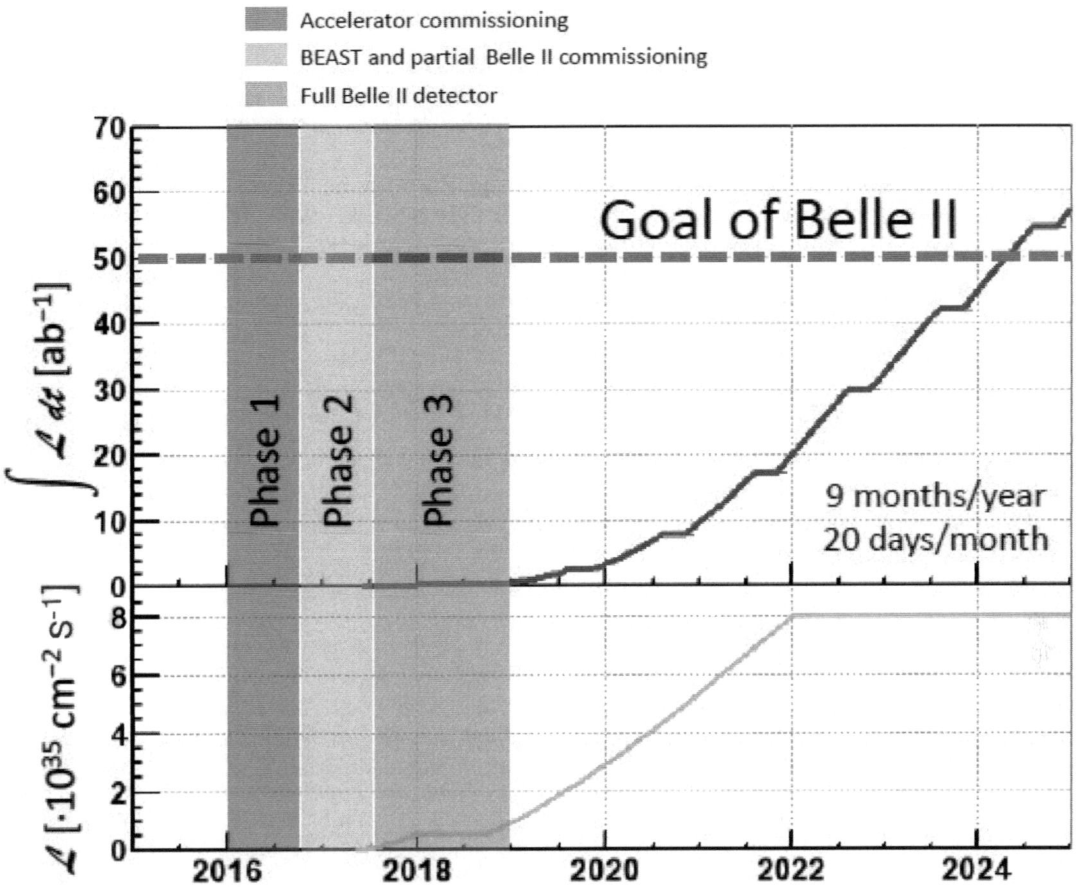


**Figure 2.** An overview of the projected luminosity and operation at SuperKEKB.

## 5. Summary

The major upgrades at KEK represent an essentially new experiment as SuperKEKB succeeds KEKB and Belle II takes the reins from the first generation $B$-factories. Many detector and electronic components have been replaced in the transition from Belle to Belle II. Significant efforts have also led to improvements in the available software and analysis tools. With these improvements, Belle II is expected to pursue a rich physics program, complementary to that of existing experiments and the energy frontier program. SuperKEKB commissioning is ongoing in 2016 and first physics are possible as early as 2017. With a goal of collecting 50 $ab^{-1}$ by 2024, Belle II is expected to make a significant impact on flavor physics and the search for new physics.

## References

[1] "The 2008 Nobel Prize in Physics - Press Release". Nobelprize.org. Nobel Media AB 2014. Web. 17 Aug 2016.
`http://www.nobelprize.org/nobel_prizes/physics/laureates/2008/press.html`
[2] CKMFitter Group (J. Charles et al.), Phys. Rev. D **91**, 073007 (2015).
[3] CKMFitter Group (J. Charles et al.), Phys. Rev. D **89**, 033016 (2014).
[4] T. Abe et al, KEK Report 2010-1 (2010), arXiv: 1011.0352v1
[5] S. Godfrey and K. Moats, Phys. Rev. D **92**, 054034 (2015).

[6] I. Adachi et al. (Belle Collaboration), Phys. Rev. Lett. **108**, 032001 (2012).
[7] A. Bondar et al. (Belle Collaboration), Phys. Rev. Lett. **108**, 122001 (2012).
[8] K. F. Chen et al. (Belle Collaboration), Phys. Rev. Lett. **98**, 031802 (2007).
[9] P. Urquijo, Nucl. Phys. B. **263-264** 15-23 (2015).
[10] Y. Ushiroda et al. (Belle Collaboration), Phys. Rev. D **74**, 111104(R) (2006).
[11] K. Hara et al. (Belle Collaboration), Phys. Rev. Lett. **110**, 131801 (2013).
[12] J. P. Lees et al. (BABAR Collaboration), Phys. Rev. D **88**, 031102(R) (2013).
[13] K. Hara et al. (Belle Collaboration), Phys. Rev. D **82**, 071101(R) (2010).
[14] B. Aubert et al. (BABAR Collaboration), Phys. Rev. D **81**, 051101(R) (2010).
[15] Heavy Flavour Averaging Group, `http://www.slac.stanford.edu/xorg/hfag/semi/winter16/winter16_dtaunu.html` (2016) [Accessed August 15, 2016]
[16] J. P. Lees et al. (BABAR Collaboration), Phys. Rev. Lett. **109**, 101802 (2012).
[17] A. Abdesselam et al. (Belle Collaboration), arXiv:1603.06711.
[18] G. Cveti, C. Dib, C. S. Kim, and J. D. Kim, Phys. Rev. D **66**, 034008 (2002).
[19] C. Yue, Y. Zhang, and L. Liu, Phys. Lett. B **547**, 252 (2002).
[20] T. Fukuyama, T. Kikuchi, and N. Okada, Phys. Rev. D **68**, 033012 (2003).
[21] J. Ellis, J. Hisano, M. Raidal, and Y. Shimizu, Phys. Rev. D **66**, 115013 (2003).
[22] A. Brignole and A. Rossi, Phys. Lett. B **566** 217 (2003).

# Particle Identification at Belle II

**S Sandilya**
**On behalf of the Belle II Collaboration**
University of Cincinnati, Cincinnati, Ohio 45221
E-mail: saurabhsandilya@gmail.com

**Abstract.** We report on the charged particle identification (PID) systems for the upcoming Belle II experiment. The time of propagation counter in the central region and the proximity focusing ring imaging Cherenkov counters with aerogel radiator in the forward region will be used as the PID devices. They are expected to provide a kaon identification efficiency of more than 94% at a low pion misidentification probability of 4%. The motivation for the upgrade, method and status of both systems are discussed.

## 1. Introduction

Reliable particle identification (PID) is essential for any high energy physics experiment, and in particular, it is crucial for the success of $B$-factories [1]. In the $B$-factories, PID is required to tag $B$-meson flavour for CP violation studies in the neutral $B$-meson system, and to suppress backgrounds in precision measurements of rare $B$ and $D$ decays.

Based on the successful operation of the Belle experiment, the KEKB [2] is being upgraded to the SuperKEKB collider, which is designed to give an instantaneous luminosity of $8 \times 10^{35}$ cm$^{-2}$s$^{-1}$. This is about 40 times larger than its predecessor. The Belle detector [3] is also being upgraded to the Belle II detector [4] in order to be able to perform in the upgraded accelerator conditions and to achieve the desired physics goals. To cope with high rates as well as with more stringent requirements on separation capabilities for rare decay channels, more efficient PID is needed for the momentum range up to 4 GeV/$c$. In Belle II, PID will be performed by the Time-Of-Propagation (TOP) counter in the central (barrel) region and the Aerogel Ring Imaging Cherenkov (ARICH) counter in the forward endcap region. Figure 1 shows a schematic of the upper half vertical cross-section of the Belle II detector in which both the PID systems are shown.

The working principle of both the TOP and ARICH counters is based on imaging the Cherenkov rings. The Cherenkov photons are emitted by a charged particle moving with a velocity ($v = \beta c$) exceeding the speed of light ($c/n$) in a dielectric medium (refractive index $n(\lambda)$) at a characteristic angle ($\theta_c$) given by, $\cos\theta_c = 1/n(\lambda)\beta$. Charged particles of the same momentum but having different masses will have different velocities, and hence the angle of Cherenkov photons will be different.

In this proceedings, we report the design, method and present status of the PID systems in both the regions of the Belle II detector.

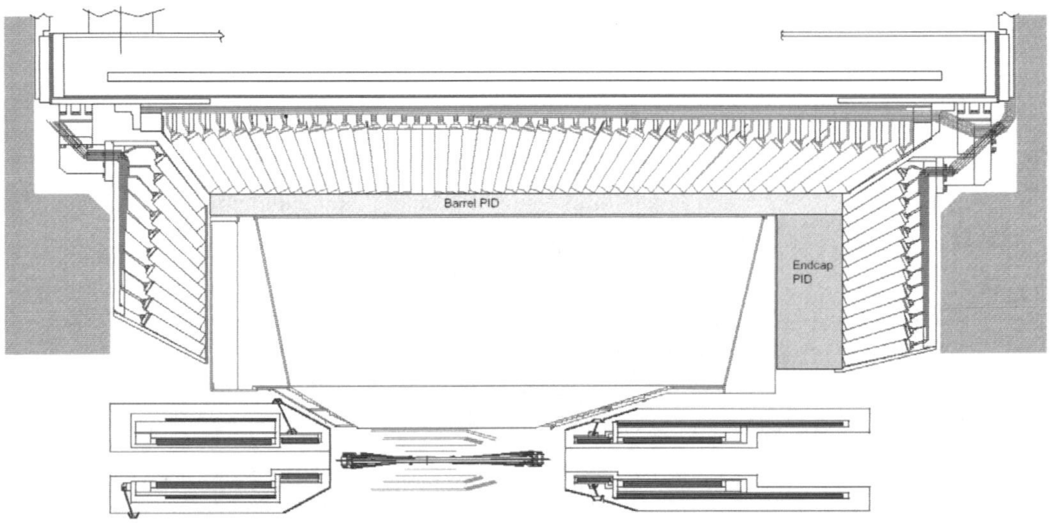


**Figure 1.** Schematic of the upper half vertical cross-section of the Belle II detector. The rectangular region shaded with cyan color represents the TOP counters in the barrel region and the orange shaded rectangular region represents ARICH counters in the forward endcap region.

## 2. TOP counters

A TOP counter module primarily consists of a quartz radiator bar, micro-channel plate photomultipliers (MCPPMTs) and a front-end readout. The Cherenkov photons emitted from a charged particle in the quartz bar are internally reflected toward the detection plane. The obtained Cherenkov ring image is characterized by its time of arrival and impact position in the detector plane [5]. Two quartz bars each having dimension $(1250 \times 450 \times 20)$ mm$^3$, a mirror with dimension $(100 \times 450 \times 20)$ mm$^3$, and a small expansion prism of dimension $(100 \times 456 \times 20 - 51)$ mm$^3$ are epoxied with an optical glue to form the quartz radiator, as shown in the Figure 2 (left). The radiator is enclosed in a box made of aluminum honeycomb panels and is supported by PEEK polymer buttons to leave an air gap between the radiator and the panels. The mirror has a radius of curvature of about 6500 mm. As such, it focuses parallel rays in the Cherenkov cone to a single point and thus removes the effect of the bar thickness (see Figure 2 (right)). Non-parallel rays are focused to different points of the detector plane, which gives a better charged particle separation and also allows to correct for chromatic dispersion. Optical components of the radiator should be of high quality to prevent any photon loss during propagation or reflections. To ensure this, several quality acceptance requirements are needed [6].

Cherenkov photons, emitted in the radiator due to passage of a charged particle, go through total internal reflections and are finally register at the expansion volume end by an array of MCPPMTs. Different path lengths are traversed by the photons depending on $\theta_c$, which results in a different time of propagation. The arrival time, including the time of flight of the charged particle and the time of propagation of emitted photons, and position in the detector plane of each detected Cherenkov photon are used to compute a likelihood for a given particle mass hypothesis.

A square-shaped MCPPMT with an effective area $(23 \times 23)$ mm$^2$ has been developed in collaboration with Hamamatsu Photonics KK for the TOP counter [7][8]. It has a multi-alkali photocathode whose average quantum efficiency is about 28% for wavelengths around 380 nm. In the MCPPMT, there are two multi-channel plates (MCP) of thickness 400 $\mu$m with a bore

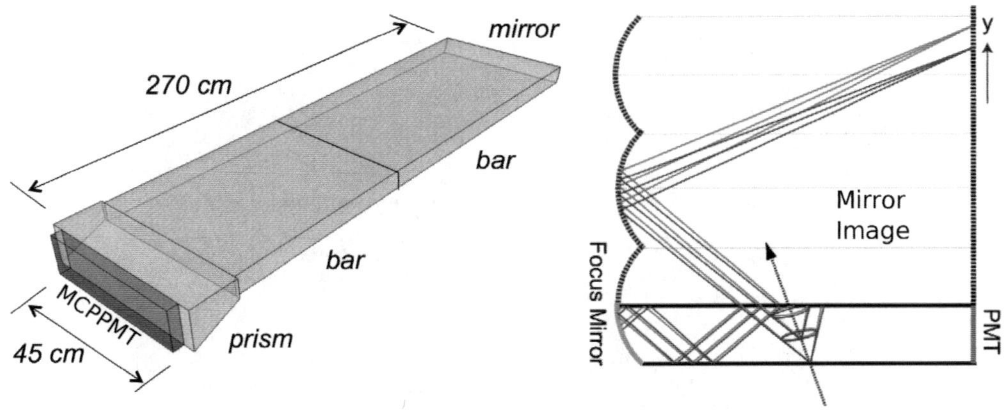


**Figure 2.** A schematic diagram of the quartz radiator of the TOP counter is shown on the left. The need of the mirror in the counter is explained on the right. Parallel (non-parallel) rays are focused at the same (different) points of the detection plane by the mirror.

size diameter of 10 $\mu$m. The bias angle of the pore is 13°. The readout from each MCPPMT anode is divided into $4 \times 4$ channels of area $(5.3 \times 5.3)$ mm$^2$. Each module contains two rows of 16 MCPPMTs, giving in total 32 MCPPMTs or 512 anode channels per module. The photo-electron emitted by the photo-cathode is multiplied in the MCP pores and then detected at one of the anode channel. The transit time spread is about 35 ps. By applying the nominal operating bias, a gain of $10^6$ is achieved.

The MCPPMTs are physically mounted on the front-end electronics modules. The main components of the front-end electronics are: front board to host the MCPPMT array; high voltage board to provide high voltages to the MCPPMTs; Application-Specific Integrated Circuits (ASICs); and Standard Control, Read-Out, and Data (SCROD) board. The ASICs consist of switched-capacitor array memory for the fast sampling and storage of time-sampled analog waveforms. The SCROD is the main controller of the entire module and acts as an entry/exit command/data point to the unit [9][10].

A small prototype of the TOP counter was tested at the 1.2 GeV/c positron beam at LEPS (Laser Electron Photon beam line at SPring-8) in June 2013. For the beam test, the positron beam was tracked by the LEPS spectrometer and the beam timing was obtained from the accelerator RF. Data were taken with the beam hitting the prototype at normal incidence and also at some angle. In the latter case, the emitted Cherenkov photons will mostly be reflected from the mirror and then reach the detector plane. For both cases of beam incidence, results in data and simulations were found to be in agreement. In a single event, the number of detected photons was 20-30. An integrated distribution in data and simulations over many events for the normal incidence case is shown in Figure 3.

## 3. ARICH counters

The ARICH counter is basically a proximity focusing RICH. Its main components are: aerogel tiles as a radiator, an array of position sensitive photon detectors, and a readout system [11][12][13]. The aerogel radiator consists of two layers with increasing refractive indices along the particle path so that the Cherenkov photons emitted by each layer overlap at the photo-detection plane, as shown in Figure 4 (left). The two aerogel tiles are of thickness 20 mm each and have refractive indices of 1.045 and 1.055 for upstream and downstream tiles, respectively. This arrangement of two layers with different refractive indices gives better performance than a single aerogel layer for the entire thickness (40 mm), as the latter arrangement leads to degradation

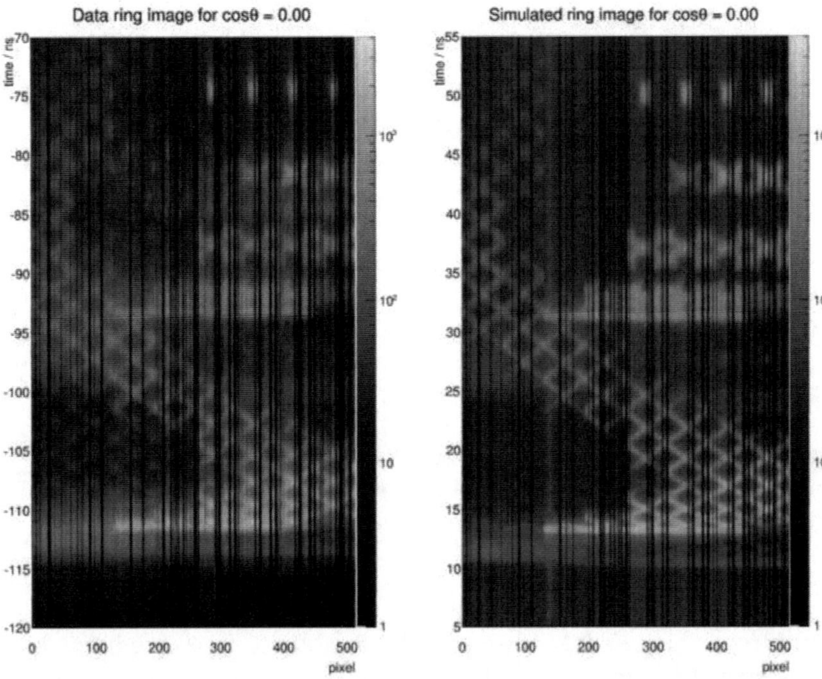


**Figure 3.** Distributions of time (ns) *vs.* MCPPMT channels, in data from the beam test at SPring-8 (left) and simulations (right) for the case of beam hitting the TOP prototype at a normal incidence.

of the single photon resolution because of an increase in emission point uncertainty. The single photon resolution ($\sigma_\theta \sim 14$ mrad) for the dual radiator is significantly better than for the case of single radiator ($\sigma_\theta \sim 21$ mrad) without affecting the number of photons detected (Figure 4).

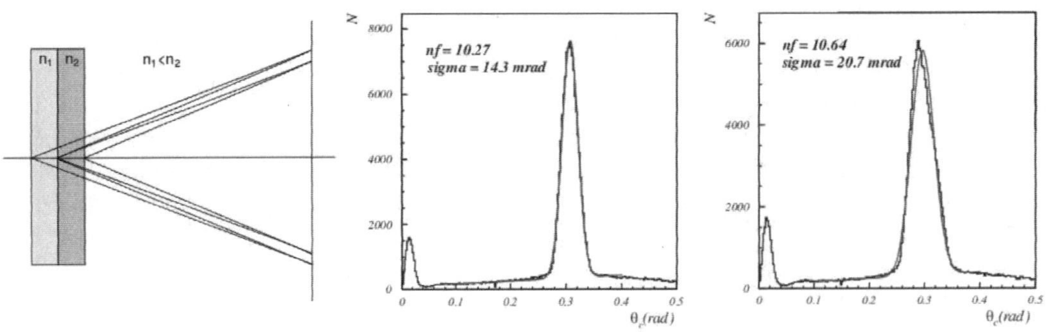


**Figure 4.** Schematic of the focusing scheme with two aerogel layers of different refractive indices (left). Number of Cherenkov photons *vs.* $\theta_c$ for the focusing configuration (center) and for a 40 mm single homogeneous aerogel layer (right) [14].

The photo-detector for the ARICH should be sensitive to single photon detection, able to provide position information, be immune to the 1.5 T magnetic field perpendicular to the photon detection plane, and be tolerant of the high radiation environment. A Hybrid Avalance Photo-

Detector (HAPD) was developed in a joint effort with Hamamatsu. The HAPD consists of a bialkali photocathode, a vacuum region with a high electric field, and four avalanche photo-diode (APD) chips that are further pixelated in $6 \times 6$ pixels giving in total 144 pixels of size ($4.9 \times 4.9$ mm$^2$). An incident photon is coverted into a photo-electron by the photocathode, which has a peak quantum efficiency of about 28% at 400 nm. The photo-electron subsequently accelerates towards the pixelated APDs via a high electric field and gets multiplied due to bombardment gain by a factor of about 1800. An additional gain of about 40 is achieved from the avalanche process of the APDs, resulting an overall gain of 70,000. The ARICH covers 3.5 m$^2$ in the forward endcap of the Belle II detector; the aerogel radiator plane and the photo-detection plane are separated by a distance of 200 mm. The radiator plane consists of 124 pairs (dual-layer, in total 248) of aerogel tiles while the photo-detection plane contains 420 HAPDs. At the outermost edge of these two planes, planar mirrors (in total, 18) are placed to redirect the outside-going photons towards the detection plane. Dedicated high gain and low noise electronics were developed for the readout. To each HAPD a front-end board with four ASICs and a field programmable gate array is attached. The digitized hit information is collected by a merger board from front-end boards, and then communicated to further stages of the data acquisition system [15].

A prototype ARICH was tested at the DESY test beam in May 2013. A photograph of the setup is shown in Figure 5 (left). The accumulated hit distribution with respect to the track position is shown in Figure 5 (right), in which a Cherenkov ring is clearly visible. The Cherenkov angle resolution is found to be 15.8 mrad, and on average 9 photons per track are detected.

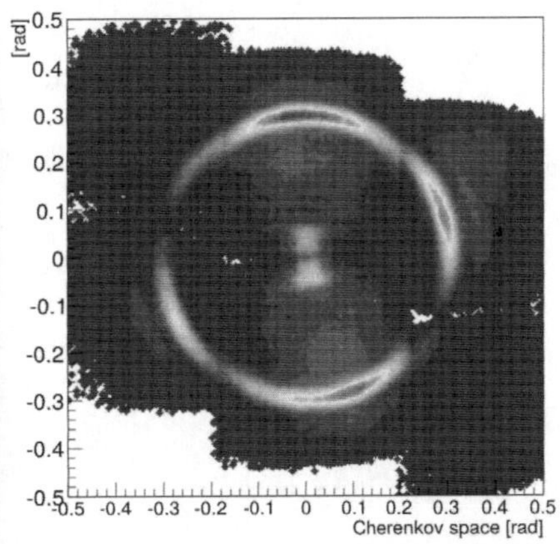


**Figure 5.** Photograph of the prototype of ARICH for the DESY beam-test(left). Accumulated photon hit distribution with respect to the track position (right).

## 4. Summary and Status

The TOP and ARICH detectors will provide PID information in the barrel and forward endcap region, respectively, of the Belle II detector. The working principle of both the detectors is based on imaging the Cherenkov ring created by the passage of charged particles. The TOP utilizes the impact position and time of arrival of the Cherenkov photons at the detection plane after total internal reflections in the quartz radiator bar to obtain PID information, whereas the ARICH is a proximity focusing RICH detector. Prototypes of both the detectors demonstrated their expected performance in test-beams. All 16 TOP counters have been successfully installed in the Belle II detector. Installation of the HAPDs and the aerogel tiles has already started and will be finished by the Summer of 2016.

Detailed simulations have been performed with the Belle II software framework, based on which we expect an excellent charged kaon efficiency ($> 90\%$) with a very small ($< 10\%$) charged pion misidentification probability.

## References

[1] Krizan P, 2009 *Journal of Instrumentation* **4**, P11015.
[2] Kurokawa S and Kikutani E 2003 *Nucl. Instrum. Methods Phys. Res., Sect.* A **499** 1 and other papers included in this volume; Abe T *et al.* 2013 *Prog. Theor. Exp. Phys.* 03A001 and following articles up to 03A011.
[3] Abashian A *et al.* (Belle Collaboration) 2002 *Nucl. Instrum. Methods Phys. Res., Sect.* A **479** 117–232; also, see the detector section in Brodzicka J *et al.* 2012 *Prog. Theor. Exp. Phys.* 04D001.
[4] Abe T*et al.* (Belle II Collaboration), arXiv:1011.0352.
[5] Ratcliff B N 2003 *Nucl. Instrum. Methods Phys. Res., Sect.* A **502** 211–221.
[6] Wang B 2013 *Nucl. Instrum. Methods Phys. Res., Sect.* A **766** 204–207.
[7] Inami K 2008 *Nucl. Instrum. Methods Phys. Res., Sect.* A **595** 96–99.
[8] Akatsu M 2004 *Nucl. Instrum. Methods Phys. Res., Sect.* A **528** 763–775.
[9] Andrew M 2012 *IEEE Realtime Conf. Rec.* 1–5.
[10] Andrew M 2014 *PoS* **(TIPP2014)** 171.
[11] Matsumoto T *et al.* 2004 *Nucl. Instrum. Methods Phys. Res., Sect.* A **521** 367–377.
[12] Iijima T *et al.* 2005 *Nucl. Instrum. Methods Phys. Res., Sect.* A **548** 383–390.
[13] Nishida S *et al.* 2014 *Nucl. Instrum. Methods Phys. Res., Sect.* A **766** 28–31.
[14] Korpar S *et al.* 2007 *Nucl. Instrum. Methods Phys. Res., Sect.* A **572** 429–431.
[15] Nishida S *et al.* 2012 *Phys. Procedia* **37** 1730–1735.

Journal of Physics: Conference Series **770** (2016) 012046 doi:10.1088/1742-6596/770/1/012046

# Search for $K^+ \to \pi^+ \nu\bar{\nu}$ at NA62

**M Perrin-Terrin on behalf of the NA62 Collaboration**[1] CERN, Switzerland

E-mail: mathieu.perrin-terrin@cern.ch

**Abstract.** The $K^+ \to \pi^+ \nu\bar{\nu}$ branching ratio is one of the key observables to test the Standard Model. The NA62 experiment has been designed to measure this branching ratio with a 10% precision. Most of the experiment systems have been commissioned during 2014 and 2015 runs and physics data have been recorded. The analysis of this data shows performances close to the nominal ones.

[1] for the NA62 Collaboration:
G. Aglieri Rinella, R. Aliberti, F. Ambrosino, R. Ammendola, B. Angelucci, A. Antonelli, G. Anzivino, R. Arcidiacono, I. Azhinenko, S. Balev, M. Barbanera, J. Bendotti, A. Biagioni, L. Bician, C. Biino, A. Bizzeti, T. Blazek, A. Blik, B. Bloch-Devaux, V. Bolotov, V. Bonaiuto, M. Boretto, M. Bragadireanu, D. Britton, G. Britvich, M.B. Brunetti, D. Bryman, F. Bucci, F. Butin, E. Capitolo, C. Capoccia, T. Capussela, A. Cassese, A. Catinaccio, A. Cecchetti, A. Ceccucci, P. Cenci, V. Cerny, C. Cerri, B. Checcucci, O. Chikilev, S. Chiozzi, R. Ciaranfi, G. Collazuol, A. Conovaloff, P. Cooke, P. Cooper, G. Corradi, E. Cortina Gil, F. Costantini, F. Cotorobai, A. Cotta Ramusino, D. Coward, G. D'Agostini, J. Dainton, P. Dalpiaz, H. Danielsson, J. Degrange, N. De Simone, D. Di Filippo, L. Di Lella, S. Di Lorenzo, N. Dixon, N. Doble, B. Dobrich, V. Duk, V. Elsha, J. Engelfried, T. Enik, N. Estrada, V. Falaleev, R. Fantechi, V. Fascianelli, L. Federici, S. Fedotov, M. Fiorini, J. Fry, J. Fu, A. Fucci, L. Fulton, S. Gallorini, S. Galeotti, E. Gamberini, L. Gatignon, G. Georgiev, A. Gianoli, M. Giorgi, S. Giudici, L. Glonti, A. Goncalves Martins, F. Gonnella, E. Goudzovski, R. Guida, E. Gushchin, F. Hahn, B. Hallgren, H. Heath, F. Herman, T. Husek, O. Hutanu, D. Hutchcroft, L. Iacobuzio, E. Iacopini, E. Imbergamo, O. Jamet, P. Jarron, E. Jones, T. Jones K. Kampf, J. Kaplon, V. Kekelidze, S. Kholodenko, G. Khoriauli, A. Khotyantsev, A. Khudyakov, Yu. Kiryushin, A. Kleimenova, K. Kleinknecht, A. Kluge, M. Koval, V. Kozhuharov, M. Krivda, Z. Kucerova, Yu. Kudenko, J. Kunze, G. Lamanna, G. Latino, C. Lazzeroni, G. Lehmann-Miotto, R. Lenci, M. Lenti, E. Leonardi, P. Lichard, R. Lietava, L. Litov, R. Lollini, D. Lomidze, A. Lonardo, M. Lupi, N. Lurkin, K. McCormick, D. Madigozhin, G. Maire, C. Mandeiro, I. Mannelli, G. Mannocchi, A. Mapelli, F. Marchetto, R. Marchevski, S. Martellotti, P. Massarotti, K. Massri, P. Matak, E. Maurice, A. Mefodev, E. Menichetti, E. Minucci, M. Mirra, M. Misheva, N. Molokanova, J. Morant, M. Morel, M. Moulson, S. Movchan, D. Munday, M. Napolitano, I. Neri, F. Newson, A. Norton, M. Noy, G. Nuessle, T. Numao, V. Obraztsov, A. Ostankov, S. Padolski, R. Page, V. Palladino, G. Paoluzzi, C. Parkinson, E. Pedreschi, M. Pepe, F. Perez Gomez, M. Perrin-Terrin, L. Peruzzo, P. Petrov, F. Petrucci, R. Piandani, M. Piccini, D. Pietreanu, J. Pinzino, I. Polenkevich, L. Pontisso, Yu. Potrebenikov, D. Protopopescu, F. Raffaelli, M. Raggi, P. Riedler, A. Romano, P. Rubin, G. Ruggiero, V. Russo, V. Ryjov, A. Salamon, G. Salina, V. Samsonov, C. Santoni, G. Saracino, F. Sargeni, V. Semenov, A. Sergi, M. Serra, A. Shaikhiev, S. Shkarovskiy, I. Skillicorn, D. Soldi, A. Sotnikov, V. Sugonyaev, M. Sozzi, T. Spadaro, F. Spinella, R. Staley, A. Sturgess, P. Sutcliffe, N. Szilasi, D. Tagnani, S. Trilov, M. Valdata-Nappi, P. Valente, M. Vasile, T. Vassilieva, B. Velghe, M. Veltri, S. Venditti, P. Vicini, R. Volpe, M. Vormstein, H. Wahl, R. Wanke, P. Wertelaers, A. Winhart, R. Winston, B. Wrona, O. Yushchenko, M. Zamkovsky, A. Zinchenko.

## 1. Introduction

The $K^+ \to \pi^+\nu\bar{\nu}$ decay occurs in the Standard Model mainly through the electroweak penguin and box diagrams such as the ones represented in Figure 1.

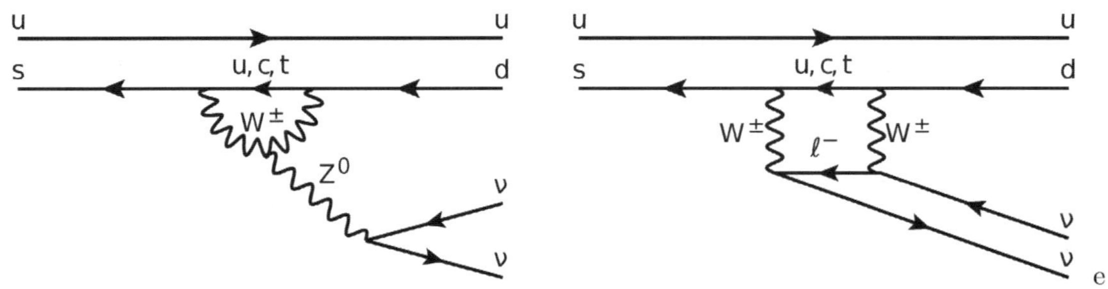


**Figure 1.** Examples of electroweak penguin and box Feynman diagrams contributing to $K^+ \to \pi^+\nu\bar{\nu}$ in the SM.

These diagrams are suppressed by the Glashow-Iliopoulos-Maiani mechanism and by the Cabibbo-Kobayashi-Maskawa (CKM) off-diagonal matrix elements $V_{is}^{\star}V_{id}$. Hence the decay is very rare in the SM, its branching ratio is [1]

$$\begin{aligned}\mathcal{B}(K^+ \to \pi^+\nu\bar{\nu}) &= (8.4 \pm 0.3) \times 10^{-11} \left(\frac{|V_{cb}|}{0.0407}\right)^{2.8} \left(\frac{\gamma}{73.2}\right)^{0.74} \\ &= (8.4 \pm 1.0) \times 10^{-11}.\end{aligned}$$

The precision of this prediction is remarkably good and dominated by the uncertainties on the external CKM parameters $|V_{cb}|$ and $\gamma$, one of the angle of the CKM triangle. The pure theoretical uncertainties is below 4%. Such a precision is achievable as the transition is dominated by perturbatively computable short-range processes, the Feynman diagrams with the top quark loop, for which the hadronic matrix elements can be derived from the well measured $K^+ \to \pi^0 e^+\nu$ branching ratio.

The $K^+ \to \pi^+\nu\bar{\nu}$ decay branching ratio allows to probe physics beyond the SM. Several recent publications [1, 2] highlighted the sensitivity of the decay up to mass scales as high as 100 TeV as well as its interplay with other observables being currently measured. For example, the correlation between the branching ratios of $K^+ \to \pi^+\nu\bar{\nu}$ and $K_L \to \pi^0\nu\bar{\nu}$ is one of the theoretically cleanest bench mark to probe the SM, as most of the CKM parametric uncertainties affecting the two branching ratios cancels in their correlation. A similar cancellation happens in the correlation between the branching ratios of $K^+ \to \pi^+\nu\bar{\nu}$ and $B_s^0 \to \mu^+\mu^-$ [1].

Experimentally, measuring such a rare decay with only one charged track in the final state is a real challenge. Searches for $K^+ \to \pi^+\nu\bar{\nu}$ have started decades ago and the most sensitive analysis, published by the E949 Collaboration [3], gives a value of

$$\mathcal{B}(K^+ \to \pi^+\nu\bar{\nu}) = (17.3^{+11.5}_{-10.5}) \times 10^{-11}. \quad (1)$$

The NA62 Collaboration aims at measuring the decay branching ratio with a 10% precision with the data that will be collected by the end of 2018. After a pilot run in 2014, the experiment is taking physics data since 2015.

## 2. Experimental Strategy

### *2.1. The NA62 Apparatus*

The NA62 experiment [4] is installed in the CERN North Area at the SPS and schematically represented in Figure 2.

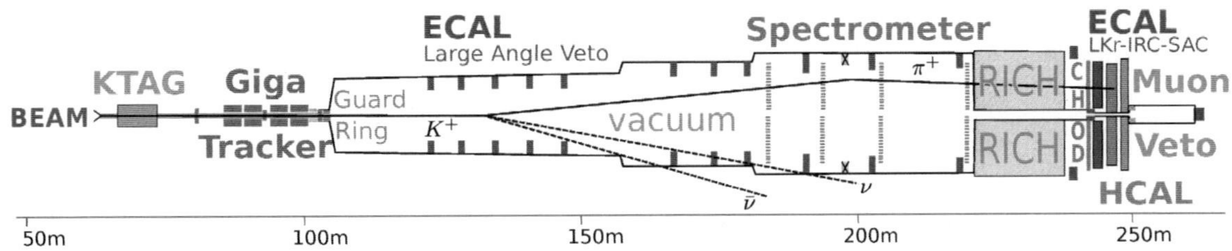


**Figure 2.** Schematic view of the NA62 experiment, excluding the beam line.

Assuming a 10% signal acceptance and a signal to background ratio of 10/1, the experiment needs to collect $10^{13}$ $\mathsf{K}^+$ decay. The beam line was designed to meet this goal using the 400 GeV protons from a SPS beam line. The beam entering the experiment is un-separated and composed at 70% of protons, 24% of $\pi^+$'s and 6% of $\mathsf{K}^+$'s with a total instantaneous nominal particle rate of 750 MHz. The beam particle momentum distribution peaks at 75 GeV and its standard deviation is 1%. The SPS extraction line delivers continuous spills during few seconds and operates in cycles with duty factors of around 20%. The experiment and the beam line compose a 300 m long evacuated tank with a pressure of $10^{-6}$ mbar reaching diameters up to almost 4 meters. In the downstream region the tank is instrumented with two sets of detectors measuring either the beam particles or the decay products properties. The beam being continuous, all detectors must also provide sub nanosecond timing information. In order to achieve the 12 orders of magnitude of background suppression needed for the measurement, the experiment employs many independent experimental techniques.

The $\mathsf{K}^+$ in the beam are identified by the KTAG, a nitrogen differential Cherenkov detector. The momentum and directions of all beam particles are measured with the GigaTracker (GTK) made of three 3×6 cm$^2$ thin planes of time-resolved silicon hybrid pixels. The beam then enters the 100 m long decay region. In order to reduce the background induced by inelastic collisions in the last GTK plane, a guard ring of scintillators is installed downstream of the GTK to collect hits from the collisions fragments.

The most abundant background source is the $\mathsf{K}^+ \to \pi^+\pi^0$ decay where the $\gamma$ from the $\pi^0$ are missed. The experimental challenge is to reduce the inefficiency on $\gamma$ from by 8 orders of magnitude. To reach this goal an electromagnetic calorimeter (ECAL) has been designed to to detect at least one of these two $\gamma$ up to 50 mrad. The ECAL is made of four subsystems. Photons with polar angles between 8.5 and 50 mrad are detected with the 12 ring lead-glass Large Angle Vetoes (LAV). The Liquid Krypton Calorimeter (LKr) covers the region between 8.5 mrad and 1 mrad. The inner edges of the LKr hole along the beam line are covered by a shashlik calorimeter called Inner Radius Calorimeter (IRC). Finally another shashlik calorimeter called Small Angle Calorimeter is installed at the end of the beam line, downstream of a bending magnet sweeping away the un-decayed beam particles and covers the angular region down to 0 mrad. The direction and momentum of the charged decay products are measured by a magnetic straw tube spectrometer operating inside the vacuum tank. The charged decay products are timed with 200 ps resolution by a fast array of scintillators called Charged Hodoscope (CHOD). The identification of the decay products are obtained by three systems: a ring imaging Cherenkov detector filled with neon at atmospheric pressure; a hadronic calorimeter (HCAL) made of two layers of iron-scintillator sandwiches, and finally a muon veto installed behind a iron wall and made of fast scintillators.

The experiment data acquisition system employs a multi-layer trigger system. At the lowest level, hits time in the CHOD, RICH and MUV and, calorimetric variables in the LKr and HCAL are computed using FPGA's.These data are sent to a central processor, operating also

Journal of Physics: Conference Series **770** (2016) 012046 doi:10.1088/1742-6596/770/1/012046

on FPGA, that builds and issues trigger requests upon which the state of all detectors, but LKr and GTK, in the time window corresponding to the trigger are readout. Based on the detectors states, algorithms running on a PC farm issue refined trigger requests upon which the remaining detectors are readout and the event assembled and stored.

In 2015 the hardware and readout of the detectors have been commissioned with a beam intensity of 10% of the nominal one. The beam line has however been commissioned up to nominal intensity. The GigaTracker was running with a partial hardware configuration. Data have been collected with a $\mathsf{K}^+ \to \pi^+\nu\bar{\nu}$ trigger using the calorimetric variables at L0 and with a minimum bias trigger based on the CHOD timing. Section 3, describes the analysis of the minimum bias sample in view of the $\mathsf{K}^+ \to \pi^+\nu\bar{\nu}$ branching ratio measurement.

### *2.2. Analysis Strategy*

The signal signature consists of an isolated $\pi^+$ track making a good vertex with a $\mathsf{K}^+$ track. Two main sources of events can mimic this signal: the incorrectly reconstructed $\mathsf{K}^+$ decays and the beam activity. The analysis is performed using the candidate square missing mass, define as:

$$m_m \quad = \quad |p_{\mathsf{K}^+} - p_{\pi^+}|^2, \tag{2}$$

where $p_{\mathsf{K}^+}$ and $p_{\pi^+}$ are the 4-momenta of the mother and daughter particles under the $\mathsf{K}^+$ and $\pi^+$ mass hypothesis. The distribution of this variable for the main $\mathsf{K}^+$ decay modes and for the signal is shown in Figure 3. The signal candidates are sought in the two regions around the $\mathsf{K}^+ \to \pi^+\pi^0$ missing mass peak and among the candidates with a downstream track momentum between 15 and 35 GeV. To suppress further the remaining backgrounds, kinematics, timing, photon rejection and particle identification are needed.

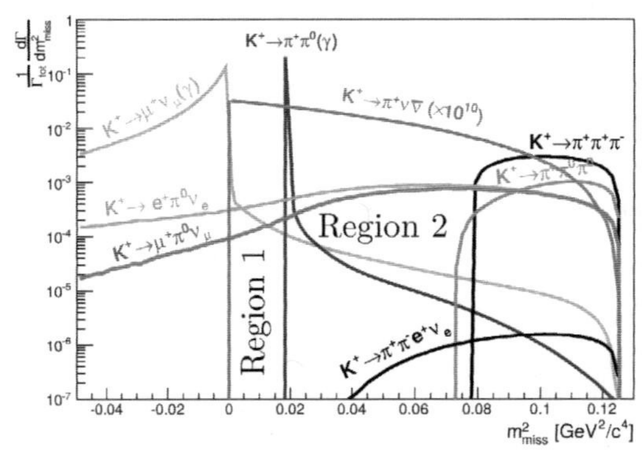


**Figure 3.** Squared missing mass theoretical distribution of the main $\mathsf{K}^+$ decays and the $\mathsf{K}^+ \to \pi^+\nu\bar{\nu}$ decay normalised by their branching ratios. The $\mathsf{K}^+ \to \pi^+\nu\bar{\nu}$ distribution is scaled by 10 orders of magnitude.

## 3. Performance in view of the 2015 minimum bias data

This section presents an analysis of data collected in 2015 with the minimum bias trigger in view of assessing the detector performance for the $\mathsf{K}^+ \to \pi^+\nu\bar{\nu}$ analysis. The kinematics, particle identification and photon rejection are assessed using a sample of single track events.

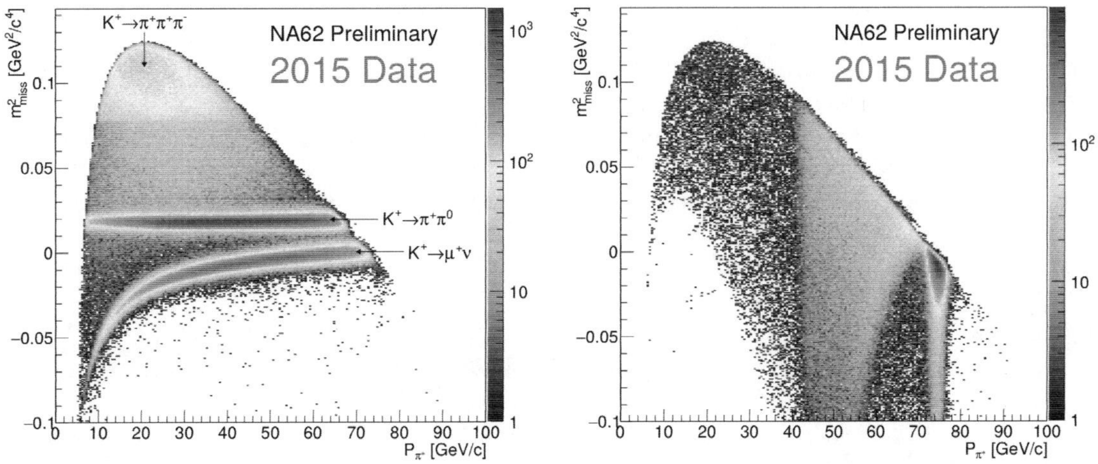


**Figure 4.** Squared missing mass distribution as a function of the straw tack momentum for the single track events with the GTK track identified (left) or not identified (right) as a $K^+$ by the KTAG.

### *3.1. Single Track Events*

The signal signature consists of a isolated $\pi^+$ track making a good vertex with a $K^+$ track. Hence signal candidates are selected asking for a good isolated spectrometer track, matched in space with energy depositions in the calorimeters and with a hit in the CHOD. The track is isolated if no other spectrometer tracks form a good vertex with it inside the fiducial decay region defined as the 65m long region between the last GTK station the the first spectrometer plane. In addition this track is requested to make a good vertex with a GTK track and its associated CHOD hit to be in time with a KTAG candidate. The squared missing mass of the selected candidates as a function of the spectrometer track momentum is shown in Figure 4 left.

Reverting the positive KTAG identification allows to study the beam activity. The distribution in Figure 4 right shows that the beam activity consists mainly of beam $\pi^+$ decay, beam particle elastic scattering and inelastic scattering in the last GTK plane. This beam activity is suppressed thanks to the good time resolution of the CHOD, KTAG and GTK measured to 200, 100 and 200 ps respectively, in agreement with the nominal design.

### *3.2. Kinematics Performance*

The control of the resolution of the $K^+ \to \pi^+\pi^0$ squared missing mass peak is essential to reduce the tails entering in the two signal regions. This resolution is measured to be around $1.2 \times 10^{-3}$ GeV$^2$/c$^4$, close to the design value ($1.0 \times 10^{-3}$ GeV$^2$/c$^4$). Note that without the GTK momentum measurement, i.e. assuming the nominal beam parameters, this resolution is three times larger.

The kinematics $K^+ \to \pi^+\pi^0$ background suppression in the signal region can be evaluated with a pure sample of $K^+ \to \pi^+\pi^0$ selected from the previous single track events with the additional requirement of having two $\pi^0$ $\gamma$'s in the LKr. A background suppression of $10^{-3}$ is found while the nominal one is $10^{-4}$. The discrepancy is understood to be caused by the partial GTK hardware setup leading to mis-reconstructed tracks.

### *3.3. Particle Identification Performance*

The particle identification system has to provide a rejection factor of $10^{-7}$ for $K^+ \rightarrow \mu^+\nu$ on top of the kinematic rejection. The $\pi^+/\mu^+$ separation is evaluated using the previous $K^+ \rightarrow \pi^+\pi^0$ sample and a $K^+ \rightarrow \mu^+\nu$ sample selected from the single track event asking a hit in the MUV. With the 2015 mirrors alignment, the RICH allows to get a $10^{-2}$ $\mu^+$ rejection for a 80% $\pi^+$ efficiency for track momenta between 15 and 35 GeV/c. Efficiencies closer to 90% are expected to be reached once the mirrors will be realigned. The background suppression from the calorimeter is under investigation. Simple cut based technique allows to reach a $10^{-4}$ to $10^{-6}$ $\mu^+$ rejection for a 90% to 50% $\pi^+$ efficiency.

### *3.4. Photon Vetoes Performance*

The photon veto system has to provide a rejection factor of $10^{-8}$ on $\pi^0$ from for $K^+ \rightarrow \pi^+\pi^0$. The momentum cut on the $\pi^+$ below 35 GeV/c insures that the $\pi^0$ has at least a momentum of 40 GeV/c and is emitted forward. Hence the single $\gamma$ rejection in the angular region below 50 mrad has only to be $10^{-5}$ for $\gamma$'s with energies above 10 GeV. The rejection factor is measured using the $K^+ \rightarrow \pi^+\pi^0$ selected within the single track event. The measurement of the $K^+ \rightarrow \pi^+\pi^0$ rejection is statistically limited and the rejection is found to be at least $10^{-6}$ at 90% CL. The effect of random veto on the signal is measured using $K^+ \rightarrow \mu^+\nu$ and scattered beam $\pi^+$'s and leads to a signal efficiency of 90%.

## 4. Conclusions and Prospects

The $K^+ \rightarrow \pi^+\nu\bar{\nu}$ branching ratio is one of the key observables to test the SM. The NA62 experiment is designed to measured its branching ratio with a 10% precision. The experiment is built and most of the systems have been commissioned. The data taken in 2015 shows performances close to the design sensitivity to measure $K^+ \rightarrow \pi^+\nu\bar{\nu}$. Refined analyses are on going and data taking has restarted in April 2016.

## 5. References

[1] A. J. Buras, D. Buttazzo, J. Girrbach-Noe, R. Knegjens, $K^+ \rightarrow \pi^+\nu\bar{\nu}$ and $K_L \rightarrow \pi^0\nu\bar{\nu}$ in the Standard Model: status and perspectives, Journal of High Energy Physics 11 (2015) 033. arXiv:1503.02693.

[2] A. J. Buras, D. Buttazzo, J. Girrbach-Noe, R. Knegjens, Can we reach the Zeptouniverse with rare $K$ and $B_{s,d}$ decays?, Journal of High Energy Physics 11 (2014) 121. arXiv:1408.0728.

[3] The E949 Collaboration, Study of the decay $K^+ \rightarrow \pi^+\nu\bar{\nu}$ in the momentum region $140 < P_\pi < 199$ MeV/c, Physical Review D79 (2009) 092004. arXiv:0903.0030.

[4] The NA62 Collaboration, NA62: Technical Design Document, NA62-10-07.

Journal of Physics: Conference Series **770** (2016) 012047 doi:10.1088/1742-6596/770/1/012047

# Prospects for Measuring $BR(K_L{}^0 \to \pi^0 \nu \nu^-)$ at CERN SPS

**P D Rubin (for the KLEVER Working Group)**

Department of Physics & Astronomy, MSN 3F3
George Mason University
Fairfax, VA 22030 USA

E-mail: prubin@gmu.edu

**Abstract.** The results of preliminary feasibility studies of a measurement of $BR(K_L^0 \to \pi^0 \nu\bar{\nu})$ with high-energy kaons from the SPS are presented.

## 1. Introduction

The unitarity of the CKM matrix implies that the sum of complex products of matched pairs of elements in two rows or columns is zero:

$$\sum_k V_{ik} V_{jk}^* = 0 \tag{1}$$

This relation can be visualized as a so-called "unitarity triangle." Over-constraining one of these triangles via measurement of its sides and angles is an example of precision tests of the Standard Model and of searches for physics beyond that model.

A program of second-order weak (flavor-changing neutral current) kaon decay branching ratio measurements can make such a precision test, as the effective Hamiltonian of these decays includes the terms $V_{cs}^* V_{cd}$ and $V_{ts}^* V_{td}$. By GIM suppression, the internal quark lines of these processes is dominated by the top quark, with, in some cases, a smaller contribution from charm. The short-distance part of $BR(K_L^0 \to \mu^+\mu^-)$ gives a measure of the base of the triangle, with an uncertainty due to the charm contribution. $BR(K^+ \to \pi^+\nu\bar{\nu})$ measures the long leg of the triangle, also subject to charm uncertainties. $BR(K_L^0 \to \pi^0\nu\bar{\nu})$ (and the short-distance part of $BR(K_L^0 \to \pi^0\ell^+\ell^-)$ gives the height (and hence the area) of the triangle, free of charm uncertainties.

The $K \to \pi\nu\bar{\nu}$ modes are particularly clean theoretically in the Standard Model, because there are no long-distance contributions from intermediate photons and contributions from hadronic elements are measured by Ke3 rates. Their Standard Model (SM) branching ratios are therefore predicted to within 20% [1]:

$$\begin{aligned} BR(K^+ \to \pi^+\nu\bar{\nu}) &= (8.4 \pm 1.0) \times 10^{-11} \\ BR(K_L^0 \to \pi^0\nu\bar{\nu}) &= (3.4 \pm 0.6) \times 10^{-11} \end{aligned}$$

Journal of Physics: Conference Series **770** (2016) 012047 doi:10.1088/1742-6596/770/1/012047

Most of this uncertainty is parametric; the intrinsic theoretical uncertainty is at the level of a few percent.

Measurements with order 10% precision, then, can probe for new physics, see Figure 1. The experimental status of these modes is now:

$$\begin{aligned} BR(K^+ \to \pi^+ \nu\bar{\nu}) &= 1.73^{+1.15}_{-1.05} \times 10^{-11} \text{ [2]} \\ BR(K^0_L \to \pi^0 \nu\bar{\nu}) &< 2.6 \times 10^{-8} \text{ (90\%) [3]} \end{aligned}$$

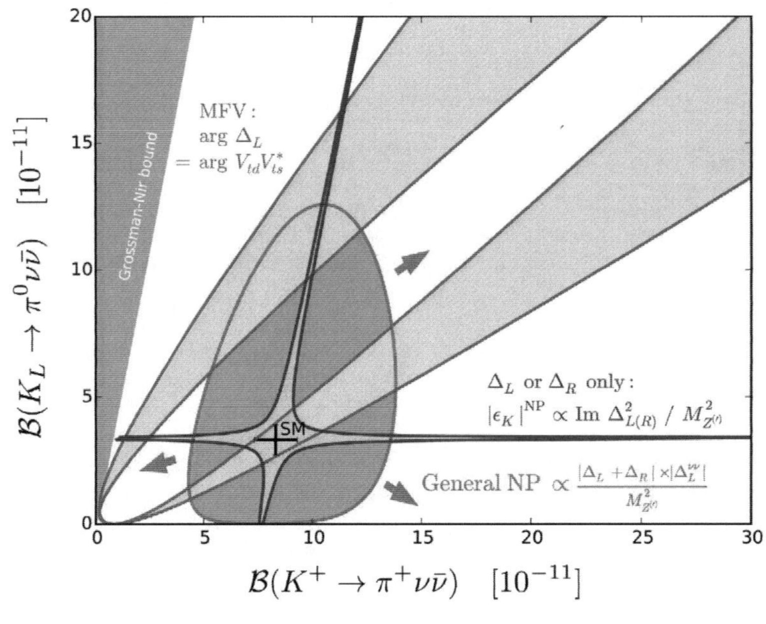


**Figure 1.** Sensitivity to new physics (NP) [1]. Measurements of both branching ratios can in principle distinguish among different classes of NP models: (diagonal swaths) CKM-like flavor structure, such as minimal flavor violation; (cross-like structure) new flavor-violating interactions in which either LH or RH couplings dominate, such as Z/Z' or T-parity Higgs; (blob) Unconstrained flavor sectors, such as Randall-Sundrum.

CERN's NA62 experiment[1] is now collecting data to measure the charged mode to nearly 10%. The neutral mode limit was set by J-PARC's E391a experiment and matched by KOTO at the same facility. KOTO intends to reach SM sensitivity by around 2019, and then to upgrade both detector and beamline so as to collect $\sim 10$ events per year.

## 2. An experiment at the SPS?

With higher energy kaons (than at J-PARC), photon vetoing–essential considering that the primary background comes from $K^0_L \to \pi^0\pi^0$–is easier, because the typical energy is greater, and polar angle coverage need only extend to 100 mr. On the other hand, a larger veto volume would be required.

[1] See the contribution by Perrin-Terrin in this volume.

### *2.1. Beam*

The primary slow-extraction proton beam at the SPS is $p_0 = 400$ GeV/c. The beam would impinge on a 400 mm-long Beryllium target ($0.94\lambda_{\rm int}$) at an angle of 2.4 mrad. This angle optimizes the $K_L^0$ to neutron and photon flux ratios. The conical angular acceptance of the secondary beam, which contains about $2.8 \times 10^{-5}$ $K_L^0$ per proton, would be collimated at 0.3 mrad, see Figure 2. Peaking at around 35 GeV/c, the $K_L^0$ spectrum would have a mean momentum of about 97 GeV/c.

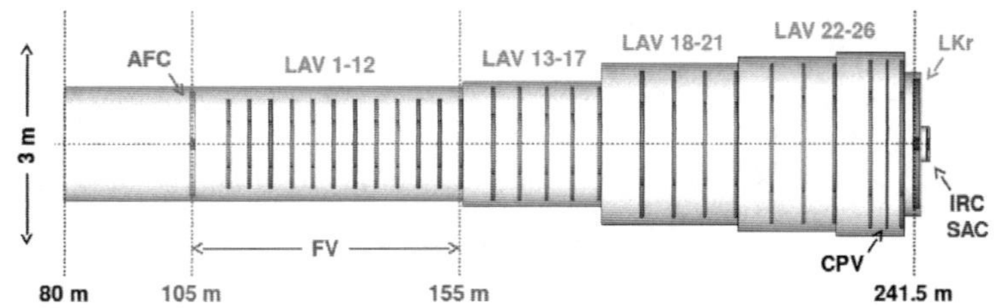


**Figure 2.** Neutral detector layout: AFC = active final collimator, LAV = large-angle veto, FV = fiducial volume, CPV = charged-particle veto, LKr = liquid Krypton calorimeter, IRC = inner-ring calorimenter, SAC = small-angle calorimeter.

For the studies reported here, a primary beam intensity of $2 \times 10^{13}$ protons pedetectorr pulse, delivered over a spill length of 4.8 s every 16.8 s (effective spill length: 3 s), was assumed. Also assumed was a 200 days/year data-taking period that is active 50% of the time. No doubt, the present beamline cavern and experimental areas would require extensive upgrades to handle the necessary rates.

### *2.2. Detector model*

A baseline assumption of the feasibility studies reported here has been that the NA48 liquid Krypton calorimeter (LKr) [4] is retained a the primary photon detector. The LKr has a photon inefficiency of $9 \times 10^{-6}$ for $E_\gamma > 10$ GeV. Clusters are resolvable to a separation of about 6 cm, and the time resolution is $\sigma_t = 2.5$ ns$/\sqrt{E}$ (GeV), which may require improvement.

High-efficiency photon vetoes would be arrayed along an 140 m-long vacuum volume in such a way as to cover out to a polar angle of 100 mrad from the beamline, see Figure 3.

Among the most daunting challenge facing such an experiment is small-angle photon detection in an environment of extremely high beam neutron and photon fluxes. An intermediate ring calorimeter (IRC) should be capable of detecting photons from downstream decays that pass through the LKr bore. While an absorber after the target can attenuate the number photons in the neutral beam, the large flux of neutrons in the neutral beam, however, creates an extremely difficult environment for vetoing photons from kaon decays that head down the beam pipe. Studies with a tungsten-crystal absorber in a small-angle calorimeter (SAC) are ongoing. Because coherent effects of the lattice can enhance pair production, such a crystal can be made very thin and consequently more transparent to neutrons.

It will also be necessary to veto charged decays, in particular $Ke3$, since, in a neutrals detector like that shown in Figure 2, $(BR \times \epsilon)_{Ke3} = 3 \times 10^{-5}$, while $(BR \times \epsilon)_{\pi^0\nu\bar{\nu}} = 8 \times 10^{-14}$.[2] Thin detectors upstream of the LKr and an hadronic calorimeter downstream can reduce this background to less than 1 event per running year. Similarly for photons from kaon decays far

[2] See Austin Bradley's contribution in this volume for details.

Journal of Physics: Conference Series **770** (2016) 012047 doi:10.1088/1742-6596/770/1/012047

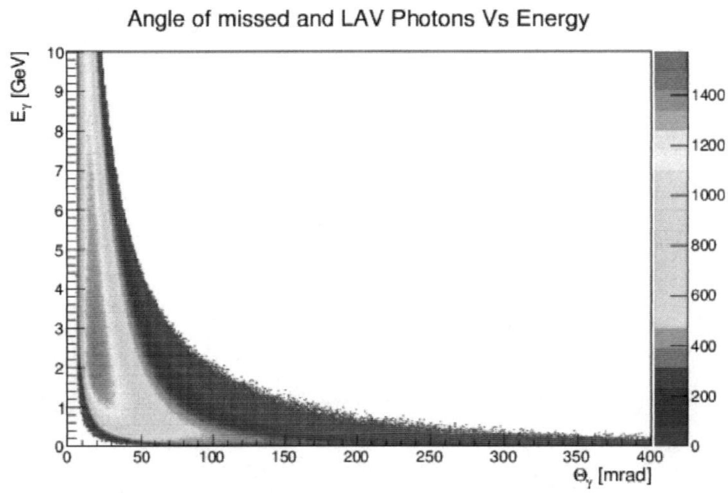


**Figure 3.** Energy vs decay angle ($E_\gamma$ vs $\theta_\gamma$) of $\pi^0$s daughter photons which pass outside the volume of the LKr. Note the extremely low energy of photons with decay angles greater than designed photon veto coverage (100 mrad).

upstream of the vacuum decay region: active collimation should be able to reduce $K_L^0 \rightarrow \pi^0\pi^0$ background to 10% or less of the total contamination from this mode.

### *2.3. Sensitivity Studies*

Two simulation packages were employed to study signal acceptance and background contamination. The first was a fast Monte Carlo algorithm with hard-coded geometry, straight-line particle propagation, and parameterized detector efficiencies. The second was Geant4-based. The two were cross-checked for consistency.

Signal identification was based on four primary selection criteria:

(i) Exactly two LKr clusters consistent with two photon interactions;

(ii) Assuming the two clusters are from the $\pi^0 \rightarrow 2\gamma$ transition, the $\pi^0$ (and therefore $K_L^0$) decay point is in the fiducial volume of the vacuum region;

(iii) Both clusters exceed a minimum distance from the beamline at the face of the LKr; and

(iv) The transverse momentum of the two photons is significantly different from zero.

Representative results for the signal can be seen in Figure 4 and Table 1. The effect of these cuts on the primary ($K_L^0 \rightarrow \pi^0\pi^0$) background can be seen in Figure 5 and also Table 1. Contamination can arise from this mode if both photons of one $\pi^0$ shower in the LKr while those of the other fail to be detected (even); if one photon from each $\pi^0$ showers in the LKr while the other photons fail to be detected (odd); or if three photons shower in the LKR, but two interact too closely to be differentiated, while the last photon fails to be detected (fused).

## 3. Conclusion

A competitive high kaon energy measurement of $K_L^0 \rightarrow \pi^0\nu\bar{\nu}$ seems feasible. Many design challenges–primarily in background suppression and small-angle activity–remain to be worked out. Alternative detector elements, particulary for charged daughters, are being considered. A reasonable time frame for such an experiment would be LHC run 4.

**Table 1.** Summary of results from a signal study (statistics generated equivalent to 100,000 years of data-taking) and a $K_L^0 \to \pi^0\pi^0$ background study (five years of data-taking generated).

| Analysis stage | $\pi^0\nu\bar{\nu}$ | $\pi^0\pi^0$ | | |
|---|---|---|---|---|
| Generated | 956.3 M (100 kyr) | | 1,215 G (5 yr) | |
| | | Even | Odd | Fused |
| $2\gamma$ on LKr | 19 M | 12 M | 104 M | 3 M |
| $105\text{m} < z_{\text{rec}} < 155$ m | 6 M | 81 k | 3 M | 607 k |
| $r_{\text{min}} > 35$ cm | 2.5 M | 23 k | 764 k | 29 k |
| $p_\perp > 0.12$ GeV | 2 M | 43 | 12 | 30 |
| All cuts, 1 yr | 19 | | 17 | |

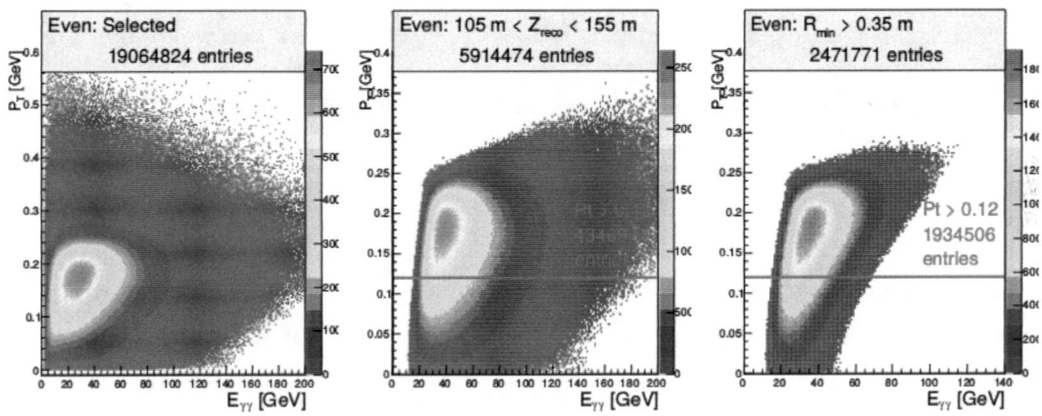


**Figure 4.** Transverse momentum vs 2-photon energy ($p_\perp$ vs $E_{\gamma\gamma}$) for simulated signal events with (left) both photons on the LKr, (center) vertex reconstructed within fiducial volume, and (right) $r > 35$ cm for both LKr clusters. The $p_\perp > 0.12$ GeV/c boundary is overlaid center and right. Note scale changes between plots.

## Acknowledgments
Presented on behalf of those who did all the work, the KLEVER working group: A. Bradley, M.B. Brunetti, F. Bucci, A. Cassese, N. Doble, D. Di Filippo, E. Gamberini, L. Gatignon, A. Gianoli, E. Imbergamo, M. Lenti, S. Martellotti, A. Mazzolari, M. Moulson, I. Neri, F. Petrucci, R. Volpe.

## References
[1] Buras A J , Buttazzo D, Girrbach-Noe J *et al* 2015 *J. High Energ. Phys* **11** 33.
[2] Artamonov A V, et al. (E949 Collaboration) 2009 *Phys. Rev. D* **79** 092004.
[3] Ahn J K, et al. (E391a Collaboration) 2010 *Phys. Rev. D* **81** 072004.
[4] Fanti V, *et al* (NA48 Collaboration) 2007 *Nucl. Instrum. Meth. A* **574** 433.

Journal of Physics: Conference Series **770** (2016) 012047 doi:10.1088/1742-6596/770/1/012047

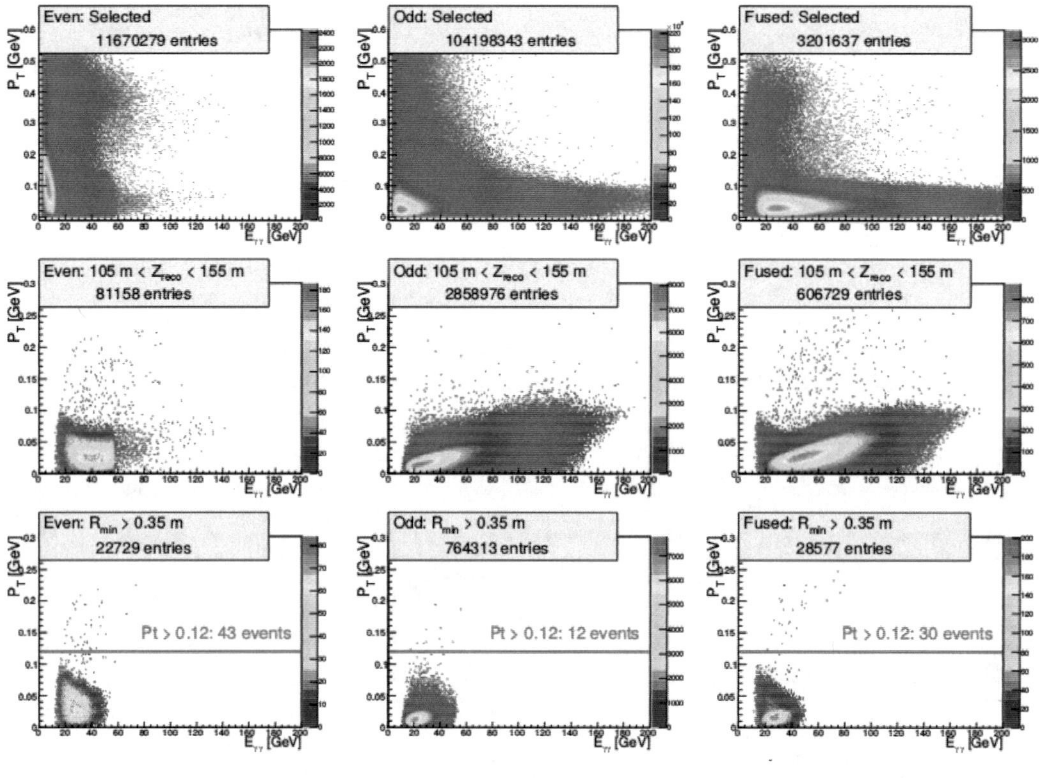


**Figure 5.** Transverse momentum vs 2-photon energy ($p_\perp$ vs $E_{\gamma\gamma}$) for simulated $K_L^0 \rightarrow \pi^0\pi^0$ events. From left to right, the columns are even, odd, and fused events, respectively. The first row shows events with both photons on the LKr, the second row shows events with vertex reconstructed within fiducial volume, and the third row shows events with $r > 35$ cm for both LKr clusters. The $p_\perp > 0.12$ GeV/c boundary is overlaid on the third row plots. Note that the vertical scale differs among the first-row plots.

Journal of Physics: Conference Series **770** (2016) 012048 doi:10.1088/1742-6596/770/1/012048

# Evaluating the Charged Background Rejection Requirement in an Experiment to Measure $\mathrm{BR}(\mathrm{K_L} \to \pi^0\nu\nu^-)$ at the CERN SPS

**A Bradley**

Department of Physics and Astronomy, George Mason University, 4400 University Drive, Fairfax, Virginia 22030, USA

E-mail: abradl12@gmu.edu

**Abstract.** Measuring the rate at which the long-lived, neutral kaon decays into a neutral pion, neutrino, and anti-neutrino allows physicists an opportunity to test precise predictions made by the Standard Model. Differences between theoretical predictions and experimental measurements may point to new physics. Not only does the Standard Model predict a very low probability at approximately 3 $K_L \to \pi^0\nu\bar{\nu}$ decay in 100 billion $K_L$ decays, but many of the common decays leave false signals in the detector that look the same as the true signal. Charged decays have been studied to determine the required detection efficiency necessary to eliminate them. The conclusion of these studies is that a reduction by a factor of $1/(3 \times 10^9)$ will be required to achieve the 10:1 signal to charged background ratio necessary for the experiment.

## 1. Introduction to the "KLEVER" feasibility study

A small collaboration within the NA62 experiment is conducting a feasibility study for measuring the very rare decay $K_L \to \pi^0\nu\bar{\nu}$. With a $\mathrm{BR}(3.4 \pm 0.6) \times 10^{-11}$ [1] predicted by the Standard Model (SM), no experiment has yet been capable of observing the decay. The goal of the "KLEVER" experiment is to observe tens of these decays while keeping the signal to background ratio at 1:1. The collaboration has published their initial findings in an NA62 technical note [2], including the findings of this study.

To accomplish this goal, long-lived, neutral kaons $K_L$ will be created by colliding 400 GeV/c protons accelerated by CERN's SPS with a beryllium target. The most common charged decays from the $K_L$, the signal decay, and their applicable BR's are shown in Table 1. The signal is two photons from the $\pi^0$ detected by the liquid krypton electromagnetic calorimeter (LKr) and no other particles detected elsewhere. Any difference between the experimental measurement and the SM branching ratio may point to new physics.[3]

## 2. Background contribution from charged decays

A 10:1 signal to charged background ratio is necessary for the experiment. The rate of charged background and the reduction required to achieve the signal to background ratio were studied. The overall 1:1 signal to background ratio includes the neutral decays, with contributions primarily from $K_L \to \pi^0\pi^0$, but is not discussed here. Additional details can be found in [2].

Journal of Physics: Conference Series **770** (2016) 012048 doi:10.1088/1742-6596/770/1/012048

### *2.1. The Detector*

In addition to the LKr ($z$ = 242 m from the target), there are 26 large angle vetoes (LAV) upstream of the LKr, and a small angle calorimeter (SAC) and inner ring calorimeter (IRC) behind the LKr. Their purpose is to detect photons and charged particles missed by the LKr.

### *2.2. Simulation and Analysis*

Charged decays of the kaon were simulated with GEANT4 [5][6]. A list of selection criteria was sequentially applied to each event in the resulting ROOT [7] file. Events were eliminated if any of the following were true: a charged particle or photon was detected by the LAVs, IRC, or SAC; fewer than two energy clusters were detected by the LKr (clusters closer than 6 cm were merged); the reconstructed $\pi^0$ decayed outside of the fiducial region ($105 < z < 155$ m); an LKr cluster was closer than 35 cm to the beamline center; or the transverse momentum of the reconstructed $\pi^0$ was less than 0.12 GeV/c. Designed primarily to eliminate background from the most problematic neutral decays, these criteria also significantly reduce charged background.

### *2.3. Results and Conclusion*

Some charged decays passed all of the selection criteria and were accepted as signal events because they were indistinguishable from the signal (e.g. a pion and electron reconstructed as two photons from a $\pi^0$) [see Table 1].

**Table 1.** Analysis results for charged and signal decay modes. $\epsilon$ is the probability a generated event was accepted as a signal. Charged BRs are from [4] and the signal BR is from [1].

| Decay Mode | Generated ($\times 10^6$) | Accepted as Signal | $\epsilon$ | BR | $\epsilon \times$ BR |
|---|---|---|---|---|---|
| $\pi^\pm e^\mp \nu_e$ | 0.24 | 18 | $7.4 \times 10^{-5}$ | $(40.55 \pm 0.11)\%$ | $3.0 \times 10^{-5}$ |
| $\pi^\pm \mu^\mp \nu_\mu$ | 44.8 | 23 | $5.1 \times 10^{-7}$ | $(27.04 \pm 0.07)\%$ | $1.4 \times 10^{-7}$ |
| $\pi^+\pi^-\pi^0$ | 153.0 | 11 | $7.2 \times 10^{-8}$ | $(12.54 \pm 0.05)\%$ | $9.0 \times 10^{-9}$ |
| $\pi^+\pi^-$ | 0.73 | 16 | $2.2 \times 10^{-5}$ | $(1.97 \pm 0.01) \times 10^{-3}$ | $4.3 \times 10^{-8}$ |
| $\pi^0\nu\bar{\nu}$ | 2.8 | 6678 | $2.4 \times 10^{-3}$ | $(3.4 \pm 0.6) \times 10^{-11}$ | $8.1 \times 10^{-14}$ |

The last column of Table 1 shows that a charged particle detection (CPD) system will need to reduce the charged background from $3 \times 10^{-5}$ to $8 \times 10^{-15}$ to achieve the 10:1 signal to charged background ratio—a reduction by a factor of $1/(3 \times 10^9)$.


## Acknowledgments

This work was supported by the U.S. National Science Foundation under Grant No. 1506088 and funding under a grant from the Italian Ministry of Education, Universities, and Research (MIUR) as the KLEVER project, a Research Project of Significant National Interest (Progetto di ricerca di Rilevante Interesse Nazionale, PRIN), call 2010-2011.


## References


[1] A.J. Buras, et al. 2015 *J. High Energy Phys.* JHEP11(2015)033
[2] A Bradley, et al. 2016 Internal Report: NA62-16-03
[3] C Smith 2014 arXiv:1409.6162
[4] Particle Data Group, K Olive, et al. 2014 *Chinese Physics* C **38**
[5] S Agostinelli, et al. 2003 *NIM* A **506** 250-303
[6] J Allison, et al. 2006 *IEEE Trans. Nucl. Sci.* **53** 270-8
[7] R Brun and F Rademakers 1997 *Nucl. Inst. & Meth. in Phys. Res.* A **389** 81-6

# Experimental Summary: Step-by-Step Towards New Physics

**A J Schwartz**

Physics Department, University of Cincinnati, P.O. Box 210011, Cincinnati, Ohio 45221 USA

E-mail: alan.j.schwartz@uc.edu

**Abstract.** We summarize some highlights from experimental results presented at the XIIth International Conference on Beauty, Charm, and Hyperons in Hadronic Interactions, held at George Mason University June 12-18, 2016.

## 1. Introduction

This year's workshop featured about fifty experimental talks covering a wide variety of results in meson and baryon decays, heavy flavor and quarkonium production, heavy ion collisions, kaon physics, neutrino physics, and searches for new physics (NP). The presentations were organized as follows: heavy flavor production (Monday); heavy flavor decays (Tuesday); neutrino physics (Wednesday); mixing and $CP$ violation (Thursday); spectroscopy (Thursday); and future experiments and facilities (Friday). Interspersed throughout the week were talks on searches for NP and theory talks. Experimental results were presented from ATLAS, ALICE, Belle, BaBar, BESIII, CDF, CMS, LHCb, PHENIX, and STAR. In this summary I discuss only a few highlights from among all these talks. For further details the reader is referred to the original presentations and these Proceedings.

## 2. Heavy flavor production

Numerous results were presented on the production of $D$ and $B$ mesons, charmonium and bottomonium mesons, $W^{\pm}$ and $Z^0$ vector bosons, and lighter pions and electrons, from LHC and RHIC experiments.

The STAR experiment (Zhang) presented measurements [1] of the correlation function $C(k^*) \equiv A(k^*)/B(k^*)$, where the variable $k^*$ is half the relative momentum between two particles produced in a collision, $A(k^*)$ is the distribution for a pair of particles in the same event, and $B(k^*)$ is the distribution for two particles produced in *different* events. If there is a net attractive interaction between the particles, $C(k^*)$ increases as $k^*$ decreases; if there is a net repulsive interaction, $C(k^*)$ decreases as $k^*$ decreases; and if there is negligible interaction, $C(k^*)$ is independent of $k^*$ and equals unity. The STAR data corresponds to Au-Au collisions at $\sqrt{s_{NN}} = 200$ GeV and is shown in Fig. 1(left). A strong attractive correlation is observed for both $p$-$p$ and $\bar{p}$-$\bar{p}$ distributions. Taking their ratio shows an equal attractive force over most of the $k^*$ range. However, at the lowest $k^*$ values the $p$-$p$ attractive force appears to be stronger. More data is needed to confirm this effect.

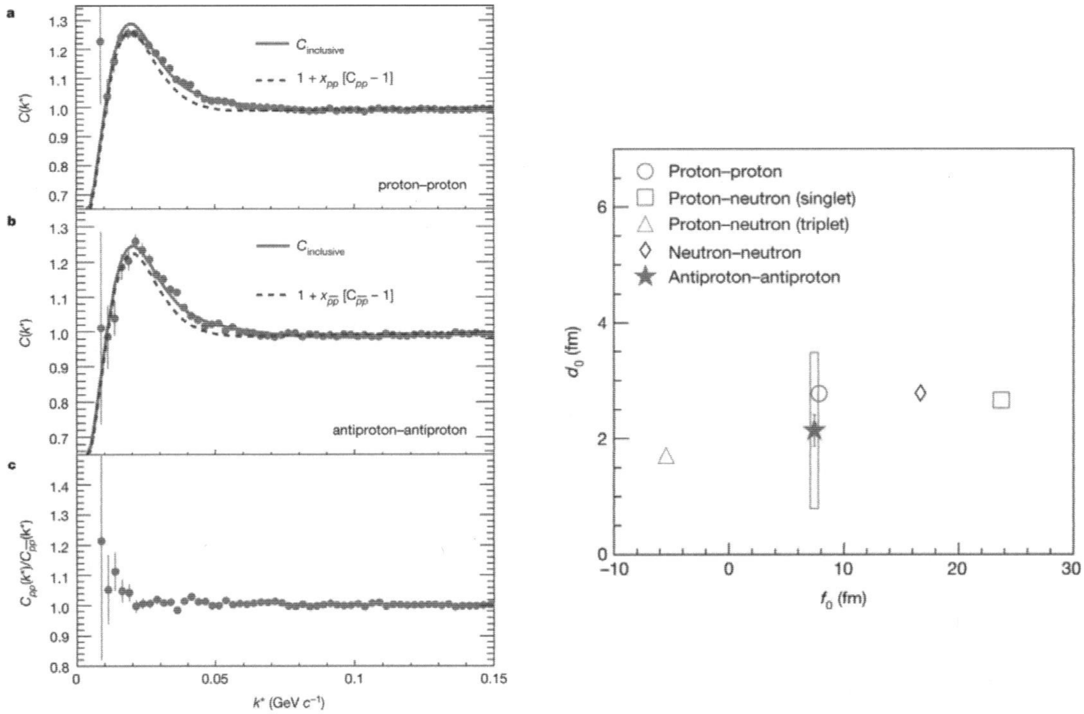


**Figure 1.** Data from Au-Au collisions at STAR. Left top: correlation function $C(k^*)_{pp}$ for $pp$ tracks; left middle: correlation function $C(k^*)_{\bar{p}\bar{p}}$ for $\bar{p}\bar{p}$ tracks; left bottom: ratio of $C(k^*)_{pp}$ to $C(k^*)_{\bar{p}\bar{p}}$. The rise at low $k^*$ for the top and middle plots indicates an attractive force. Right: values of the range $d_0$ and scattering length $f_0$ resulting from fitting the correlation functions $C(k^*)_{pp}$ and $C(k^*)_{\bar{p}\bar{p}}$.

To measure a quantitative difference between $p$-$p$ and $\bar{p}$-$\bar{p}$ interactions and in this manner test $CPT$, STAR fits the $p$-$p$ and $\bar{p}$-$\bar{p}$ spectra for $C(k^*)$ using the Lednicky and Lyuboshitz model [2]. The fit yields two parameters: the effective range of the interaction $d_0$, and the scattering length $f_0$. The latter indicates whether the interaction corresponds to an attractive bound state ($f_0 < 0$), an attractive but unbound state ($f_0 > d_0$), or is repulsive ($0 < f_0 < d_0$). The fit result is plotted in Fig. 1(right), which shows that $d_0$ and also $f_0$ are the same for $p$-$p$ and $\bar{p}$-$\bar{p}$ interactions, within errors. In both cases the range $d_0$ is approximately 2.5 fm, and $f_0$ corresponds to an attractive unbound state.

The LHCb experiment (Szumlak) presented measurements of differential cross sections for prompt and secondary $J/\psi$ production, $\Upsilon(1S)$, $\Upsilon(2S)$, and $\Upsilon(3S)$ production, and $D$ and $\Lambda_b$ production. All cross sections are measured as a function of $p_T$ and rapidity $y$ and correspond to $pp$ collisions at $\sqrt{s} = 7, 8$ TeV. From measurements of $D^0$, $D^{(*)+}$, and $D_s^+$ production, LHCb obtains a $c\bar{c}$ total production cross section of $(2940 \pm 3 \pm 180 \pm 160)$ $\mu$b, where the first error is statistical, the second is systematic, and the last error is due to the $c \rightarrow D$ fragmentation model. From measurements of non-prompt $J/\psi$ production, i.e., events in which the $J/\psi$ candidate forms a secondary (rather than primary) vertex, LHCb calculates a $b\bar{b}$ total production cross section of $(515 \pm 2.0 \pm 53.0)$ $\mu$b. This value is plotted in Fig. 2 along with measurements of $\sigma_{b\bar{b}}$ presented by PHENIX (Haseler). The latter results are from three independent analyses: one using same-sign $\mu^{\pm}\mu^{\mp}$ pairs; one using opposite-sign $e^+e^-$ pairs; and one using electron-hadron correlations. All $\sigma_{b\bar{b}}$ measurements show remarkable agreement with next-to-leading-

Journal of Physics: Conference Series **770** (2016) 012049 doi:10.1088/1742-6596/770/1/012049

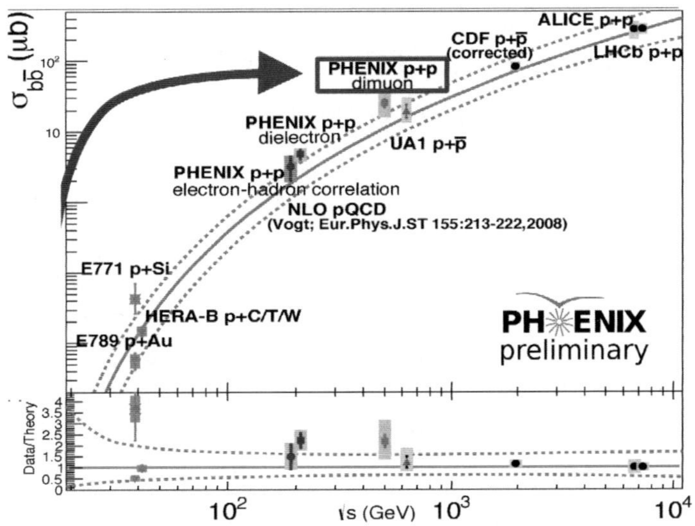


**Figure 2.** Measurements of the $b\bar{b}$ production cross section, and next-to-leading-order pQCD predictions [3].

order pQCD [3] over four orders of magnitude.

LHCb also measured the production of $B_c^+$ mesons by reconstructing Cabibbo-favored $B_c^+ \to J/\psi\,\pi^+$ decays [4]. The signal yield is normalized to the number of color-suppressed $B^+ \to J/\psi\,K^+$ decays reconstructed, and the result is $R \equiv [\sigma_{B_c^+} \cdot B(B_c^+ \to J/\psi\,\pi^+)]/[\sigma_{B^+} \cdot B(B^+ \to J/\psi\,K^+)] = (0.683 \pm 0.018 \pm 0.009)\%$. This corresponds to the kinematic range $2.5 < y < 4.5$ and $4 < p_T < 20$ GeV/$c$. The value indicates that, within the kinematic range, $c\bar{c}$ pairs are produced via fragmentation at a rate $\sim 1/9$ that of $d\bar{d}$ pairs.

The ALICE experiment (De) presented results for meson production in proton-nucleus and nucleus-nucleus collisions. The energy density of such collisions corresponds to the environment of a quark-gluon plasma (QGP), and partons produced in collisions interact with the QGP when escaping and subsequently lose energy. This parton energy loss is expected to decrease as the parton mass increases. The parameter quantifying parton interactions with the QGP is the "nuclear modification factor:"

$$R_{AA}(p_T) \;\equiv\; \frac{1}{N_{\rm coll}}\frac{(dN_{AA}/dp_T)}{(dN_{pp}/dp_T)}\,, \tag{1}$$

where $N_{\rm coll}$ is the number of nucleons participating in the $A$-$A$ collision. The greater the interaction with the QGP and subsequent energy loss, the lower $R_{AA}(p_T)$; this is referred to as "suppression." The amount of suppression is found to increase with $p_T$.

Figure 3(left) shows $R_{AA}$ plotted as a function of $p_T$ for $D^0$ and $D^{(*)+}$ production as measured by ALICE. The highest points correspond to $p$-Pb collisions and show little suppression; presumably the incoming proton does not generate a sufficient QGP energy density. The other points plotted correspond to Pb-Pb collisions and show large suppression. Figure 3(middle) shows $R_{AA}$ for electrons in ALICE that have a large impact parameter with respect to the primary interaction point, i.e., they originate from heavy flavor decays. This data also shows significant suppression. Figure 3(right) shows $R_{AA}$ as measured by ALICE for $D$ mesons along with $R_{AA}$ as measured by CMS [5] for $J/\psi$ mesons having a large impact parameter with respect to the primary interaction; these $J/\psi$ decays originate from $B$ decays. The plot shows that $R_{AA}^{D} < R_{AA}^{B}$, as expected due to $m_b > m_c$. This is an important confirmation of this relationship. Also superimposed on the plot are several theoretical predictions [6], which are consistent with the data.

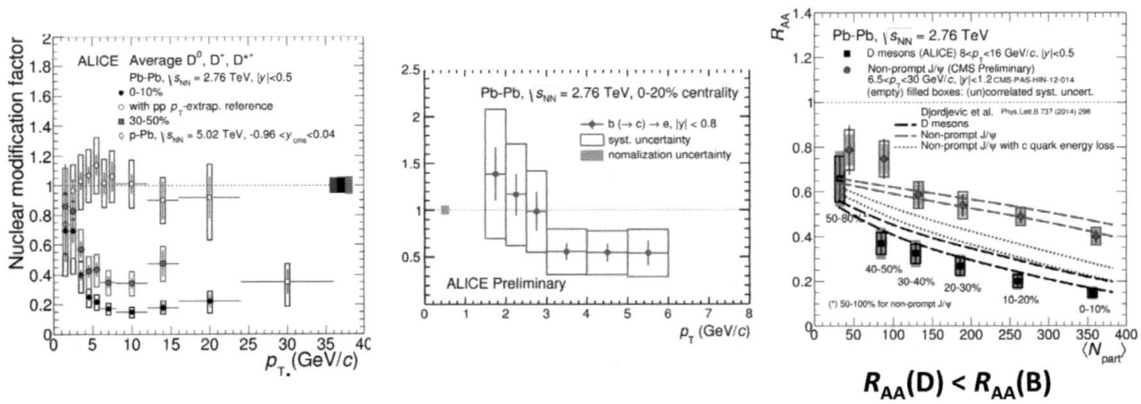


**Figure 3.** Left: $R_{AA}^{D}$ measured by ALICE for $D^0$ and $D^{(*)+}$ production in Pb-Pb and $p$-Pb collisions. Middle: $R_{AA}^{B}$ measured by ALICE for high-impact-parameter electrons in Pb-Pb collisions. Right: $R_{AA}^{D}$ measured by ALICE as compared to $R_{AA}^{B}$ measured by CMS using non-prompt $J/\psi$ decays [5], in Pb-Pb collisions. Also shown are theoretical predictions [6].

## 3. Heavy flavor decays

The BESIII experiment (Ke) presented new measurements of branching fractions for the $\Lambda_c^+$ baryon decaying into a dozen hadronic final states; see Table 1. With the exception of $B(\Lambda_c^+ \to pK^-\pi^+)$ [7], the BESIII results are the world's most precise and represent a significant improvement over previous results. BESIII also presented $\Lambda_c^+$ semileptonic branching fractions:

$$\begin{aligned} B(\Lambda_c^+ \to \Lambda\mu^+\nu_\mu) &= (3.49 \pm 0.46 \pm 0.26)\% \\ B(\Lambda_c^+ \to \Lambda e^+\nu_e) &= (3.63 \pm 0.38 \pm 0.20)\% \\ B(\Lambda\mu^+\nu_\mu)/B(\Lambda e^+\nu_e) &= 0.96 \pm 0.16 \pm 0.04\,. \end{aligned}$$

The last result constitutes a test of lepton universality in $\Lambda_c^+$ decays.

A higher precision test of lepton universality was presented by LHCb (Hamilton), which measured the ratio of branching fractions $R_K \equiv B(B \to K\mu^+\mu^-)/B(B \to Ke^+e^-)$. Within the Standard Model (SM), this ratio is within 0.1% of unity [8]. The measurement is challenging for LHCb due to the electrons. Photons reconstructed in the electromagnetic calorimeter that lie close to the candidate electron's trajectory are added to the electron's four-momentum to improve resolution; as a consequence, the signal shape for the $Ke^+e^-$ mass distribution must be treated separately for one, two, or three "recovered" photons. In addition, the background shape depends on whether the event was electron-triggered, kaon-triggered, or passed some other trigger criterion; thus the different trigger streams are fitted separately. The result is $R_K = 0.745\,^{+0.090}_{-0.074} \pm 0.036$ for $q^2 < (6\text{ GeV})^2$, which is $2.6\sigma$ below unity. This measurement has notably higher precision than similar measurements made by Belle [9] and BaBar [10].

The Belle experiment (King) presented a new result for the ratio $R(D^*) \equiv B(B \to D^*\tau\nu_\tau)/B(B \to D^*\ell\nu)$ ($\ell = e, \mu$) that uses "leptonic tagging," i.e., events in which the opposite-side $B$ is required to decay semileptonically, producing an electron or muon. This requirement significantly reduces backgrounds. The analysis obtains $231 \pm 23$ signal ($D^*\tau\nu$) decays and

**Table 1.** $\Lambda_c^+$ branching fractions to hadronic final states, as measured by BESIII.

| Mode | BESIII | 2014 PDG | Belle |
|---|---|---|---|
| $pK_S^0$ | $1.52 \pm 0.08 \pm 0.03$ | $1.15 \pm 0.30$ | |
| $pK^-\pi^+$ | $5.84 \pm 0.27 \pm 0.23$ | $5.0 \pm 1.3$ | $6.84 \pm 0.24\,^{+0.21}_{-0.27}$ |
| $pK_S^0\pi^0$ | $1.87 \pm 0.13 \pm 0.05$ | $1.65 \pm 0.50$ | |
| $pK_S^0\pi^+\pi^-$ | $1.53 \pm 0.11 \pm 0.09$ | $1.30 \pm 0.35$ | |
| $pK^-\pi^+\pi^0$ | $4.53 \pm 0.23 \pm 0.30$ | $3.4 \pm 1.0$ | |
| $\Lambda\pi^+$ | $1.24 \pm 0.07 \pm 0.03$ | $1.07 \pm 0.28$ | |
| $\Lambda\pi^+\pi^0$ | $7.01 \pm 0.37 \pm 0.19$ | $3.6 \pm 1.3$ | |
| $\Lambda\pi^+\pi^-\pi^+$ | $3.81 \pm 0.24 \pm 0.18$ | $2.6 \pm 0.7$ | |
| $\Sigma^0\pi^+$ | $1.27 \pm 0.08 \pm 0.03$ | $1.05 \pm 0.28$ | |
| $\Sigma^+\pi^0$ | $1.18 \pm 0.10 \pm 0.03$ | $1.00 \pm 0.34$ | |
| $\Sigma^+\pi^+\pi^-$ | $4.25 \pm 0.24 \pm 0.20$ | $3.6 \pm 1.0$ | |
| $\Sigma^+\omega$ | $1.56 \pm 0.20 \pm 0.07$ | $2.7 \pm 1.0$ | |

$2800 \pm 57$ normalization ($D^*\ell\nu$) decays; the result is $R(D^*) = 0.302 \pm 0.030 \pm 0.011$, which, like a previous measurement by Belle using a hadronic tagging method [11], and also an LHCb measurement [12] (Hamilton), is significantly higher than the SM prediction of 0.25 [13]. Such a difference was also seen for the ratio $R(D) = B(B \to D\tau\nu_\tau)/B(B \to D\ell\nu)$, i.e., the measured values were higher than the SM prediction. All measurements and the SM predictions are summarized in Fig. 4.

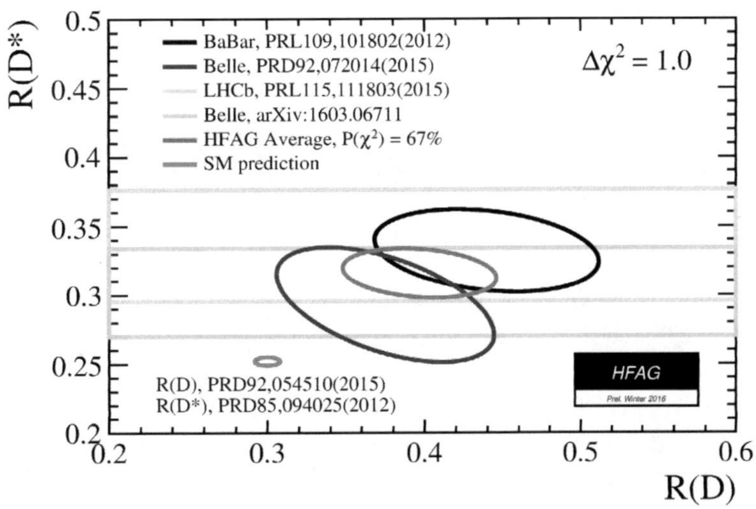


**Figure 4.** Measurements and theoretical predictions for $R(D^*)$ and $R(D)$, as compiled by the Heavy Flavor Averaging Group [14].

Both Belle (King) and LHCb (Coutinho) presented results for the angular distribution of $B^0 \to K^{*0}\mu^+\mu^-$ decays. The angular distribution can be parameterized as [15]:

$$\frac{d^4\Gamma}{dq^2 d\Omega} \quad \propto \quad \frac{3}{4}(1-F_L)\sin^2\theta_k + F_L\cos^2\theta_k * \frac{1}{4}(1-F_L)\sin^2\theta_k\cos 2\theta_\ell - F_L\cos^2\theta_k\cos 2\theta_\ell +$$

$$S_3 \sin^2\theta_k \sin^2\theta_\ell \cos 2\phi + S_4 \sin 2\theta_k \sin 2\theta_\ell \cos\phi + S_5 \sin 2\theta_k \sin\theta_\ell \cos\phi +$$
$$\frac{4}{3} A_{FB} \sin^2\theta_k \cos\theta_\ell + S_7 \sin 2\theta_k \sin\theta_\ell \sin\phi + S_8 \sin 2\theta_k \sin 2\theta_\ell \sin\phi +$$
$$S_9 \sin^2\theta_k \sin^2\theta_\ell \sin 2\phi \qquad (2)$$

where $\theta_k$ is the helicity angle of the $K^{0*} \to K^+\pi^-$ decay, $\theta_\ell$ is the helicity angle of the $\mu^+\mu^-$ system, and $\phi$ is the azimuthal angle between the $K^+\pi^-$ plane and the $\mu^+\mu^-$ plane. There are eight underlying parameters, among which $F_L$ is the longitudinal polarization of the final state, and $A_{FB}$ is the forward-backward asymmetry of the $\mu^+\mu^-$ system. Calculations of these parameters have large theoretical uncertainties, but for the ratio $P_5' \equiv S_5/\sqrt{F_L(1-F_L)}$ the leading form factor uncertainties cancel [16]. Figure 5 shows measurements of $P_5'$ from both Belle and LHCb as a function of $q^2$, which is the invariant mass of the $\mu^+\mu^-$ system. Superimposed on the data points are SM predictions [17]. There is good agreement between the measured values from the two experiments, but these measurements disagree with the SM prediction for two $q^2$ bins. If these differences were statistical fluctuations, it is notable that both experiments observe fluctuations in the same direction for the same $q^2$ bins. More data from LHCb and the future Belle II experiment is needed to better understand this difference.

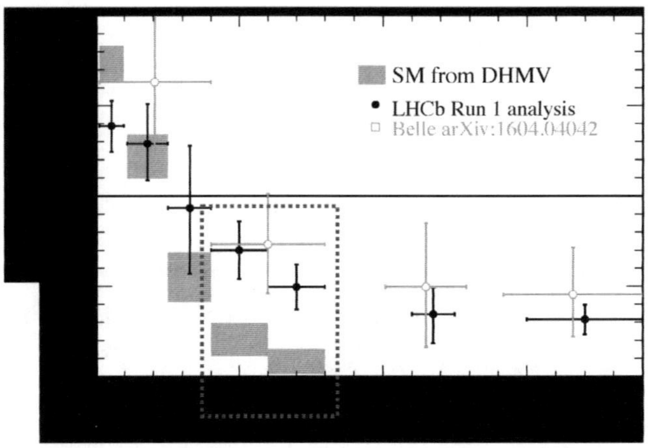


**Figure 5.** Measurements and theoretical predictions [17] for the parameter $P_5'$ measured in $B^0 \to K^{*0}\mu^+\mu^-$ decays.

## 4. Mixing and $CP$ violation

The LHCb experiment (Carbone) presented three measurements of $D^0$-$\overline{D}^0$ mixing and $CP$ violation: measurements of $A_\Gamma$ and $A_{CP}$ in $D^0 \to K^+K^-/\pi^+\pi^-$ decays, and a measurement of mixing in doubly Cabibbo-suppressed $D^0 \to K^+\pi^-\pi^+\pi^-$ decays.

The parameter $A_\Gamma$ is defined as the asymmetry in lifetimes between $D^0$ and $\overline{D}^0$ decays: $A_\Gamma = (\tau_{\overline{D}^0} - \tau_{D^0})/(\tau_{\overline{D}^0} + \tau_{D^0})$. One way to determine $A_\Gamma$ is by fitting the decay time distribution of the $CP$ asymmetry $A_{CP}$ for decays to a self-conjugate final state $f$:

$$A_{CP}(t) = \frac{dN(D^0 \to f)/dt - dN(\overline{D}^0 \to f)/dt}{dN(D^0 \to f)/dt + dN(\overline{D}^0 \to f)/dt} \approx A_{CP}^{\rm dir} - A_\Gamma \frac{t}{\tau}. \qquad (3)$$

LHCb performs this measurement for $D^0 \to K^+K^-$ and $D^0 \to \pi^+\pi^-$ decays. The flavor of the decaying $D^0$ or $\overline{D}^0$ is identified by reconstructing $D^{*+} \to D^0\pi^+$ decays. The resulting

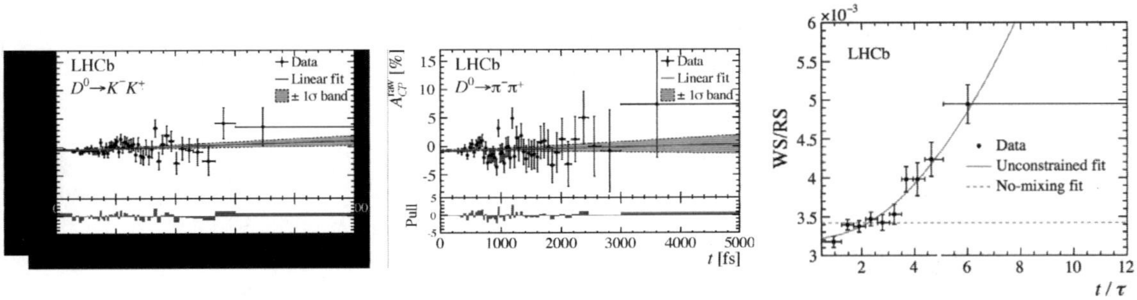

**Figure 6.** LHCb measurements of the $CP$ asymmetry in $D^0 \to K^+K^-$ decays (left); the $CP$ asymmetry in $D^0 \to \pi^+\pi^-$ decays (middle); and the decay time distribution of $D^0 \to K^+\pi^-\pi^+\pi^-$ decays (right).

$A_{CP}$ distributions are shown in Fig. 6. Performing a simultaneous fit to the $K^+K^-$ and $\pi^+\pi^-$ distributions gives $A_\Gamma = (-0.125 \pm 0.073)\%$, which is the world's most precise measurement.

Using time-*integrated* (rather than time-dependent) samples of $D^0 \to K^+K^-$ and $D^0 \to \pi^+\pi^-$ decays, LHCb measures $CP$-violating parameters $a_{CP}^{\rm ind}$ and $\Delta a_{CP}^{\rm dir} = a_{CP}^{\rm dir}(K^+K^-) - a_{CP}^{\rm dir}(\pi^+\pi^-)$. For the analysis presented here the flavor of the $D^0$ is determined by requiring that the $D^0$ originate from semileptonic $\overline{B}^0 \to D^0\mu^-\nu_\mu$ decays; the charge of the accompanying $\mu^\pm$ then identifies the charm meson as $D^0$ or $\overline{D}^0$. The result is $a_{CP}^{\rm ind} = (0.058 \pm 0.044)\%$ and $\Delta a_{CP}^{\rm dir} = (-0.061 \pm 0.076)\%$, which are both consistent with no $CP$ violation.

Finally, LHCb fits the decay time distribution of $D^0 \to K^+\pi^-\pi^+\pi^-$ decays to search for mixing in this doubly Cabibbo-suppressed decay mode. The decay time distribution, normalized to that for the Cabibbo-favored decay $D^0 \to K^-\pi^+\pi^+\pi^-$, is given by

$$\frac{dN}{dt} \approx R_{K3\pi}^2 - \kappa^{K3\pi} \cdot R_{K3\pi} \cdot y'_{K3\pi} \cdot \left(\frac{t}{\tau}\right) + \frac{x^2+y^2}{4} \cdot \left(\frac{t}{\tau}\right)^2 \qquad (4)$$

where $R_{K3\pi}$ is the ratio of the $D^0 \to K^+\pi^-\pi^+\pi^-$ amplitude integrated over phase space to the $D^0 \to K^-\pi^+\pi^+\pi^-$ amplitude integrated over phase space; $\kappa^{K3\pi}$ is the coherence factor for $D^0 \to K^+\pi^-\pi^+\pi^-$ decays; $y'_{K3\pi} = y\cos\delta_{K3\pi} - x\sin\delta_{K3\pi}$, and $\delta_{K3\pi}$ is the average strong phase difference between $D^0 \to K^+\pi^-\pi^+\pi^-$ and $D^0 \to K^-\pi^+\pi^+\pi^-$ amplitudes. The decay time distribution is shown in Fig. 6(right). Fitting this distribution to Eq. (4) gives $(x^2+y^2)/4 = (4.8 \pm 1.8) \times 10^{-5}$, where the error includes systematic uncertainties. Due to a large correlation between the fitted terms $\kappa^{K3\pi}y'_{K3\pi}$ and $(x^2+y^2)/4$, the no-mixing hypothesis $x = y = y'_{K3\pi} = 0$ is rejected with a relatively high significance: $8.2\sigma$.

LHCb (Whitehead) presented three measurements of the CKM phase $\phi_3$ (or $\gamma$). The first measurement is based on the Atwood-Dunietz-Soni method [18], in which one compares the rate for $B^+ \to (D^0, \overline{D}^0)K^+, (D^0, \overline{D}^0) \to K^-\pi^+$ with that for the charge-conjugate decay $B^- \to (D^0, \overline{D}^0)K^-, (D^0, \overline{D}^0) \to K^+\pi^-$. These decays proceed through the interference of two amplitudes: a Cabibbo-favored $B$ decay followed by a doubly Cabibbo-suppressed $D$ decay, and a Cabibbo-suppressed $B$ decay followed by a Cabibbo-favored $D$ decay. The phase difference between the overall $B^+$ and $B^-$ decay amplitudes is $2\phi_3$, which results in a difference in decay rates. Previous measurements by Belle [19] and BaBar [20] had relatively low statistics. The LHCb data is shown in Fig. 7(top). Systematic uncertainties are determined by repeating the measurement with control samples of $B^+ \to (D^0, \overline{D}^0)_{[K^-\pi^+]}\pi^+$ and $B^- \to (D^0, \overline{D}^0)_{[K^+\pi^-]}\pi^-$ decays, which should exhibit no difference in decay rates. A total signal yield of $553 \pm 54$ events is obtained, and the $CP$ asymmetry is clearly visible; the statistical significance of the asymmetry is almost $8\sigma$. LHCb subsequently applied this method to four-body $D^0 \to K^+\pi^-\pi^+\pi^-$ decays and observed a similar $CP$ asymmetry; see Fig. 7(bottom). However, the statistics is lower

(159 ± 17 events) and the difference seen is not statistically significant. More data should establish a $CP$ asymmetry in this mode also.

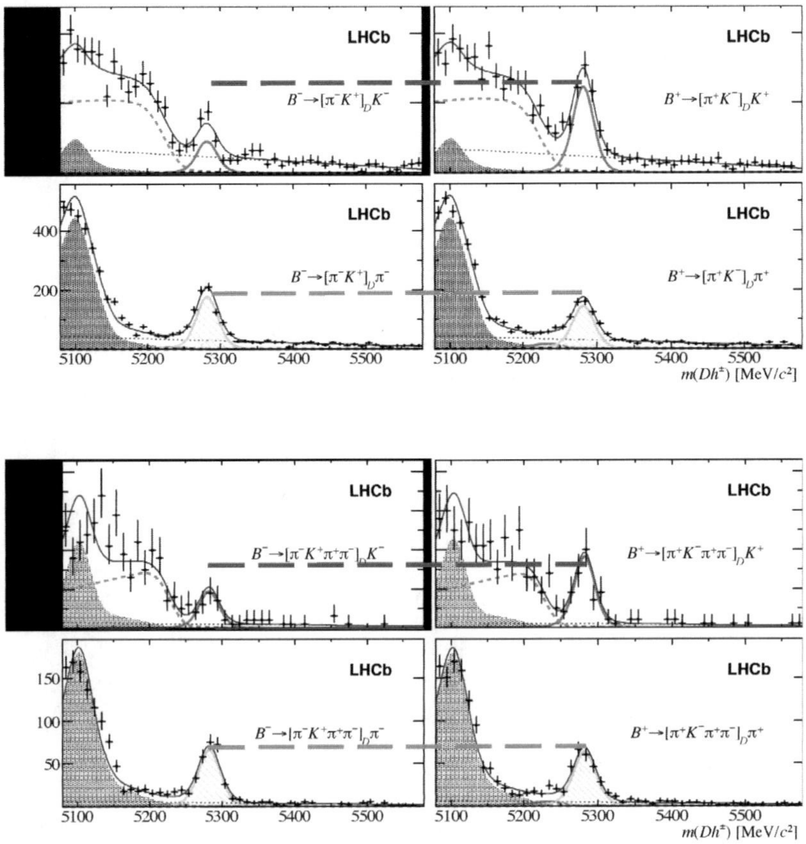


**Figure 7.** LHCb measurements of $B^- \to (D^0, \overline{D}^0)K^-, (D^0, \overline{D}^0) \to K^+\pi^-$ decays (top left) and $B^+ \to (D^0, \overline{D}^0)K^+, (D^0, \overline{D}^0) \to K^-\pi^+$ decays (top right). A clear $CP$ asymmetry is seen. The control samples $B^\mp \to D^0\pi^\mp$ are also shown. LHCb measurements of $B^- \to (D^0, \overline{D}^0)K^-, (D^0, \overline{D}^0) \to K^+\pi^-\pi^+\pi^-$ decays (bottom left) and $B^+ \to (D^0, \overline{D}^0)K^+, (D^0, \overline{D}^0) \to K^-\pi^+\pi^+\pi^-$ decays (bottom right). A clear $CP$ asymmetry is also seen, although the statistics are low. The control samples $B^\mp \to D^0\pi^\mp$ are also shown.

The ATLAS experiment (Barton) has measured the $B^0$-$\overline{B}^0$ mixing parameter $\Delta\Gamma_d/\Gamma_d$, where $\Delta\Gamma_d$ is the difference in decay widths between the two mass eigenstates, and $\Gamma_d$ is the mean decay width. For this measurement ATLAS reconstructs $B^0 \to J/\psi K^0_S$ decays and fits the decay time distribution. This distribution contains four terms:

$$\frac{dN}{dt} \propto e^{-\Gamma t}\left[\cosh\frac{\Delta\Gamma t}{2} + A_P A_{CP}^{\rm dir}\cos(\Delta M t) + A_{\Delta\Gamma}\sinh\frac{\Delta\Gamma t}{2} + A_P A_{CP}^{\rm indir}\sin(\Delta M t)\right], \quad (5)$$

where $A_P$ is the production asymmetry between $B^0$ and $\overline{B}^0$ mesons. This parameter is measured by fitting the decay time distribution of flavor-specific $B^0 \to J/\psi\, K^{*0}, K^{*0} \to K^+\pi^-$ decays, which should be purely exponential; the result is $A_P = (0.25 \pm 0.48 \pm 0.05)\%$. Inserting this into Eq. (5) along with the theoretical values $A_{CP}^{\rm dir} = 0$, $A_{\Delta\Gamma} = \cos(2\phi_1)$, and $A_{CP}^{\rm mix} = -\sin(2\phi_1)$, ATLAS obtains $\Delta\Gamma_d/\Gamma_d = (-0.1 \pm 1.1 \pm 0.9)\%$. This result is (surprisingly) more precise than

**Table 2.** Fitted parameters of exotic states used by LHCb to fit the $M(J/\psi\,\phi)$ invariant mass distribution of $B^+ \to J/\psi\,\phi$ decays.

| State | $J^{PC}$ | significance | Mass | Width | Fit fraction |
|---|---|---|---|---|---|
| X(4140) | $1^{++}$ | 8.4 | $4165 \pm 4.5\,^{+4.6}_{-2.8}$ | $83 \pm 21\,^{+21}_{-14}$ | $13.0 \pm 3.2\,^{+4.8}_{-2.0}$ |
| X(4274) | $1^{++}$ | 6.0 | $4273.3 \pm 8.3\,^{+17.2}_{-3.6}$ | $56 \pm 11\,^{+8}_{-11}$ | $7.1 \pm 2.5\,^{+3.5}_{-2.4}$ |
| X(4500) | $0^{++}$ | 6.1 | $4506 \pm 11\,^{+12}_{-15}$ | $92 \pm 21\,^{+21}_{-20}$ | $6.6 \pm 2.4\,^{+2.5}_{-2.3}$ |
| X(4700) | $0^{++}$ | 6.1 | $4704 \pm 10\,^{+14}_{-24}$ | $120 \pm 31\,^{+42}_{-33}$ | $12 \pm 5\,^{+9}_{-5}$ |

measurements made at the $B$ factories and at LHCb, narrowly eclipsing the world's previous best measurement (from Belle [21]) of $(1.7 \pm 1.8 \pm 1.1)\%$. It is consistent with the SM prediction of $(0.042 \pm 0.008)\%$ [22].

## 5. Spectroscopy

There were numerous talks on spectroscopy; here we discuss two recent results.

The LHCb experiment (Dey) reconstructed a large sample of $B^+ \to J/\psi\,\phi K^+$ decays and studied the $J/\psi$-$\phi$ invariant mass distribution for unusual structure. This distribution is shown in Fig. 8(left) and exhibits four prominent peaks corresponding to the known states $X(4140)$, $X(4274)$, $X(4500)$, and $X(4700)$. Aside from these states no additional structure is apparent. Fitting the $M(J/\psi\,\phi)$ distribution with these states plus background gives a satisfactory goodness-of-fit: the $p$-value is 0.22. The fit projection is shown in Fig. 8(right), and the fit results are listed in Table 2.

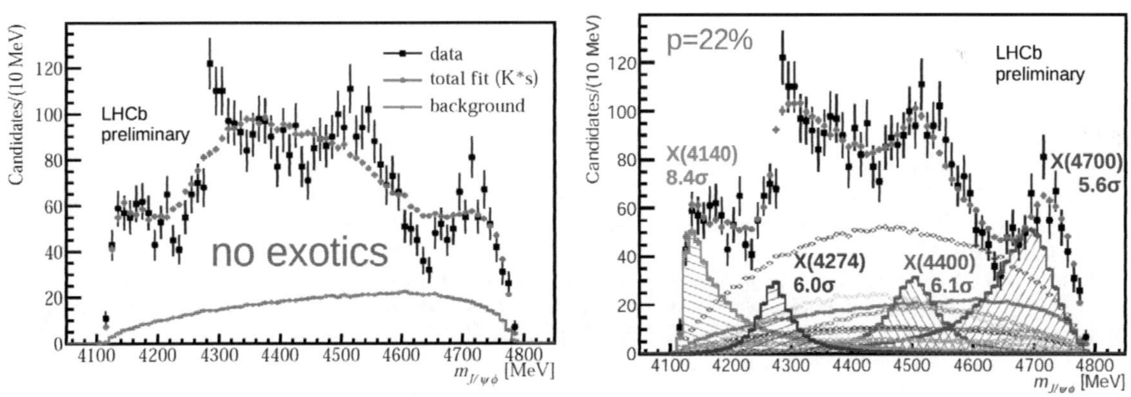


**Figure 8.** LHCb sample of $B^+ \to J/\psi\,\phi K^+$ decays. The $M(J/\psi\,\phi)$ invariant mass distribution is fitted without (left) and with (right) states $X(4140)$, $X(4274)$, $X(4500)$, and $X(4700)$.

The BESIII experiment (Pelizaus) reconstructed an especially large sample of $J/\psi \to \eta'\pi^+\pi^-\gamma$ decays with the goal of identifying intermediate $J/\psi \to X(1835)\gamma$ decays followed by $X(1835) \to \eta'\pi^+\pi^-$. The $\eta'$ is reconstructed in both $\eta' \to \rho^0\gamma$ and $\eta' \to \eta\pi^+\pi^-$ modes, where $\eta \to \gamma\gamma$. The resulting $M(\eta'\pi^+\pi^-)$ invariant mass distribution shows a clear peak near $M \approx 1835$ MeV$/c^2$, as expected, but it also shows a sharp drop at the $p\bar{p}$ threshold, which was unexpected. The observed lineshape (including this drop) is subsequently modeled in two ways: with a broad $X(1835)$ state and narrow $X(1920)$ state (the latter being just above $p\bar{p}$ threshold); and with a broad $X(1835)$ state and narrow $X(1870)$ state (the latter being just below $p\bar{p}$ threshold). Both parameterizations give satisfactory fits, which are shown in Fig. 9.

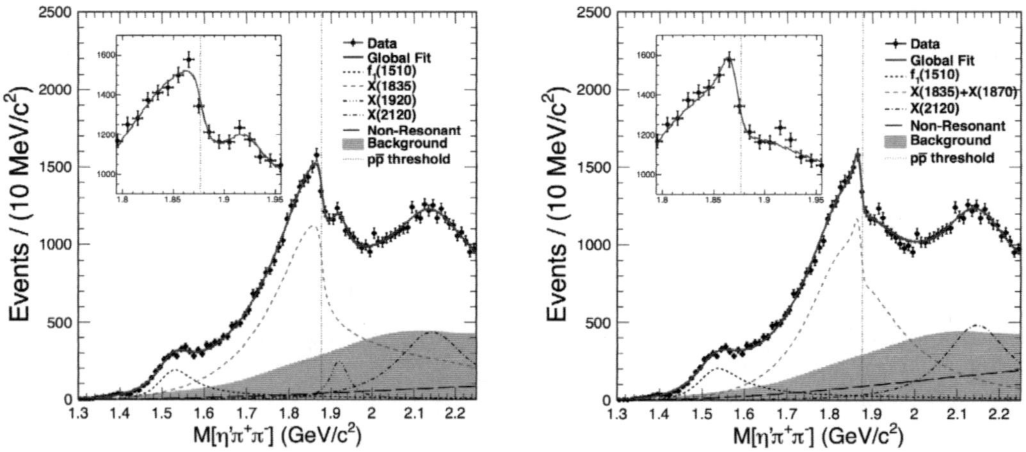


**Figure 9.** BESIII sample of $J/\psi \to \eta'\pi^+\pi^-\gamma$ decays. Left: fit result with a broad $X(1835)$ and narrow $X(1920)$ state. Right: fit result with a broad $X(1835)$ and a narrow $X(1870)$ state. Both fits are satisfactory.

## Acknowledgments

The authors thank the BEACH 2016 organizers for a well-run workshop and excellent hospitality. This research is supported by the U.S. Department of Energy.

## References

[1] Adamczyk L *et al* (STAR Collaboration) 2015 *Nature* **527** 345348
[2] Lednicky R, Lyuboshitz V L and Lyuboshitz V V 2004 *Phys. Part. Nucl.* **35** S50-S53
Lednicky R and Lyuboshitz V 2001 *Phys. Lett.* B **508** 146-154
[3] Vogt R 2008 *Eur. Phys. J.* ST **155** 213-222
[4] Charge-conjugate modes are implicitly included unless noted otherwise
[5] CMS Collaboration 2012 *Preprint* CMS-PAS-HIN-12-014
[6] Djordjevic M, Djordjevic M and Blagojevic B 2014 *Phys. Lett.* B **737** 298-302
[7] Zupanc A *et al* (Belle Collaboration) 2014 *Phys. Rev. Lett.* **113** 042002
[8] Bobeth C, Hiller G and Piranishvili G 2007 *J. High Energy Phys.* JHEP12(2007)040
Bouchard C, Lepage G P, Monahan C, Na H and Shigemitsu J 2013 *Phys. Rev. Lett.* **111** 162002
[9] Wei J-T *et al* (Belle Collaboration) 2009 *Phys. Rev. Lett.* **103** 171801
[10] Lees J P *et al* (BaBar Collaboration) 2012 *Phys. Rev.* D **86** 032012
[11] Huschle M *et al* (Belle Collaboration) 2015 *Phys. Rev.* D **92** 072014
[12] Aaij R *et al* (LHCb Collaboration) 2015 *Phys. Rev. Lett.* **115** 111803
Erratum 2015 *Phys. Rev. Lett.* **115** 159901
[13] Fajfer S, Kamenik J F and Niandi I 2012 *Phys. Rev.* D **85** 094025
[14] Amhis Y *et al* (Heavy Flavor Averaging Group) 2014 *Preprint* arXiv:1412.7515 and online update at www.slac.stanford.edu/xorg/hfag/
[15] Aaij R *et al* (LHCb Collaboration) 2016 *J. High Energy Phys.* JHEP02(2016)104
[16] Descotes-Genon S, Matias J, Ramon M and Virto J 2013 *J. High Energy Phys.* JHEP01(2013)048
[17] Descotes-Genon S, Hofer L, Matias J and Virto J 2014 *J. High Energy Phys.* JHEP12(2014)125
[18] Atwood D, Dunietz I and Soni A 1997 *Phys. Rev. Lett.* **78** 3257-3260
[19] Horii Y *et al* (Belle Collaboration) 2011 *Phys. Rev. Lett.* **106** 231803
[20] del Amo Sanchez P *et al* (BaBar Collaboration) 2010 *Phys. Rev.* D **82** 072006
[21] Higuchi T *et al* (Belle Collaboration) 2012 *Phys. Rev.* D **85** 071105
[22] Bobeth C, Haisch U, Lenz A, Pecjak B and Tetlalmatzi-Xolocotzi G 2014 *J. High Energy Phys.* JHEP06(2014)040

# A cosmic Ray Muon Experiment: a Way to Teach Standard Model of Particles at Community Colleges

**C Barazandeh, A Gutarra-Leon, R Rivas, H Glaser and W Majewski**

Physics Department, Northern Virginia Community College, Annandale, US
Email: wmajewski@nvcc.edu

**Abstract.** This experiment is an example of research for early undergraduate students and of its benefits and challenges as an accessible strategy for community colleges, in the spirit of the report on improving undergraduate STEM education from the US President's Council of Advisors on Science and Technology. The goals of this project include measuring average low-energy muon flux, day/night flux difference, time dilation, energy spectra of electrons and muons in arbitrary units, muon decay curve, average lifetime of muons. From the lifetime data we calculate the weak coupling constant $g_w$, electric charge e and the Higgs energy density.

## 1. Experiment
In our first experiment, we measured the muon average lifetime.

*1.1. Our system.*

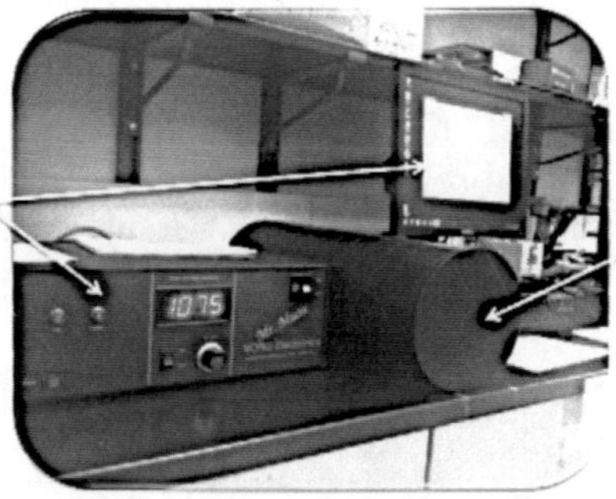


Single cylindrical plastic scintillation detector coupled to a 5" diameter PMT; its power supply voltage (set at 1060 V)/control box; an ADC card PCI-DAS4020/1 sitting in the computer, and the associated data analysis software. The muon trigger is sent out only if muon has decayed in the detector. Dead time set to 230 ns (not to have two triggers within that time). Ramp reset time is set at 35 $\mu$s: if a second (coincidence) event happens within this time, it is considered a muon decay event. Our detector only stops and identifies - by their emitted electron - muons of kinetic energy <160 MeV

Journal of Physics: Conference Series **770** (2016) 012050 doi:10.1088/1742-6596/770/1/012050

(calculated from the stopping power of 2 MeV/(g/cm$^2$) typical for relativistic muons at minimum ionization). Due to low rate of cosmic muons and small volume of the detector to collect enough data we ran the data acquisition each time for a period of one week, collecting simultaneously muon decay times, low-energy stoppable muon fluxes, and the fluxes of high-energy non-differentiated cosmic particles (muons making up more than half of this overall flux, according to the existing literature data). Energy spectra in arbitrary units of stoppable muons and of decay electrons were also collected.

## 2. Results

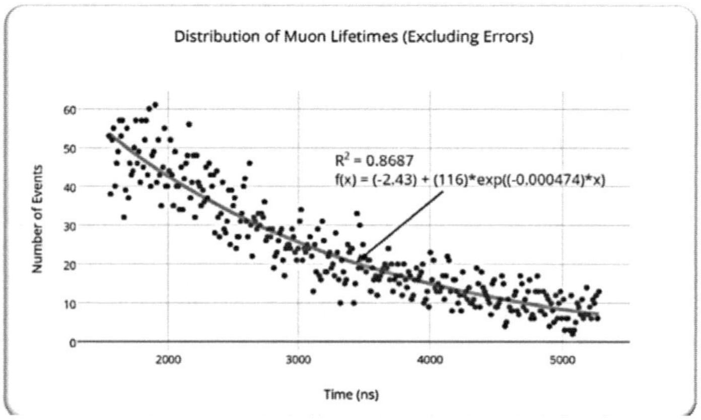


### *2.1. Mean lifetime.*

We fitted the data to the exponential decay formula

$$N = N_0 e^{-\lambda t} \tag{1}$$

where λ is the decay rate. This yielded 0.542 ns$^{-1}$ for λ. So τ=1/λ gives the average lifetime τ of a muon to be 1.85$\mu$s. We also analysed our smoothed-out data fitting them to a linear graph of: lnN = lnN$_0$ −λt. The value for λ appeared to be 0.474, yielding a τ of 2.11 (+/-) 0.07$\mu$s, with an error of 4% from the accepted value in vacuum of 2.20$\mu$s [1]. An independent run of the experiment gave

$$\tau=2.165\ (+\text{ or }-)\ 0.403\mu s. \tag{2}$$

### 2.2. *Charge Ratio.*

With observed muon lifetime, positive muon lifetime $\tau^+$ as the accepted value of free muon lifetime, and negative muon lifetime $\tau^-$ as known lifetime in carbon, detector material, we estimated the ratio

$$\varrho = N^+ / N^- \tag{3}$$

of fluxes of positive and negative muons, where $\tau_{obs}$ is our result (including muons of both signs). We obtained the charge ratio of muons in our detector's energy detection range, $\varrho = N^+/N^- = 0.764$ [2]. Negative muons both decay and are absorbed by atomic nuclei during muon capture. Positive muons only decay, so they have longer lifetime [3]. Free negative and positive muons have the same lifetime.

## 3. Calculated Results

### *3.1. Energy and Time of Flight.*

Using measurements of the muon flux at two different altitudes from similar experiments in literature, was the calculated the factor by which lifetime has been dilated - the Lorenz γ factor.

$$\gamma = \frac{1}{\sqrt{1-\frac{v^2}{c^2}}} \tag{4}$$

From that, the average kinetic energy was

Journal of Physics: Conference Series **770** (2016) 012050 doi:10.1088/1742-6596/770/1/012050

$$K_{av} = (\gamma - 1) m_{\mu} c^2 \tag{5}$$

The results for v = 2.206 x $10^8$ m/s, $\gamma$ =1.48 and $K_{av}$ = 50.3 MeV fit with expectations, as our detector can only record muon events of energies < 160 MeV. With this speed, the average time of flight is 5.789 μs, meaning that a fraction of $e^{-5.789/2.165}$ = 0.069, about 7% of the muons survive the descent [4].

*3.2. Flux.*

Our result for the flux of low-energy (< 160 MeV) muons is $\sigma$ = 0.78 muons/s*$m^2$ [5]. We found no systematic difference between day and night muon fluxes. Our result of $\Phi_D/\Phi_N = 1.04 \pm 0.04$ indicates that the sun is not a significant source of cosmic rays.

## 4. Standard Model (SM) -Derived Relations

From the lifetime formula and our result for $\tau$, we can, assuming other quantities have known values, calculate general SM parameters, such as the ratio of $g_w/M_wc^2$. Our calculations show interdependence of fundamental constants within the SM. The most surprising is an ability to calculate electric charge value from an apparently unrelated weak decay process. Our results agree with accepted values [6].

*4.1. Weak Coupling Constant.*

After solving the SM lifetime formula [7] for the ratio $g_w^2/M_Wc^2$ and using experimental W-boson mass of $M_Wc^2$ = 80.4 GeV and our value for $\tau$, we get as the dimensionless weak coupling constant $g_w$ = 0.680. This has a 4% difference from the accepted value of 0.653 [8].

*4.2. Elementary Electric Charge.*

In the SM, the electroweak force contains both electromagnetic force, mediated by the photon exchange, and weak force mediated by the heavy bosons $W^+$, $W^-$ and $Z^0$. Using predictions of the SM, we calculate elementary electric charge from our experiment. According to the SM,

$$e = g_w \sin\theta \sqrt{\hbar c \varepsilon_0} = 1.72 \times 10^{-19}\,C \tag{6}$$

if we use our experimental value of $g_w$ and take the accepted value of the weak angle mixing photon with the $Z^0$ boson as 29° [1]. As a result, we have an error of 7.5% with the textbook value of the elementary electric charge e=1.6x$10^{-19}$C [1].

*4.3. The vacuum expectation value, vev, of the Higgs field*

It determines the masses of all particles of SM; as calculated from our muon experiment by the SM vev formula [7], it is

$$(\hbar c)^{1/2}\,(\text{vev}) = (2M_wc^2/g_w) = 236\ \text{GeV}. \tag{7}$$

This value differs from the accepted value of 247 GeV by 4%.

## References


[1] NIST Physical Measurement Laboratory. n.d. http://www.physics.nist.gov (accessed 2015)

[2] Reitner R A, Romanowski T A, Sutton R B and Chidley B G 1960 Precise measurements of the mean lives of $\mu$ + and $\mu$ - mesons in carbon. *Physical Review Letters* **5.1** 22-3

[3] Ye J and Coan T E n.d. *Muon Physics* (Report of Rutgers University)

[4] Pierce M and Tekniska K April 2003 *Measuring the Lifetime of Cosmic Ray Muons* (Report, Section of Experimental Particle Physics)

[5] L3 Collaboration et al 2004 Measurement of the atmospheric muon spectrum from 20 to 3000 GeV *Physics Letters B* **598** 1-2; 15-32

[6] Schwartz M D 2014 *Quantum Field Theory and the Standard Model*. (Cambridge: Cambridge University Press)

[7] Giffith D 2008 *Introduction to Elementary Particles*. (Verlag: Wiley-VCH)

[8] Quigg C 2013 *Gauge Theories of the Strong, Weak and Electromagnetic Interactions*. (Princeton: Princeton University Press)